122

Mechanisms and Machine Science

Volume 104

This book series establishes a well-defined forum for monographs, edited Books, and proceedings on mechanical engineering with particular emphasis on MMS (Mechanism and Machine Science). The final goal is the publication of research that shows the development of mechanical engineering and particularly MMS in all technical aspects, even in very recent assessments. Published works share an approach by which technical details and formulation are discussed, and discuss modern formalisms with the aim to circulate research and technical achievements for use in professional, research, academic, and teaching activities.

This technical approach is an essential characteristic of the series. By discussing technical details and formulations in terms of modern formalisms, the possibility is created not only to show technical developments but also to explain achievements for technical teaching and research activity today and for the future.

The book series is intended to collect technical views on developments of the broad field of MMS in a unique frame that can be seen in its totality as an Encyclopaedia of MMS but with the additional purpose of archiving and teaching MMS achievements. Therefore, the book series will be of use not only for researchers and teachers in Mechanical Engineering but also for professionals and students for their formation and future work.

The series is promoted under the auspices of International Federation for the Promotion of Mechanism and Machine Science (IFToMM).

Prospective authors and editors can contact Mr. Pierpaolo Riva (publishing editor, Springer) at: pierpaolo.riva@springer.com

Indexed by SCOPUS and Google Scholar.

More information about this series at http://www.springer.com/series/8779

Marc Gouttefarde · Tobias Bruckmann ·
Andreas Pott
Editors

Cable-Driven Parallel Robots

Proceedings of the 5th International Conference on Cable-Driven Parallel Robots

Springer

Editors
Marc Gouttefarde
LIRMM, Univ Montpellier,
CNRS, Montpellier
France

Tobias Bruckmann
Chair of Mechatronics
University of Duisburg-Essen
Duisburg, Germany

Andreas Pott
University of Stuttgart
Stuttgart, Germany

ISSN 2211-0984 ISSN 2211-0992 (electronic)
Mechanisms and Machine Science
ISBN 978-3-030-75788-5 ISBN 978-3-030-75789-2 (eBook)
https://doi.org/10.1007/978-3-030-75789-2

This Springer imprint is published by the registered company Springer Nature Switzerland AG
The registered company address is: Gewerbestrasse 11, 6330 Cham, Switzerland

Preface

In 2012, leading experts from three continents gathered during the "First International Conference on Cable-Driven Parallel Robots (CableCon 2012)" in Stuttgart, Germany. This conference initiated a forum for the cable robot community. Due to the great success, the event was continued by the "Second International Conference on Cable-Driven Parallel Robots (CableCon 2014)" at the University Duisburg-Essen in 2014, establishing CableCon as a home for cable robot researchers. In 2017, the "Third International Conference on Cable-Driven Parallel Robots (CableCon 2017)" was organized by the Université Laval in Québec, Canada. This time, the conference went across the Atlantic and emphasized its role as an international event, where researchers from all over the world find a place to connect with the most recognized experts from the field. Back in Europe, the "Fourth International Conference on Cable-Driven Parallel Robots" took place in 2019 as part of the "15th IFToMM World Congress" in Krakow, Poland, a very special event which celebrated the 50th anniversary of IFToMM.

All four previous conferences on cable-driven parallel robots were organized under the patronage of International Federation for the Promotion of Mechanism and Machine Science (IFToMM), and this "Fifth International Conference on Cable-Driven Parallel Robots" is no exception.

Because of the COVID-19 pandemic, this fifth edition could not be organized as a physical event in Montpellier, France, as initially planned. However, we believe that the researchers and engineers working on cable-driven parallel robots will soon have the occasion to meet in person again.

Meanwhile, practical investigations on cable-driven parallel robots are attracting the focus of research teams around the world. At the same time, fundamental research still continues to provide new insights into the physical understanding of cable-driven parallel robots. This broad variety of research activities is reflected by the content of this book.

The editors would like to thank the authors for their valuable contributions. Noteworthy, the strict schedule in the preparation of this book would not have been feasible without the support of the reviewers and the scientific committee.

We would like to express our gratefulness to Springer for smoothly processing the material for this book.

April 2021

Marc Gouttefarde
Tobias Bruckmann
Andreas Pott

Organization

International Scientific Committee

Sunil K. Agrawal	Columbia University, USA
Marc Arsenault	Laurentian University, Canada
Tobias Bruckmann	University of Duisburg-Essen, Germany
Philippe Cardou	Laval University, Canada
Stephane Caro	CNRS, LS2N, France
Marco Carricato	University of Bologna, Italy
Clement Gosselin	Laval University, Canada
Marc Gouttefarde	LIRMM, Univ Montpellier, France
Darwin Lau	The Chinese University of Hong Kong, China
Jean-Pierre Merlet	Inria, France
Leila Notash	Queen's University, Canada
Andreas Pott	University of Stuttgart, Germany
Dieter Schramm	University of Duisburg-Essen, Germany
Alexander Verl	University of Stuttgart, Germany

Contents

Applications

Kinematics

Efficient Kinematics of a 2-1 and 3-1 CDPR with Non-elastic Sagging Cables

Jean-Pierre Merlet(✉)

HEPHAISTOS Project, INRIA Sophia -Antipolis, Sophia-Antipolis, France
Jean-Pierre.Merlet@inria.fr

Abstract. Solving the kinematics of CDPR is complex as soon as cable sagging is taken into account. We are considering here CDPRs having 2 cables whose extremities are attached at the same point on the platform (i.e. CDPRs allowing only translational motion). Regarding the cables we assume that they are non-elastic but have a mass so that they will exhibit sagging. We first show that the inverse and direct kinematics (IK and DK) amount to solve a square system of equations. We then show that these systems of equations may be reduced to solving an equation in a single variable that cannot be solved analytically but can easily be solved numerically. Using this result we show that the sagging will play a role on the result of the IK/FK only if the load mass is lower than a threshold. We then present some preliminary results regarding the case of 3 cables which is much more complex: the IK may be reduced to solving an equation in a single variable but this solving is numerically difficult. Here again it seems that sagging may be neglected if the load mass is high enough. We also present a preliminary analysis of the 3-1 case.

Keywords: Cable-driven parallel robot · Sagging cable · Inverse kinematics · Direct kinematics

1 Introduction

Solving the inverse kinematics and direct kinematics of CDPRs is an essential element for the design and control. Inverse kinematics is straightforward if the cable sagging is not taken into account but becomes complex as soon as sagging is considered with possibly several solutions if both the cable mass and its elasticity are taken into account [1]. Direct kinematics (DK) is even more complex whatever cable model is used. There are several variants of the DK problem:

1. finding the current platform pose using an estimation of the cable lengths but being given the platform pose a short time before the current one (this is often called *real-time kinematics*)
2. finding all solution poses based only on an estimation of the cable lengths
3. finding the current platform pose based on a combination of partial measurements of the pose components [2–5] and/or other information on the CDPR (e.g. cable angles [6–8] or force sensors [9]) and of the cable lengths

M. Gouttefarde et al. (Eds.): CableCon 2021, MMS 104, pp. 3–12, 2021.
https://doi.org/10.1007/978-3-030-75789-2_1

We are considering that real-time kinematics is no more an issue as there are very efficient and certified algorithms (implementable in a distributed manner) that are able to solve it whatever the cable model is [10,11]. Many works have addressed point 3 but there are still improvements to be gained such as fusing additional measurements beside cable lengths and pose estimation (quite often obtained at a lower rate than the cable lengths) or mixing IK and DK to manage more efficiently slack cables [12,13]. Regarding point 2 the DK problem of CDPRs with ideal cable (no mass, no elasticity) has been addressed in several works [14–17] but there are much less works on the DK with sagging cables [18–22]. The proposed algorithms either cannot guarantee to find all solutions or are relatively expensive in term of computation time. This is mostly a consequence that sagging cable models are not algebraic so that classical methods such as elimination or Gröebner basis, which have been proven effective for classical parallel robots, cannot be used. In this paper we will address the inverse and direct kinematics of two specific classes of CDPR (Fig. 1):

- The planar CDPR 2-1 with 2 cables attached at the same point on the platform
- The spatial CDPR 3-1 with 3 cables attached at the same point on the platform

These CDPRs allow only for translational degrees-of-freedom and we will assume that the cables are non-elastic but have a mass. We will show that for both the 2-1 and 3-1 CDPR the inverse kinematics may be reduced to solve a single equation in one unknown. This property will also be verified for the direct kinematics of the 2-1 CDPR.

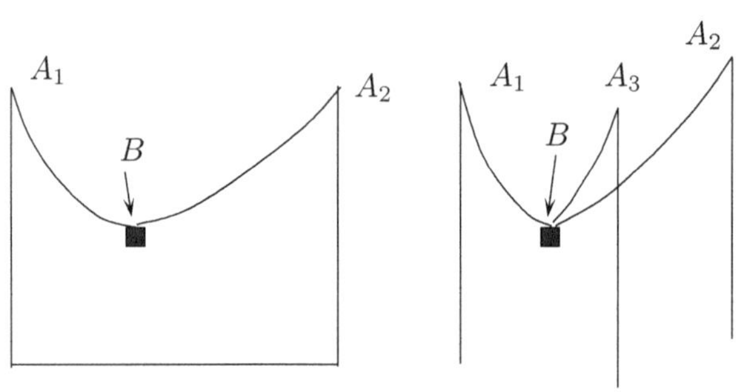

Fig. 1. The two CDPRs that will be studied: with 2 cables we have a planar motion and with 3 cables a spatial one.

2 Cable Model

Beside the ideal and elastic-ideal mode, which assume that the cable shape is a line, multiple models that take into account elasticity and/or the cable mass have been proposed [23–27] but many of them have been simplified in order to be able to manage the kinematic issues. There is however a model usually called the Irvine model [28] that is extensively used in the analysis of cabled structures and it has been shown experimentally to be quite realistic for CDPRs [21]. This model is planar and takes into account both the elasticity and the mass of the cable. In this paper we will use a formulation of this model derived from [29] in which the elastic part is neglected. The motivation of neglecting elasticity is that the CDPRs we are considering are assumed to carry a relatively light load so that the cable tensions are relatively small.

We assume that a reference frame $\mathscr{R} = (0, \mathbf{x}, \mathbf{y}, \mathbf{z})$ has been defined with the $\mathbf{z}$ axis being the local vertical. The cables are attached at a common point B on the platform whose coordinates in $\mathscr{R}$ are (x_b, y_b, z_b). For a given cable we assume that the cable exits from the winch at a known fixed point A with coordinates (x_a, y_a, z_a) in $\mathscr{R}$. The platform exerts at B a force $\mathbf{F}$ whose components in $\mathscr{R}$ are (F_x, F_y, F_z). The cable tensions at A, B will be respectively denoted τ_A, τ_B, L_0 is the cable length and μ is the linear density of the cable material, that is assumed to be identical for all cables. With this notation the cable model we will use is:

$$x_A = x_B + \frac{F_x}{\mu g} ln(\frac{\tau_A + F_z - \mu g L_0}{\tau_B + F_z}) \tag{1}$$

$$y_A = y_B + \frac{F_y}{\mu g} ln(\frac{\tau_A + F_z - \mu g L_0}{\tau_B + F_z}) \tag{2}$$

$$z_A = z_B + \frac{\tau_A - \tau_B}{\mu g} \tag{3}$$

We have also

$$\tau_B^2 = F_x^2 + F_y^2 + F_z^2 \qquad \tau_A^2 = F_x^2 + F_y^2 + (F_z - \mu g L_0)^2 \tag{4}$$

In the sequel of this paper F_x, F_z, L_0 for cable i will be denoted $F_{x_i}, F_{z_i}, L_{0_i}$.

3 Analysis of the 2-1 CDPR

The 2-1 CDPR is a planar robot with 2 cables and therefore 2 winch output points A_1, A_2. For the sake of simplicity we will assume that A_1, A_2 have the same height so that their coordinates are $(0, 0), (d, 0)$. For the planar case we have to use only Eqs. (1, 3, 4) of the cable model.

3.1 Inverse Kinematics

For the inverse kinematics we have to determine the cable lengths L_{0_1}, L_{0_2} for given x_B, z_B. The unknowns of this problem are the $F_x, F_z, \tau_A, \tau_B, L_0$ for each cable so that we have a total of 10 unknowns. We have the set of Eqs. (1, 3, 4) for each cable which provide 8 constraints. At B we have a load of mass m and the mechanical equilibrium imposes the 2 constraints:

$$F_{x_1} + F_{x_2} = 0 \qquad F_{z_1} + F_{z_2} = -mg \tag{5}$$

so that we have a total of 10 constraints and hence the IK admits usually a finite number of solutions. Such a system may be solved with interval analysis but this method requires to have bounds for the unknowns but we have no systematic mean to determine bounds for the F_x, F_z, L_0. We define $U_i = e^{\mu g(x_{a_i} - x_B)/F_{x_i}}$ so that we have $U_1 = e^{-x_B \mu g/F_{x_1}}$ and according to our notation we have $x_B > 0, F_{x_1} > 0$. As for U_2 taking into account that $F_{x_2} = -F_{x_1}$ we have $U_2 = e^{-(d-x_B)\mu g/F_{x_1}} = e^{-\mu g d/F_{x_1}}/U1$. We note that both U_1 and U_2 lies in the range [0,1] and we have $F_{x_1} = -\mu g x_B / ln(U_1)$. Equation (1) is written as

$$\frac{(x_{a_i} - x_b)\mu g}{F_{x_i}} = ln(\frac{\tau_{A_i} + F_{z_i} - \mu g L_{0_i}}{\tau_{B_i} + F_{z_i}}) \tag{6}$$

Taking the exponential of both terms leads to

$$\tau_{A_i} + F_{z_i} - \mu g L_{0_i} - U_i(\tau_{B_i} + F_{z_i}) = 0 \tag{7}$$

while Eq. (3) is written as

$$\mu g(z_{A_i} - z_B) - (\tau_{A_i} - \tau_{B_i}) = 0 \tag{8}$$

These equations are linear in τ_{A_i}, τ_{B_i} and by solving this system we obtain τ_{A_i}, τ_{B_i} as functions of L_{0_i}, F_{z_i} and F_{x_1} as

$$\tau_{A_i} = \frac{\mu g((z_{A_i} - z_b)U_i - L_{0_i})}{U_i - 1} - F_{z_i} \qquad \tau_{B_i} = \frac{\mu g((z_{A_i} - z_b) - L_{0_i})}{U_i - 1} - F_{z_i} \tag{9}$$

Reporting this result in (4) leads for each cable to 2 equations F_1, F_2 that are algebraic in L_{0_i}. Taking the resultant of these 2 equations leads to an equation whose unknowns are U_i, F_{z_i}, F_{x_1}. As $F_{z_2} = -mg - F_{z_1}$ we get for the 2 cables 2 equations G_1, G_2 whose unknowns are $U_1, U_2, F_{z_1}, F_{x_1}$. These equations are too large to be given here but they are written as

$$G_i = a_i + b_i F_{x_1}^2 + c_i F_{z_1} + d_i F_{z_1}^2 \tag{10}$$

where the coefficients a_i, b_i, c_i, d_i are only functions of the U_i. These equations are algebraic in F_{z_1} and their resultant H in F_{z_1} is an equation in U_1, U_2, F_{x_1} only. As U_1, U_2 are functions of F_{x_1} only, then H is a function of the single unknown F_{x_1}. The equation $H = 0$ may be written as

$$H = \sum_{j=0}^{j=2} A_j(U_1, U_2) F_{x_1}^{2j} \tag{11}$$

As we have in H both terms in $F_{x_1}^j$ and in $e^{A/F_{x_1}}$ there is no closed-form of the roots in F_{x_1} of $H = 0$. Note that using the expression of F_{x_1} as function of U_1 we get an expression o fH in U_1, U_2: a multi-Taylor expansion of H at 1-ϵ, for U_1, U_2, where ϵ is supposed to be small, leads to $H > 0$ meaning that there is a limit value V for F_{x_1} such that $H(\forall F_{x_1} > V) > 0$ and therefore V is an upper bound for the roots in F_{x_1} of $H = 0$. The value of V may be safety set to $V = 1000 x_B \mu g$ as in this case $U_1 = 0.999$. Interval analysis is then used to solve $H = 0$ with a range for F_{x_1} set to $[1e - 3, V]$. For each root F_{x_1} of $H = 0$ we calculate the common roots of $G_1 = 0, G_2 = 0$ which provide the possible value of F_{z_1}. For a pair F_{x_1}, F_{z_1} the length(s) L_{0_i} are obtained as the common root of $F_1 = 0, F_2 = 0$.

3.2 Example

As example we consider the IK of a CDPR with $d = 20$ m for the pose $x_B = 7$ m, $z_B = -2$ m with a load mass of 1 kg or 50 kg. For this pose the distances ρ_1, ρ_2 between the A, B points of cable 1, 2 are 7.280109 m and 13.15294 m. The cables are made of Dyneema synthetic fiber with various cable diameter d_c and Fig. 2 shows the values of L_{0_1}, L_{0_2} and τ_{B_1} as a function of d_c for a mass of 10 kg and of 50 kg. A typical running time for the IK solver is about 3 ms. As may be seen from the figure the differences between the ideal cable lengths is very small if $m = 50$ kg (in the worst case 0.38 mm and 2.5 mm for cable 1 and 2) while the tension at B for cable 1 exhibits a more significant difference going from 1100 N to 1300 N. If $m = 10$ kg there is a more significant difference in the L_0 (0.5 cm for L_{0_1} and 2.5 cm for L_{0_2}). There is also a significant change in the tension at B, going from 32 N to 180 N. However being given the breaking point of the Dyneema cable a small diameter cable can be used in both cases with negligible effect on the kinematics solution compared to ideal cables. The influence of the load mass will be more extensively studied in the next section.

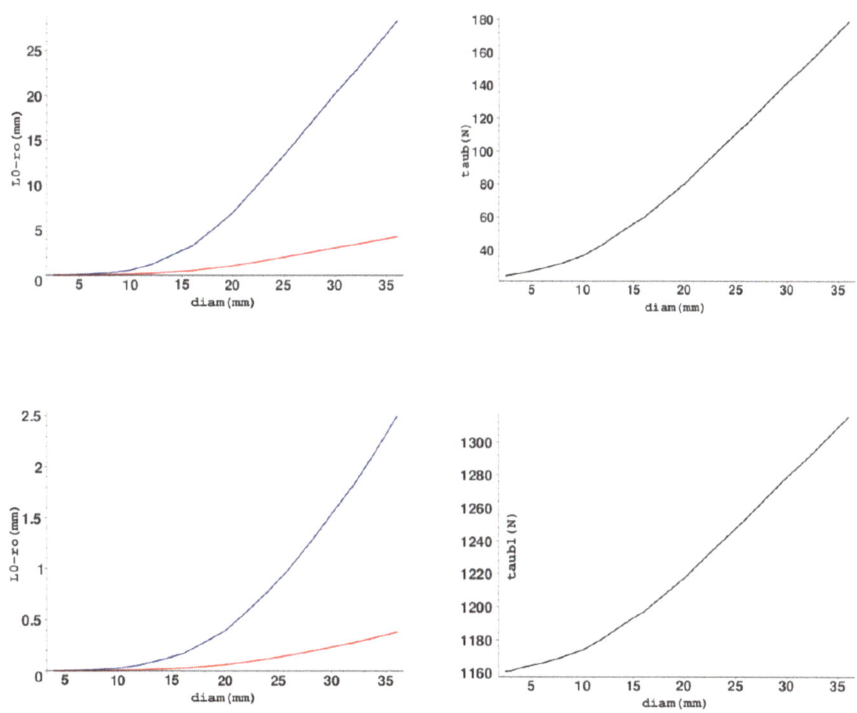

Fig. 2. For the IK the value of the ideal cable lengths are ρ_1, ρ_2 while with sagging the values are L_{0_1}, L_{0_2}. This figure presents the values of $L_{0_1} - \rho_1$ (red), $L_{0_2} - \rho_2$ (blue) and τ_{B_1} as a function of the cable diameter for a mass of 10 kg and of 50 kg

3.3 Direct Kinematics

For this problem L_{0_1}, L_{0_2} are given and we have to determine x_B, z_B. The solving process starts as for the IK by determining τ_{A_i}, τ_{B_i} by solving Eqs. (1, 3). We then report the result in the Eq. (4). Subtracting the left-side equation presented in (4) obtained for both cables and doing the same for the second equation leads to a system that is linear in z_B, F_{z_1}. This system is solved and the result is reported i n the 2 Eq. (4) for the first cable. The unknowns in these equations are F_{x_1}, U_1, U_2 but as $U_2 = e^{-d\mu g/F_{x_1}}/U_1 = A/U_1$ we get 2 equations $D_1(F_{x_1}, U_1) = 0, D_2(F_{x_1}, U_1) = 0$ that are algebraic in U_1. The resultant of these equations in U_1 leads to the constraint $T(F_{x_1}) = 0$. The function T may be written as

$$T = \sum_{j=0}^{j=12} u_j F_{x_1}^j \tag{12}$$

where the coefficients u_j are algebraic function of A, which is restricted to lie in the range [0,1]. Furthermore we have $F_{x_1} = -d\mu g/ln(A)$ so that function T leads to a constraint $S(A) = 0$. This equation is solved numerically to obtain the solution in A which lead to the solution in F_{x_1}. Note that there is a symmetry in the T constraint as $T(F_{x_1}, A) = T(-F_{x_1}, 1/A)$.

For solving $S(A) = 0$ we use interval analysis and the solutions provide the possible value of F_{x_1}. For a given F_{x_1} D_1, D_2 are algebraic function of U_1 only and their common roots are the possible values of U_1. Then x_B is calculated as $x_B = F_{x_1} ln(U_1)/(\mu g)$ while F_{z_1}, z_B have been obtained in the process as functions of A, U_1. The total computation time is around 5 ms and may probably be reduced.

3.4 Example

We consider a CDPR with $d = 20$ m, $\mu = 0.023$ kg/m which correspond to a 6 mm Dyneema cable). For the cable lengths we use as cable lengths the distance between A_1, A_2 and the pose $x_B = 7$ m, $z_B = -2$ m. We consider the following load masses: 0.1, 1 and 10 kg with the following differences between the calculated pose and x_B, z_B in cm: (4.04, 18.95) (0.36, 1.595) (0.0053,0.023).

As analyzed in the IK and DK it appears that if the load mass is sufficiently large, then the sagging is negligible and the ideal cable models may be used. It is therefore of interest to determine the minimal load mass that will lead to a small difference $\Delta\mathbf{X}$ between the pose obtained with ideal cables and the one obtained with the sagging cables. Figure 3 presents the minimal mass that leads to a $\Delta\mathbf{X} = 0.5$ cm as function of the cable diameter. If the load has a larger mass than this minimum, then the sagging effect may be neglected.

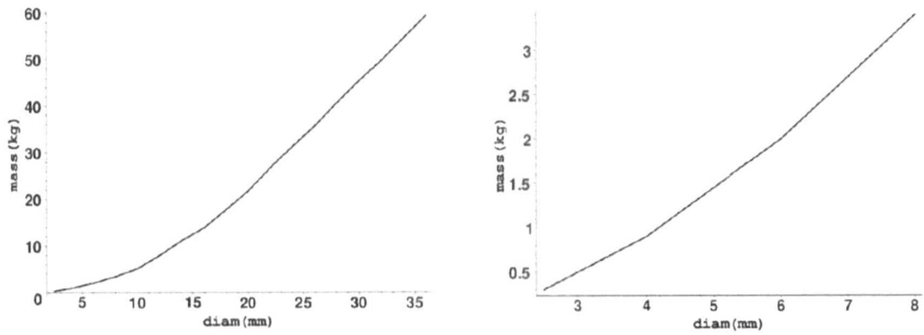

Fig. 3. We show here the load mass that leads to a distance of 5 mm between the 2-cables DK solution pose obtained for ideal cable and the one for sagging cable as function of the cable diameter If the load mass is larger than this mass, then the sagging may be neglected. On the right a detail for small diameter cables.

4 Preliminary Analysis of the 3-1 CDPR

As will be seen in the next section the inverse kinematics of the 3-1 CDPR may also be reduced to solve a single equation, while we have not be able to obtain a similar result for the direct kinematics. Still we believe that the inverse kinematics process may be improved while reducing the complexity of the direct kinematics may be possible.

4.1 Inverse Kinematics

Using Eqs. (1, 2) we get $F_{y_i} = F_{x_i}(ya_i - yb)/(xa_i - xb)$. The mechanical equilibrium equations are $\sum_{j=1}^{j=3} F_{x_i} = \sum_{j=1}^{j=3} F_{y_i} = 0$ and $\sum_{j=1}^{j=3} F_{z_i} = -mg$. The 2 first equations are linear in the F_{x_i} so that we can get F_{x_2}, F_{x_3} as functions of F_{x_1}. As in the 2-1 cable section we use the 2 Eq. (8) to get the value of the τ_{A_i}, τ_{B_i}. We then set $F_{z_3} = -mg - F_{z_1} - F_{z_2}$ and consider for each cable the 2 equations P_{1_i}, P_{2_i} of (4). For a given cable j the unknowns in these 2 equations are $F_{x_1}, F_{z_1}, F_{z_2}, L_{0_j}, U_j$ and are algebraic in L_{0_j}. We compute the resultant H_j of the 2 equations in L_{0_j}. For cable 2 and 3 H_j is algebraic in F_{z_2} so that may calculate the resultant R_{23} of H_2, H_3 in F_{z_2}. Finally we compute the resultant W of H_1, R_{23} in F_{z_1} which is a function of F_{x_1}, U_1, U_2, U_3 where $U_j = e^{\mu g(x_{A_i} - x_B)/F_{x_j}}$. As F_{x_2}, F_{x_3} has been obtained as function of F_{x_1} the resultant W is a function of F_{x_1} may be written as

$$W = \sum_{j=0}^{j=4} v_j F_{x_1}^{2j} \tag{13}$$

whose coefficients v_j are functions of the U_j. If W can be solved numerically for a given root of W we get F_{z_1}, F_{z_2} from H_1, H_2 which allow to calculate F_{z_3}. The 2 equations of (4) for each cable are now only algebraic functions of L_{0_j} and their common roots are the possible values of L_{0_j}. However it appears that W is difficult to solve as soon as the load mass is low.

4.2 Direct Kinematics

Regarding the DK it has a single solution for ideal cable so that we can claim that it will also have a single solution for sagging cables as soon as the DK equations are not singular [20]. Indeed the cable model is continuous in terms of E, μ with the deal cable being obtained for $E \rightarrow \infty, \mu = 0$. Hence starting from the single solution obtained for the DK with ideal cable we may use a continuation scheme on E, μ to obtain the DK solution for given E, μ. If the DK equations do not become singular during the continuation there will be a single solution. But if a singularity occurs, then the followed branch may split in several branches, thereby possibly leading to several DK solutions.

However we have not yet been able to reduce the DK system to a single variable equation. WE use an interval analysis based approach: bounds for x_B, y_B, z_B are easily obtained and we consider very large intervals for $F_{x_1}, F_{z_1}, F_{z_2}, \tau_{A_i}, \tau_{B_i}$. The trick here is to assume that there is a single solution of the DK and to run a few iterations of the Newton scheme with as initial guess the center of the boxes obtained during the interval analysis algorithm and to stop the algorithm as soon as the Newton scheme converge. However the Newton scheme may converge to an incorrect solution. Indeed there is a symmetry in the DK equations so that if $(x_B, y_B, z_B, F_{x_1}, F_{z_1}, F_{z_2}, U_i, \tau_{A_i}, \tau_{B_i})$ is a solution, then $(x_B, y_B, -z_B, -F_{x_1}, F_{z_1}, F_{z_2}, 1/U_i, -\tau_{A_i}, -\tau_{B_i})$ is also a solution: hence only solution(s) with positive τ_{A_i}, τ_{B_i} have to be retained. Here again the sagging has an effect only for small load masses. For example we consider the CDPR with $x_{A_1} = y_{A_1} = y_{A_2} = x_{A_3} = 0$, $x_{A_2} = 20$, $y_{A_3} = 10$, the cable lengths equal to the distance from the A_i to the pose $\mathscr{I} = (10, 4, -3)$. For a load mass of 0 kg the pose will be $9.78, 3.615, -2.56$ i.e. at a distance of 62.47 cm from $\mathscr{I}$. The difference between the reached pose and $\mathscr{I}$ decreases with the load mass: an error of 5 mm is obtained for a mass of 6.8 kg and an error of 1 mm for a load mass of 15.5 kg. We have also investigated the convergence of the Newton scheme with as guess the DK solution obtained for ideal cables starting for various cable diameter and load mass. Curiously it appears that for cable diameter from 2.5 mm to 10 mm Newton diverges if the load mass is lower than 2 kg while this mass increases for larger cables (3 kg for a 12 mm diameter, 4.1 kg for 14 mm, 6.2 kg for 16 mm, 9.7 kg for 20 mm, 11.8 kg for 22 mm). Hence for most existing prototypes the Newton scheme may be used to solve the DK.

5 Conclusion

The first contribution of this paper are very efficient FK and DK algorithms for CDPR with 2 sagging cables that are based on the solving of a function in one single variable. The second contribution is to show that the sagging of non-elastic synthetic fiber may be neglected as soon as the load mass is sufficiently large. At the opposite for a small mass the sagging may induce a very significant error in the CDPR positioning. The algorithms hence allows one to determine a lower bound m_m for the load mass so that sagging may be neglected. Future work will address the case of elastic cables and it may be expected that the effect of elasticity will play a significant role only for a mass that is larger than a threshold M_M. If this is the case we will be able to determine a range $[m_m, m_M]$ for the load mass so that cable may be considered as ideal for the IK and DK, which greatly simplify the analysis.

Acknowledgment. the author want to thank the reviewers for their interesting comments.

References

1. Merlet, J.-P.: A new generic approach for the inverse kinematics of cable-driven parallel robot with 6 deformable cables. In: ARK, Grasse, 27–30 June 2016
2. Zavatta, M., et al.: A vision-based referencing procedure for cable-driven parallel manipulators. Journal of Mechanisms and Robotics, vol. 12, no. 4, August 2020
3. Korayem, M.H., et al.: A novel method for recording the position and orientation of the end effector of a spatial cable-suspended robot and using for closed-loop control. Int. J. Adv. Manuf. Technol. **72**, 739–755 (2014)
4. Mousavi, M., et al.: Sensory feedback performance improvement on RoboCab: an experimental approach to wire-driven parallel manipulator. In: 4th International Conference on Robotics and Mechatronics, Teheran, 26–28 October 2016
5. Zi, B., Lin, J., Quian, S.: Localization, obstacle avoidance planning and control of a cooperative cable parallel robot for multiple mobile cranes. Robot. Comput.-Integr. Manuf. **34**, 105–123 (2015)
6. Fortin-Côté, A., Cardou, P., Campeau-Lecours, A.: Improving cable driven parallel robot accuracy through angular position sensors. In: IEEE International Conference on Intelligent Robots and Systems (IROS), pp. 4350–4355. Daejeon, 9–14 October 2016
7. Garant, X., et al.: Improving the forward kinematics of cable-driven parallel robots through cable angle sensors. In: 3rd International Conference on Cable-driven Parallel Robots (CableCon), Québec (2017)
8. Merlet, J.-P.: Direct kinematics of CDPR with extra cable orientation sensors: the 2 and 3 cables case with perfect measurement and sagging cables. In: IEEE International Conference on Intelligent Robots and Systems (IROS), Vancouver, 24–28 September 2017
9. Oftadeh, R., Aref, M.M., Taghirad, H.D.: Forward kinematic analysis of a planar cable driven redundant parallel manipulator using force sensors. In: IEEE International Conference on Intelligent Robots and Systems (IROS), pp. 2295–2300. Taipei, 18–22 October 2010
10. Merlet, J.-P.: On the real-time calculation of the forward kinematics of suspended cable-driven parallel robots. In: 14th IFToMM World Congress on the Theory of Machines and Mechanisms, Taipei, 27–30 October 2015
11. Pott, A.: An algorithm for real-time forward kinematics of cable-driven parallel robots. In: ARK, pp. 529–538. Piran, June 28–July 1 2010
12. Zake, Z., et al.: Vision-based control and stability analysis of a cable-driven parallel robot. IEEE Robot. Autom. Lett. **4**(2), 1029–1036 (2019)
13. Merlet, J.-P., Papegay, Y., Gasc, A.-V.: The Prince's tears, a large cable-driven parallel robot for an artistic exhibition. In: IEEE International Conference on Robotics and Automation, Paris, May 31–August 31 2020

14. Abbasnejad, G., Carricato, M.: Direct geometrico-static problem of underconstrained cable-driven parallel robots with n cables. IEEE Trans. Robot. **31**(2), 468–478 (2015)
15. Berti, A., Merlet, J.-P., Carricato, M.: Solving the direct geometrico-static problem of underconstrained cable-driven parallel robots by interval analysis. Int. J. Robot. Res. **35**(6), 723–739 (2016)
16. Husty, M., Zsombor-Murray, P.: Geometric contribution to the analysis of 2-2 wire driven cranes. In: International Conference on Interdisciplinary Applications of Kinematics, pp. 1–8. Lima, 9–11 September 2013
17. Pott, A., Tempel, P.: A unified approach to forward kinematics for cable-driven parallel robots based on energy. In: ARK, Bologna, 1–5 July 2018
18. Gouttefarde, M., Nguyen, D.Q., Baradat, C.: Kinetostatics analysis of cable-driven parallel robots with consideration of sagging and pulleys. In: ARK, pp. 213–221. Ljulbjana, June 29–July 3 2014
19. Merlet, J.-P.: The forward kinematics of cable-driven parallel robots with sagging cables. In: 2nd International Conference on Cable-driven Parallel Robots (CableCon), pp. 3–16. Duisburg, 24–27 August 2014
20. Merlet, J.-P.: A generic numerical continuation scheme for solving the direct kinematics of cable-driven parallel robot with deformable cables. In: IEEE International Conference on Intelligent Robots and Systems (IROS), Daejeon, 9–14 October 2016
21. Riehl, N., et al.: Effects of non-negligible cable mass on the static behavior of large workspace cable-driven parallel mechanisms. In: IEEE International Conference on Robotics and Automation, pp. 2193–2198. Kobe, 14–16 May 2009
22. Such, M., et al.: An approach based on the catenary equation to deal with static analysis of three dimensional cable structures. Eng. Struct. **31**(9), 2162–2170 (2009)
23. Baklouti, S., Caro, S., Courteille, E.: Dynamic and oscillatory motions of cable-driven parallel robots based on a non-linear cable tension model. Journal of Mechanisms and Robotics, 9 September 2017
24. Ferravante, V., et al.: Dynamic analysis of high precision construction cable-driven parallel robots. Mechanism and Machine Theory **135**, 54–64 (2019)
25. Miermeister, P., et al.: An elastic cable model for cable-driven parallel robots including hysteresis effect. In: 2nd International Conference on Cable-driven Parallel Robots (CableCon), pp. 17–28. Duisburg, 24–27 August 2014
26. Miyasaka, M., et al.: Hysteresis model of longitudinally loaded cable for cable driven robots and identification of the parameters. In: IEEE International Conference on Robotics and Automation, Stockholm, 16–20 May 2016
27. Piao, J., et al.: Open-loop position control of a polymer cable-driven parallel robot via a viscoelastic cable model for high payload workspaces. Adv. Mech. Eng. **9**(12), 1687814017737199 (2017)
28. Irvine, H.M.: Cable Structures. MIT Press, Cambridge (1981)
29. Nguyen, D.Q., Gouttefarde, M.: Stiffness matrix of 6-dof cabl-driven parallel robots and its homogenization. In: ARK, Ljulbjana, June 29–July 3 2014

Kinematics of a 2-DOF Planar Suspended Cable-Driven Parallel Robot

Dinh Quan Nguyen(✉)

Le Quy Don Technical University, Hanoi, Vietnam
ndquan@lqdtu.edu.vn

Abstract. This paper discusses the kinematics of a two degrees of freedom (2-DOF) planar suspended cable-driven parallel robot (CDPR) with massless cables. The solutions for the inverse and direct geometric as well as the inverse and direct kinematic problems are derived while fully considering the influence of pulleys and cable winding drums.

Keywords: Inverse kinematics · Direct kinematics · Suspended CDPR

1 Introduction

The studies of CDPRs have been carried on for several decades. Due to their appealing advantages over rigid-body manipulators such as simple in design, having large workspace and low construction cost, there are more and more prototypes of CDPRs have been developed. However, one significant drawback of CDPRs compared to rigid-body manipulators is the low accuracy in their kinematic modeling which lead to their limits in industrial applications. Simple design CDPRs tend to bring more complexity in solving their kinematic problems, especially when all the main factors of the CDPR systems are considered, for example, pulleys and cable winding drums or the cable sag (in the case of large workspace and heavy payload CDPRs). The simpler in design, the greater difficulty there is in order to solve the kinematic modeling problem which is the well-known trade-off of CDPRs.

Over the past years, many research works have been focusing on dealing with these problems of CDPR in order to improve their performance in control applications. Basic solutions to solve the inverse and direct kinematic problems of CDPR (where the influence of pulleys and winding drums are ignored) can be seen in [1–3]. Recent studies on CDPRs that consider the involvement of pulleys can be seen in [4–6]. The works considered both the pulleys and drums of CDPR systems can be seen in [7,8], however, kinematic modeling solutions has not been developed explicitly.

Nevertheless, there is the fact that it is difficult to archive the accuracy in CDPR kinematic modeling similar to that o f rigid-body manipulators. For very large dimension CDPRs, in most applications, simplified kinematic modeling solutions might be used to get satisfying results. However, for relatively small workspace CDPRs, in order to increase their applicability, it is important to

M. Gouttefarde et al. (Eds.): CableCon 2021, MMS 104, pp. 13–23, 2021.
https://doi.org/10.1007/978-3-030-75789-2_2

have a good CDPR kinematic modeling solution. We will try to find a solution regarding this problem in this paper, starting with the most basic CDPR system: a 2-DOF planar suspended CDPR with massless cables.

2 Geometric Models

Figure 1 shows the geometric structure of the 2-DOF suspended CDPR. It consists of a point mass driven by two massless cables. For this model, we have several assumptions as follows:

- The cables are massless. The cable segments that do not mount on a pulley or drum are considered as straight lines.
- The end-effector point is always lies on the plane $(\mathbf{x}_0, \mathbf{z}_0)$ of the global coordinate frame $\langle O_0 \rangle$.
- All the cables connect to the point mass at point B.
- The position of the pulleys and drums are fixed.
- The last two pulleys have the same coordinate in z-axis ($z_{P1} = z_{P2}$).
- The total lengths of the cables are the same.
- The deformations of the cable length due to friction (between pulleys or drums and cables) or due to cable twisting or ovalization are neglected.
- There is no deformation of the base frame of the CDPR system (the mounting positions of the pulleys and drums are fixed in space).
- The distance between the drums and first pulleys are large enough so that the cables lie properly in corresponding grooves of the drums.

2.1 Inverse Geometric Problem

When fully considering the influence of all the components in the cable driving system, we will find the equations where the inputs are the coordinates of point B and the output is the position of the cable exit point on the drum (or the number of cable turns on the drums).

Figure 2 describes the cable-driving system for one cable. For simplicity, the cable index $i\,(i = 1, 2)$ will be omitted in all the terms in this section. For each cable, the total cable length is constant:

$$L_T = L^{(D)} + L^{(D,E)} + L^{(E,P)} + L^{(P,B)} \tag{1}$$

where:

- L_T: total cable length.
- $L^{(D)}$: length of cable segment on the drum.
- $L^{(D,E)}$: length of cable segment from the cable exit point on the drum to the tangent point of the first pulley.
- $L^{(E,P)}$: length of cable segment from the input tangent point of the first pulley to the input tangent point of the last pulley.

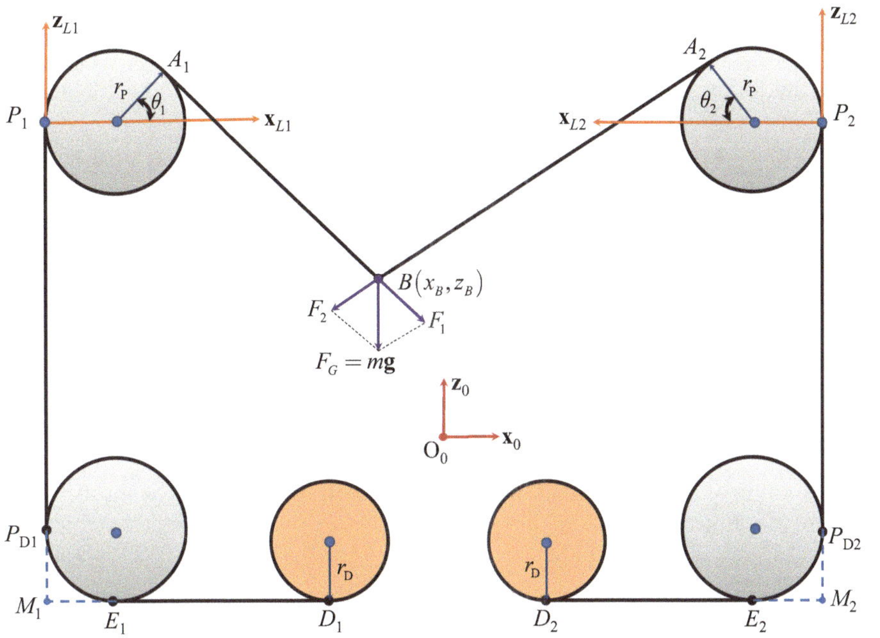

Fig. 1. Geometric structure of 2-DOF suspended CDPR.

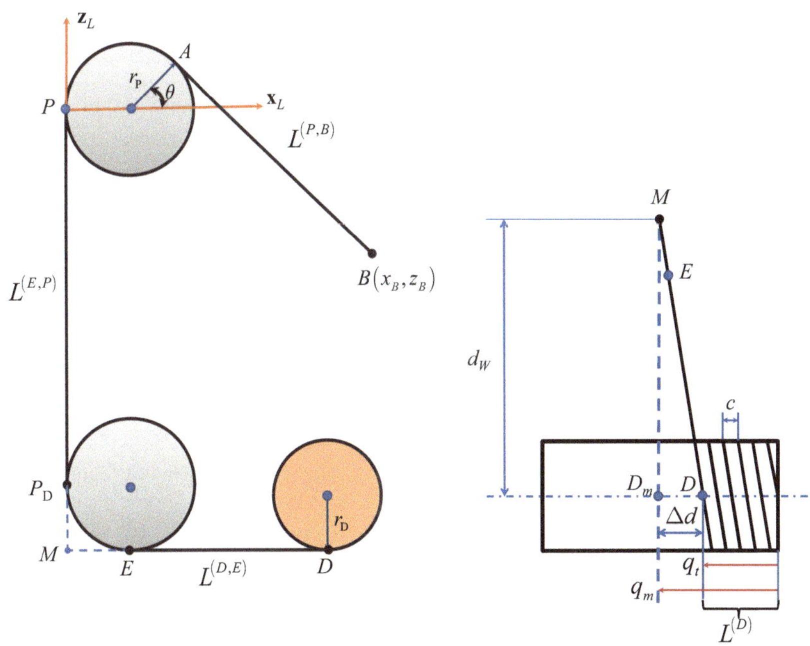

Fig. 2. Cable-driving system for one cable.

- $L^{(P,B)}$: length of cable segment from the input tangent point of the last pulley to the end-effector point.

In the following, we will compute each part of the cable. The cable segment on the drum is calculated as follows:

$$L^{(D)} = 2\pi\sqrt{r_D^2 + c^2} \cdot q_t = r_q \cdot q_t \tag{2}$$

with:

- r_D: drum radius
- c: helical step on the drum (distance between two consecutive grooves on the drum).
- q_t: current number of cable turns on the drum at time t.
- $r_q = 2\pi\sqrt{r_D^2 + c^2}$: conversion ratio between the cable length on the drum and the number of cable turns.

The cable segment from the cable exit point on the drum to the input tangent point of the first pulley:

$$\begin{aligned} L^{(D,E)} &= MD - ME \\ &= \sqrt{d_w^2 + \Delta d^2} - r_p \\ &= \sqrt{d_w^2 + c^2\left(q_t - q_m\right)^2} - r_p \end{aligned} \tag{3}$$

with:

- d_w: distance from M to the orthogonal cable exit point D_m on the drum (Fig. 2).
- $\Delta d = c \cdot \Delta q$: distance from the cable exit point D to the orthogonal cable exit point D_m on the drum.
- r_P: pulley radius.
- $\Delta q = (q_t - q_m)$: difference in cable turns from the point D_m to D.
- q_m: number of cable turns on the drum with respect to the cable exit point at D_m.

The cable segment from the input tangent point of the first pulley to the input tangent point of the last pulley:

$$L^{(E,P)} = L^{(P_D,P)} + \frac{\pi}{2} r_P \tag{4}$$

The term $L^{(P_D,P)}$ is determined since all the pulleys are fixed at the pivot points (P_D, P).

The cable segment from the input tangent point of the last pulley to the end-effector point:

$$L^{(P,B)} = L_t + (\pi - \theta_t)\, r_P \tag{5}$$

The term L_t is calculated as follows:

$$L_t = \|AB\| = \sqrt{\left(x_B - x_A\right)^2 + \left(z_B - z_A\right)^2} \tag{6}$$

In Eq. (6), the coordinates of point A need to be computed through the tangent equation of the line AB to the pulley inner circle, expressed in the local cable coordinate frame at point P. The subscript "L" is used to indicate that all the terms are expressed in the cable local coordinate frame $\langle P \rangle$. The coordinates of point B in frame $\langle P \rangle$ are:

$$\begin{aligned} x_{LB} &= |x_B - x_P| \\ z_{LB} &= z_B - z_P \end{aligned} \tag{7}$$

The coordinates of point A in the frame $\langle P \rangle$ are computed as follows:

$$\begin{aligned} z_{LA} &= \frac{z_{LB} r_P^2 + r_P \left| x_{LB} - r_P \right| \sqrt{z_{LB}^2 + \left(x_{LB} - r_P\right)^2 - r_P^2}}{z_{LB}^2 + \left(x_{LB} - r_P\right)^2} \\ x_{LA} &= r_P + \sqrt{r_P^2 - z_{LA}^2} \end{aligned} \tag{8}$$

The coordinates of point A in the global coordinate frame $\langle O_0 \rangle$ are:

$$\begin{aligned} x_A &= x_P + x_{LA} \cos(\gamma) \\ z_A &= z_P + z_{LA} \end{aligned} \tag{9}$$

with the angle γ is computed as follows:

$$\gamma = \operatorname{atan2}(0, x_B - x_P) \tag{10}$$

To determine the term q_t, let us rewrite the Eq. (1) after substituting all the calculated terms:

$$L_T = r_q q_t + \sqrt{d_w^2 + c^2 \left(q_t - q_m\right)^2} + L^{(P_D, P)} + L_t + \left(\frac{3\pi}{2} - \theta_t - 1\right) r_P \tag{11}$$

In Eq. (11), we have:

- L_t is determined from Eqs. (6)–(10).
- θ_t can be computed from Eq. (7):

$$\theta_t = \operatorname{asin}\left(\frac{z_{LA}}{r_P}\right) \tag{12}$$

- q_m is determined from the CDPR design.

Deriving Eq. (11), we get:

$$\sqrt{d_w^2 + c^2 \left(q_t - q_m\right)^2} = M - r_q q_t \tag{13}$$

where:

$$M = L_T - L^{(P_D, P)} - L_t - \left(\frac{3\pi}{2} - \theta_t - 1\right) r_P \tag{14}$$

Equation (13) has two roots:

$$q_t = \frac{-\left(c^2 q_m - r_q M\right) \pm \sqrt{\triangle}}{r_q^2 - c^2} \tag{15}$$

with:

$$\triangle = \left(r_q^2 - c^2\right) d_w^2 + c^2 \left(r_q q_m - M\right)^2 \tag{16}$$

Note that q_t always has positive value. Besides, from Eq. (13) we have:

$$M - r_q q_t > 0 \tag{17}$$

If we choose $q_t = \frac{-\left(c^2 q_m - r_q M\right) + \sqrt{\triangle}}{r_q^2 - c^2}$, substituting this solution in condition (17) we have:

$$M - r_q \frac{-\left(c^2 q_m - r_q M\right) + \sqrt{\triangle}}{r_q^2 - c^2} = \frac{-c^2 \left(M - r_q q_m\right) - \sqrt{\triangle}}{r_q^2 - c^2} > 0 \tag{18}$$

which is not valid. Therefore, Eq. (13) has only one root:

$$q_t = \frac{-\left(c^2 q_m - r_q M\right) - \sqrt{\triangle}}{r_q^2 - c^2} \tag{19}$$

Equation (19) is the solution of the inverse geometric problem: given the coordinates of the end-effector point, find the number of cable turns on the drum.

Also, from Eq. (19) one can see that, in order to calculate exactly the value of q_t, we need to know exactly the values of the following terms: q_m, d_w, L_T and the coordinates of each pulley (in this study case are the coordinates of points P and P_D). The determination of these terms depends on the accuracy of machining and assembling all the parts of the CDPR.

2.2 Direct Geometric Problem

For the CDPR in this work, we will derive the constraint equations that describe the relations between the coordinates of the end-effector point and the number of cable turns on each drum q_{ti} $(i = 1, 2)$.

Derive Eq. (11) for two cables, we have:

$$L_i + \theta_i r_P = L_{ti} \qquad (i = 1, 2) \tag{20}$$

with:

$$L_i = L_{Ti} - r_q q_{ti} - \sqrt{d_{wi}^2 + c^2 \left(q_{ti} - q_{mi}\right)^2} - L^{(P_{Di}, P_i)} - \left(\frac{3\pi}{2} - 1\right) r_P \qquad (i = 1, 2) \tag{21}$$

Therefore, if $q_{ti}\,(i=1,2)$ are known, one can compute $L_i\,(i=1,2)$. From Eq. (20), we have:

$$(L_i+\theta_i r_P)^2=(x_{LAi}-x_{LBi})^2+(z_{LAi}-z_{LBi})^2 \qquad (i=1,2) \tag{22}$$

All the coordinates in Eq. (22) are expressed in the local cable coordinate frames that have their origins at the mounting points of respective pulleys, and are computed as follows $(i=1,2)$:

$$\begin{cases} x_{LAi}=r_P\,(C_i+1)\\ z_{LAi}=r_P S_i\\ x_{LBi}=|x_B-x_{Pi}|\\ z_{LBi}=z_B-z_{Pi}\end{cases} \tag{23}$$

with $C_i=\cos(\theta_i)\,, S_i=\sin(\theta_i)\quad(i=1,2)$. All the terms that have the subscripts "L" and "i" are expressed in the local cable coordinate frames that have their origins at P_i. All the terms without the subscript "L" are expressed in the global coordinate frame $\langle O_0\rangle$. The tangent equations of the line BA_i expressed in the local cable coordinate frames:

$$(x_{LAi}-r_P)(x_{LBi}-r_P)+z_{LAi}z_{LBi}=r_P^2 \qquad (i=1,2) \tag{24}$$

which leads to:

$$\begin{aligned} & r_P C_i\,(x_{LBi}-r_P)+r_P S_i z_{LBi}=r_P^2\\ \Rightarrow & \begin{cases} z_{LBi}=\dfrac{r_P-C_i\,(x_{LBi}-r_P)}{S_i}\\ x_{LBi}=\dfrac{r_P-S_i z_{LBi}}{C_i}+r_P\end{cases} \qquad (i=1,2)\end{aligned} \tag{25}$$

Note that the angles $\theta_i\ \ (i=1,2)$ are strictly constrained by the following condition:

$$0<\theta_i<\frac{\pi}{2} \qquad (i=1,2) \tag{26}$$

Therefore, $S_i>0,\ C_i>0\ \ (i=1,2)$, thus $z_{LBi},\ x_{LBi}$ are always exist. The condition (26) only happens when the end-effector point lies within the feasible workspace which satisfies the conditions of the tension distribution problem where the end-effector is at its static equilibrium. In the cases the end-effector moves, there could be situations where the angles θ_i fall outside the range in (26) but the cable tensions are still positive. These cases are not considered in this study.

Substituting Eqs. (23), (25) into Eq. (22), we have:

$$(L_i+\theta_i r_P)^2=\frac{1}{C_i^2}(z_{LAi}-z_{LBi})^2 \qquad (i=1,2) \tag{27}$$

For suspended CDPR, we have the condition:

$$z_{LAi}>z_{LBi} \qquad (i=1,2) \tag{28}$$

Thus, Eq. (27) becomes:

$$\begin{aligned} L_i + \theta_i r_P &= \frac{1}{C_i}\left(z_{LAi} - z_{LBi}\right) \\ \Leftrightarrow z_{LBi} &= r_P S_i - C_i\left(L_i + \theta_i r_P\right) \qquad (i = 1, 2) \end{aligned} \tag{29}$$

From Eq. (23) and Eq. (29) we have:

$$z_B = z_{Pi} + r_P S_i - C_i\left(L_i + \theta_i r_P\right) \qquad (i = 1, 2) \tag{30}$$

Substituting Eq. (29) into Eq. (25) we have:

$$\begin{aligned} x_{LBi} &= \frac{r_P - S_i\left[r_P S_i - C_i\left(L_i + \theta_i r_P\right)\right]}{C_i} + r_P \\ \Rightarrow x_{LBi} &= r_P\left(C_i + 1\right) + S_i\left(L_i + \theta_i r_P\right) \qquad (i = 1, 2) \end{aligned} \tag{31}$$

From Eq. (23) and Eq. (31) we have:

$$\left|x_B - x_{Pi}\right| = r_P\left(C_i + 1\right) + S_i\left(L_i + \theta_i r_P\right) \qquad (i = 1, 2) \tag{32}$$

Combining Eq. (30) and Eq. (32) and note the condition $x_{P1} < x_B < x_{P2}$, we have a system of four equations with four unknown terms $(x_B, z_B, \theta_1, \theta_2)$:

$$\begin{cases} x_B = x_{P1} + r_P\left(C_1 + 1\right) + S_1\left(L_1 + \theta_1 r_P\right) \\ x_B = x_{P2} - r_P\left(C_2 + 1\right) - S_2\left(L_2 + \theta_2 r_P\right) \\ z_B = z_{P1} + r_P S_1 - C_1\left(L_1 + \theta_1 r_P\right) \\ z_B = z_{P2} + r_P S_2 - C_2\left(L_2 + \theta_2 r_P\right) \end{cases} \tag{33}$$

The system of equations (33) does not have analytical solution. That is why we have to use numerical methods to compute the unknown terms $(x_B, z_B, \theta_1, \theta_2)$ in which (x_B, z_B) is the solution of the direct geometric problem.

Derive the Eq. (33) and note that $z_{P1} = z_{P2}$, we can transform the system of equations to:

$$\begin{cases} S_{12}\left(L_1 + r_P\theta_1\right) - \left(x_{P2} - x_{P1} - 2r_P\right)C_2 + r_P\left(C_{12} + 1\right) = 0 \\ S_{12}\left(L_2 + r_P\theta_2\right) - \left(x_{P2} - x_{P1} - 2r_P\right)C_1 + r_P\left(C_{12} + 1\right) = 0 \\ x_B = x_{P1} + r_P\left(C_1 + 1\right) + S_1\left(L_1 + \theta_1 r_P\right) \\ z_B = z_{P1} + r_P S_1 - C_1\left(L_1 + \theta_1 r_P\right) \end{cases} \tag{34}$$

where $S_{12} = \sin\left(\theta_1 + \theta_2\right), C_{12} = \cos\left(\theta_1 + \theta_2\right)$.

Equation (34) can be solved through two steps as follows.

Step 1: *Using numerical methods to compute* θ_1, θ_2 from the first two equations:

$$\begin{cases} f_1\left(\boldsymbol{\theta}\right) = S_{12}\left(L_1 + r_P\theta_1\right) - \left(x_{P2} - x_{P1} - 2r_P\right)C_2 + r_P\left(C_{12} + 1\right) = 0 \\ f_2\left(\boldsymbol{\theta}\right) = S_{12}\left(L_2 + r_P\theta_2\right) - \left(x_{P2} - x_{P1} - 2r_P\right)C_1 + r_P\left(C_{12} + 1\right) = 0 \end{cases} \tag{35}$$

with $\boldsymbol{\theta} = \left[\theta_1 \quad \theta_2\right]^T$.

Let us define $\mathbf{F}(\boldsymbol{\theta}) = [f_1(\boldsymbol{\theta}) \quad f_2(\boldsymbol{\theta})]^T$. Using the Newton-Raphson method, the solution of (35) is updated as follows:

$$\begin{aligned} \boldsymbol{\theta}^{(k)} &= \boldsymbol{\theta}^{(k-1)} + \triangle\boldsymbol{\theta}^{(k-1)} \\ \triangle\boldsymbol{\theta}^{(k-1)} &= -\mathbf{J}_a\left(\boldsymbol{\theta}^{(k-1)}\right)^{-1}\mathbf{F}\left(\boldsymbol{\theta}^{(k-1)}\right) \end{aligned} \tag{36}$$

with:

$$\begin{aligned} \mathbf{J}_a(\boldsymbol{\theta}) &= \begin{bmatrix} \dfrac{\partial f_1}{\partial \theta_1} & \dfrac{\partial f_1}{\partial \theta_2} \\ \dfrac{\partial f_2}{\partial \theta_1} & \dfrac{\partial f_2}{\partial \theta_2} \end{bmatrix} \\ \frac{\partial f_1}{\partial \theta_1} &= C_{12}(L_1 + r_P\theta_1) \\ \frac{\partial f_1}{\partial \theta_2} &= C_{12}(L_1 + r_P\theta_1) + (x_{P2} - x_{P1} - 2r_P)S_2 - r_P S_{12} \\ \frac{\partial f_2}{\partial \theta_1} &= C_{12}(L_2 + r_P\theta_2) + (x_{P2} - x_{P1} - 2r_P)S_1 - r_P S_{12} \\ \frac{\partial f_2}{\partial \theta_2} &= C_{12}(L_2 + r_P\theta_2) \end{aligned} \tag{37}$$

For a general case, one can start with the initial condition:

$$\begin{aligned} \boldsymbol{\theta}^{(0)} &= [\pi/4 \quad \pi/4]^T \\ \triangle\boldsymbol{\theta}^{(0)} &= -\mathbf{J}_a\left(\boldsymbol{\theta}^{(0)}\right)^{-1}\mathbf{F}\left(\boldsymbol{\theta}^{(0)}\right) \end{aligned} \tag{38}$$

We reach a solution when $\|\triangle\boldsymbol{\theta}\| \to 0$.

In the trajectory tracking control problem, where we have to continuously update $\boldsymbol{\theta} = \boldsymbol{\theta}_t$, the initial value $\boldsymbol{\theta}_t^{(0)}$ can be chosen as its previous values: $\boldsymbol{\theta}_t^{(0)} = \boldsymbol{\theta}_{t-1}$.

Step 2: *Update the coordinates of the end-effector point* by the last two equations in Eq. (34).

3 Kinematic Models

The kinematic models of the robot describe the relation between the velocity vector of the end-effector point B with respect to the angular velocity of the drums in term of $\dot{q}_{t1}, \dot{q}_{t2}$. In order to derive the kinematic equations, we will consider Eqs. (21), (33).

Taking the time derivatives of both sides of Eq. (21), we have:

$$\dot{L}_i = r_i \dot{q}_{ti} \qquad (i = 1, 2) \tag{39}$$

with:

$$r_i = -\left[r_q + \frac{c^2(q_{ti} - q_{mi})}{\sqrt{d_{wi}^2 + c^2(q_{ti} - q_{mi})^2}}\right] \qquad (i = 1, 2) \tag{40}$$

Taking the time derivatives of both sides of equations in Eq. (33), we have:

$$\begin{cases} \dot{x}_B = r_1 S_1 \dot{q}_{t1} + C_1 \left(L_1 + r_P \theta_1\right) \dot{\theta}_1 \\ \dot{x}_B = -r_2 S_2 \dot{q}_{t2} - C_2 \left(L_2 + r_P \theta_2\right) \dot{\theta}_2 \\ \dot{z}_B = -r_1 C_1 \dot{q}_{t1} + S_1 \left(L_1 + r_P \theta_1\right) \dot{\theta}_1 \\ \dot{z}_B = -r_2 C_2 \dot{q}_{t2} + S_2 \left(L_2 + r_P \theta_2\right) \dot{\theta}_2 \end{cases} \tag{41}$$

Rewrite Eq. (41) in the matrix form:

$$\begin{bmatrix} \dot{x}_B \\ \dot{x}_B \\ \dot{z}_B \\ \dot{z}_B \end{bmatrix} = \begin{bmatrix} C_1 \left(L_1 + r_P \theta_1\right) & 0 & r_1 S_1 & 0 \\ 0 & -C_2 \left(L_2 + r_P \theta_2\right) & 0 & -r_2 S_2 \\ S_1 \left(L_1 + r_P \theta_1\right) & 0 & -r_1 C_1 & 0 \\ 0 & S_2 \left(L_2 + r_P \theta_2\right) & 0 & -r_2 C_2 \end{bmatrix} \cdot \begin{bmatrix} \dot{\theta}_1 \\ \dot{\theta}_2 \\ \dot{q}_{t1} \\ \dot{q}_{t2} \end{bmatrix} \tag{42}$$

The solution for Eq. (42) is:

$$\begin{cases} \dot{\theta}_1 = \dfrac{1}{L_1 + r_P \theta_1} \left(C_1 \dot{x}_B + S_1 \dot{z}_B\right) \\ \dot{\theta}_2 = \dfrac{1}{L_2 + r_P \theta_2} \left(C_2 \dot{x}_B - S_2 \dot{z}_B\right) \\ \dot{q}_{t1} = \dfrac{1}{r_1} \left(S_1 \dot{x}_B - C_1 \dot{z}_B\right) \\ \dot{q}_{t2} = \dfrac{1}{r_2} \left(S_2 \dot{x}_B + C_2 \dot{z}_B\right) \end{cases} \tag{43}$$

From the last two equations in Eq. (43), we get the direct and inverse kinematic equations of the CDPR:

$$\begin{matrix} \dot{\mathbf{X}} = \mathbf{J} \dot{\mathbf{q}} \\ \dot{\mathbf{q}} = \mathbf{J}^{-1} \dot{\mathbf{X}} \end{matrix} \tag{44}$$

where $\mathbf{X} = \left[x_B, z_B\right]^T, \dot{\mathbf{X}} = \left[\dot{x}_B, \dot{z}_B\right]^T, \mathbf{q} = \left[q_{t1}, q_{t2}\right]^T, \dot{\mathbf{q}} = \left[\dot{q}_{t1}, \dot{q}_{t2}\right]^T$ (with unit in *rps - round per second*) and $\mathbf{J}, \mathbf{J}^{-1}$ are the Jacobian matrix and its inverse:

$$\mathbf{J} = \frac{1}{S_{12}} \begin{bmatrix} r_1 C_2 & -r_2 C_1 \\ -r_1 S_2 & -r_2 S_1 \end{bmatrix}, \qquad \mathbf{J}^{-1} = \begin{bmatrix} \dfrac{S_1}{r_1} & -\dfrac{C_1}{r_1} \\ -\dfrac{S_2}{r_2} & -\dfrac{C_2}{r_2} \end{bmatrix} \tag{45}$$

Note that we have $r_i > 0 \, (i = 1, 2)$ and $S_{12} > 0$ $\left(\text{because } 0 < \theta_1, \theta_2 < \dfrac{\pi}{2}\right)$, thus there is no singularity in the CDPR feasible work space.

4 Conclusion

This paper presented a solution in solving the direct and inverse geometric and kinematic problems of a 2-DOF planar suspended CDPR. Analytical solution can be found for the inverse geometric problem whereas only numerical solution can be found for the direct geometric problem of the CDPR.

The found solutions help to improve the accuracy of the CDPR kinematic modeling since the influence of the pulleys and cable winding drums are both considered. These results can be used in the kinematic analysis as well as in the kinematic-based control of CDPRs with a similar structure to the one in this work.

References

1. Merlet, J.-P.: The kinematics of redundant N-1 wire driven parallel robot. In: IEEE International Conference on Robotics and Automation, USA (2012)
2. Abbasnejad, G., Carricato, M.: Direct geometrico-static problem of underconstrained cable-driven parallel robots with n cables. IEEE Trans. Robot. **31**, 468–478 (2015)
3. Pott, A., Schmidt, V.: On the forward kinematics of cable-driven parallel robots. In: IEEE International Conference on Intelligent Robots and Systems, pp. 928–934 (2015)
4. Paty, T., Binaud, N., Caro, S., Segonds, S.: Cable-driven parallel robot modelling considering pulley kinematics and cable elasticity. Mech. Mach. Theory **159**(5), 104263 (2021)
5. Santos, J.C., Gouttefarde, M.: A real-time capable forward kinematics algorithm for cable-driven parallel robots considering pulley kinematics. In: Advances in Robot Kinematics 2020, Slovenia (2020)
6. Jin, X., Jung, J., Piao, J., Choi, E., Park, J.-O., Kim, C.-S.: Solving the pulley inclusion problem for a cable-driven parallel robotic system: Extended kinematics and twin-pulley mechanism. J. Mech. Sci. Technol. **32**(6), 2829–2838 (2018)
7. Schmidt, V.L.: Modelling techniques and reliable real-time implementation of kinematics for cable-driven parallel robots using polymer fiber cables. Ph.D. thesis, Fraunhofer Verlag, Stuttgart (2017)
8. Mousavi, M.R., Ghanbari, M., Moosavian, S.A.A., et al.: Explicit dynamics of redundant parallel cable robots. Nonlinear Dyn. **94**, 2077–2096 (2018)

A New Performance Index for Underactuated Cable-Driven Parallel Robots

Edoardo Idà(✉) and Marco Carricato

Department of Industrial Engineering, University of Bologna, Bologna, Italy
{edoardo.ida2,marco.carricato}@unibo.it
http://www.irmalab.org/

Abstract. Cable-driven parallel robots (*CDPRs*) equipped with less cables than the end-effector (*EE*) degrees of freedom (*DoFs*) are underactued and underconstrained by design. These characteristics imply that only a subset of the *EE DoFs* can be assigned for planning purposes, and that the *EE* can freely move when actuators are locked. The performance of this class of manipulators has yet to be fully analyzed, since common performance indices for fully or redundantly actuated robots may not directly apply. In this paper, a novel index is proposed, which is tailored for *suspended* underactuated *CDPRs*. This index analyses cable-tension sensitivity to cable-length variation, and aims at determining if a static equilibrium configuration is attainable under bounded cable-length control errors. When 4- or 5-cable *UACDPRs* have to perform positioning tasks, the optimization of such an index may be used to determine the safest *EE* orientation with respect to actuation errors.

Keywords: Underactuated cable-driven parallel robots · Underconstrained cable-driven parallel robot · Performance index · Trajectory optimization

1 Introduction

Underactuated Cable-driven parallel robots (underactuated *CDPRs*, or *UACDPRs* in short) control the end-effector (*EE*) pose by means of n extendable cables, whose number is smaller than the *EE* degrees of freedom (*DoFs*). *UACDPRs* are intrinsically underconstrained [1], since their *EE* is subject to less constaint actions than the number of its *DoFs*. As a consequence, only a subset of the *EE* coordinates can be directly controlled, and the *EE* preserves some freedoms once the actuators are locked. Given the reduced mechanical complexity and increased workspace accessibility of *UACDPRs*, a growing number of studies are being conducted on these systems in the field of trajectory planning [2–5], geometrico-static modeling [6], control [7], calibration [8], and workspace analysis [9]. On the other hand, *UACDPRs* performance evaluation is still an open problem.

M. Gouttefarde et al. (Eds.): CableCon 2021, MMS 104, pp. 24–36, 2021.
https://doi.org/10.1007/978-3-030-75789-2_3

Robot performance indices have been vastly investigated [10], and a few specific studies exist on their application to suspended, completely actuated *CDPRs*: the *global conditioning index* was used in [11] to evaluate the kinematic dexterity performance of a generic 6-cable *CDPR*; *rotation* and *displacement sensitivity to actuator displacement errors* were investigated in [12] for translational *CDPRs* and in [13] for rotational *CDPRs*. As far as redundantly constrained *CDPRs* are concerned, the ratio between maximum and minimum cable tensions was introduced in [14] (called *tension factor* in [15]); the *maximum acceptable horizontal distance between the platform reference point and the center of mass of the set composed of the platform and a payload* was introduced in [16] to determine whether a handling task was feasible; and the maximum wrench that can be applied in a given direction, the *wrench exertion capability*, was explored in [17].

In this paper, a novel index for the evaluation of the geometrico-static performance of *UACDPRs* is introduced. This index, called *maximum tension variation under a cable displacement error*, aims at analysing the influence of actuator errors on *UACDPRs* static cable tensions. This analysis allows one to determine whether an equilibrium configuration may be safely attained under bounded cable-length control errors, with cables remaining likely taut. In addition, if 4- or 5-cable *UACDPRs* have to perform positioning tasks and the additional cables are used only to widen their static reachable workspace, the optimization of such an index could be used to determine the safest *EE* orientation with respect to actuation errors.

This paper is structured as follows. Section 2 recalls the geometrico-static model of *UACDPRs* with massless straight cables. Section 3 introduces the *maximum tension variation under a cable displacement error* index, based on the differential analysis of the geometrico-static model. An application is proposed in Sect. 4, where the orientation of a 4-cable *UACDPR* for a given reference position is determined as the one that minimizes the new index. In the end, conclusions and future works are discussed in Sect. 5.

2 Geometrico-Static Model

A *UACDPR* consists of a mobile platform coupled to the base by $n < 6$ cables, which can be coiled and uncoiled by motorized winches. In the following, $Oxyz$ is an inertial frame, whereas $Gx'y'z'$ is a mobile frame attached to the moving platform center of mass. The *EE* pose is described by the position vector $\mathbf{p}$ of G, and the rotation matrix $\mathbf{R}$ (Fig. 1). In this paper, $\mathbf{R}$ is parametrized by a minimal set of angles $\boldsymbol{\epsilon} = [\phi, \theta, \chi]^T$. The platform generalized coordinates are thus $\boldsymbol{\zeta} = \left[\mathbf{p}^T\, \boldsymbol{\epsilon}^T\right]^T$.

2.1 Geometric Model

For the sake of simplicity, we assume that the i-th cable is massless and rigid, and is guided into the workspace through an eyelet at point B_i, described in the inertial frame by vector $\mathbf{b}_i$, and attached to the platform in point A_i, described in the moving frame by (the constant) ${}^P\mathbf{a}_i'$, and in the inertial frame by $\mathbf{a}_i = \mathbf{p} + \mathbf{a}_i' = \mathbf{p} + \mathbf{R}\,{}^P\mathbf{a}_i'$, where $\mathbf{a}_i' = \mathbf{R}\,{}^P\mathbf{a}_i'$. The cable vector connecting points A_i and B_i is $\boldsymbol{\rho}_i = \mathbf{a}_i - \mathbf{b}_i$ and, if l_i is the length of the i-th cable and $(\cdot)$ indicates the scalar product between vectors, the geometric constraint imposed by the cable on the platform is:

$$\boldsymbol{\rho}_i \cdot \boldsymbol{\rho}_i - l_i^2 = 0 \tag{1}$$

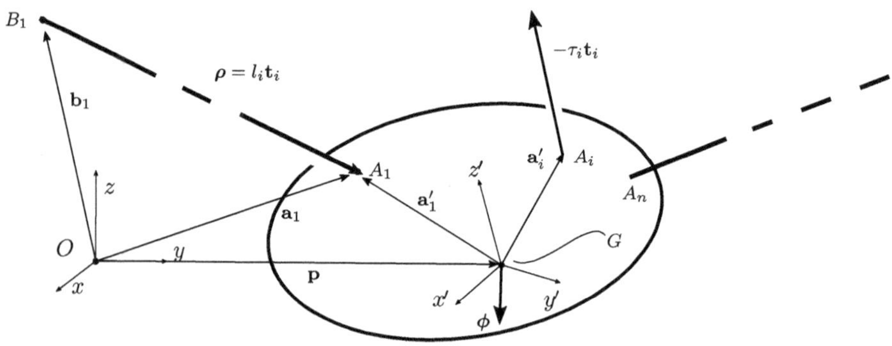

Fig. 1. Geometrico-static model of the *UACDPR*

2.2 Kinematic Model

If $\boldsymbol{\omega}$ is the *EE* angular velocity, $\mathbf{v} = [\dot{\mathbf{p}}^T \boldsymbol{\omega}^T]^T$ the *EE* twist, $\mathbf{t}_i = \boldsymbol{\rho}_i / l_i$ the cable direction, and $(\times)$ indicates the vector product between vectors, differentiating Eq. (1) with respect to time yields:

$$\boldsymbol{\xi}_i \cdot \mathbf{v} - \dot{l}_i = 0, \qquad \boldsymbol{\xi}_i = \begin{bmatrix} \mathbf{t}_i \\ \mathbf{a}_i' \times \mathbf{t}_i \end{bmatrix}, \qquad i = 1, \ldots, n \tag{2}$$

The n relations in Eq. (2) can be grouped as:

$$\boldsymbol{\Xi}\mathbf{v} - \dot{\mathbf{l}} = \mathbf{0}_{n\times 1}, \qquad \dot{\mathbf{l}} = [\dot{l}_1 \cdots \dot{l}_i \cdots \dot{l}_n]^T, \qquad \boldsymbol{\Xi} = [\boldsymbol{\xi}_1 \cdots \boldsymbol{\xi}_i \cdots \boldsymbol{\xi}_n]^T \in \mathbb{R}^{n\times 6} \tag{3}$$

The *EE* twist $\mathbf{v}$ is related to the pose time derivative $\dot{\boldsymbol{\zeta}}$ through a matrix $\mathbf{D}$, whose formulation depends on the specific orientation parametrization employed and whose elements depend on the *EE* pose [4]:

$$\mathbf{v} = \mathbf{D}(\boldsymbol{\zeta})\dot{\boldsymbol{\zeta}} = \mathbf{D}\dot{\boldsymbol{\zeta}} \tag{4}$$

If Eq. (4) is substituted in Eq. (3), one gets:

$$\boldsymbol{\Xi}\mathbf{D}\dot{\boldsymbol{\zeta}} - \dot{\mathbf{l}} = \mathbf{J}\dot{\boldsymbol{\zeta}} - \dot{\mathbf{l}} = \mathbf{0}_{n\times 1}, \qquad \mathbf{J} = \boldsymbol{\Xi}\mathbf{D} \in \mathbb{R}^{n\times 6} \tag{5}$$

Matrices $\boldsymbol{\Xi}$ and $\mathbf{J}$ are the kinematic and analytical Jacobian matrices of the system and, since the robot is underactuated, they are rectangular; in case the manipulator is not in a singular configuration (whether a geometric or a representation one), rank($\boldsymbol{\Xi}$) = rank($\mathbf{J}$) = $n < 6$, which is strictly less than the number of *EE DoFs*. Thus, only n coordinates of the *EE* pose can be controlled by varying the *UACDPR* cable lengths, while the remaining $\lambda = 6 - n$ are to be determined according to the mechanical equilibrium of the platform. In addition, even if the actuators are locked and cable lengths are kept constant, λ freedoms remains. The n *controlled coordinates* are denoted as $\boldsymbol{\zeta}_c \in \mathbb{R}^n$, whereas the non-controllable coordinates will be referred to as *free coordinates* and denoted as $\boldsymbol{\zeta}_f \in \mathbb{R}^\lambda$. For simplicity sake, we will assume that the controlled coordinates are the first in the array $\boldsymbol{\zeta}$, that is, $\boldsymbol{\zeta} = [\boldsymbol{\zeta}_c{}^T \, \boldsymbol{\zeta}_f{}^T]^T$.

This coordinate partition is particularly useful in *trajectory planning* of *UACDPRs*, but it attains additional kinematic meaning. In fact, the *EE* twist $\mathbf{v}$ can be decomposed into two contributions, namely a *free twist*[1] $\mathbf{v}_f$ and a *controlled twist* $\mathbf{v}_c$, so that:

$$\mathbf{v} = \mathbf{v}_f + \mathbf{v}_c \tag{6}$$

The *free twist* is defined as the *EE* twist when the platform is in free motion, and it can be derived as the solution of Eq. (3) when $\dot{\mathbf{l}} = \mathbf{0}_{n\times 1}$, namely:

$$\boldsymbol{\Xi}\mathbf{v}_f = \mathbf{0}_{n\times 1} \tag{7}$$

The solution to Eq. (7) is readily obtained by considering the right nullspace $\boldsymbol{\Xi}^\perp$ of matrix $\boldsymbol{\Xi}$. By definition, the right nullspace of a $(n \times 6)$ matrix is a $(6 \times \lambda)$ matrix such that $\boldsymbol{\Xi}\boldsymbol{\Xi}^\perp = \mathbf{0}_{n\times\lambda}$, thus its columns define a basis for the free twist $\mathbf{v}_f$:

$$\mathbf{v}_f = \boldsymbol{\Xi}^\perp \mathbf{c} \qquad \text{for some } \mathbf{c} \in \mathbb{R}^\lambda \tag{8}$$

If $\mathbf{J}^\perp$ is the right nullspace of matrix $\mathbf{J}$, we also have:

$$\boldsymbol{\Xi}\mathbf{v}_f = \boldsymbol{\Xi}\,\underbrace{\mathbf{D}\dot{\boldsymbol{\zeta}}}_{\mathbf{v}_f} = \underbrace{\boldsymbol{\Xi}\mathbf{D}}_{\mathbf{J}}\,\dot{\boldsymbol{\zeta}} = \mathbf{0}_{n\times 1} \qquad \Longrightarrow \qquad \dot{\boldsymbol{\zeta}} = \mathbf{J}^\perp \mathbf{c}' \;\; \text{for some } \mathbf{c}' \in \mathbb{R}^\lambda \tag{9}$$

so that:

$$\mathbf{v}_f = \mathbf{D}\dot{\boldsymbol{\zeta}} = \mathbf{D}\mathbf{J}^\perp \mathbf{c}' \tag{10}$$

By comparing Eqs. (8) and (10) and by choosing $\mathbf{c} = \mathbf{c}'$, we have:

$$\boldsymbol{\Xi}^\perp = \mathbf{D}\mathbf{J}^\perp \tag{11}$$

The coefficients $\mathbf{c}$ coincide with the free coordinates derivative $\dot{\boldsymbol{\zeta}}_f$, if $\boldsymbol{\Xi}^\perp$ (and thus $\mathbf{J}^\perp$, cf. Eq. (11)) is computed according to the following procedure. First, we partition the analytical Jacobian matrix $\mathbf{J}$ as:

$$\mathbf{J} = \boldsymbol{\Xi}\mathbf{D} = \boldsymbol{\Xi}\left[\mathbf{D}_c \, \mathbf{D}_f\right] = \left[\boldsymbol{\Xi}\mathbf{D}_c \, \boldsymbol{\Xi}\mathbf{D}_f\right] = \left[\mathbf{J}_c \, \mathbf{J}_f\right] \tag{12}$$

where $\mathbf{D}_c \in \mathbb{R}^{6\times n}$, $\mathbf{D}_f \in \mathbb{R}^{6\times\lambda}$, $\mathbf{J}_c = \boldsymbol{\Xi}\mathbf{D}_c \in \mathbb{R}^{n\times n}$, and $\mathbf{J}_f = \boldsymbol{\Xi}\mathbf{D}_f \in \mathbb{R}^{n\times\lambda}$. Matrix $\mathbf{J}^\perp$ can be symbolically computed under the assumption that rank($\mathbf{J}_c$) =

[1] The free twist is derived in [18] and reported here in order to make the paper self contained.

n and, thus, invertible. By assuming so[2], matrix $\mathbf{J}^{\perp}$ can be derived from Eq. (5) by setting $\dot{\mathbf{l}} = \mathbf{0}_{n\times 1}$:

$$\mathbf{J}\dot{\boldsymbol{\zeta}} = \mathbf{J}_c\dot{\boldsymbol{\zeta}}_c + \mathbf{J}_f\dot{\boldsymbol{\zeta}}_f = \mathbf{0}_{n\times 1} \qquad \Longrightarrow \qquad \dot{\boldsymbol{\zeta}}_c = -\mathbf{J}_c^{-1}\mathbf{J}_f\dot{\boldsymbol{\zeta}}_f \tag{13}$$

and finally:

$$\dot{\boldsymbol{\zeta}} = \begin{bmatrix}\dot{\boldsymbol{\zeta}}_c\\ \dot{\boldsymbol{\zeta}}_f\end{bmatrix} = \begin{bmatrix}-\mathbf{J}_c^{-1}\mathbf{J}_f\\ \mathbf{I}_{\lambda\times\lambda}\end{bmatrix}\dot{\boldsymbol{\zeta}}_f = \mathbf{J}^{\perp}\dot{\boldsymbol{\zeta}}_f, \qquad \mathbf{J}^{\perp} = \begin{bmatrix}-\mathbf{J}_c^{-1}\mathbf{J}_f\\ \mathbf{I}_{\lambda\times\lambda}\end{bmatrix} \tag{14}$$

$$\mathbf{v}_f = \mathbf{D}\dot{\boldsymbol{\zeta}} = \mathbf{D}\mathbf{J}^{\perp}\dot{\boldsymbol{\zeta}}_f = \boldsymbol{\Xi}^{\perp}\dot{\boldsymbol{\zeta}}_f, \qquad \boldsymbol{\Xi}^{\perp} = \mathbf{D}\mathbf{J}^{\perp} \tag{15}$$

where $\mathbf{c} = \dot{\boldsymbol{\zeta}}_f$. It should be noted that the expression of $\dot{\boldsymbol{\zeta}}$ provided in Eq. (14) is valid *only when* $\dot{\mathbf{l}} = \mathbf{0}_{n\times 1}$.

The *controlled twist* is defined as the *EE* twist due to cable actuation only, that is, the twist resulting from a zero free-coordinate derivative, $\dot{\boldsymbol{\zeta}}_f = \mathbf{0}_{\lambda\times 1}$. The expression of $\mathbf{v}_c$ is determined by considering Eq. (4) and setting $\dot{\boldsymbol{\zeta}}_f = \mathbf{0}_{\lambda\times 1}$:

$$\mathbf{v}_c = \mathbf{D}\dot{\boldsymbol{\zeta}} = \begin{bmatrix}\mathbf{D}_c\ \mathbf{D}_f\end{bmatrix}\begin{bmatrix}\dot{\boldsymbol{\zeta}}_c\\ \mathbf{0}_{\lambda\times 1}\end{bmatrix} = \begin{bmatrix}\mathbf{D}_c\dot{\boldsymbol{\zeta}}_c\\ \mathbf{0}_{\lambda\times 1}\end{bmatrix} \tag{16}$$

In addition, by recalling position (12):

$$\dot{\mathbf{l}} = \mathbf{J}\dot{\boldsymbol{\zeta}} = \begin{bmatrix}\mathbf{J}_c\ \mathbf{J}_f\end{bmatrix}\begin{bmatrix}\dot{\boldsymbol{\zeta}}_c\\ \mathbf{0}_{\lambda\times 1}\end{bmatrix} \qquad \Longrightarrow \qquad \dot{\boldsymbol{\zeta}}_c = \mathbf{J}_c^{-1}\dot{\mathbf{l}} \tag{17}$$

and thus:

$$\mathbf{v}_c = \begin{bmatrix}\mathbf{D}_c\dot{\boldsymbol{\zeta}}_c\\ \mathbf{0}_{\lambda\times 1}\end{bmatrix} = \begin{bmatrix}\mathbf{D}_c\mathbf{J}_c^{-1}\dot{\mathbf{l}}\\ \mathbf{0}_{\lambda\times 1}\end{bmatrix} = \begin{bmatrix}\mathbf{D}_c\mathbf{J}_c^{-1}\\ \mathbf{0}_{\lambda\times n}\end{bmatrix}\dot{\mathbf{l}} = \boldsymbol{\Xi}^{\parallel}\dot{\mathbf{l}}, \qquad \boldsymbol{\Xi}^{\parallel} = \begin{bmatrix}\mathbf{D}_c\mathbf{J}_c^{-1}\\ \mathbf{0}_{\lambda\times n}\end{bmatrix} \in \mathbb{R}^{6\times n} \tag{18}$$

while an expression of $\dot{\boldsymbol{\zeta}}$ which is valid *only when* $\dot{\boldsymbol{\zeta}}_f = \mathbf{0}_{\lambda\times 1}$ is deduced from Eq. (17):

$$\dot{\boldsymbol{\zeta}} = \begin{bmatrix}\mathbf{J}_c^{-1}\\ \mathbf{0}_{\lambda\times n}\end{bmatrix}\dot{\mathbf{l}} = \mathbf{J}^{\parallel}\dot{\mathbf{l}}, \qquad \mathbf{J}^{\parallel} = \begin{bmatrix}\mathbf{J}_c^{-1}\\ \mathbf{0}_{\lambda\times n}\end{bmatrix}, \qquad \boldsymbol{\Xi}^{\parallel} = \mathbf{D}\mathbf{J}^{\parallel} \tag{19}$$

The newly defined matrix $\boldsymbol{\Xi}^{\parallel}$ has the property to be a right inverse for the kinematic Jacobian $\boldsymbol{\Xi}$, namely $\boldsymbol{\Xi}\boldsymbol{\Xi}^{\parallel} = \mathbf{I}_{n\times n}$, when $\dot{\boldsymbol{\zeta}}_f = \mathbf{0}_{\lambda\times 1}$. This can be verified by substituting Eq. (18) in (3):

$$\boldsymbol{\Xi}\mathbf{v}_c - \dot{\mathbf{l}} = \boldsymbol{\Xi}\boldsymbol{\Xi}^{\parallel}\dot{\mathbf{l}} - \dot{\mathbf{l}} = \mathbf{0}_{n\times 1} \qquad \Longrightarrow \qquad \boldsymbol{\Xi}\boldsymbol{\Xi}^{\parallel} = \mathbf{I}_{n\times n} \tag{20}$$

Finally, one has:

$$\mathbf{v} = \mathbf{v}_f + \mathbf{v}_c = \boldsymbol{\Xi}^{\perp}\dot{\boldsymbol{\zeta}}_f + \boldsymbol{\Xi}^{\parallel}\dot{\mathbf{l}} \tag{21}$$

[2] By suitably choosing controlled and free coordinates, it can be proven that this is always true for *UACDPR* [18], but the demonstration is omitted due to space limitation.

It can be verified that the expression of $\mathbf{v}$ in Eq. (21) verifies the *EE* first-order kinematics expressed in Eq. (3):

$$\boldsymbol{\Xi}\mathbf{v} - \dot{\mathbf{l}} = \boldsymbol{\Xi}\left(\boldsymbol{\Xi}^{\perp}\dot{\boldsymbol{\zeta}}_f + \boldsymbol{\Xi}^{\parallel}\dot{\mathbf{l}}\right) - \dot{\mathbf{l}} = \underbrace{\boldsymbol{\Xi}\boldsymbol{\Xi}^{\perp}}_{\mathbf{0}_{n\times\lambda}}\dot{\boldsymbol{\zeta}}_f + \underbrace{\boldsymbol{\Xi}\boldsymbol{\Xi}^{\parallel}}_{\mathbf{I}_{n\times n}}\dot{\mathbf{l}} - \dot{\mathbf{l}} = \mathbf{0}_{n\times 1} \tag{22}$$

At last, the complete expression of the pose derivative can be written as:

$$\dot{\boldsymbol{\zeta}} = \mathbf{J}^{\perp}\dot{\boldsymbol{\zeta}}_f + \mathbf{J}^{\parallel}\dot{\mathbf{l}} \tag{23}$$

2.3 Static Modelling

The static model of the *EE* is determined by considering that cable tensions counteract the external actions applied onto the *EE*. If we consider gravity as the only load applied to the *EE* center of mass G, the static equilibrium yields (see Fig. 1):

$$\boldsymbol{\Xi}^T\boldsymbol{\tau} = \mathbf{f}, \qquad \mathbf{f} = \begin{bmatrix} \boldsymbol{\phi} \\ \mathbf{0}_{3\times 1} \end{bmatrix} \tag{24}$$

where $\boldsymbol{\phi} = m\mathbf{g}$, m is the *EE* mass, $\mathbf{g}$ is the gravitational acceleration, and $\boldsymbol{\tau} = [\tau_1, \cdots, \tau_n]^T$ is the array containing the cable tensions.

The static constraint that the *EE* coordinates must statify at equilibrium (regardless of the sign of cable tensions), is obtained by pre-multiplying Eq. (24) by ${\boldsymbol{\Xi}^{\perp}}^T$:

$${\boldsymbol{\Xi}^{\perp}}^T\boldsymbol{\Xi}^T\boldsymbol{\tau} = \mathbf{0}_{\lambda\times 1} = {\boldsymbol{\Xi}^{\perp}}^T\mathbf{f} \tag{25}$$

where ${\boldsymbol{\Xi}^{\perp}}^T\boldsymbol{\Xi}^T = \mathbf{0}_{\lambda\times n}$ by definition of nullspace. Cable tensions can be computed by pre-multiplying Eq. (24) by $\boldsymbol{\Xi}^{\parallel T}$, instead:

$$\boldsymbol{\Xi}^{\parallel T}\boldsymbol{\Xi}^T\boldsymbol{\tau} = \boldsymbol{\tau} = \boldsymbol{\Xi}^{\parallel T}\mathbf{f} \tag{26}$$

The geometrico-static model is established by considering Eq. (1) for $i = 1, \ldots, n$ and Eq. (25):

$$\begin{cases} \boldsymbol{\rho}_1 \cdot \boldsymbol{\rho}_1 - l_1^2 = 0 \\ \vdots \\ \boldsymbol{\rho}_\mathbf{n} \cdot \boldsymbol{\rho}_n - l_n^2 = 0 \\ {\boldsymbol{\Xi}^{\perp}}^T\mathbf{f} = \mathbf{0}_{\lambda\times 1} \end{cases} \tag{27}$$

The non-linear system (27) has 6 equations in $6+n$ unknowns, namely the 6 *EE* coordinates $\boldsymbol{\zeta}$ and the n cable lengths $\mathbf{l}$. By assigning $\mathbf{l}$, the forward geometrico-static problem is established; if instead $\boldsymbol{\zeta}_c$ is assigned, the inverse geometrico-static problem is determined. Both problems are well-posed, since the system (27) has 6 equations in 6 unknowns, and they may admit multiple real solutions: a solution is acceptable if cable tensions, obtained from Eq. (26), are strictly positive, and the equilibrium is stable [1].

3 Maximum Tension Variation Under a Cable Displacement Error

In this section, the *maximum tension variation under a cable displacement error*, denoted by $\sigma_{\tau\%,\infty}$, is formulated by considering how cable length variations, or errors, influence cable tensions around equilibrium.

The equilibrium of a *UACDPR* may vary for two reasons, namely a change of cable lengths or external wrench. In this paper, we only consider the former. By differentiating Eq. (26), we evaluate how tensions vary due to pose variations:

$$\mathrm{d}\boldsymbol{\tau} = \mathrm{d}\left(\boldsymbol{\Xi}^{\parallel T}\mathbf{f}\right) = \frac{\partial\left(\boldsymbol{\Xi}^{\parallel T}\mathbf{f}\right)}{\partial\boldsymbol{\zeta}}\mathrm{d}\boldsymbol{\zeta} \tag{28}$$

If we consider $\boldsymbol{\Xi}^T\boldsymbol{\tau} = \mathbf{f}$, the partial derivative in Eq. (28) is:

$$\frac{\partial(\boldsymbol{\Xi}^{\parallel T}\mathbf{f})}{\partial\boldsymbol{\zeta}} = \left(\frac{\partial\boldsymbol{\Xi}^{\parallel T}}{\partial\boldsymbol{\zeta}}\mathbf{f} + \boldsymbol{\Xi}^{\parallel T}\frac{\partial\mathbf{f}}{\partial\boldsymbol{\zeta}}\right) = \left(\frac{\partial\boldsymbol{\Xi}^{\parallel T}}{\partial\boldsymbol{\zeta}}\boldsymbol{\Xi}^T\boldsymbol{\tau} + \boldsymbol{\Xi}^{\parallel T}\frac{\partial\mathbf{f}}{\partial\boldsymbol{\zeta}}\right) \tag{29}$$

$$\frac{\partial(\boldsymbol{\Xi}^{\parallel T}\mathbf{f})}{\partial\boldsymbol{\zeta}} = -\left(\boldsymbol{\Xi}^{\parallel T}\frac{\partial\boldsymbol{\Xi}^T}{\partial\boldsymbol{\zeta}}\boldsymbol{\tau} - \boldsymbol{\Xi}^{\parallel T}\frac{\partial\mathbf{f}}{\partial\boldsymbol{\zeta}}\right) = -\boldsymbol{\Xi}^{\parallel T}\left[\sum_{i=1}^{n}\left(\tau_i\frac{\partial\boldsymbol{\xi}_i}{\partial\boldsymbol{\zeta}}\right) - \frac{\partial\mathbf{f}}{\partial\boldsymbol{\zeta}}\right] \tag{30}$$

where we have taken into consideration that, since $\boldsymbol{\Xi}^{\parallel T}\boldsymbol{\Xi}^T = \mathbf{I}_{n\times n}$:

$$\frac{\partial\boldsymbol{\Xi}^{\parallel T}}{\partial\boldsymbol{\zeta}}\boldsymbol{\Xi}^T + \boldsymbol{\Xi}^{\parallel T}\frac{\partial\boldsymbol{\Xi}^T}{\partial\boldsymbol{\zeta}} = \mathbf{0}_{n\times n} \tag{31}$$

It can be shown by computation that Eq. (30) can be rewritten as[3]:

$$\frac{\partial(\boldsymbol{\Xi}^{\parallel T}\mathbf{f})}{\partial\boldsymbol{\zeta}} = -\boldsymbol{\Xi}^{\parallel T}\mathbf{K}\mathbf{D} \tag{32}$$

$$\mathbf{K} = \sum_{i=1}^{n}\frac{\tau_i}{l_i}\begin{bmatrix}\mathbf{T}_i & -\mathbf{T}_i\tilde{\mathbf{a}}'_i \\ \tilde{\mathbf{a}}'_i\mathbf{T}_i & -\tilde{\mathbf{a}}'_i\mathbf{T}_i\tilde{\mathbf{a}}'_i\end{bmatrix}, \qquad \mathbf{T}_i = \mathbf{I}_{3\times 3} - \mathbf{t}_i\mathbf{t}_i^T \tag{33}$$

where $(\tilde{\cdot})$ denotes the skew-symmetric representation of a vector. Matrix $\mathbf{K}$ is referred to as *Geometric Stiffness* of the *CDPR* [18,19], because it depends on geometry, and it is fundamentally different from the so-called *Passive Stiffness* generated by cable deformations (not considered in this paper).

From Eq. (23), we infer that the pose variation $\mathrm{d}\boldsymbol{\zeta}$ in Eq. (28) is produced by a variation of either cable lengths or free-pose coordinates, namely:

$$\mathrm{d}\boldsymbol{\zeta} = \mathbf{J}^{\perp}\mathrm{d}\boldsymbol{\zeta}_f + \mathbf{J}^{\parallel}\mathrm{d}\mathbf{l} \tag{34}$$

[3] Detailed computation of $\mathbf{K}$ is reported in [18] for a more general case, i.e. when cables are guided by swivel pulleys and an arbitrary external wrench is applied to the *EE*.

On the other hand, the variation of free coordinates is not independent, since Eq. (25) must hold in the newly attained equilibrium. Thus, the free-coordinate variation upon a cable-length change can be determined by differentiating Eq. (25):

$$\mathrm{d}\left(\boldsymbol{\Xi}^{\perp^T}\mathbf{f}\right) = \frac{\partial\left(\boldsymbol{\Xi}^{\perp^T}\mathbf{f}\right)}{\partial\boldsymbol{\zeta}}\mathrm{d}\boldsymbol{\zeta} = \mathbf{0}_{\lambda\times 1} \tag{35}$$

If we substitute Eq. (34) in Eq. (35), we obtain:

$$\mathrm{d}\left(\boldsymbol{\Xi}^{\perp^T}\mathbf{f}\right) = \frac{\partial\left(\boldsymbol{\Xi}^{\perp^T}\mathbf{f}\right)}{\partial\boldsymbol{\zeta}}\mathbf{J}^{\perp}\mathrm{d}\boldsymbol{\zeta}_f + \frac{\partial\left(\boldsymbol{\Xi}^{\perp^T}\mathbf{f}\right)}{\partial\boldsymbol{\zeta}}\mathbf{J}^{\parallel}\mathrm{d}\mathbf{l} = \mathbf{0}_{\lambda\times 1} \tag{36}$$

Since $\boldsymbol{\Xi}^T\boldsymbol{\tau} = \mathbf{f}$ at equilibrium, the partial derivative in Eq. (36) is:

$$\frac{\partial\left(\boldsymbol{\Xi}^{\perp^T}\mathbf{f}\right)}{\partial\boldsymbol{\zeta}} = \left(\frac{\partial\boldsymbol{\Xi}^{\perp^T}}{\partial\boldsymbol{\zeta}}\mathbf{f} + \boldsymbol{\Xi}^{\perp^T}\frac{\partial\mathbf{f}}{\partial\boldsymbol{\zeta}}\right) = \left(\frac{\partial\boldsymbol{\Xi}^{\perp^T}}{\partial\boldsymbol{\zeta}}\boldsymbol{\Xi}^T\boldsymbol{\tau} + \boldsymbol{\Xi}^{\perp^T}\frac{\partial\mathbf{f}}{\partial\boldsymbol{\zeta}}\right) \tag{37}$$

$$\frac{\partial\left(\boldsymbol{\Xi}^{\perp^T}\mathbf{f}\right)}{\partial\boldsymbol{\zeta}} = -\left(\boldsymbol{\Xi}^{\perp^T}\frac{\partial\boldsymbol{\Xi}^T}{\partial\boldsymbol{\zeta}}\boldsymbol{\tau} - \boldsymbol{\Xi}^{\perp^T}\frac{\partial\mathbf{f}}{\partial\boldsymbol{\zeta}}\right) = -\boldsymbol{\Xi}^{\perp^T}\left(\frac{\partial\boldsymbol{\Xi}^T}{\partial\boldsymbol{\zeta}}\boldsymbol{\tau} - \frac{\partial\mathbf{f}}{\partial\boldsymbol{\zeta}}\right) \tag{38}$$

where we have taken into consideration that, since $\boldsymbol{\Xi}^{\perp^T}\boldsymbol{\Xi}^T = \mathbf{0}_{\lambda\times n}$:

$$\frac{\partial\boldsymbol{\Xi}^{\perp^T}}{\partial\boldsymbol{\zeta}}\boldsymbol{\Xi}^T + \boldsymbol{\Xi}^{\perp^T}\frac{\partial\boldsymbol{\Xi}^T}{\partial\boldsymbol{\zeta}} = \mathbf{0}_{\lambda\times n} \tag{39}$$

and ultimately (cf. Eqs. (30) and (32)):

$$\frac{\partial\left(\boldsymbol{\Xi}^{\perp^T}\mathbf{f}\right)}{\partial\boldsymbol{\zeta}} = -\boldsymbol{\Xi}^{\perp^T}\mathbf{KD} \tag{40}$$

By substituting Eq. (40) in Eq. (36), and considering Eqs. (19) and (15), one obtains:

$$\boldsymbol{\Xi}^{\perp^T}\mathbf{KD}\mathbf{J}^{\perp}\mathrm{d}\boldsymbol{\zeta}_f + \boldsymbol{\Xi}^{\perp^T}\mathbf{KD}\mathbf{J}^{\parallel}\mathrm{d}\mathbf{l} = \mathbf{K}_f^{\perp}\mathrm{d}\boldsymbol{\zeta}_f + \mathbf{K}_l^{\perp}\mathrm{d}\mathbf{l} = \mathbf{0}_{\lambda\times 1} \tag{41}$$

where:

$$\mathbf{K}_f^{\perp} = \boldsymbol{\Xi}^{\perp^T}\mathbf{K}\boldsymbol{\Xi}^{\perp} \in \mathbb{R}^{\lambda\times\lambda}, \qquad \mathbf{K}_l^{\perp} = \boldsymbol{\Xi}^{\perp^T}\mathbf{K}\boldsymbol{\Xi}^{\parallel} \in \mathbb{R}^{\lambda\times n} \tag{42}$$

Note that, if the equilibrium is stable, $\mathbf{K}_f^{\perp}$ is positive definite, thus invertible [18].

Then, the free-coordinate variation upon a variation of cable lengths is:

$$\mathrm{d}\boldsymbol{\zeta}_f = -\mathbf{K}_f^{\perp^{-1}}\mathbf{K}_l^{\perp}\mathrm{d}\mathbf{l} \tag{43}$$

and substituting Eq. (43) in Eq. (34) the overall pose variation is determined as:

$$\mathrm{d}\boldsymbol{\zeta} = \left(-\mathbf{J}^{\perp}\mathbf{K}_f^{\perp^{-1}}\mathbf{K}_l^{\perp} + \mathbf{J}^{\parallel}\right)\mathrm{d}\mathbf{l} \tag{44}$$

In the end, the variation of cable tension upon a change of the cable lengths is evaluated by substituting Eq. (32) and (44) into Eq. (28):

$$\mathrm{d}\boldsymbol{\tau} = -\boldsymbol{\Xi}^{\parallel T}\mathbf{KD}\left(-\mathbf{J}^{\perp}\mathbf{K}_f^{-\perp}\mathbf{K}_l^{\perp} + \mathbf{J}^{\parallel}\right)\mathrm{d}\mathbf{l} = \mathbf{K}_\tau \mathrm{d}\mathbf{l} \tag{45}$$

$$\mathbf{K}_\tau = -\boldsymbol{\Xi}^{\parallel T}\mathbf{K}\left(-\boldsymbol{\Xi}^{\perp}\mathbf{K}_f^{-\perp}\mathbf{K}_l^{\perp} + \boldsymbol{\Xi}^{\parallel}\right) \tag{46}$$

Finally, the *maximum tension variation under a cable displacement error* $\sigma_{\tau\%,\infty}$ is defined as:

$$\sigma_{\tau\%,\infty} = \max_{\|\mathrm{d}\mathbf{l}\|_\infty = 1} \|\mathrm{d}\boldsymbol{\tau}_{\%}\|_\infty \tag{47}$$

where $\mathrm{d}\boldsymbol{\tau}_{\%} = 100[(\mathrm{d}\tau_1)/\tau_1, \cdots, (\mathrm{d}\tau_n)/\tau_n]$ is the percentage tension error, and $\|\cdot\|_\infty$ indicates the vector ∞-norm. As highlighted in [20], the use of the infinity-norm attains the clearest physical meaning, since it is consistent with a realistic actuation error model, namely $-\mathrm{d}l_{i,\max} \leq \mathrm{d}l_i \leq \mathrm{d}l_{i,\max}$ for $i = 1, \ldots, n$. The definition of the index based on the percentage variation of cable tensions, and not on their absolute variation, is justified by the fact that a small tension variation on an almost slack cable is comparably relevant to a large tension variation on a well-taut cable.

By considering Eq. (45) and the definitions of matrix norms [21], we have:

$$\sigma_{\tau\%,\infty} = \max_{\|\mathrm{d}\mathbf{l}\|_\infty = 1} \|\mathrm{d}\boldsymbol{\tau}_{\%}\|_\infty = \|\mathbf{K}_{\tau\%}\|_\infty \tag{48}$$

where $\mathbf{K}_\tau\%$ is, by accounting for Eq. (46):

$$\mathbf{K}_\tau\% = 100\mathbf{T}\mathbf{K}_\tau, \qquad \mathbf{T} = \mathrm{diag}(\boldsymbol{\tau})^{-1} \tag{49}$$

Note that the computation of the infinity norm of a matrix is straightforward, as it consists of computing the 1-norms of its row vectors and choosing the largest.

Notice that $\sigma_{\tau\%,\infty}$ is not dimensionless. If cable lengths are expressed in millimeters, $\sigma_{\tau\%,\infty}$ is the percentage error on the estimated cable tension if a control error of at the most 1mm occurs in cable lengths.

4 Application

As an application of the proposed performance index, we optimize the static orientation of a 4-cable *UACDPR EE* that has to perform a positioning task. Since only 3 of the 4 controllable coordinates of the robot needs to be assigned to perform the task, i t is reasonable to investigate which orientation among the possibly infinite ones attainable by *EE* would locally optimize the performance of the robot. Then, for an assigned *EE* position, its orientation may be determined as the one minimizing $\sigma_{\tau\%,\infty}$, so that, should cable lengths not be perfectly controlled by the robot, the risk of cables becoming slack is minimized.

Consider a *UACDPR* with the following parameters:

$${}^P\mathbf{a}_1 = \begin{bmatrix} 0.2 \\ 0.3 \\ 0.3 \end{bmatrix} \mathrm{m}, \quad {}^P\mathbf{a}_2 = \begin{bmatrix} 0.2 \\ -0.3 \\ 0.3 \end{bmatrix} \mathrm{m}, \quad {}^P\mathbf{a}_3 = \begin{bmatrix} -0.2 \\ -0.3 \\ 0.3 \end{bmatrix} \mathrm{m}, \quad {}^P\mathbf{a}_4 = \begin{bmatrix} -0.2 \\ 0.3 \\ 0.3 \end{bmatrix} \mathrm{m},$$

$$\mathbf{b}_1 = \begin{bmatrix} 1.5 \\ 1 \\ 0 \end{bmatrix} \text{m}, \quad \mathbf{b}_2 = \begin{bmatrix} 1.5 \\ -1 \\ 0 \end{bmatrix} \text{m}, \quad \mathbf{b}_3 = \begin{bmatrix} -1.5 \\ -1 \\ 0 \end{bmatrix} \text{m}, \quad \mathbf{b}_4 = \begin{bmatrix} -1.5 \\ 1 \\ 0 \end{bmatrix} \text{m}, \quad m = 1\text{Kg}$$

and the position $\mathbf{p}_0 = [0\,0\, -2]^T$ m that the *EE* has to reach. Orientation $\boldsymbol{\epsilon}$ is parametrized with *Roll-Pitch-Yaw* Euler angles, namely the *EE* rotation matrix is $\mathbf{R}(\boldsymbol{\epsilon}) = \mathbf{R}_z(\phi)\mathbf{R}_y(\theta)\mathbf{R}_x(\chi)$, with $\mathbf{R}_i$, $i = x, y, z$ being elementary rotation matrices.

The optimal orientation of the *EE* is determined by solving the non-linear constrained optimization:

$$\boldsymbol{\epsilon}_{opt} = \min_{\boldsymbol{\epsilon}} \sigma_{\boldsymbol{\tau}\%,\infty}(\boldsymbol{\epsilon}) \tag{50}$$

subject to:

$$\boldsymbol{\Xi}^{\perp^T}\mathbf{f} = \mathbf{0}_{\lambda\times 1} \qquad \text{and} \qquad \boldsymbol{\tau} \succeq \tau_{min} \tag{51}$$

where $\succeq$ denotes element-wise inequality, and $\tau_{min} = 2$N is the inferior tension limit. The model presented in this paper was implemented in Matlab, and Eq. (50) was solved by the *fmincon* function. The results are reported in Table 1.

Equilibrium orientations in the neighbourhood of the optimal one were also investigated, in order to understand the entity of the variation of $\sigma_{\boldsymbol{\tau}\%,\infty}$ in the assigned $\mathbf{p}_0$. Accordingly, angle ϕ was continuously varied around ϕ_{opt}, while θ and χ were determined as the solutions of the inverse geometrico-static problem in Eq. (27). The results of this analysis are reported in Fig. 2a, and, in the authors' opinion, one detail is particularly interesting. The worst (highest) values of $\sigma_{\boldsymbol{\tau}\%,\infty}$ are obtained for those equilibrium orientations that results in one cable reaching the lower tension limit τ_{min}: even though $\sigma_{\boldsymbol{\tau}\%,\infty}$ does not vary considerably in the assigned position of the *EE*, the highest value of this index, where one cable is the least tense, implies that those configurations are intrinsically not safe to reach if cable lengths are not perfectly controlled, since one cable may become slack.

Table 1. Optimization results in $\mathbf{p}_0$

$\boldsymbol{\epsilon}$ [rad]	$\mathbf{l}$ [m]	$\boldsymbol{\tau}$ [N]	$\sigma_{\boldsymbol{\tau}\%,\infty}$ [%/m]
$\left[0, 0, 0\right]^T$	$\left[2.252, 2.252, 2.252, 2.252\right]^T$	$\left[3.25, 3.25, 3.25, 3.25\right]^T$	2117

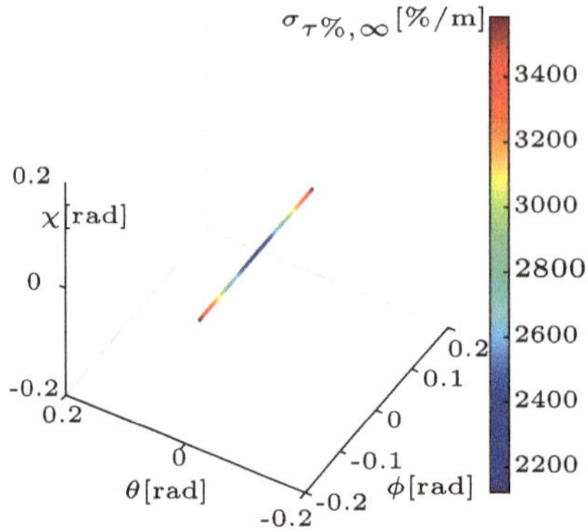

(a) $\sigma_{\tau\%,\infty}$ [%/m] *for* $\mathbf{p}_0$

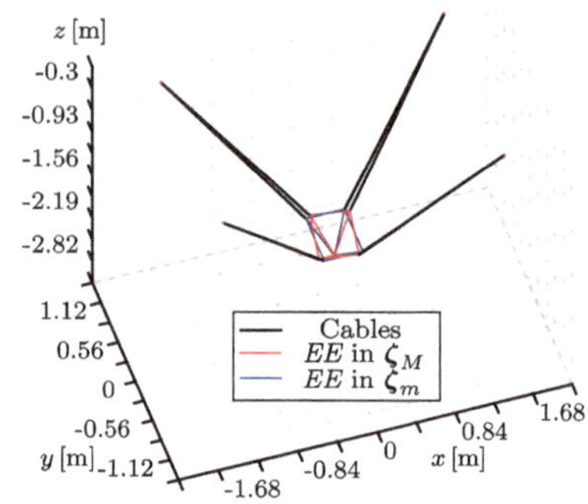

(b) EE *configurations in* $\mathbf{p}_0$

Fig. 2. Simulation results for $\mathbf{p}_0$

In order to emphasize why this result is critical, consider two equilibrium orientations $\boldsymbol{\epsilon}_M$ and $\boldsymbol{\epsilon}_m$ attainable by the *UACDPR* in $\mathbf{p}_0$, respectively maximizing and minimizing $\sigma_{\tau\%,\infty}$ (see Fig. 2b, and note that the subscript M and m will be used to denote all variables associated with such orientations):

$$\boldsymbol{\zeta}_M = \begin{bmatrix} 0 \\ 0 \\ -2 \\ -0.161 \\ 0 \\ 0 \end{bmatrix} [\text{m}, \text{rad}], \quad \mathbf{l}_M = \begin{bmatrix} 2.237 \\ 2.273 \\ 2.237 \\ 2.273 \end{bmatrix} \text{m}, \quad \boldsymbol{\tau}_M = \begin{bmatrix} 4.48 \\ 2 \\ 4.48 \\ 2 \end{bmatrix} \text{N}, \quad \sigma_{\tau\%,\infty,M} = 3615\,\%/\text{m}$$

$$\boldsymbol{\zeta}_m = \begin{bmatrix} 0 \\ 0 \\ -2 \\ 0 \\ 0 \\ 0 \end{bmatrix} [\text{m}, \text{rad}], \quad \mathbf{l}_m = \begin{bmatrix} 2.252 \\ 2.252 \\ 2.252 \\ 2.252 \end{bmatrix} \text{m}, \quad \boldsymbol{\tau}_m = \begin{bmatrix} 3.25 \\ 3.25 \\ 3.25 \\ 3.25 \end{bmatrix} \text{N}, \quad \sigma_{\tau\%,\infty,m} = 2117\,\%/\text{m}$$

Suppose now that the second and the fourth cable are 1 cm longer than their values in $\mathbf{l}_M$ and $\mathbf{l}_m$: these new cable lengths are denoted as $\mathbf{l}_M{}^\star$ and $\mathbf{l}_m{}^\star$, respectively. The solutions to the forward geometrico-static problem for $\mathbf{l}_M{}^\star$ and $\mathbf{l}_m{}^\star$ are:

$$\boldsymbol{\zeta}_M{}^\star = \begin{bmatrix} 0 \\ 0 \\ -2.004 \\ -0.207 \\ 0 \\ 0 \end{bmatrix} [\text{m}, \text{rad}], \quad \mathbf{l}_M{}^\star = \begin{bmatrix} 2.237 \\ 2.283 \\ 2.237 \\ 2.283 \end{bmatrix} \text{m}, \quad \boldsymbol{\tau}_M{}^\star = \begin{bmatrix} 4.85 \\ 1.63 \\ 4.85 \\ 1.63 \end{bmatrix} \text{N}$$

$$\boldsymbol{\zeta}_m{}^\star = \begin{bmatrix} 0 \\ 0 \\ -2.006 \\ -0.045 \\ 0 \\ 0 \end{bmatrix} [\mathrm{m}, \mathrm{rad}], \quad \mathbf{l}_m{}^\star = \begin{bmatrix} 2.252 \\ 2.262 \\ 2.252 \\ 2.262 \end{bmatrix} \mathrm{m}, \quad \boldsymbol{\tau}_m{}^\star = \begin{bmatrix} 3.59 \\ 2.90 \\ 3.59 \\ 2.90 \end{bmatrix} \mathrm{N}$$

It is clear that for the pose associated with $\mathbf{l}_M{}^\star$ cable tensions drop below the tension limit, while for the pose associated with $\mathbf{l}_m{}^\star$ this issues does not occur.

5 Conclusions

In this paper, a novel index $\sigma_{\tau\%,\infty}$ for the evaluation of geometrico-static performances of *UACDPRs*, called *maximum tension variation under a cable displacement error*, was introduced and analytically derived. Its relevance was demonstrated on a 4-cable *UACDPR*, by determining the orientations minimizing and maximizing said index when the position of the *EE* is assigned. It was shown that, in case cable lengths corresponding to minimal and maximal values of $\sigma_{\tau\%,\infty}$ exhibit errors, the risk of loosing tension in one cable is the highest in the latter configuration.

In the future, this index will be used for the characterizing and optimizing *UACDPRs* reachable workspace.

References

1. Carricato, M., Merlet, J.: Stability analysis of underconstrained cable-driven parallel robots. IEEE Trans. Rob. **29**(1), 288–296 (2013)
2. Fink, J., Michael, N., Kim, S., Kumar, V.: Planning and control for cooperative manipulation and transportation with aerial robots. Int. J. Robot. Res. **30**(3), 324–334 (2011)
3. Zoso, N., Gosselin, C.: Point-to-point motion planning of a parallel 3-DoF underactuated cable-suspended robot. In: 2012 IEEE International Conference on Robotics and Automation, pp. 2325–2330, May 2012
4. Idà, E., Bruckmann, T., Carricato, M.: Rest-to-rest trajectory planning for underactuated cable-driven parallel robots. IEEE Trans. Rob. **35**(6), 1338–1351 (2019)
5. Idà, E., Briot, S., Carricato, M.: Robust trajectory planning of under-actuated cable-driven parallel robot with 3 cables. In: Lenarčič, J., Siciliano, B. (eds.) Advances in Robot Kinematics 2020, pp. 65–72. Springer, Cham (2021)
6. Abbasnejad, G., Carricato, M.: Direct geometrico-static problem of underconstrained cable-driven parallel robots with n cables. IEEE Trans. Rob. **31**(2), 468–478 (2015)
7. Zarei, M., Aflakian, A., Kalhor, A., Masouleh, M.T.: Oscillation damping of nonlinear control systems based on the phase trajectory length concept: An experimental case study on a cable-driven parallel robot. Mech. Mach. Theory **126**, 377–396 (2018)
8. Idá, E., Merlet, J.-P., Carricato, M.: Automatic self-calibration of suspended underactuated cable-driven parallel robot using incremental measurements. In: Pott, A., Bruckmann, T. (eds.) Cable-Driven Parallel Robots, pp. 333–344. Springer, Cham (2019)

9. Kumar, A.A., Antoine, J.-F., Zattarin, P., Abba, G.: Workspace analysis of a 4 cable-driven spatial parallel robot. In: Arakelian, V., Wenger, P. (eds.) ROMANSY 22 - Robot Design, Dynamics and Control, pp. 204–212. Springer, Cham (2019)
10. Patel, S., Sobh, T.: Manipulator performance measures - a comprehensive literature survey. J. Intell. Robot. Syst. **77**(3), 547–570 (2015)
11. Pusey, J., Fattah, A., Agrawal, S., Messina, E.: Design and workspace analysis of a 6–6 cable-suspended parallel robot. Mech. Mach. Theory **39**(7), 761–778 (2004)
12. Mottola, G., Gosselin, C., Carricato, M.: Effect of actuation errors on a purely-translational spatial cable-driven parallel robot. In: 2019 IEEE 9th Annual International Conference on CYBER Technology in Automation, Control, and Intelligent Systems (CYBER), pp. 701–707 (2019)
13. Idà, E., Marian, D., Carricato, M.: A deployable cable-driven parallel robot with large rotational capabilities for laser-scanning applications. IEEE Robot. Autom. Lett. **5**(3), 4140–4147 (2020)
14. Verhoeven, R., Hiller, M.: Tension distribution in tendon-based stewart platforms. In: Advances in Robot Kinematics, pp. 117–124. Springer, Dordrecht (2002)
15. Pham, C.B., Yeo, S.H., Yang, G., Chen, I.-M.: Workspace analysis of fully restrained cable-driven manipulators. Robot. Auton. Syst. **57**(9), 901–912 (2009)
16. Gouttefarde, M., Collard, J., Riehl, N., Baradat, C.: Geometry selection of a redundantly actuated cable-suspended parallel robot. IEEE Trans. Rob. **31**(2), 501–510 (2015)
17. Boschetti, G., Trevisani, A.: Cable robot performance evaluation by wrench exertion capability. Robotics **7**(2), 15 (2018)
18. Idà, E., Briot, S., Carricato, M.: Natural oscillations of underactuated cable-driven parallel robots. IEEE Access (2021). https://doi.org/10.1109/ACCESS.2021.3071014
19. Surdilovic, D., Radojicic, J., Krüger, J.: Geometric stiffness analysis of wire robots: a mechanical approach. In: Bruckmann, T., Pott, A. (eds.) Cable-Driven Parallel Robots, pp. 389–404. Springer, Heidelberg (2013)
20. Merlet, J.P.: Jacobian, manipulability, condition number, and accuracy of parallel robots. J. Mech. Des. **128**(1), 199–206 (2005)
21. Golub, G.H., Van Loan, C.F.: Matrix Computations, 4th edn. Johns Hopkins (2013)

CDPR Forward Kinematics with Error Covariance Bounds for Unconstrained End-Effector Attitude Parameterizations

Vinh Le Nguyen and Ryan James Caverly(✉)

Department of Aerospace Engineering and Mechanics, University of Minnesota, Minneapolis, MN 55455, USA
{nguy2876,rcaverly}@umn.edu

Abstract. This paper presents two methods to solve the forward kinematics of an overconstrained six degree-of-freedom cable-driven parallel robot (CDPR) that explicitly account for cable length measurement noise and are applicable to any unconstrained attitude parameterization of the CDPR's end-effector. Nonlinear weighted least-squares optimization is used to solve the CDPR's forward kinematics and determine covariance bounds on the pose estimation error using loop-closure equations based on either the magnitude of the CDPR's cable lengths or the square of the CDPR's cable lengths. It is shown through numerical simulations that the error covariance bounds obtained when using the cable length loop-closure equations are significantly more accurate than those found when using the cable length squared loop-closure equations.

Keywords: Cable-driven parallel robots · Pose estimation · Forward kinematics · Attitude parameterizations

1 Introduction

Cable-driven parallel robots (CDPRs) make use of cables actuated by winches to manipulate an end-effector. Accurate knowledge of the end-effector's pose is necessary in order to accurately control the motion of the end-effector. The computational backbone of CDPR pose estimation is forward kinematics (i.e., direct kinematics), where measurements of the length of each CDPR cable are taken, typically with winch encoders, and loop-closure equations are solved to determine the associated end-effector pose. There are multiple challenges associated with CDPR forward kinematics computations, including the fact that the loop-closure equations are nonlinear in the pose parameters, making them difficult to solve for in general and leading to multiple pose solutions, and the cable length measurements (or winch encoder measurements) are noisy, which leads to error in the pose estimate. The iterative Levenberg-Marquardt algorithm [13], an energy-based method [14], or an interval method [1] can be used to overcome the challenge of solving the nonlinear forward kinematics equations. Realistic effects can also be added to the forward kinematics computations, including pulley

M. Gouttefarde et al. (Eds.): CableCon 2021, MMS 104, pp. 37–49, 2021.
https://doi.org/10.1007/978-3-030-75789-2_4

kinematics, cable sag, and cable elasticity, to improve the pose estimate [15]. The effect of measurement noise on the error in the pose estimate is typically quantified through Monte-Carlo simulations [12,13] or large batches of experiments [1,13,15].

Additional sensors have been proposed to improve the CDPR pose estimate [11], including cable orientation sensors [7], exteroceptive vision systems [3,5], end-effector-mounted cameras [17], and end-effector-mounted inertial measurement units (IMUs) [2,9,12,16]. Although these sensors show promise in obtaining pose estimates that are more accurate than the use of forward kinematics alone, many of them rely on the fusion of a forward kinematics solution with the additional sensor measurements. Moreover, Kalman-filter-based sensor fusion techniques [2,9,12] require statistical information regarding the error associated with the forward kinematics solution in order to obtain an optimal fused pose estimate with meaningful covariance error bounds. This information may be difficult to obtain *a priori* using simulation or experimental techniques, especially since the error in the forward kinematics solution is pose dependent.

This paper addresses the need for pose estimation error covariance bounds in the calculation of CDPR forward kinematics. The first novel contribution is the development of two forward kinematics computation methods that can be performed with any choice of unconstrained end-effector attitude (orientation) parameterization. Both methods make use of the general identities derived in [6] to determine analytic expressions for the Jacobians involved in the forward kinematics computations. Method 1 develops loop-closure equations based on the square of the cable lengths (similar to [12–14]), while the loop-closure equations in Method 2 are based on the magnitude of the cable lengths (similar to [7,15]). The paper's second novel contribution is the calculation of covariance bounds on the error in the forward kinematics pose estimate. These bounds are computed using least-squares estimation theory to ensure the cable length or winch encoder noise covariance is explicitly accounted for in the forward kinematics computations. There are similarities between the proposed forward kinematics methods and the weighted least-squares method used in [7] to optimally fuse cable length and cable orientation measurements to obtain a pose estimate. The proposed Method 2 is distinct from the method in [7], in the sense that (1) covariance bounds on the pose estimation error are computed, which allow for the forward kinematics to be accurately fused with any additional sensors; and (2) any unconstrained attitude parameterization of the end-effector can be implemented. The remainder of this section proceeds with an introduction to the notation used throughout the paper, along with a description of the CDPR architecture considered and a discussion on attitude parameterizations.

1.1 Notation

In this paper, the identity and zero matrices are $\mathbf{1}$ and $\mathbf{0}$, respectively. The cross operator, $(\cdot)^\times : \mathbb{R}^3 \rightarrow \mathfrak{so}(3)$, satisfies $\mathbf{a}^\times \mathbf{b} = -\mathbf{b}^\times \mathbf{a}$ for all $\mathbf{a}, \mathbf{b} \in \mathbb{R}^3$, where $\mathfrak{so}(3) = \{\mathbf{S} \in \mathbb{R}^{3\times 3} \,|\, \mathbf{S}+\mathbf{S}^\mathsf{T} = \mathbf{0}\}$ [8, pp. 525–526]. A reference frame, $\mathcal{F}_i$, is defined by three dextral, orthonormal basis vectors, $\underrightarrow{i}^1$, $\underrightarrow{i}^2$, $\underrightarrow{i}^3$. The physical vector

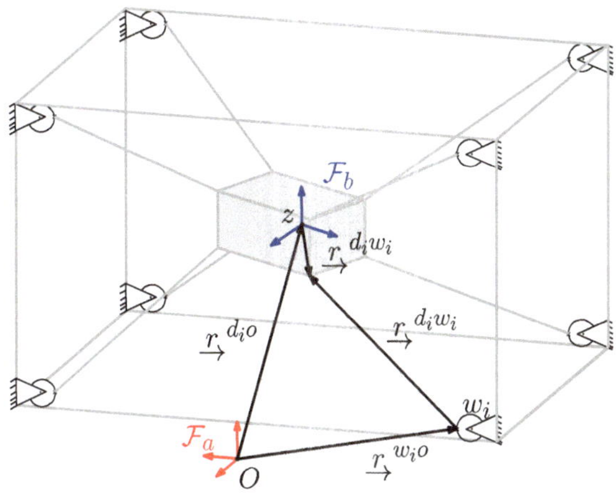

Fig. 1. An overconstrained 6 DOF CDPR with a rigid-body end-effector payload.

$\underset{\rightarrow}{r}$ resolved in $\mathcal{F}_i$ is denoted as $\mathbf{r}_i$. The continuous random variable $\mathbf{v} \sim \mathcal{N}(\bar{\mathbf{v}}, \mathbf{Q}_v)$ satisfies a normal distribution with mean $\bar{\mathbf{v}}$ and covariance $\mathbf{Q}_v = \mathbf{Q}_v^{\mathsf{T}} \geq 0$.

1.2 System Description

Consider an overconstrained 6 degree-of-freedom (DOF) CDPR with m cables, shown in Fig. 1. The reference frame $\mathcal{F}_a$ is an inertial frame with point o at its origin, while $\mathcal{F}_b$ is an end-effector-fixed frame. The end-effector's center of mass is at point z, and the position of z relative to o is $\underset{\rightarrow}{r}^{zo}$. The ith cable is anchored at point w_i, which may represent a routing pulley, and attached to the end-effector at point d_i. The position of d_i relative to z is $\underset{\rightarrow}{r}^{d_i z}$, the position of d_i relative to w_i is $\underset{\rightarrow}{r}^{d_i w_i}$, and the position of w_i relative to o is $\underset{\rightarrow}{r}^{w_i o}$. It is assumed that $\mathbf{r}_b^{d_i z}$ and $\mathbf{r}_a^{w_i o}$ are known constants, and that the cables remain in tension at all times, do not feature any sag, and are inelastic.

1.3 Attitude Parameterizations

The direction cosine matrix (DCM), $\mathbf{C}_{ba} \in SO(3)$, describes the attitude (orientation) of $\mathcal{F}_b$ relative to $\mathcal{F}_a$, where $SO(3) = \{\mathbf{C} \in \mathbb{R}^{3\times 3} \,|\, \mathbf{C}^{\mathsf{T}}\mathbf{C} = \mathbf{1}, \det(\mathbf{C}) = +1\}$. The DCM can be used to change the frame of reference that a physical vector is resolved in (e.g., $\mathbf{r}_b = \mathbf{C}_{ba}\mathbf{r}_a$). The DCM (or orientation matrix) is related to the rotation matrix $\mathbf{R}_{ba}$ that describes the rotation from frame $\mathcal{F}_a$ to $\mathcal{F}_b$ by $\mathbf{R}_{ba} = \mathbf{C}_{ba}^{\mathsf{T}}$. The angular velocity of $\mathcal{F}_b$ relative to $\mathcal{F}_a$, resolved in $\mathcal{F}_b$, is $\boldsymbol{\omega}_b^{ba}$ and satisfies Poisson's equation, $\dot{\mathbf{C}}_{ba} = -\boldsymbol{\omega}_b^{ba^\times}\mathbf{C}_{ba}$ [6].

A number of attitude parameterizations exist that can be used rather than directly computing the DCM. In this paper, $\mathbf{q}^{ba} \in \mathbb{R}^3$ denotes an unconstrained attitude parameterization. Examples of unconstrained attitude parameterizations include Euler angle sequences, the rotation vector, Rodrigues parameters,

Table 1. Examples of unconstrained attitude parameterizations with corresponding DCM representation, $\mathbf{C}_{ba}(\mathbf{q}^{ba})$, and kinematic mapping matrix, $\mathbf{S}(\mathbf{q}^{ba})$.

Name	Parameter	$\mathbf{C}_{ba}(\mathbf{q}^{ba})$, $\mathbf{S}(\mathbf{q}^{ba})$
3-2-1 Euler Angle Sequence [8]	$\boldsymbol{\theta} = \begin{bmatrix} \theta_1 \\ \theta_2 \\ \theta_3 \end{bmatrix}$	$\mathbf{C}_{ba}(\boldsymbol{\theta}) = \mathbf{C}_1(\theta_1)\mathbf{C}_2(\theta_2)\mathbf{C}_3(\theta_3)$ $\mathbf{S}(\boldsymbol{\theta}) = \begin{bmatrix} 1 & 0 & -\sin(\theta_2) \\ 0 & \cos(\theta_1) & \sin(\theta_1)\cos(\theta_2) \\ 0 & -\sin(\theta_1) & \cos(\theta_1)\cos(\theta_2) \end{bmatrix}$
Rotation Vector [6,8,10]	$\boldsymbol{\psi} = \phi\mathbf{a}$	$\mathbf{C}_{ba}(\boldsymbol{\psi}) = \mathbf{1} - \frac{\sin(\psi)}{\lVert\boldsymbol{\psi}\rVert_2}\boldsymbol{\psi}^\times + \frac{1-\cos(\psi)}{\lVert\boldsymbol{\psi}\rVert_2^2}\boldsymbol{\psi}^\times\boldsymbol{\psi}^\times$ $\mathbf{S}(\boldsymbol{\psi}) = \mathbf{1} - \frac{1-\cos(\psi)}{\lVert\boldsymbol{\psi}\rVert_2^2}\boldsymbol{\psi}^\times + \frac{\psi-\sin(\psi)}{\lVert\boldsymbol{\psi}\rVert_2^3}\boldsymbol{\psi}^\times\boldsymbol{\psi}^\times$
Rodrigues Parameters [10]	$\mathbf{p} = \mathbf{a}\tan\left(\frac{\phi}{2}\right)$	$\mathbf{C}_{ba}(\mathbf{p}) = \mathbf{1} + \frac{2}{1+\mathbf{p}^\mathsf{T}\mathbf{p}}\left(\mathbf{p}^\times\mathbf{p}^\times - \mathbf{p}^\times\right)$ $\mathbf{S}(\mathbf{p}) = \frac{2}{1+\mathbf{p}^\mathsf{T}\mathbf{p}}\left(\mathbf{1} - \mathbf{p}^\times\right)$
Modified Rodrigues Parameters (MRPs) [10]	$\mathbf{s} = \mathbf{a}\tan\left(\frac{\phi}{4}\right)$	$\mathbf{C}_{ba}(\mathbf{s}) = \mathbf{1} + \frac{4}{(1+\mathbf{s}^\mathsf{T}\mathbf{s})^2}\left(2\mathbf{s}^\times\mathbf{s}^\times - \left(1-\mathbf{s}^\mathsf{T}\mathbf{s}\right)\mathbf{s}^\times\right)$ $\mathbf{S}(\mathbf{s}) = \frac{4}{1+\mathbf{s}^\mathsf{T}\mathbf{s}}\left(\mathbf{1} + 2\left(\frac{\mathbf{s}^\times\mathbf{s}^\times - \mathbf{s}^\times}{1+\mathbf{s}^\mathsf{T}\mathbf{s}}\right)\right)$

and modified Rodrigues parameters (MRPs). All unconstrained attitude parameterizations feature kinematic singularities, although this is typically not an issue with CDPRs due to their limited rotational workspace. A useful identity derived in [6] for any unconstrained attitude parameterization is given by

$$\frac{\partial}{\partial \mathbf{q}^{ba}}\left(\mathbf{C}_{ba}^\mathsf{T}\mathbf{u}_b\right) = -\mathbf{C}_{ba}^\mathsf{T}\mathbf{u}_b^\times\mathbf{S}(\mathbf{q}^{ba}), \ \forall \mathbf{u}_b \in \mathbb{R}^3, \tag{1}$$

where $\mathbf{q}^{ba}$ is an unconstrained attitude parameterization, $\mathbf{S}(\mathbf{q}^{ba})$ satisfies the attitude kinematics $\boldsymbol{\omega}_b^{ba} = \mathbf{S}(\mathbf{q}^{ba})\dot{\mathbf{q}}^{ba}$, and $\mathbf{u}_b$ is any 3×1 column matrix. This identity will prove useful in the forward kinematics derivation in Sect. 2. A summary of common unconstrained attitude parameterizations, their relationship to the DCM, and their associated kinematic mapping matrix satisfying $\boldsymbol{\omega}_b^{ba} = \mathbf{S}(\mathbf{q}^{ba})\dot{\mathbf{q}}^{ba}$ is included in Table 1, where the rotation vector, Rodrigues parameters, and modified Rodrigues parameters are described using an axis-angle parameterization with unit rotation axis $\mathbf{a}$ and rotation angle ϕ [8, pp. 10–14].

2 Forward Kinematics Formulation

CDPR forward kinematics involves calculating the CDPR's end-effector pose using knowledge of the CDPR's cable lengths, ℓ_i, $i = 1, \ldots, m$, which are typically measured using winch encoders. It is assumed that the measurement noise associated with each cable length measurement satisfies $v_{\ell_i} \sim \mathcal{N}(0, \sigma_{\ell_i}^2)$, where σ_{ℓ_i} is the standard deviation of the noise. Although realistic effects such as cable sag and pulley kinematics can be included in forward kinematics computations, they are not considered in this work. The pose of the CDPR's end-effector is described by $\boldsymbol{\rho}^\mathsf{T} = \left[\mathbf{r}_a^{zo^\mathsf{T}} \ \mathbf{q}^{ba^\mathsf{T}}\right]$, where $\mathbf{q}^{ba}$ is any unconstrained attitude parameterization of $\mathbf{C}_{ba}$.

The position of d_i, the ith cable's end-effector attachment point, relative to w_i, the anchor point of the ith cable, in $\mathcal{F}_a$ is

$$\mathbf{r}_a^{d_i w_i}(\boldsymbol{\rho}) = \mathbf{r}_a^{zo} + \mathbf{C}_{ba}^{\mathsf{T}}(\mathbf{q}^{ba})\mathbf{r}_b^{d_i z} - \mathbf{r}_a^{w_i o}.$$

The cable length measurements must satisfy the loop-closure equation $\ell_i = \left\|\mathbf{r}_a^{d_i w_i}\right\|_2 + v_{\ell_i} = g_i(\boldsymbol{\rho}) + v_{\ell_i}$, where $g_i(\boldsymbol{\rho}) = \sqrt{\mathbf{r}_a^{d_i w_i^{\mathsf{T}}}(\boldsymbol{\rho})\mathbf{r}_a^{d_i w_i}(\boldsymbol{\rho})}$. The same procedure is performed for all m cables to obtain

$$\boldsymbol{\ell} = \mathbf{g}(\boldsymbol{\rho}) + \mathbf{v}_\ell, \tag{2}$$

where $\boldsymbol{\ell}^{\mathsf{T}} = [\ell_1 \;\cdots\; \ell_m]$ contains all the CDPR cable length measurements, $\mathbf{g}^{\mathsf{T}}(\boldsymbol{\rho}) = [g_1(\boldsymbol{\rho}) \;\cdots\; g_m(\boldsymbol{\rho})]$, and $\mathbf{v}_\ell^{\mathsf{T}} = [v_{\ell_1} \;\cdots\; v_{\ell_2}]$ contains the cable length measurement noise. Alternatively, the loop-closure equation for each cable can be written as $\ell_i^2 = (g_i(\boldsymbol{\rho}) + v_{\ell_i})^2$ and concatenated for the m cables as

$$\mathbf{0} = \mathbf{f}(\boldsymbol{\rho}, \mathbf{v}_\ell), \tag{3}$$

where $\mathbf{f}^{\mathsf{T}}(\boldsymbol{\rho}, \mathbf{v}_\ell) = [f_1(\boldsymbol{\rho}, v_{\ell_1}) \;\cdots\; f_m(\boldsymbol{\rho}, v_{\ell_m})]$ and

$$\begin{aligned} f_i(\boldsymbol{\rho}, v_{\ell_i}) &= g_i^2(\boldsymbol{\rho}) + 2g_i(\boldsymbol{\rho})v_{\ell_i} + v_{\ell_i}^2 - \ell_i^2 \\ &= \mathbf{r}_a^{d_i w_i^{\mathsf{T}}}(\boldsymbol{\rho})\mathbf{r}_a^{d_i w_i}(\boldsymbol{\rho}) + 2\sqrt{\mathbf{r}_a^{d_i w_i^{\mathsf{T}}}(\boldsymbol{\rho})\mathbf{r}_a^{d_i w_i}(\boldsymbol{\rho})}v_{\ell_i} + v_{\ell_i}^2 - \ell_i^2. \end{aligned}$$

Two forward kinematics methods are presented next that explicitly account for cable length measurement noise and provide covariance bounds on the pose estimation errors. The first method uses the cable length squared loop-closure equations in (3), while the second method make use of the cable length loop-closure equations in (2).

2.1 Method 1: Cable Length Squared Loop-Closure Equations

Linearizing the cable length measurement model in (3) about the pose $\bar{\boldsymbol{\rho}}$ and zero mean measurement noise $\bar{\mathbf{v}}_\ell = \mathbb{E}(\mathbf{v}_\ell) = \mathbf{0}$ yields

$$\mathbf{0} = \mathbf{f}(\boldsymbol{\rho}, \mathbf{v}_\ell) \approx \mathbf{f}(\bar{\boldsymbol{\rho}}, \mathbf{0}) + \mathbf{J}_\rho(\bar{\boldsymbol{\rho}}, \mathbf{0})\delta\boldsymbol{\rho} + \mathbf{J}_v(\bar{\boldsymbol{\rho}}, \mathbf{0})\mathbf{v}_\ell, \tag{4}$$

where $\delta\boldsymbol{\rho} = \boldsymbol{\rho} - \bar{\boldsymbol{\rho}}$, $\mathbf{J}_\rho(\boldsymbol{\rho}, \mathbf{0}) = \frac{\partial \mathbf{f}(\boldsymbol{\rho}, \mathbf{0})}{\partial \boldsymbol{\rho}} = \left[\frac{\partial \mathbf{f}(\boldsymbol{\rho}, \mathbf{0})}{\partial \mathbf{r}_a^{zo}} \; \frac{\partial \mathbf{f}(\boldsymbol{\rho}, \mathbf{0})}{\partial \mathbf{q}^{ba}}\right]$, and $\mathbf{J}_v(\boldsymbol{\rho}, \mathbf{0}) = \frac{\partial \mathbf{f}}{\partial \mathbf{v}_\ell} = \operatorname{diag}\{2g_1(\boldsymbol{\rho}), \ldots, 2g_m(\boldsymbol{\rho})\}$. The entries of $\left(\frac{\partial \mathbf{f}(\boldsymbol{\rho}, \mathbf{0})}{\partial \mathbf{r}_a^{zo}}\right)^{\mathsf{T}} = \left[\frac{\partial f_1(\boldsymbol{\rho}, \mathbf{0})}{\partial \mathbf{r}_a^{zo}} \;\cdots\; \frac{\partial f_m(\boldsymbol{\rho}, \mathbf{0})}{\partial \mathbf{r}_a^{zo}}\right]$ and $\left(\frac{\partial \mathbf{f}(\boldsymbol{\rho}, \mathbf{0})}{\partial \mathbf{q}^{ba}}\right)^{\mathsf{T}} = \left[\frac{\partial f_1(\boldsymbol{\rho}, \mathbf{0})}{\partial \mathbf{q}^{ba}} \;\cdots\; \frac{\partial f_m(\boldsymbol{\rho}, \mathbf{0})}{\partial \mathbf{q}^{ba}}\right]$ are computed as

$$\begin{aligned} \frac{\partial f_i(\boldsymbol{\rho}, 0)}{\partial \mathbf{r}_a^{zo}} &= \frac{\partial f_i(\boldsymbol{\rho}, 0)}{\partial \mathbf{r}_a^{d_i w_i}} \frac{\partial \mathbf{r}_a^{d_i w_i}}{\partial \mathbf{r}_a^{zo}} = 2\mathbf{r}_a^{d_i w_i^{\mathsf{T}}}(\boldsymbol{\rho}), \\ \frac{\partial f_i(\boldsymbol{\rho}, 0)}{\partial \mathbf{q}^{ba}} &= \frac{\partial f_i(\boldsymbol{\rho}, 0)}{\partial \mathbf{r}_a^{d_i w_i}} \frac{\partial \mathbf{r}_a^{d_i w_i}}{\partial \mathbf{q}^{ba}} = 2\mathbf{r}_a^{d_i w_i^{\mathsf{T}}}(\boldsymbol{\rho}) \frac{\partial}{\partial \mathbf{q}^{ba}} \left(\mathbf{C}_{ba}^{\mathsf{T}}(\mathbf{q}^{ba})\mathbf{r}_b^{d_i z}\right). \end{aligned} \tag{5}$$

Since $\mathbf{q}^{ba} \in \mathbb{R}^3$ is an unconstrained attitude parameterization, the identity in (1) yields $\frac{\partial}{\partial \mathbf{q}^{ba}}\left(\mathbf{C}_{ba}^{\mathsf{T}}(\mathbf{q}^{ba})\mathbf{r}_b^{d_i z}\right) = -\mathbf{C}_{ba}^{\mathsf{T}}(\mathbf{q}^{ba})\mathbf{r}_b^{d_i z^\times}\mathbf{S}(\mathbf{q}^{ba})$. Substituting this into (5), the Jacobian $\mathbf{J}_\rho(\boldsymbol{\rho}, \mathbf{0})$ can be written as

$$\mathbf{J}_\rho(\boldsymbol{\rho}, \mathbf{0}) = \begin{bmatrix} 2\mathbf{r}_a^{d_1 w_1^{\mathsf{T}}}(\boldsymbol{\rho}) \begin{bmatrix} \mathbf{1} & -\mathbf{C}_{ba}^{\mathsf{T}}(\mathbf{q}^{ba})\mathbf{r}_b^{d_1 z^\times}\mathbf{S}(\mathbf{q}^{ba}) \end{bmatrix} \\ \vdots \\ 2\mathbf{r}_a^{d_m w_m^{\mathsf{T}}}(\boldsymbol{\rho}) \begin{bmatrix} \mathbf{1} & -\mathbf{C}_{ba}^{\mathsf{T}}(\mathbf{q}^{ba})\mathbf{r}_b^{d_m z^\times}\mathbf{S}(\mathbf{q}^{ba}) \end{bmatrix} \end{bmatrix}.$$

With the linearized cable measurement model in (4), the CDPR forward kinematics is solved through the nonlinear weighted least-squares optimization problem

$$\min_{\boldsymbol{\rho}} \mathbf{f}^{\mathsf{T}}(\boldsymbol{\rho}, \mathbf{0})\bar{\mathbf{V}}^{-1}(\bar{\boldsymbol{\rho}})\mathbf{f}(\boldsymbol{\rho}, \mathbf{0}),$$

where $\bar{\mathbf{V}}(\bar{\boldsymbol{\rho}}) = \mathbf{J}_v(\bar{\boldsymbol{\rho}}, \mathbf{0})\mathbf{V}\mathbf{J}_v^{\mathsf{T}}(\bar{\boldsymbol{\rho}}, \mathbf{0}) = \text{diag}\{4\sigma_{\ell_1}^2 g_1^2(\bar{\boldsymbol{\rho}}), \ldots, 4\sigma_{\ell_m}^2 g_m^2(\bar{\boldsymbol{\rho}})\}$ and the cable length measurement noise covariance is $\mathbf{V} = \mathbb{E}(\mathbf{v}_\ell \mathbf{v}_\ell^{\mathsf{T}}) = \text{diag}\{\sigma_{\ell_1}^2, \ldots, \sigma_{\ell_m}^2\}$. The solution to this optimization problem corresponds to the best linear unbiased estimator through the Gauss-Markov Theorem [4, pp. 63–67]. This optimization problem is solved using the Levenberg-Marquardt algorithm, which begins with an initial estimate of the CDPR's end-effector pose, $\boldsymbol{\rho}_0$ and then proceeds at the jth iteration with the update

$$\boldsymbol{\rho}_{j+1} = \boldsymbol{\rho}_j - \left(\mathbf{J}_\rho^{\mathsf{T}}(\boldsymbol{\rho}_j, \mathbf{0})\bar{\mathbf{V}}^{-1}(\boldsymbol{\rho}_j)\mathbf{J}_\rho(\boldsymbol{\rho}_j, \mathbf{0}) + \eta\mathbf{1}\right)^{-1}\mathbf{J}_\rho^{\mathsf{T}}(\boldsymbol{\rho}_j, \mathbf{0})\bar{\mathbf{V}}^{-1}(\boldsymbol{\rho}_j)\mathbf{f}(\boldsymbol{\rho}_j, \mathbf{0}),$$

where the parameter $\eta > 0$ provides numerical damping to the iterative calculations [4, pp. 48–49]. The iteration is terminated when a specified stopping criterion is satisfied, such as achieving convergence in terms of $\|\boldsymbol{\rho}_j - \boldsymbol{\rho}_{j-1}\|_2 < \epsilon_1$ or $\|\mathbf{f}(\boldsymbol{\rho}_j, \mathbf{0})\|_2 < \epsilon_2^2$, where $\epsilon_1 > 0$ and $\epsilon_2 > 0$ are numerical tolerances. The covariance of the pose estimate at the jth iteration is given by [4, p. 84]

$$\mathbf{P}_j = \mathbb{E}[(\boldsymbol{\rho}_{\text{true}} - \boldsymbol{\rho}_j)(\boldsymbol{\rho}_{\text{true}} - \boldsymbol{\rho}_j)^{\mathsf{T}}] = \left(\mathbf{J}_\rho^{\mathsf{T}}(\boldsymbol{\rho}_j, \mathbf{0})\bar{\mathbf{V}}^{-1}(\boldsymbol{\rho}_j)\mathbf{J}_\rho(\boldsymbol{\rho}_j, \mathbf{0})\right)^{-1}. \quad (6)$$

2.2 Method 2: Cable Length Loop-Closure Equations

Linearizing the cable length measurement model in (2) about the pose $\bar{\boldsymbol{\rho}}$ yields

$$\boldsymbol{\ell} \approx \mathbf{g}(\bar{\boldsymbol{\rho}}) + \mathbf{H}(\bar{\boldsymbol{\rho}})\delta\boldsymbol{\rho} + \mathbf{v}_\ell,$$

where $\mathbf{H}(\boldsymbol{\rho}) = \frac{\partial \mathbf{g}}{\partial \boldsymbol{\rho}} = \begin{bmatrix} \frac{\partial \mathbf{g}}{\partial \mathbf{r}_a^{zo}} & \frac{\partial \mathbf{g}}{\partial \mathbf{q}^{ba}} \end{bmatrix}$ and $\delta\boldsymbol{\rho} = \boldsymbol{\rho} - \bar{\boldsymbol{\rho}}$. The entries of $\mathbf{H}(\boldsymbol{\rho})$ are $\left(\frac{\partial \mathbf{g}}{\partial \mathbf{r}_a^{zo}}\right)^{\mathsf{T}} = \begin{bmatrix} \frac{\partial g_1}{\partial \mathbf{r}_a^{zo}} & \cdots & \frac{\partial g_m}{\partial \mathbf{r}_a^{zo}} \end{bmatrix}$ and $\left(\frac{\partial \mathbf{g}}{\partial \mathbf{q}^{ba}}\right)^{\mathsf{T}} = \begin{bmatrix} \frac{\partial g_1}{\partial \mathbf{q}^{ba}} & \cdots & \frac{\partial g_m}{\partial \mathbf{q}^{ba}} \end{bmatrix}$, where

$$\begin{aligned} \frac{\partial g_i}{\partial \mathbf{r}_a^{zo}} &= \frac{\partial g_i}{\partial \mathbf{r}_a^{d_i w_i}}\frac{\partial \mathbf{r}_a^{d_i w_i}}{\partial \mathbf{r}_a^{zo}} = \left(\mathbf{r}_a^{d_i w_i^{\mathsf{T}}}(\boldsymbol{\rho})\mathbf{r}_a^{d_i w_i}(\boldsymbol{\rho})\right)^{-\frac{1}{2}}\mathbf{r}_a^{d_i w_i^{\mathsf{T}}}(\boldsymbol{\rho}), \\ \frac{\partial g_i}{\partial \mathbf{q}^{ba}} &= \frac{\partial g_i}{\partial \mathbf{r}_a^{d_i w_i}}\frac{\partial \mathbf{r}_a^{d_i w_i}}{\partial \mathbf{q}^{ba}} = \left(\mathbf{r}_a^{d_i w_i^{\mathsf{T}}}(\boldsymbol{\rho})\mathbf{r}_a^{d_i w_i}(\boldsymbol{\rho})\right)^{-\frac{1}{2}}\mathbf{r}_a^{d_i w_i^{\mathsf{T}}}(\boldsymbol{\rho})\frac{\partial}{\partial \mathbf{q}^{ba}}\left(\mathbf{C}_{ba}^{\mathsf{T}}(\mathbf{q}^{ba})\mathbf{r}_b^{d_i z}\right). \end{aligned} \quad (7)$$

The attitude parameterization identity in (1) is substituted into (7) to yield

$$\mathbf{H}(\boldsymbol{\rho}) = \begin{bmatrix} \left(\mathbf{r}_a^{d_1 w_1^\mathsf{T}}(\boldsymbol{\rho})\mathbf{r}_a^{d_1 w_1}(\boldsymbol{\rho})\right)^{-\frac{1}{2}} \mathbf{r}_a^{d_1 w_1^\mathsf{T}}(\boldsymbol{\rho}) \left[\mathbf{1} \;\; -\mathbf{C}_{ba}^\mathsf{T}(\mathbf{q}^{ba})\mathbf{r}_b^{d_1 z^\times}\mathbf{S}(\mathbf{q}^{ba})\right] \\ \vdots \\ \left(\mathbf{r}_a^{d_m w_m^\mathsf{T}}(\boldsymbol{\rho})\mathbf{r}_a^{d_m w_m}(\boldsymbol{\rho})\right)^{-\frac{1}{2}} \mathbf{r}_a^{d_m w_m^\mathsf{T}}(\boldsymbol{\rho}) \left[\mathbf{1} \;\; -\mathbf{C}_{ba}^\mathsf{T}(\mathbf{q}^{ba})\mathbf{r}_b^{d_m z^\times}\mathbf{S}(\mathbf{q}^{ba})\right] \end{bmatrix}.$$

The forward kinematics of the CDPR is solved through the nonlinear weighted least-squares optimization problem

$$\min_{\boldsymbol{\rho}} \left(\boldsymbol{\ell} - \mathbf{g}(\boldsymbol{\rho})\right)^\mathsf{T} \mathbf{V}^{-1} \left(\boldsymbol{\ell} - \mathbf{g}(\boldsymbol{\rho})\right),$$

where $\mathbf{V} = \mathbb{E}(\mathbf{v}_\ell \mathbf{v}_\ell^\mathsf{T}) = \operatorname{diag}\{\sigma_{\ell_1}^2, \ldots, \sigma_{\ell_m}^2\}$. The Levenberg-Marquardt algorithm is used to solve this optimization problem, beginning with an initial CDPR end-effector pose estimate $\boldsymbol{\rho}_0$, and proceeding at the jth iteration with the update

$$\boldsymbol{\rho}_{j+1} = \boldsymbol{\rho}_j + \left(\mathbf{H}^\mathsf{T}(\boldsymbol{\rho}_j)\mathbf{V}^{-1}\mathbf{H}(\boldsymbol{\rho}_j) + \eta\mathbf{1}\right)^{-1} \mathbf{H}^\mathsf{T}(\boldsymbol{\rho}_j)\mathbf{V}^{-1}\left(\boldsymbol{\ell} - \mathbf{g}(\boldsymbol{\rho}_j)\right),$$

where $\eta > 0$ provides numerical damping to the iterative calculations and the iteration is terminated when a specified stopping criterion is satisfied. The covariance of the pose estimate at the jth iteration is given by [4, p. 84]

$$\mathbf{P}_j = \mathbb{E}[(\boldsymbol{\rho}_{\text{true}} - \boldsymbol{\rho}_j)(\boldsymbol{\rho}_{\text{true}} - \boldsymbol{\rho}_j)^\mathsf{T}] = \left(\mathbf{H}^\mathsf{T}(\boldsymbol{\rho}_j)\mathbf{V}^{-1}\mathbf{H}(\boldsymbol{\rho}_j)\right)^{-1}. \qquad (8)$$

2.3 Discussion

The main difference between the two methods presented in this section is in the linearization of the loop-closure equations. Since Method 1 is based on the square of the cable lengths, the cable length measurement noise appears nonlinearly in the loop-closure equations of (3), which requires a linearization when solving the optimization problem. The cable length measurement noise appears linearly in the loop-closure equations of Method 2 in (2), which allows for the effect of measurement noise on the pose estimation error to be captured more accurately. Note that there is no guarantee that the proposed iterative methods converge to a solution or that the solution is near the one found at the previous time step due to the fact that in general there are multiple forward kinematics solutions.

3 Numerical Results

Numerical inverse-kinematics-based simulations are performed on an overconstrained 6-DOF CDPR with 8 cables, shown in Fig. 2(a), whose dimensions are provided in Table 2. The simulations use a discrete time step $T = 0.005$ s and an end-effector reference trajectory that is solved for using the kinematic equations $\mathbf{r}_{a_k}^{zo} = \mathbf{r}_{a_{k-1}}^{zo} + T\mathbf{v}_{a_{k-1}}^{zo/a}$, $\mathbf{v}_{a_k}^{zo/a} = \mathbf{v}_{a_{k-1}}^{zo/a} + T\dot{\mathbf{v}}_{a_{k-1}}^{zo/a}$, and $\mathbf{C}_{b_k a} = \exp(-T\boldsymbol{\omega}_{b_{k-1}}^{ba^\times})\mathbf{C}_{b_{k-1} a}$,

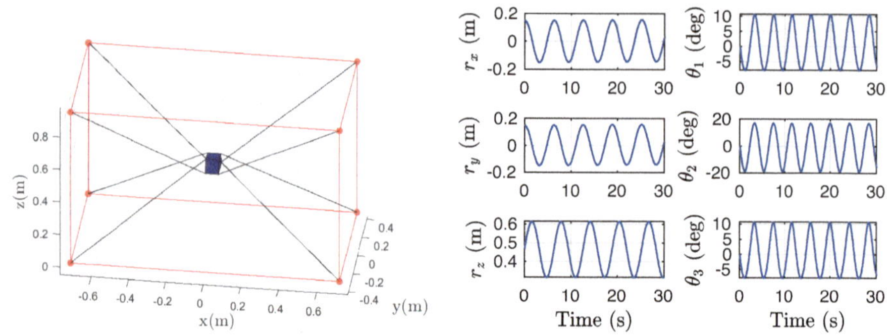

(a) (b)

Fig. 2. (a) Depiction of the 6-DOF overconstrained CDPR used in the numerical simulations of Sect. 3. (b) Plots of the simulated end-effector pose trajectory versus time, where $\mathbf{r}_a^{zo} = [r_x \ r_y \ r_z]^\mathsf{T}$ and $\boldsymbol{\theta}^{ba} = [\theta_1 \ \theta_2 \ \theta_3]^\mathsf{T}$.

Table 2. CDPR cable anchor locations and cable end-effector attachment points.

Cable	Winch position, $\mathbf{r}_{a,\mathrm{ref}}^{w_i o \mathsf{T}}$ (m)	Cable End-Effector Attachment Point, $\mathbf{r}_b^{d_i z \mathsf{T}}$ (cm)
1	$[0.71 \ \ 0.38 \ \ 0.93]$	$[3.75 \ \ 7.5 \ \ -3.75]$
2	$[-0.71 \ \ 0.38 \ \ 0.93]$	$[-3.75 \ \ 7.5 \ \ -3.75]$
3	$[-0.71 \ \ -0.38 \ \ 0.93]$	$[-3.75 \ \ -7.5 \ \ -3.75]$
4	$[0.71 \ \ -0.38 \ \ 0.93]$	$[3.75 \ \ -7.5 \ \ -3.75]$
5	$[-0.71 \ \ -0.38 \ \ 0]$	$[-1.875 \ \ -7.5 \ \ 3.75]$
6	$[0.71 \ \ 0.38 \ \ 0]$	$[1.875 \ \ 7.5 \ \ 3.75]$
7	$[-0.71 \ \ 0.38 \ \ 0]$	$[-1.875 \ \ 7.5 \ \ 3.75]$
8	$[0.71 \ \ -0.38 \ \ 0]$	$[1.875 \ \ -7.5 \ \ -3.75]$

where $\mathbf{v}_a^{zo/a} = \dot{\mathbf{r}}_a^{zo}$, $\dot{\mathbf{v}}_{a_k}^{zo/a\mathsf{T}} = -1.5\left[\cos(kT)\ \cos(kT)\ \sin(kT)\right]$ m/s^2, and $\boldsymbol{\omega}_{b_k}^{ba\mathsf{T}} = -13.5\cos(1.5kT)\left[1\ 2\ 1\right]$ deg/s, with initial conditions $\mathbf{r}_{a_0}^{zo\mathsf{T}} = \left[0.15\ 0.15\ 0.465\right]$ m, $\mathbf{v}_{a_0}^{zo\mathsf{T}} = \left[0\ 0\ 0.15\right]$ m/s, and $\mathbf{C}_{b_0 a} = \mathbf{1}$. The resulting end-effector pose trajectory is given in Fig. 2(b), where the end-effector attitude is shown using a 3-2-1 Euler angle sequence for ease of interpretation. The CDPR's inverse kinematics are solved at each time step to find the cable lengths associated with the end-effector pose. Artificial zero mean Gaussian measurement noise with standard deviation $\sigma_{\ell_i} = 1$ mm, $i = 1, \ldots, 8$ is added to each cable length before proceeding with the forward kinematics computations.

Forward kinematics Methods 1 and Method 2 are implemented at each time step of the simulation, where the end-effector attitude is parameterized first using a 3-2-1 Euler angle sequence, followed by the rotation vector. In all cases, the initial pose estimate used in the Levenberg-Marquardt algorithm at time step k is the pose estimate obtained at the previous time step $k-1$. The damping and termination tolerances are chosen as $\eta = 0.001$ and $\epsilon_1 = \epsilon_2 = 5 \times 10^{-4}$. The pose estimation errors obtained with the Euler angle and rotation vector

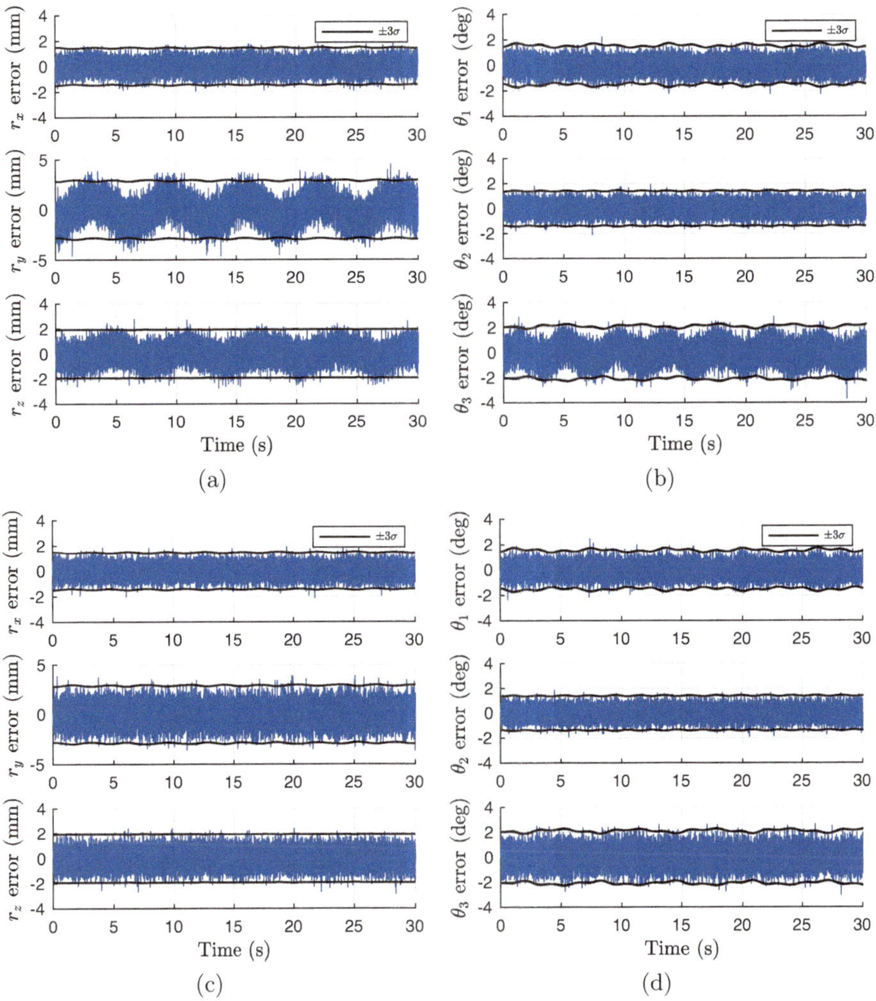

Fig. 3. Plots of the pose estimation error and $\pm 3\sigma$ bounds obtained using a 3-2-1 Euler angle sequence attitude parameterization with the proposed forward kinematics Method 1 in (a) and (b), and forward kinematics Method 2 in (c) and (d).

parameterizations are shown in Figs. 3 and 4, respectively. The end-effector position is described by $\mathbf{r}_a^{zo} = [r_x \; r_y \; r_z]^\mathsf{T}$ and the end-effector attitude is given by $\boldsymbol{\theta}^{ba} = [\theta_1 \; \theta_2 \; \theta_3]^\mathsf{T}$ for Euler angles and $\boldsymbol{\psi}^{ba} = [\psi_1 \; \psi_2 \; \psi_3]^\mathsf{T}$ for the rotation vector. The pose estimation error plots in Fig. 3 and Fig. 4 include $\pm 3\sigma$ bounds that are computed at each time step using the covariance matrix associated with the weighted least-squares solution in Method 1 and Method 2. Specifically, the variance associated with each component of the error is found from the diagonal elements of the covariance matrix $\mathbf{P}_j$ computed using (6) or (8). Figures 3 and 4 demonstrate that although there is little difference in the magnitude of the pose

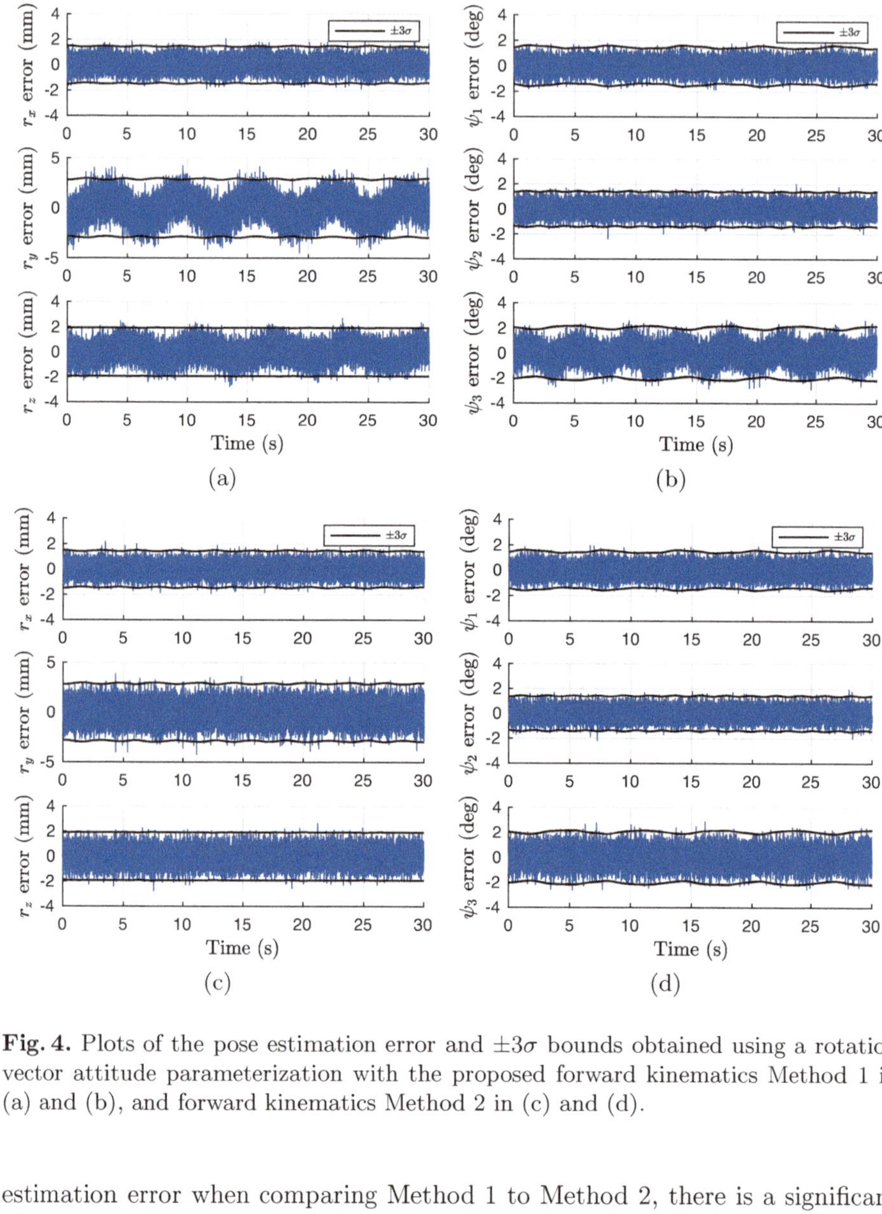

Fig. 4. Plots of the pose estimation error and $\pm 3\sigma$ bounds obtained using a rotation vector attitude parameterization with the proposed forward kinematics Method 1 in (a) and (b), and forward kinematics Method 2 in (c) and (d).

estimation error when comparing Method 1 to Method 2, there is a significant difference in the accuracy of the computed covariance bounds. The $\pm 3\sigma$ bounds in Figs. 3 and 4 are more representative of the true pose estimation error when using Method 2 compared with Method 1. This is emphasized in Figs. 5 and 6, where the root mean square error (RMSE) in the pose estimates for the four simulations are summarized, along with the percent error between the mean of the standard deviation in pose estimation error computed using (6) or (8) over the entire time series and the exact standard deviation of the pose estimation error over the duration of the simulation. The standard deviation obtained using

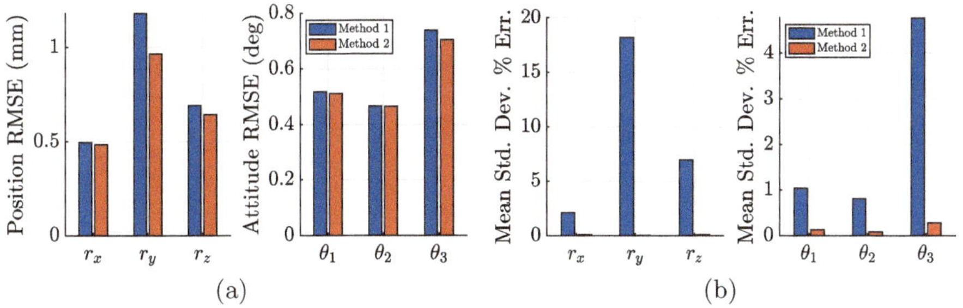

Fig. 5. Plots of (a) the RMSE in the pose estimate and (b) the percent error between the mean standard deviation of the pose estimation error computed using (6) and the true standard deviation for Methods 1 and 2 with the Euler angle parameterization.

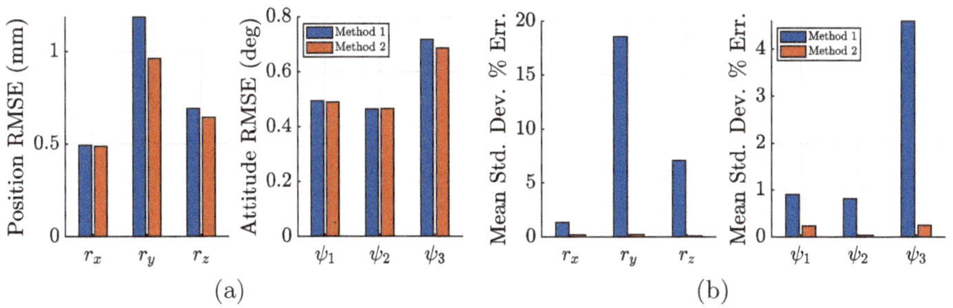

Fig. 6. Plots of (a) the RMSE in the pose estimate and (b) the percent error between the mean standard deviation of the pose estimation error computed using (8) and the true standard deviation for Methods 1 and 2 with the rotation vector parameterization.

the covariance matrix in (8) is accurate within a fraction of a percent of the true standard deviation, while using (6) with Method 1 does not provide a reliable statistical characterization of the pose estimation error.

4 Conclusion

Two novel forward kinematics formulations were presented in this paper that can be performed with any unconstrained end-effector attitude parameterization and explicitly account for cable length measurement noise using nonlinear weighted least-squares optimization. A key novelty of the proposed methods is the computation of covariance bounds on the pose estimation error. While both methods result in similar pose estimation errors, the covariance bounds computed using the cable length loop-closure equations in Method 2 greatly outperform those computed using the cable length squared loop-closure equations in Method 1. This makes Method 2 more suitable in situations where accurate covariance bounds on the forward kinematics pose estimation error are required, such as when performing sensor fusion or Kalman filtering. Future work will compare the computational complexity of the two approaches, examine the effect of calibration errors and measurement biases, and implement the proposed approaches experimentally on a CDPR that is being built at the University of Minnesota.

References

1. Berti, A., Merlet, J.P., Carricato, M.: Solving the direct geometrico-static problem of underconstrained cable-driven parallel robots by interval analysis. Int. J. Robot. Res. **35**(6), 723–739 (2016)
2. Caverly, R.J., Forbes, J.R.: State estimator design for a single degree of freedom cable-actuated system. J. Frankl. Inst. **353**(18), 4845–4869 (2016)
3. Chellal, R., Cuvillon, L., Laroche, E.: A kinematic vision-based position control of a 6-DoF cable-driven parallel robot. In: Pott, A., Bruckmann, T. (eds.) Cable-Driven Parallel Robots, Mechanisms and Machine Science, vol. 32, pp. 213–225. Springer, Cham, Switzerland (2015)
4. Crassidis, J.L., Junkins, J.L.: Optimal Estimation of Dynamic Systems. CRC Press, Boca Raton (2004)
5. Dallej, T., Gouttefarde, M., Andreff, N., Hervé, P.E., Martinet, P.: Modeling and vision-based control of large-dimension cable-driven parallel robots using a multiple-camera setup. Mechatronics **61**, 20–36 (2019)
6. de Ruiter, A.H.J., Forbes, J.R.: General identities for parameterizations of SO(3) with applications. ASME. J. Appl. Mech. **81**(7) (2014). 16 pages
7. Garant, X., Campeau-Lecours, A., Cardou, P., Gosselin, C.: Improving the forward kinematics of cable-driven parallel robots through cable angle sensors. In: Gosselin, C., Cardou, P., Bruckmann, T., Pott, A. (eds.) Cable-Driven Parallel Robots, Mechanisms and Machine Science, vol. 53, pp. 167–179. Springer International Publishing, Cham, Switzerland (2018)
8. Hughes, P.C.: Spacecraft Attitude Dynamics. Dover, Mineola (2004)
9. Korayem, M.H., Yousefzadeh, M., Kian, S.: Precise end-effector pose estimation in spatial cable-driven parallel robots with elastic cables using a data fusion method. Measurement **130**, 177–190 (2018)
10. Markley, F.L., Crassidis, J.L.: Fundamentals of Spacecraft Attitude Determination and Control. Springer, New York (2014)
11. Merlet, J.P.: An experimental investigation of extra measurements for solving the direct kinematics of cable-driven parallel robots. In: IEEE International Conference on Robotics, pp. 6947–6952 (2018)
12. Nguyen, V.L., Caverly, R.J.: Cable-driven parallel robot pose estimation using extended Kalman filtering with inertial payload measurements. IEEE Rob. Autom. Lett. **6**(2), 3615–3622 (2021)
13. Pott, A.: An algorithm for real-time forward kinematics for cable-driven parallel robots. In: Lenarčič, J., Stanišić, M.M. (eds.) Advances in Robot Kinematics: Motion in Man and Machine, pp. 529–538. Springer, Dordrecht (2010)
14. Pott, A., Schmidt, V.: On the forward kinematics of cable-driven parallel robots. In: IEEE International Conference on International Robotics, pp. 3182–3187 (2015)
15. Santos, J.C., Gouttefarde, M.: A real-time capable forward kinematics algorithm for cable-driven parallel robots considering pulley kinematics. In: Lenarčič, J., Siciliano, B. (eds.) Advances in Robot Kinematics 2020, Springer Proceedings in Advanced Robotics, vol. 15, pp. 199–208. Springer, Cham (2021)

16. Schenk, C.T.: Modelling and control of a cable-driven parallel robot methods for vibration reduction and motion quality improvement. Ph.D. thesis, University of Stuttgart, Stuttgart, Germany (2019)
17. Zake, Z., Chaumette, F., Pedemonte, N., Caro, S.: Vision-based control and stability analysis of a cable-driven parallel robot. IEEE Rob. Autom. Lett. **4**(2), 1029–1036 (2019)

Neural Network-Based Inverse Kineto-Static Analysis of Cable-Driven Parallel Robot Considering Cable Mass and Elasticity

Ishan Chawla[1], Pushparaj Mani Pathak[1](✉), Leila Notash[2], Arun Kumar Samantaray[3], Qingguo Li[2], and Umesh Kumar Sharma[1]

[1] Indian Institute of Technology Roorkee, Roorkee 247667, India
pushparaj.pathak@me.iitr.ac.in
[2] Queen's University, Kingston, Canada
[3] Indian Institute of Technology Kharagpur, Kharagpur 721302, India

Abstract. Inverse kinematic problem of a cable-driven parallel robot (CDPR) is a straightforward problem for massless inextensible cables. However, the problem becomes a non-linear kineto-static problem with the consideration of cable mass and elasticity. This problem is conventionally solved using numerical methods. These methods are iterative in nature, have slow convergence speed, and are highly dependent on initial values, which makes them difficult to be used in real-time applications. Therefore, this paper proposes a deep neural network-based approach to obtain the inverse kineto-static (IKS) solution of CDPR considering cable mass and elasticity. The training dataset has been obtained from IKS solutions attained by using numerical methods. Then, appropriate parameters as well as training algorithm is utilized for off-line training of the network. The performance of the proposed approach is validated by the simulation results for a redundantly constrained planar CDPR. From the simulation results, it has been observed that the proposed approach shows an 83% reduction in computational time in comparison to conventional numerical methods. In addition, the proposed neural network-based approach has performed satisfactorily with a minor error in the IKS solution. Therefore, the proposed approach can be employed for real-time applications due to its low and bounded computational time.

Keywords: Neural network · Cable-driven parallel robot · Parallel robot · Inverse Kineto-static · Parallel robots · Sagging · Kinematics

1 Introduction

Cable-driven parallel robots belong to the class of parallel manipulators that utilize flexible cables instead of rigid links to manipulate the end-effector. Since cables are lighter in weight and can be extended to much larger lengths, cable robots possess a much larger workspace with high payload to weight ratio. Therefore, CDPRs have been extensively investigated by researchers for various applications such as logistics,

M. Gouttefarde et al. (Eds.): CableCon 2021, MMS 104, pp. 50–62, 2021.
https://doi.org/10.1007/978-3-030-75789-2_5

production, construction [1], motion-simulators [2], and large-scale telescopes [3], to name a few.

In the past, the work on CDPR was mostly done based on the assumption that cables are of negligible mass and elasticity. The inverse kinematic (IK) problem with massless inextensible cables is a straightforward problem where the cable lengths can be easily obtained using closed-loop kinematic relations for the given mobile platform pose. However, the forward kinematics (FK) problem is generally complex for CDPR (like other parallel robots). The FK problem [4] has been extensively studied for real-time applications [5, 6]. A neural network-based approach was proposed in [7, 8] for real-time FK analysis of CDPR considering massless inextensible cables.

Since cable-driven parallel robots are known for their large-scale applications, cable mass and elasticity cannot be ignored for large-scale robots. Therefore, it becomes crucial to consider cable mass and elasticity in the analysis. However, the consideration of cable mass in the analysis leads to sagging in the cable, where the amount of sagging depends upon the tension in the cables. Therefore, the kinematics of the robot is coupled with the statics of the robot. The resulting inverse kinematic problem becomes a non-linear kineto-static problem. Thus, the static balance equations, along with non-linear kinematic constraints are required to be solved to obtain the solution. This makes the inverse kinematic problem a complex problem that is impossible to solve analytically.

From the last decade, many researchers have shifted their interest from massless cables to cables with non-negligible mass and elasticity for the analysis of CDPRs. Initial work on static analysis of CDPR considering non-negligible cable mass and elasticity was presented in [9]. This work is based on Irvine's static cable model [10] and utilizes numerical methods to solve the inverse kineto-static (IKS) problem. A study on the effect of considering non-negligible cable mass on the kineto-static of large-scale redundant CDPRs was conducted in [11]. The authors concluded that the mass of the cable has a significant effect on the cable forces and cable lengths and hence, cannot be ignored for large-scale robots. Notash studied the effect of various parameters on the deflection of a CDPR [12]. A numerical method is employed in this work to solve the IKS problem. Recently, the author employed artificial neural network for the prediction of deflection maps of CDPR [13]. The encouraging results show the capability of neural network in learning the deflection maps by utilizing limited amount of data for training.

Merlet proposed an interval analysis-based algorithm to solve the inverse kineto-static problem [14] of CDPR with sagging cables. The key feature of this algorithm is its ability to provide all possible solutions. Furthermore, the author recently conducted a detailed study on kineto-static equations in order to enhance the computational speed of the algorithm [15]. However, the algorithm is computationally intensive and is not intended for real-time applications. For applications in real-time, a simplified cable model has been proposed in the literature that considers the effect of cable mass and elasticity to give an approximate IKS solution [16]. However, these models can only be applied to limited cases where the cable tensions should be low enough to restrict large elastic elongations and high enough to limit large cable sagging.

Thus, it has been observed that the IKS problem for CDPR have been solved in the literature either numerically or by utilizing an interval analysis-based algorithm. As mentioned above, the interval analysis-based algorithms are computationally intensive and hence, are not suitable for real-time applications. To date, numerical algorithms are the only alternative to solve the IKS problem of CDPR. However, the computational speed of these algorithms highly depends on the initial value, which is generally chosen as an inverse kinematic and static solution for inextensible massless cables. If the solution is far away from the initial value (in case of heavy cables with large sagging), the algorithm may not converge and if it does, it may require large computational time. The solution to this problem can however be computed by solving the kineto-static problem for low-density cable initially. Then, the density is increased iteratively and the kineto-static problem is solved using the previous solution as the next initial value in each iteration until the cable density reaches the desired level [9]. Although this solution may provide a good convergence, the computational speed is slow due to large iterations. Hence, the computational time of the numerical algorithms is unbounded due to its dependency on the initial value.

Therefore, the primary objective of this work is to provide a faster and computationally bounded approach to solve the inverse kineto-static problem of CDPR, which could prove to be beneficial for real-time applications. To achieve this objective, a neural network-based approach for a real-time IKS analysis of CDPR has been proposed by taking cables mass and elasticity into consideration. The performance of the proposed approach is validated by the simulation results for a redundant planar CDPR. The obtained results are further compared against the conventional numerical method-based approach in terms of computational time and error in the IKS solution to show the real-time applicability of the proposed approach.

This paper is organized as follows: The inverse kineto-static problem is briefly discussed in Sect. 2. Section 3 explains the proposed neural network-based approach in detail. Simulation results for redundantly constrained planar cable-driven parallel robot are presented in Sect. 4. Finally, Sect. 5 draws the conclusions from the proposed work.

2 Inverse Kineto-Static Problem

The presented formulation assumes the cables to be homogeneous and elastic with negligible flexural stiffness. Furthermore, the mobile platform is assumed to be moving slowly enough to be considered static. This assumption is valid for many large-scale applications such as telescope, 3D/concrete printing where the speed of the mobile platform is relatively slow.

A cable of unstrained length L_0, unstrained cross-sectional area A_0, unstrained linear density ρ_0 and elastic modulus E fixed at one end and is pulled by the cable force vector $\vec{\mathbf{F}}$ at the other end is considered. Due to the non-negligible mass of the cable and the effect of gravitational force, the cable undergoes sagging, as shown in Fig. 1. Here, ΔL is the elastic elongation induced in the cable as a result of cable elasticity and forces acting on the cable. The strained length L of the cable can be obtained by adding L_0 and ΔL. The length variable p represents the strained length of the cable from the fixed end of the cable such that $p \in [0, L]$ and s denotes another length variable representing the unstrained length of the cable such that $s \in [0, L_0]$.

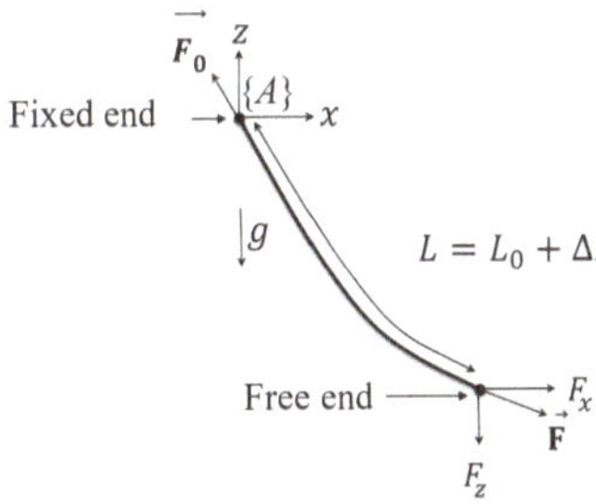

Fig. 1. Schematic diagram of cable static profile

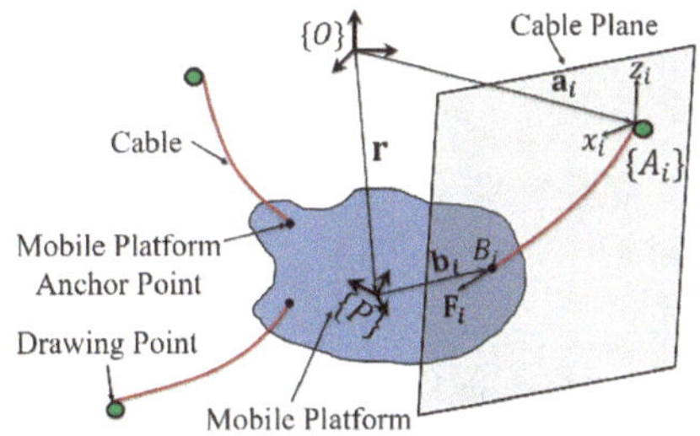

Fig. 2. Schematic diagram of a cable-driven parallel robot

The static profile of the cable in the cable frame $\{A\}$ located at the fixed end of the cable is derived by Irvine [10] (and validated experimentally in [17]) and is given by

$$x(s) = \frac{F_x s}{EA_0} + \frac{F_x}{\rho_0 g}\left[\sinh^{-1}\left(\frac{F_z + \rho_0 g(s - L_0)}{F_x}\right) - \sinh^{-1}\left(\frac{F_z - \rho_0 g L_0}{F_x}\right)\right] \quad (1)$$

$$z(s) = \frac{F_z s}{EA_0} + \frac{\rho_0 g}{EA_0}\left(\frac{s^2}{2} - L_0 s\right) + \frac{1}{\rho_0 g}\left(\sqrt{F_x^2 + (F_z + \rho_0 g(s - L_0))^2} - \sqrt{F_x^2 + (F_z - \rho_0 g L_0)^2}\right) \quad (2)$$

Here, $x(s)$ and $z(s)$ in the above equations give the static profile of cable with reference to cable frame $\{A\}$ as a function of an unstrained length variable s for the given force vector $\vec{\mathbf{F}}$ on the free end of the cable. The variable F_x and F_z represent the components of the force vector $\vec{\mathbf{F}}$ in x and z-directions, respectively.

A CDPR containing a mobile platform of mass m held by n_c number of cables is considered, as shown in Fig. 2. The mobile platform can be manipulated within the workspace of the robot by changing the cable lengths with the application of actuators (generally called winches). Here, $\mathbf{a}_i$ represent the position of ith cable drawing point (origin of cable frame $\{A_i\}$) with respect to the global coordinate frame $\{O\}$ and the rotation matrix $\mathbf{R}_i$ represent the orientation of the cable frame $\{A_i\}$ with respect to the global coordinate frame $\{O\}$. The mobile platform vector $\mathbf{b_i}$ represent the position of ith cable anchor point on the mobile platform with respect to the local coordinate frame $\{P\}$ attached to the center of mass of the mobile platform. The position vector $\mathbf{r}$ and rotation matrix $\mathbf{R}$ represent the position and orientation of the mobile platform with respect to the global coordinate frame $\{O\}$, respectively. To statically balance the mobile platform of mass m, the following equations must be satisfied:

$$\sum_{i=1}^{n_c} -\mathbf{R}_i\mathbf{F}_i + m\mathbf{g} = 0 \quad (3)$$

$$\sum_{i=1}^{n_c} \mathbf{R}\mathbf{b}_i \times (-\mathbf{R}_i\mathbf{F}_i) = 0 \quad (4)$$

where, $\mathbf{F}_i$ denotes the force applied by the mobile platform on the ith cable and $\mathbf{g} = [0, 0, -9.8]^T$ represents the gravitational vector. In addition, the cable attachment point

B_i corresponding to the mobile platform must satisfy the following closed-loop kinematic constraints given by:

$$\mathbf{r} + \mathbf{R}\mathbf{b}_i = \boldsymbol{a}_i + \mathbf{R}_i \begin{bmatrix} x_i(L_{0_i}) \\ 0 \\ z_i(L_{0_i}) \end{bmatrix} \tag{5}$$

where, $x_i(L_{0_i})$ and $z_i(L_{0_i})$ represent the mobile platform attachment points of ith cable with respect to cable frame $\{A_i\}$ and are obtained by incorporation $s = L_{0_i}$ in Eq. (1) and Eq. (2), respectively. Therefore, from Eq. (5), the ith cable attachment point to the mobile platform must satisfy the following non-linear kinematic constraints in the cable frame $\{A_i\}$:

$$\begin{bmatrix} x_i(L_{0_i}) \\ 0 \\ z_i(L_{0_i}) \end{bmatrix} = \mathbf{R}_i^{-1}(\mathbf{r} + \mathbf{R}\mathbf{b}_i - \mathbf{a}_i) \tag{6}$$

Equation (6) represents 2 non-linear kinematic constraints for each cable, which corresponds to $2n_c$ non-linear constraints for n_c number of cables. Therefore, Eq. (3) and Eq. (4) represent the linear static constraints while Eq. (6) denotes the non-linear kinematic constraints which are to be satisfied for solving the IKS problem.

In general, the primary purpose of the inverse kinematic problem is to obtain the cable lengths for the given mobile platform position $\mathbf{r}$ and orientation $\mathbf{R}$. However, the inverse kinematic of a CDPR considering non-negligible cable mass is a kineto-static problem as the sag in the cable is dependent on the force acting on the cable. Therefore, obtaining the inverse kineto-static solution, in this case, corresponds to finding the unstrained length L_0 of the cable and the cable force components $[F_x F_z]$ for all cables in order to statically balance the mobile platform at the given pose (position and orientation).

The unknown parameters to be obtained are the unstrained length L_{0_i} of the cable and the cable force components $\left[F_{x_i} F_{z_i}\right]$, which correspond to $3n_c$ unknowns for a CDPR with n_c cables. Therefore, for a n DOF CDPR system with n_c cables, a total of $n + 2n_c$ equations (n static balance equations and $2n_c$ kinematic equations) must be solved for $3n_c$ unknowns to attain the solution to inverse kineto-static problem. For *redundantly constrained* CDPRs, the number of cables n_c are larger than the degrees of freedom n of the system. Hence, the number of unknowns are larger than the number of equations, thus making the inverse kineto-static problem an under-determined non-linear problem. This is why more than one solution is possible for this problem. To determine the unique solution out of multiple solutions, this problem is conventionally solved using constrained optimization with an objective to minimize/maximize the specific cost function along with satisfying the constraints given by static and kinematic equations.

The initial value for the optimization problem is attained by solving for cable lengths and forces considering massless cables. The initial value of cable lengths is obtained using closed-loop kinematic relations. However, multiple solutions to cable forces are possible due to the redundancy of the robot. Therefore, a unique initial value of cable forces is determined using tension distribution algorithms, which may lead to an increase in the computational time for solving the IKS problem. Thus, it can be stated that the inverse kineto-static problem for redundant robots is relatively more computationally intensive in comparison to IKS analysis for minimally constrained robots. The calculation of real-time IKS is important in providing reference cable lengths for motion control of a mobile platform.

3 Neural Network-Based Inverse Kineto-Static Analysis

Artificial neural network (ANN), initially introduced by McCulloch and Pitts in 1943, is an information processing paradigm that is inspired by networks of biological neurons. It comprises of a large number of interconnected perceptron/neurons that work in unison to provide a solution to a particular problem. The mathematical model of the perceptron/neuron is based on a biological neuron.

Deep neural network is a class of artificial neural networks with more than one hidden layer. The basic architecture of a deep neural network is given in Fig. 3. Deep neural networks have made a significant breakthrough through tremendous performance in numerous applications, such as computer vision, natural language processing, face recognition, adaptive testing, object detection, speech recognition, stock market analysis and many more [18]. In this paper, the deep neural network has been used for the inverse kineto-static analysis of a cable-driven parallel robot. The procedure to train the deep neural network for IKS analysis of a CDPR is generally divided into three steps given in the following subsections.

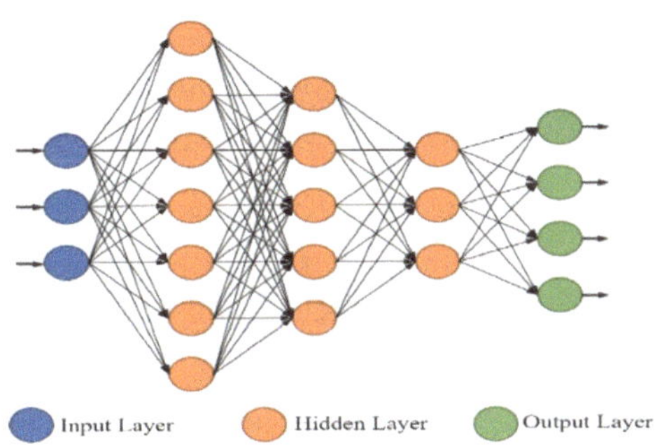

Fig. 3. Architecture of deep neural network

3.1 Dataset Creation

For training the deep neural network, it is important to have an appropriate dataset containing the information of inputs and their corresponding outputs. For the inverse kineto-static problem, the input data corresponds to the desired pose (position and orientation) of the mobile platform while the output data corresponds to the unstrained cable lengths as well as cable forces for the desired pose of the mobile platform. To create an input-output dataset, it is important to know the cable lengths and cable force components for the adequate number of mobile platform poses within the workspace of the robot. This can be attained experimentally using appropriate sensors to measure the cable lengths and forces for a large number of mobile platform poses within the workspace of the robot. Alternatively, the forward kineto-static or the inverse kineto-static relations can be employed to obtain the input-output dataset. In this paper, the inverse kineto-static solution using a numerical/constrained-optimization algorithm has been used to attain the input-output dataset.

3.2 Defining the Parameters of Deep Neural Network

The parameters of deep neural network significantly affect the efficiency of the neural network in learning a relation between the given input-output dataset. Various parameters such as the number of hidden layers, number of neurons in each layer, activation function, learning algorithms, number of epochs and error tolerance are to be selected for training the neural network. Appropriate parameters for the neural network are generally selected based upon the user's experience. However, there are certain guidelines that can be followed to decide appropriate parameters. The parameters (such as the number of hidden layers and number of neurons in each layer) are generally selected by keeping an appropriate balance between the desired error tolerance and the computational time for the training. The complexity of the system to be approximated also plays a major role in selecting the parameters of the neural network. Selecting a small number of layers or neurons may result in underfitting and can cause large training error. In contrast, a large number of layers/neurons may lead to overfitting resulting in poor generalization and hence large testing error. Therefore, proper care must be taken while deciding the parameters of the deep neural network. An appropriate activation function out of various activation functions such as *purelin*, *tansig*, *logsig*, *relu*, *radbas* and *sigmoid,* etc. can be chosen based on the nature (linear/non-linear or positive/real) of the input-output dataset. The learning algorithm also plays a significant role in the training of the neural network. Gauss-Newton algorithm, Gradient descent algorithm and Levenberg-Marquardt algorithm are some of the well-established algorithms employed for the training of deep neural networks. In this paper, the appropriate neural network architectures, activation function and training algorithm is selected by trial and error.

3.3 Training and Validation

Based on the defined parameters, the training of the neural network is done using an appropriate training algorithm. The input-output dataset is randomly divided into a training dataset, validation dataset and testing dataset. In general, the large percentage of the dataset is used for the training dataset, a few percentages for validation and testing dataset. The training dataset is loaded into the learning algorithm, which optimizes the weights of the neural network in order to reduce the error between the desired output and the predicted output of the neural network. Once this process is completed for all the samples in the training dataset, it is termed as one epoch. After each epoch, the neural network is tested on the validation dataset and the mean square error for the validation dataset is calculated, which is termed as a validation error. The validation error is used to keep a check on the overfitting while training the neural network. A large number of epochs are performed until the algorithm meets one of the stopping criteria. Some of the common stopping criteria are:

1) If the training error reaches the desired error tolerance.
2) If the number of epochs exceeds the defined maximum number of epochs.
3) If a negligible reduction in validation error is observed for defined number of consecutive epochs.

4 Results and Discussion

A 3 DOF planar cable-driven parallel robot consisting of 4 cables lying in a vertical plane is considered, as shown in Fig. 4. The geometry of the robot is defined by a square base frame (with drawing points at the corner of the frame) of the side $s_b = 50\,\text{m}$ and a square mobile platform (mass $m = 100\,\text{kg}$) of side $s_m = 2\,\text{m}$ with the cables anchored in a crossed configuration around the mobile platform. The orientation θ of the mobile-platform is constrained to $[-20°, +20°]$ to prevent cable-platform collisions. This kind of robot can be potentially used in large storage retrieval systems or for glass cleaning/painting of a large building. The dimensions of the robot are intentionally taken large to test the accuracy of the neural network-based approach for large-scale robots. The stainless steel cable of unstrained linear cable density $\rho_0 = 0.402\,\text{kg/m}$ and unstrained cable cross-sectional area $A_0 = 7.8 \times 10^{-5}\text{m}^2$ is utilized for the analysis. The cables are assumed to be linearly elastic with Young's modulus $E = 200\,\text{GPa}$.

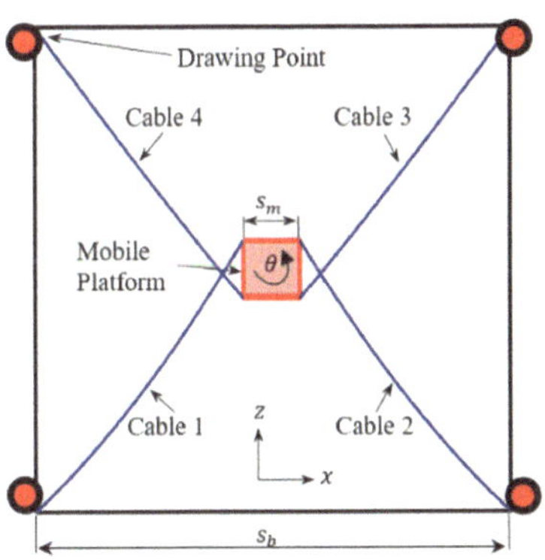

Fig. 4. Schematic diagram of redundantly constrained planar CDPR

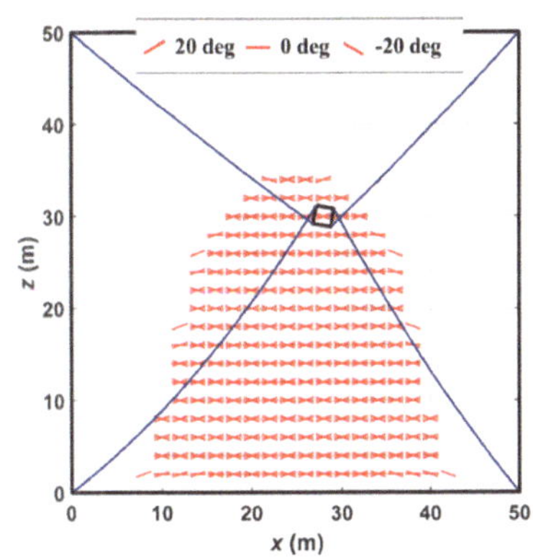

Fig. 5. WFW workspace of considered CDPR for three different orientations

The robot is a redundant robot with one degree of redundancy as the number of cables $n_c = 4$ is larger than the DOF $n = 3$, i.e., (x, z, θ) of the robot. Thus, there are multiple solutions possible for the tension distribution in the cables to balance a specific wrench (in this case, weight of the platform) and hence, multiple solutions of the inverse kineto-static problem are possible considering cable mass and elasticity.

The wrench-feasible workspace (WFW) of the CDPR for 3 different orientations is given in Fig. 5. The tension in the cables is constraint to be between $T_{min} = 100$ N to $T_{max} = 1500$ N. The lower bound on cable tension is selected based on minimum allowable sagging in the cable while the upper bound is selected to take care of actuators and mechanical limitations with an appropriate factor of safety. The collision of the lower cables with the ground is also considered as one of the constraints for plotting the workspace of the robot. This is done by obtaining cable profile using Eqs. (1) and (2) and checking the collision with the line $z = 0$.

Table 1. Error in NNIKS solution for the test dataset

Error in cable unstrained lengths L_{0_i} (m)		L_{0_1}	L_{0_2}	L_{0_3}	L_{0_4}
	max	0.0441	0.0356	0.0279	0.0433
	rmse	0.0131	0.0115	0.0106	0.0117
	min	−0.0384	−0.0340	−0.0269	−0.0407
Error in cable force F_{x_i} (N)		F_{x_1}	F_{x_2}	F_{x_3}	F_{x_4}
	max	9.1730	11.6913	2.4640	3.8955
	rmse	1.6745	1.8296	0.8690	0.7642
	min	−11.0195	−10.7183	−5.7087	−2.3207
Error in cable force F_{z_i} (N)		F_{z_1}	F_{z_2}	F_{z_3}	F_{z_4}
	max	5.4456	4.2460	3.1829	4.1929
	rmse	0.8170	0.7829	0.9481	0.8647
	min	−3.8337	−3.6601	−5.2833	−4.6545

The next step is to formulate the input-output dataset for the training and validation of the neural network. A large number of poses (both position and orientation) are selected within the workspace of the robot and the corresponding cable lengths and forces are attained using numerical optimization algorithms. Since multiple IKS solutions are possible for a single pose, a unique IKS solution for each pose is obtained by using the constrained optimization with an objective to minimize the cable sagging. Sagging can be minimized by minimizing the difference between the Euclidean distance L_{e_i} between the two anchor points (i.e., the drawing point and the mobile platform anchor point) and the strained length L_i of the cable. The objective function for the constrained optimization is given as:

$$\min f = \sqrt{\sum_{i=1}^{n_c} \left(L_i - L_{e_i}\right)^2} \tag{7}$$

The constraints of the optimization are given by static constraints (given by Eq. (3) and Eq. (4)) and the kinematic constraints (given by Eq. (6)). As the considered robot is redundant, a large training dataset is required to learn the complex kineto-statics of the robot. Therefore, a large input-output dataset containing 219560 samples is formulated. This dataset is randomly divided such that 75% of the dataset is assigned as the training dataset, 15% as validation dataset and the remaining 15% as testing dataset. An architecture of $3 \times 48 \times 24 \times 12$ is chosen for the training of the deep neural network. The position and orientation of the mobile platform, i.e., (x, z, θ) acts as inputs to the neural network while the cable lengths and force components correspond to 12 outputs of the neural network. The *tansig* function is chosen as an activation function for the hidden layers, while *purelin* function is chosen as an activation function for the output layers. The Levenberg-Marquardt algorithm is employed for the training of the neural network. The appropriate neural network architectures and parameters are selected after performing experiments on large number of architectures and parameters. The training of the network is completed in 168 h with 500000 epochs. The simulations are carried out in MATLAB on a system with an Intel Core i7-2600 CPU @3.40 GHz processor and a 16 GB of RAM. The trained neural network is further tested on a test dataset. The obtained results are given in Table 1, which shows the deviation of neural network based inverse kineto-static solution (NNIKS) from the numerical methods based inverse kineto-static solution (NIKS) in terms of the maximum error, root mean square error (rmse) and minimum error.

From the results, the maximum error of 0.0441 m, 11.6913 N and 5.4456 N is obtained for unstrained cable lengths L_{0_i}, the cable force component F_{x_i} and a cable force component F_{z_i}, respectively. This leads to a maximum relative error (i.e., $\max\limits_{i=1..n_c} \frac{\tau_i - \tau_i^{'}}{\tau_i}$ where τ_i and $\tau_i^{'}$ represents the tension in the ith cable attained for numerical and neural network-based approach, respectively) of 0.8101% in cable tension using the proposed approach. However, the error is small as compared to the scale of the CDPR.

The comparison of the cable profile for the neural network solution and the conventional numerical/optimization solution is visualized for 4 random poses in Fig. 6. A minor deviation in cable profiles can be seen between NNIKS and NIKS for the top left pose, while for the remaining poses, the cable profile is analogous to each other. Thus, after testing, it can be concluded that the deep neural network performed satisfactorily in approximating the IKS of CDPR with non-negligible cable mass and elasticity.

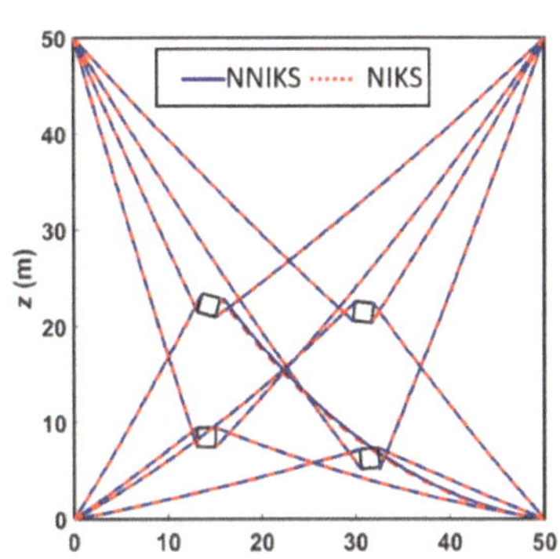

Fig. 6. IKS solution of 4 random test samples using NNIKS and NIKS

Furthermore, the computational efficiency of the proposed neural network-based approach is tested against the numerical approach. The comparison is done for the various number of runs and is compiled in Table 2. The obtained results show an approximately 83% reduction in computational time using a neural network-based approach as compared to the conventional numerical algorithm-based approach. The average computational time of the proposed neural network-based approach to compute the inverse kineto-static solution is 0.0070 s/pose. On the other hand, the average computational time to obtain IKS using numerical methods is 0.043 s/pose. Therefore, the proposed approach can significantly improve the computational speed of IKS analysis (compared to numerical approach) at the cost of a minor error in the IKS solution.

Table 2. Comparison of computational time between NNIKS and NIKS

No. of runs	100	1000	10000
Numerical solution (s)	6.93	48.34	430.81
Neural network solution (s)	0.74	7.44	70.34
Percentage reduction	89.45%	84.61%	83.67%

The computational time of the neural network can be further improved by pruning the pre-trained large neural network into a small network with an insignificant reduction in the performance [19]. One of the major advantages of using a neural network-based approach over the numerical approach is its non-dependency on the initial values. This makes the approach computationally bounded and reliable for real-time purposes.

5 Conclusions

In this paper, a deep neural network-based approach for the real-time inverse kineto-static analysis of cable-driven parallel robots with non-negligible cable mass and elasticity has been proposed. The methodology to obtain the inverse kineto-static solution of the robot using deep neural networks has been proposed. To validate the performance of the proposed approach, the simulation results are presented for a redundantly constrained planar CDPR. From the simulation results, a significant reduction of 83% in computational time has been observed for the considered redundantly constrained CDPR. Also, the proposed neural network-based approach has performed satisfactorily in terms of low error in the IKS solution. Therefore, the proposed approach is recommended for real-time applications due to its low and bounded computational time.

The requirement of large training time and large training dataset is one of the key limitations of the proposed approach. However, as the training is required for one time only, the large training time can be disregarded. In contrast, large training time can be significantly reduced using parallel computing for training the neural network. On the other hand, transfer learning [20] can be employed to reduce the requirement of large dataset in the case when the dataset is generated from the sensors of physical robot. In this case, a pre-trained neural network trained using large dataset constructed from simulation can be transferred to learn the IKS with relatively small dataset obtained from physical robot for the training of neural network. Transfer learning can also speed-up the training process of the neural network for different robot parameters or configurations.

Acknowledgments. This research was undertaken through the SPARC project funded by MHRD, Government of India, under Project No. SPARC/2018-2019/P713/SL. IIT Roorkee (India), IIT Kharagpur (India) and Queen's University (Canada) are the collaborating institutions.

References

1. Bosscher, P., Robert, L.W., Bryson, L.S., Castro-Lacouture, D.: Cable-suspended robotic contour crafting system. Autom. Constr. **17**, 45–55 (2007)
2. Miermeister, P., et al.: The cable robot simulator large scale motion platform based on cable robot technology. In: International Conference on Intelligent Robots and Systems (IROS), pp. 3024–3029. IEEE, Daejeon, South Korea (2016)
3. Li, D., Pan, Z.: The five-hundred-meter aperture spherical radio telescope project. Radio Sci. **51**(7), 1060–1064 (2016)
4. Carricato, M., Merlet, J.P.: Direct geometrico-static problem of under-constrained cable-driven parallel robots with three cables. In: International Conference on Robotics and Automation (ICRA), pp. 3011–3017. IEEE, Shanghai, China (2011)
5. Merlet, J.P.: On the real-time calculation of the forward kinematics of suspended cable-driven parallel robots. In: 14th IFToMM World Congress on the Theory of Machines and Mechanisms, pp. 1–10. Taipei, Taiwan (2015)
6. Cavalcanti, S.J., Gouttefarde, M.: A real-time capable forward kinematics algorithm for cable-driven parallel robots considering pulley kinematics. In: International Symposium on Advances in Robot Kinematics, pp. 199–208. Springer, Cham (2020)
7. Ghasemi, A., Eghtesad, M., Farid, M.: Neural network solution for forward kinematics problem of cable robots. J. Intell. Rob. Syst. **60**, 201–215 (2010)
8. Schmidt, V., Müller, B., Pott, A.: Solving the forward kinematics of cable-driven parallel robots with neural networks and interval arithmetic. In: Thomas, F., Pérez Gracia, A. (eds.) Computational Kinematics. MMS, vol. 15, pp. 103–110. Springer, Dordrecht (2014). https://doi.org/10.1007/978-94-007-7214-4_12
9. Kozak, K., Zhou, Q., Wang, J.: Static analysis of cable-driven manipulators with non-negligible cable mass. IEEE Trans. Rob. **22**(3), 425–433 (2006)
10. Irvine, H.M.: Cable Structures. The MIT Press, Cambridge (1981)
11. Chawla, I., Pathak, P.M., Notash, L., Samantaray, A.K., Li, Q., Sharma, U.K.: Effect of selection criterion on the kineto-static solution of a redundant cable-driven parallel robot considering cable mass and elasticity. Mech. Mach. Theory **156**, (2021)
12. Notash, L.: Deflection maps of planar elastic catenary cable-driven robots. J. Mech. Robot. **12**(2), 1–17 (2020)
13. Notash, L.: Artificial neural network prediction of deflection maps for cable-driven robots. In: International Design Engineering Technical Conferences and Computers and Information in Engineering Conference, p. V010T10A094. ASME (2020)
14. Merlet, J.P.: On the inverse kinematics of cable-driven parallel robots with up to 6 sagging cables. In: IEEE International Conference on Intelligent Robots and Systems, pp. 4356–4361. IEEE, Hamburg (2015)
15. Merlet, J.P.: Some properties of the Irvine cable model and their use for the kinematic analysis of cable-driven parallel robots. Mech. Mach. Theory **135**, 271–280 (2019)

16. Nguyen, D.Q., Gouttefarde, M., Company, O., Pierrot, F.: On the simplifications of cable model in static analysis of large-dimension cable-driven parallel robots. In: IEEE International Conference on Intelligent Robots and Systems, pp. 928–934. IEEE, Tokyo (2013)
17. Tang, L., Tang, X., Jiang, X., Gosselin, C.: Dynamic trajectory planning study of planar two-dof redundantly actuated cable-suspended parallel robots. Mechatronics **30**, 187–197 (2015)
18. Dargan, S., Kumar, M., Ayyagari, M.R., Kumar, G.: A survey of deep learning and its applications: a new paradigm to machine learning. Arch. Comput. Methods Eng. **27**, 1071–1092 (2019)
19. Frankle, J., Carbin, M.: The lottery ticket hypothesis: finding sparse, trainable neural networks. In: 7th International Conference on Learning Representations, pp. 1–42 (2019)
20. Tang, H., Notash, L.: Neural network-based transfer learning of manipulator inverse displacement analysis. J. Mech. Robot. **13**(3), 035004 (2021)

Unsupervised Neural Network Based Forward Kinematics for Cable-Driven Parallel Robots with Elastic Cables

Utkarsh A. Mishra[1] and Stéphane Caro[2(✉)]

[1] Indian Institute of Technology Roorkee, Roorkee 247667, Uttarakhand, India
umishra@me.iitr.ac.in

[2] CNRS, Laboratoire des Sciences du Numérique de Nantes, UMR CNRS 6004, 1, Rue de la Noë, 44321 Nantes, France
stephane.caro@ls2n.fr

Abstract. Forward Kinematic (FK) analysis of under-constrained Cable-Driven Parallel Robots (CDPRs) deals with the inherent coupling between the loop-closure and static equilibrium equations. The non-linearity of the problem is magnified with the addition of the coupling between the cable lengths and their tensions based on the elastic cable model. The paper proposes an unsupervised neural network algorithm to perform fast forward geometrico-static analysis for CDPRs in a suspended configuration with elastic cables. The formulation determines a non-linear function approximation to model the FK and proves to be efficient in solving for consecutive and close waypoints in a path. The methodology is applied on a simulated six-degree-of-freedom (6-DOF) spatial under-constrained suspended cable-driven parallel robot. Specific comparison results to show the effectiveness of the proposed method in tracking a given path are presented against the results obtained from least-square non-linear optimization.

Keywords: Unsupervised neural networks · Elastic cables · Forward kinematics

1 Introduction

Cable-Driven Parallel Robots (CDPRs) belong to a class of parallel manipulators where cables are used to manipulate a Moving Platform (MP) and connects it to a fixed base frame. CDPRs are inherently accompanied with various advantages of low inertia, high payload to weight ratio, and a significantly large workspace as compared to their serial and parallel counterparts. Such advantages have been exploited for various tasks like large scale 3D printing [2], in rehabilitation mechanisms [8], as transfer robots for the elderly [11] and rescue robots [12] as well as in large-scale telescopes [22]. The kinematic analysis for CDPRs play a vital role in realizing various fundamental characteristics consisting of the size of the static workspace, maintaining static equilibrium and achieving appropriate

M. Gouttefarde et al. (Eds.): CableCon 2021, MMS 104, pp. 63–76, 2021.
https://doi.org/10.1007/978-3-030-75789-2_6

stiffness. However, such an analysis turns out to be more complex as compared to parallel manipulators with rigid links, driven by the fact that cables can only apply unilateral forces, i.e., only pull the MP and not push it. Thus, constraining its capability to account for static equilibrium. Furthermore, such an analysis is critically dependent on the configuration of the complete manipulator as well as on the nature of the cables used.

The robot configuration is generally based on the number of cables modulating the intended degrees of freedom (DOFs) and on the availability of the cables to pull the MP in both upward and downward directions. In this context, we are particularly interested in suspended configurations i.e. when cables can only pull the MP upward. With the reduction in the number of cables, the system becomes under constrained and typically has n-DOFs controlled by $m(< n)$ cables. Thus, allowing only m-DOFs to be controlled. The kinematic analysis of such robots is inherently coupled with static analysis and cannot be solved without solving the combined kinetostatic problem [5]. For such systems, even when the cable lengths are fixed, the MP still moves and behaves according to the external wrench. Accordingly, the geometrico-static [4] problem becomes more complex and may have several solutions. The cable lengths tend to be the most feasible proprioceptive measurement for CDPRs as compared to cable tensions or orientations. If only cable lengths, i.e., pulley motor position measurements, are considered, the kinematic analysis is significantly influenced by the cable model used. Cables can be considered mass-less and non-elastic, mass-less but elastic, or with non-negligible mass and elastic [6]. In this paper, the cables are supposed to be massless and elastic.

Solving the forward kinematics for under-constrained suspended CDPRs has been approached as finding all the possible solutions to the geometrico-static equations or incorporating various iterative strategies to solve in real-time. Interval analysis [3,10] is one possible approach to solve for the complete analysis taking into account that some cables may be slack and have been explored for underconstrained systems. Various works have led to optimization problems for real-time analysis based on minimizing the potential energy [18], finding the lowest equilibrium pose [7] or minimizing cable tensions [14]. Iterative (usually, Levenberg-Marquadt) algorithms have found themselves useful in locally minimizing pose errors [18], using Hessian matrix to construct convex problem [17] and Jacobian matrix to solve the linearized approximation of the FK problem [21] at hand.

The optimization problem discussed for the real-time analysis can be seen to be a highly non-linear system and can be easily correlated to neural network architecture. Artificial Neural Network (ANN) has been used extensively in recent research to solve the inverse analysis of serial robots and forward analysis of parallel robots. Supervised methods are employed by collecting ground truth data and training to find a function approximation. Such practices have been extensively applied to serial [23], and cable [16] robots. Furthermore, as most controllers directly work on joint space applying direct control over motor winches to modulate cable lengths, path planning in joint space [15] is a better option. In such a situation, processing Cartesian poses for accurate tracking of a planned path can take advantage of a real-time forward kinematics module.

2 Contributions and Paper Organization

In this paper, we propose an unsupervised feed-forward neural network algorithm based forward kinematics of under-constrained suspended CDPRs with elastic cables. On the one hand, the under-constrained configuration couples the kinematics and statics of the robot, whereas, on the other hand, the elastic cable model couples the cable lengths with their tensions. This makes the problem formulation highly non-linear and coupled. The algorithm is motivated by the aim of performing forward kinematics for such situations while tracking a pre-planned path and is fulfilled using the unsupervised neural network weight adaptation strategy. Assuming that the waypoints belonging to the path are close, our model does not require significant weight updates. Thus, the convergence turns out to be slower for the starting position and faster for the waypoints as compared to the popular least-square non-linear (*lsqnonlin* by ®MATLAB) optimization framework. Finally, a pre-planned path is tracked using a simulated spatial CDPR (with 6-DOFs, 4 cables). The overall performance for the proposed algorithm and *lsqnonlin* is compared based on the time efficiency and accuracy.

The model of the manipulator and the associated nomenclature are discussed in Sect. 3. Section 4 describes the elastic cable model used Sect. 5 formulates the geometrico-static problem of the CDPR under study. The unsupervised neural network is described in Sect. 6. Section 7 gives the simulation results obtained using the proposed approach. Those results are discussed in Sect. 8. Conclusions and future work are drawn in Sect. 9.

3 Parametrization

We consider a CDPR consisting of m cables and n-DOFs. Its ith closed-loop being represented in Fig. 1. The two frames of reference, namely, $\mathcal{F}_b$ and $\mathcal{F}_p$ are attached to the fixed base frame and the MP frame respectively. $\mathbf{a}_i$ represents the Cartesian coordinates vector of exit point A_i on the fixed frame, expressed in $\mathcal{F}_b$. The cable attachment with the MP, B_i is defined by the Cartesian coordinate vector $\mathbf{b}_i$ expressed in $\mathcal{F}_p$. The MP pose is defined by the vector $\mathbf{p}$ from O to P and the corresponding orientation of the MP w.r.t. the fixed frame is defined by the rotation matrix ${}^b\mathbf{R}_p \in SO(3)$ from $\mathcal{F}_b$ to $\mathcal{F}_p$.

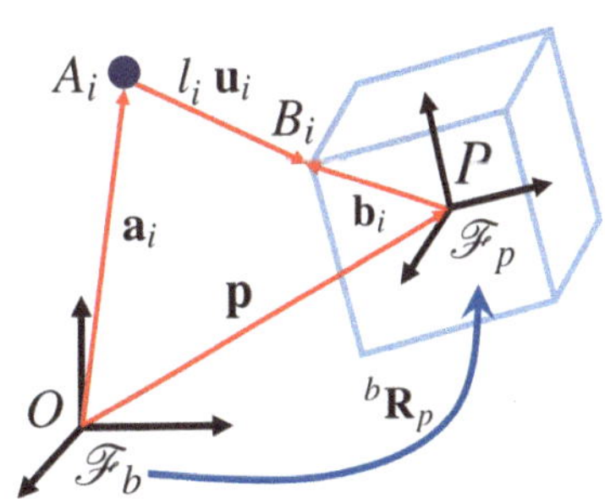

Fig. 1. i^{th}-closed loop of the CDPR

The ith loop closure equation is expressed as:

$$
{}^b\mathbf{a}_i + \ell_i^b \mathbf{u}_i - {}^b\mathbf{R}_p {}^p\mathbf{b}_i -{}^b \mathbf{p} = \mathbf{0}_3 \tag{1}
$$

$\mathbf{u}_i$ is the unit vector along the ith cable pointing from A_i to B_i and $\mathbf{0}_3$ is a three-dimensional zero vector. The subscript ^{b} (^{p}, resp.) means that the corresponding vector is expressed in frame $\mathcal{F}_b$ ($\mathcal{F}_p$, resp.).

The pre-planned path used for verification purpose is chosen to be within robot workspace (based on CRAFT [13] with an upper and lower bound on cable tensions equal to $\bar{t} = 86\,\mathrm{N}$ and $\underline{t} = 1\,\mathrm{N}$ respectively). The MP mass is named m_E.

4 Elastic Cable Model

We consider the mass-less elastic cable model wherein the cables are modelled as linear springs instead of rigid links. Their elongation is a function of their elasticiy and the tension. This eventually couples the kinematic analysis of the robot with the statics. Naturally, for a cable to get into a stable elongation, it is a logarithmic function of time [1]. However, for simplicity, we consider solving for the final elongation and use the stress-strain relationship [6]. For a cable, let the parameters be A_0, E_0, t_c, ℓ_a and ℓ_e defined as the undeformed cross-sectional area, elastic modulus, cable tension, length from the actuator-encoder and the final elongated length, respectively. Then, the elastic elongation, $\Delta\ell$, is given by:

$$\Delta\ell = t_c \frac{\ell_a}{E_0\ A_0} \tag{2}$$

and along with the relation, $\ell_e = \ell_a + \Delta\ell$, the undeformed length can be obtained as a function of cable tension and elongated length, given by:

$$\ell_a = \frac{\ell_e}{1 + \frac{t_c}{E_0\ A_0}} \tag{3}$$

5 Geometrico-Static Modeling

After realizing the first coupling between kinematics and statics based on elastic cable model in Sect. 4, this section formulates the second coupling between them based on the under-constrained suspended configuration. Specifically, this section presents the static equilibrium formulations with the inverse and forward kinematics. It should be noted that all occurrences of ℓ_i correspond to the elongated cable length.

5.1 Static Equilibrium

Let an external force $\mathbf{f}_p$ and moment $\boldsymbol{\tau}_p$ be applied on the moving platform. For the moving platform to remain in equilibrium, there must exist a m-dimensional vector of cable tensions, $\mathbf{t} = [t_1, t_2, \ldots, t_m]$, satisfying the Newton-Euler equation given by,

$$-\mathbf{W}\ \mathbf{t} + \mathbf{w_e} = \mathbf{0}_n \tag{4}$$

where $\mathbf{W}$ is the normalized Wrench matrix of the mechanism at this particular MP pose, given by

$$\mathbf{W} = \begin{bmatrix} {}^b\mathbf{u}_1 & \dots & {}^b\mathbf{u}_m \\ (1/r){}^b\mathbf{R}_p{}^p\mathbf{b}_1 \times {}^b\mathbf{u}_1 & \dots & (1/r){}^b\mathbf{R}_p{}^p\mathbf{b}_m \times {}^b\mathbf{u}_m \end{bmatrix} \tag{5}$$

and $\mathbf{w}_e = [\mathbf{f}_p^T \; (1/r)\boldsymbol{\tau}_p^T]^T$ as dimensionless homogenized matrices with the help of a characteristic length r which is defined as $r^2 = 1/m \sum_{i=1}^{m} \left\|{}^{\mathcal{F}}\mathbf{b}_i\right\|_2^2$ (refer [19,20]). Hereafter, we denote $\mathbf{W}$ as $\mathbf{W}(\mathbf{s})$, because wrench matrix is a function of the MP pose, $\mathbf{s}$ (defined in the next subsection).

Now, as the wrench matrix formulation is similar for both massless rigid and elastic cables [9], the tension distribution obtained from Eq. (4) holds true for our case.

5.2 Inverse Kinematics

The inverse kinematics (IK) formulation for CDPRs is the mapping from the Cartesian space to the cable space. For a given For a given pose (position + orientation) of the moving-platform (MP) of a n-DOF CDPR, IK computes the undeformed lengths of the m cables.

The MP pose, $\mathbf{s} = [{}^b\mathbf{p}, {}^b\varphi]^T$, is given by the position vector of the geometric center of the moving platform ${}^b\mathbf{p}$ and its orientation with respect to the base frame ${}^b\varphi = [\phi, \theta, \psi]$. Thus, ${}^b\mathbf{R}_p = \mathbf{R}_z(\psi)\mathbf{R}_y(\theta)\mathbf{R}_x(\phi)$ and from Eq. (1),

$$\ell_i^b\mathbf{u}_i =^b \mathbf{p} + {}^b\mathbf{R}_p{}^p\mathbf{b}_i -^b \mathbf{a}_i \tag{6}$$

$$\ell_i = \left\|{}^b\mathbf{p} + {}^b\mathbf{R}_p{}^p\mathbf{b}_i -^b \mathbf{a}_i\right\|_2 \tag{7}$$

where $\boldsymbol{\ell}_a = [\ell_{a1}, \ell_{a2}, \dots, \ell_{am}]$ is the vector of undeformed actuated cable lengths obtained using the corresponding elongated cable lengths, ℓ_i (from Eq. (7)), and cable tensions, $\mathbf{t}$ (from Eq. (4) using MP pose $\mathbf{s}$) through Eq. (3). $\|.\|_2$ denotes the two-norm of a vector. This IK mapping from $\mathbf{s}$, $\mathbf{t}$ to $\boldsymbol{\ell}_a$ is referred to the mapping function defined as,

$$\xi : \mathbf{s} \in \mathbb{R}^n, \mathbf{t} \in \mathbb{R}^m \mapsto \boldsymbol{\ell}_a \in \mathbb{R}^m$$

5.3 Forward Kinematics

The forward kinematics (FK) of a cable-driven manipulator consists of obtaining the platform pose, $\mathbf{s}$, based on given cable lengths $\boldsymbol{\ell}_a$ and external wrench. For the under-actuated CDPR studied in this paper, the kinematic model is under-determined. Thus, the FK problem is set up as a minimization problem where the relative error between the given actuated cable lengths, $\hat{\boldsymbol{\ell}}_a$, and the lengths obtained from the IK at the current pose, $\boldsymbol{\ell}_a = \xi(\mathbf{s},\ \mathbf{t})$ is a minimum. The error required to be minimized is given by:

$$e(\mathbf{s}) = \left\|\hat{\boldsymbol{\ell}}_a - \xi(\mathbf{s},\ \mathbf{t})\right\|_2 \tag{8}$$

Thus, for a given $\hat{\boldsymbol{\ell}}_a$, the FK formulation is expressed as:

$$\begin{gathered}\zeta : \hat{\boldsymbol{\ell}}_a \in \mathbb{R}^m \mapsto \mathbf{s} \in \mathbb{R}^n \\ \text{such that} \\ \zeta(\hat{\boldsymbol{\ell}}_a) = \arg\min_{\mathbf{s}} \left\| \hat{\boldsymbol{\ell}}_a - \xi(\mathbf{s},\ \mathbf{t}) \right\|_2\end{gathered}$$

6 Neural-Network Formulation

The components of robot kinetostatics play a vital role in defining the solution of the forward kinematics and the feasibility of the solution as a measure of satisfaction of the static equilibrium. Unlike serial robots, solving the forward kinematics is more difficult than solving the inverse kinematics for such parallel robots. The proposed neural network (NN) formulation is guided by an unsupervised iterative strategy to solve the surrogate objective of finding a suitable MP pose for given cable lengths.

The complete NN framework is built to take the cable lengths from the trajectory in joint space along with the external wrench being applied on the MP. Thus, the input vector is given by $\mathbf{x} = [\hat{\boldsymbol{\ell}}_a^T\ \mathbf{w_e}^T]^T$. Finally, the framework solves for the MP pose and required cable tensions to maintain the MP in a static equilibrium while considering elastic cable elongations. Accordingly, the output vector is given by $\mathbf{y} = [\mathbf{s}^T\ \mathbf{t}^T]^T$. The overall proposed strategy is formulated as solving:

$$\begin{gathered}NN : \mathbf{x} \in \mathbb{R}^{m+n} \mapsto \mathbf{y} \in \mathbb{R}^{m+n} \\ \text{s.t. for } \hat{\boldsymbol{\ell}}_a, \mathbf{w_e} \in \mathbf{x} \text{ and } \mathbf{s}, \mathbf{t} \in \mathbf{y} \\ NN(\mathbf{x}) = \mathbf{y} = \arg\min_{\mathbf{s},\ \mathbf{t}} \left\| \hat{\boldsymbol{\ell}}_a - \xi(\mathbf{s},\ \mathbf{t}) \right\|_2 \\ -\mathbf{W}(\mathbf{s})\ \mathbf{t} + \mathbf{w_e} = \mathbf{0}_n\end{gathered}$$

The above quadratic optimization problem was formulated as a surrogate objective ($\mathcal{L}$) based on penalty formation from the constraints such that

$$\mathcal{L}(\mathbf{y} = [\mathbf{s}^T, \mathbf{t}^T]^T) = \left\| \hat{\boldsymbol{\ell}}_a - \xi(\mathbf{s},\ \mathbf{t}) \right\|_2^2 + \mu \left\| -\mathbf{W}(\mathbf{s})\ \mathbf{t} + \mathbf{w_e} \right\|_2^2 \tag{9}$$

However, it can be observed that the summation holds two different measurement units, namely, square meter and Newton, respectively. In order to homogenize the surrogate objective, the characteristic length (meter) and the weight (Newton) of the MP were used respectively. Hence, the homogenized objective function is expressed as:

$$\hat{\mathcal{L}}(\mathbf{y}) = \frac{1}{r^2} \left\| \hat{\boldsymbol{\ell}}_a - \xi(\mathbf{s},\ \mathbf{t}) \right\|_2^2 + \mu \left(\frac{\left\| -\mathbf{W}(\mathbf{s})\ \mathbf{t} + \mathbf{w_e} \right\|_2}{m_E\ g} \right)^2 \tag{10}$$

where $g = 9.81\,\mathrm{m/s^2}$ is the acceleration due to gravity. Such a penalty function formulation can be solved using Gradient Descent, Newton or Quasi-Newton method and Trust Region methods.

6.1 Network Architecture

The formulated input-output problem is solved by means of an unsupervised neural network scheme where the error is calculated using the cable lengths corresponding to the predicted MP pose. The gradients are calculated based on the Stochastic Gradient Descent (SGD) algorithm and the weights are updated based on those gradients. Consider a Neural Network (NN), as shown in Fig. 2, consisting of two hidden layers with weights W of size $h \times (m+n)$, V of size $(m+n) \times h$ and biases B of size $h \times 1$, respectively. Here, h is the number of hidden nodes in the hidden layers. The activation functions used for the hidden layers is the sigmoid function, $\sigma(x)$, defined by:

$$\sigma(x) = \frac{1}{1+e^{-x}} \tag{11}$$

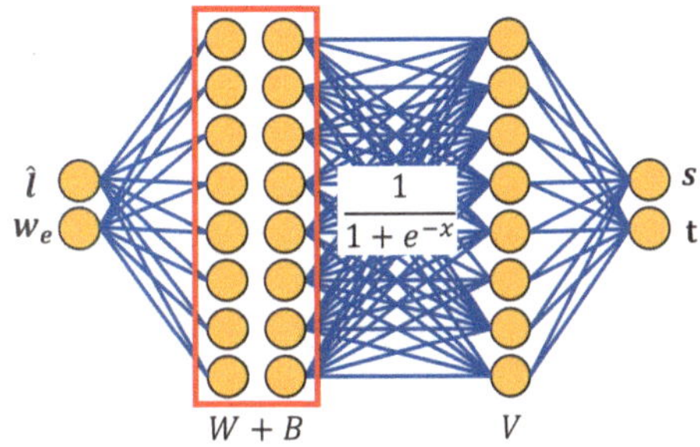

Fig. 2. Neural network architecture and activation function

The back-propagation error is defined using $\hat{\mathcal{L}}$ from Eq. (8). From SGD formulation, the weights and biases are optimized with reference to calculated change in weights δ_W of size $h \times (m+n)$, δ_V of size $(m+n) \times h$ and biases δ_B of size $h \times 1$ given by:

$$\delta_W(j,i) = \frac{\partial \hat{\mathcal{L}}}{\partial W(j,i)} \quad \forall \quad i = [1\ (m+n)], \quad j = [1\ h] \tag{12}$$

$$\delta_V(k,j) = \frac{\partial \hat{\mathcal{L}}}{\partial V(k,j)} \quad \forall \quad j = [1\ h], \quad k = [1\ (m+n)] \tag{13}$$

$$\delta_B(j,1) = \frac{\partial \hat{\mathcal{L}}}{\partial B(j,1)} \quad \forall \quad j = [1\ h] \tag{14}$$

The updated weights and biases ($\Box_{new}$) are obtained by using the above $\delta_\Box$ values obtained for the corresponding current weights and biases ($\Box_{current}$). Here. $\Box$ represents all of W, V (weights) and B (biases). The neural network parameter update equations are given by:

$$W_{new} = W_{current} - \alpha\delta_W \tag{15}$$

$$V_{new} = V_{current} - \alpha\delta_V \tag{16}$$

$$B_{new} = B_{current} - \alpha\delta_B \tag{17}$$

where α is defined as the learning rate, typically in the range of 10^{-4}.

6.2 Derivatives and Gradient Descent

The final problem gets reduced to calculation of the gradients, i.e., the changes in weights. To solve this problem, some pre-calculations are done to represent the predicted output vector $\mathbf{y}^{nn}$ from the input $\mathbf{x}$, such that,

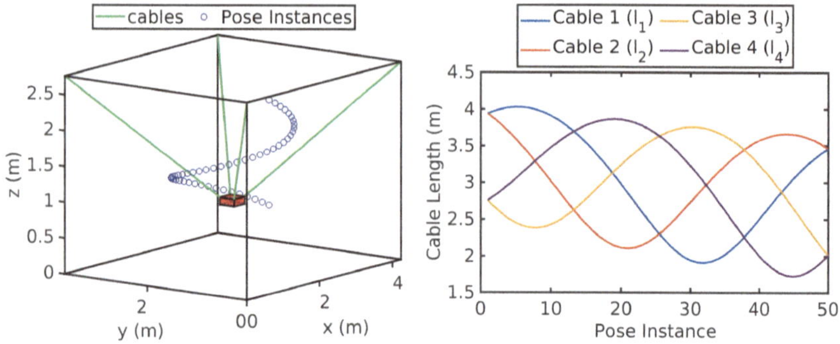

(a) Model and Cartesian Path (b) Pre-planned path in Cable Space

Fig. 3. CDPR configuration and reference path

$$\mathbf{y}^{nn}{}_{(m+n)\times 1} = V_{(m+n)\times h}\sigma(W_{h\times(m+n)}\mathbf{x}_{(m+n)\times 1} + B_{h\times 1}) \tag{18}$$

Each uth individual term in the output vector is a function of the components of the input vector:

$$\mathbf{y}_u^{nn} = \sum_{j=1}^{h}\left(V(u,j)\sigma\left(\sum_{i=1}^{m+n} W(j,i)\mathbf{x}(i) + B(j,1)\right)\right) \tag{19}$$

As a result, the gradients defined in Eqs. (10) to (12) take the form:

$$\frac{\partial \hat{\mathcal{L}}}{\partial W(j,i)} = \sum_{u=1}^{m+n} \frac{\partial \hat{\mathcal{L}}}{\partial \mathbf{y}_u^{nn}} \frac{\partial \mathbf{y}_u^{nn}}{\partial W(j,i)} \tag{20}$$

$$\frac{\partial \hat{\mathcal{L}}}{\partial V(k,j)} = \sum_{u=1}^{m+n} \frac{\partial \hat{\mathcal{L}}}{\partial \mathbf{y}_u^{nn}} \frac{\partial \mathbf{y}_u^{nn}}{\partial V(k,j)} \tag{21}$$

$$\frac{\partial \hat{\mathcal{L}}}{\partial B(j,1)} = \sum_{u=1}^{m+n} \frac{\partial \hat{\mathcal{L}}}{\partial \mathbf{y}_u^{nn}} \frac{\partial \mathbf{y}_u^{nn}}{\partial B(j,1)} \tag{22}$$

The individual segments can be calculated based on derivative of the error with respect to the outputs and the derivatives of outputs with respect to the weights. Hereafter, we refer this complete algorithm as the Unsupervised Forward Kinematics Neural Network (UFKNN).

7 Simulation Results

UFKNN algorithm is implemented on a 6-DOF spatial robot with four cables. The primary motivation of the methodology is to develop re-usability by making a module to learn the FK for a particular manipulator. Thus, the results are analyzed by making the manipulator follow a planned path in the joint space and are compared against the results obtained with *lsqnonlin*, a non-linear optimizer using the Levenberg-Marquardt algorithm. It is expected that the weights estimated for one instance do not vary much when the MP moves to the next waypoint, decreasing the time taken to compute FK for each instance. This makes tracking faster using only proprioceptive sensors. The results are present thereafter considering the elastic cable model.

The suspended underconstrained configuration as shown in Fig. 3a is considered. The setup with a size of 4.24 m × 3.67 m × 2.76 m (l × b × h) has a moving platform of size 0.28 m × 0.28 m × 0.20 m. The appropriate learning rate (α) was obtained iteratively by observing the quality of the results. Table 1 gives the parameters that are used in this illustrative example.

Table 1. Parameters for spatial CDPR-UFKNN algorithm, *lsqnonlin* and elastic cable model

Parameter	Value	Parameter	Value	Parameter	Value
m_E	5.6 Kg	t_1	15 N	itr^{lsq}_{max}	5×10^4
r	0.396 m	t_2	18 N	tol^{lsq}	10^{-8}
μ	50	t_3	25 N	E_0	70 GPa
h	64	t_4	20 N	d_0	5 mm
α	10^{-5}	$\hat{\mathcal{L}}_{threshold}$	0.0001	A_0	$\pi\ d_0^2/4$

The reference path for this case is chosen to be a vertical spiral in the center of the fixed frame, as in Fig. 3a, and the corresponding path in joint space is shown in Fig. 3b. Such a path considers the extreme points at varying heights. Now, to track the accuracy of the algorithm in finding an appropriate solution, we prepare ground truth outputs using the Cartesian data and fixed tension vector equal to $\mathbf{t} = [t_1, t_2, t_3, t_4]$, cable tension values being given in Table 1. The wrench to be applied on the MP at every pose instance is calculated by considering the fixed tension vector and the wrench matrix corresponding to that MP pose.

The comparison setup for both the algorithms was based on an iterative rule based on the fact that the right initial solution should speed up the convergence of the algorithms. The ground truth output belonging to the first pose instance was fed into the algorithms. During this initial setup process, while *lsqnonlin* takes only 4 s to converge, the neural network formulation requires about 1.5-min to tune the initial network weights. All the calculations were performed using ©MATLAB with CPU computations on an Intel ®i7-7500U CPU@2.70 GHz.

After both the algorithms perfectly fit into the starting pose, the previous outputs from the algorithms were used as initial guesses (for *lsqnonlin*) or weights (for neural network) for each of the next pose instances. The performance and accuracy are measured on a variety of factors as described below. Here, performance is referred to the overall computational time objective which is the primary objective. Accuracy, on the other hand, refers to the relative deviation of the cable lengths, cable tensions and Cartesian MP poses from the desired values.

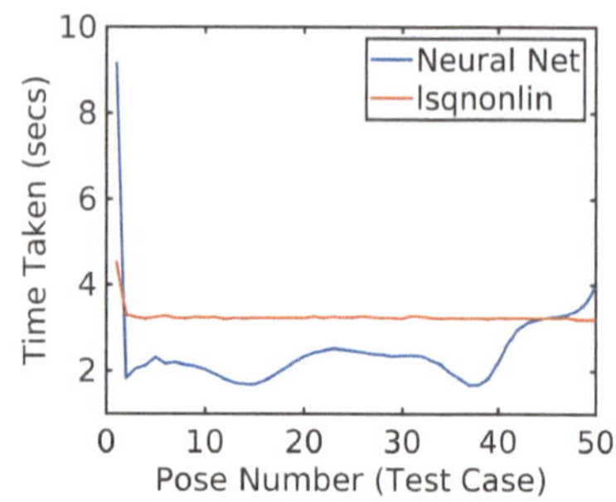

Fig. 4. Computational time efficiency comparison

As a performance measure, the overall performance for both the algorithms was compared based on the computation time for each of the 50 pose instances. Figure 4 clearly justifies the learning of the neural network. It should be noted that the pre-iterations for the initial pose for 1.5 min can be considered as the time taken by the algorithm to learn the model-specific kineto-statics. Though *lsqnonlin* gets the privilege of having a very close initial guess (as the waypoints are very close to each other), it has to perform a black-box optimization every time due to its model-free approach, and takes nearly the same time (≈ 3.5 s). The UFKNN algorithm does not store the guesses, but a non-linear approximation of the kineto-static model itself and progressively decreases the computational time. This is to be noted that the test conducted in the presented work maintains constant tension targets for all the test poses in the reference path. With this consideration, the final points on the reference path lie at the boundary of the wrench feasible workspace and hence the computation time increases in the later stages.

8 Discussion

As we deal with coupled kinematics and statics of the under-constrained CDPR model, the pose can be changed to satisfy wrench feasibility and decrease the wrench satisfaction error. Thus, it is essential to judge the formulation's accuracy by comparing it with the ground truth reference poses. The metrics for comparison are the cable lengths, cable tensions, and the predicted MP poses. Figures 5 and 6 show the accuracy comparison of both the algorithms.

For cable lengths, the desired cable lengths obtained from the planned path are a part of the inputs to the algorithms. After predicting the MP pose by the UFKNN algorithm, the obtained cable lengths are calculated through IK. The final error is the difference between the calculated and the desired cable lengths. The comparison of these cable errors is given in Fig. 5 for all cables. The maximum deviation in cable length is equal to 0.6 mm, which is negligible with respect to four-meter frame length. It is apparent that UFKNN reaches the same accuracy as *lsqnonlin* in terms of cable lengths, but in lesser time.

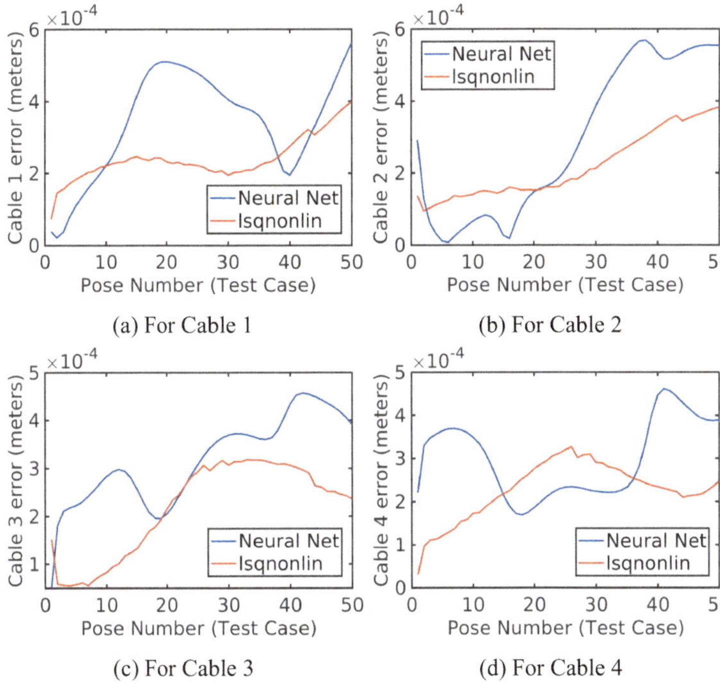

Fig. 5. Comparison of cable length errors

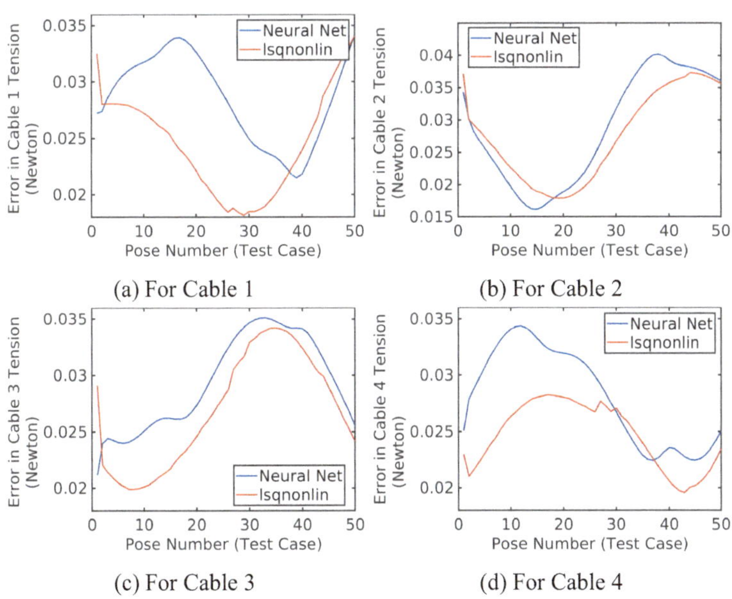

Fig. 6. Comparison of cable tension errors

For cable tensions, the cable tension values were fixed for all MP pose instances, whereas the external wrench, which is also an input to the algorithm, is varied such that the wrench feasibility condition holds. The outputs of the algorithm correspond to the MP pose and cable tensions. The error between the obtained and fixed desired tensions is shown in Fig. 6. The maximum absolute error in cable tensions is 0.035 N for *lsqnonlin* and 0.04 N for the UFKNN algorithm. The corresponding mean errors are 0.020 N and 0.025 N respectively. All these cable tensions usually result in a very high degree of wrench satisfaction i.e. $\frac{\|-\mathbf{W(s)}\ \mathbf{t}+\mathbf{w_e}\|_2}{m_E\ g} \approx 10^{-5}$.

For predicted MP poses, the mean Cartesian error for the MP poses of the complete path is 0.6 mm for UFKNN algorithm and 0.5 mm for *lsqnonlin*. The deviations are mostly due to angular displacements, which should be null, but such a high precision cannot be expected from numerical computations.

9 Conclusions and Future Work

This paper presented an unsupervised neural network weight adaptation framework according to the kineto-statics of under-constrained Cable-Driven Parallel Robots (CDPRs) with elastic cables and solves the FK problem for a suspended configuration. The approach is validated with simulated results in a 6-DOF spatial CDPR setup and compared with a non-linear least-square optimization-based *lsqnonlin* algorithm. The comparison demonstrates the advantages of the model-based UFKNN algorithm to approximate the robot's kineto-statics compared with model-free black-box optimization techniques currently in practice. This method is beneficial to track paths in cable space, namely in the actuation space. The method is computation time-efficient and works at the same accuracy as other related optimization techniques. Future work includes implementing the proposed algorithm on hardware and conducting simulated experiments to fully leverage its capabilities. Finally, the algorithm will be extended to non-negligible mass and sagging-based cable models to determine the solutions to the forward kinematics of more complex and realistic CDPRs.

Acknowledgement. This work was supported by the ANR CRAFT project, grant ANR-18-CE10-0004, https://anr.fr/Project-ANR-18-CE10-0004.

References

1. Baklouti, S., Courteille, E., Caro, S., Dkhil, M.: Dynamic and oscillatory motions of cable-driven parallel robots based on a nonlinear cable tension model. J. Mech. Robot. (2017). https://doi.org/10.1115/1.4038068
2. Barnett, E., Gosselin, C.: Large-scale 3D printing with a cable-suspended robot. Addit. Manufact. (2015). https://doi.org/10.1016/j.addma.2015.05.001
3. Berti, A., Merlet, J.P., Carricato, M.: Solving the direct geometrico-static problem of underconstrained cable-driven parallel robots by interval analysis. Int. J. Robot. Res. (2016). https://doi.org/10.1177/0278364915595277

4. Carricato, M.: Direct geometrico-static problem of underconstrained cable-driven parallel robots with three cables. J. Mech. Robot. **5**(3) (2013). https://doi.org/10.1115/1.4024293
5. Carricato, M., Merlet, J.P.: Stability analysis of underconstrained cable-driven parallel robots. IEEE Trans. Robot. 1 (2012). https://doi.org/10.1109/vr.2012.6180941
6. Chawla, I., Pathak, P.M., Notash, L., Samantaray, A.K., Li, Q., Sharma, U.K.: Effect of selection criterion on the kineto-static solution of a redundant cable-driven parallel robot considering cable mass and elasticity. Mech. Mach. Theory (2021). https://doi.org/10.1016/j.mechmachtheory.2020.104175
7. Collard, J.F., Cardou, P.: Computing the lowest equilibrium pose of a cable-suspended rigid body. Optim. Eng. (2013). https://doi.org/10.1007/s11081-012-9191-5
8. Homma, K., Fukuda, O., Sugawara, J., Nagata, Y., Usuba, M.: A wire-driven leg rehabilitation system: Development of a 4-DOF experimental system. In: IEEE/ASME International Conference on Advanced Intelligent Mechatronics, AIM (2003). https://doi.org/10.1109/AIM.2003.1225463
9. Maier, T., Woernle, C.: Inverse kinematics for an underconstrained cable suspension manipulator. In: Advances in Robot Kinematics: Analysis and Control (1998). https://doi.org/10.1007/978-94-015-9064-8_10
10. Merlet, J.P.: Solving the forward kinematics of a gough-type parallel manipulator with interval analysis. Int. J. Robot. Res. (2004). https://doi.org/10.1177/0278364904039806
11. Merlet, J.P.: MARIONET, a family of modular wire-driven parallel robots. In: Advances in Robot Kinematics: Motion in Man and Machine (2010). https://doi.org/10.1007/978-90-481-9262-5_6
12. Merlet, J.P., Daney, D.: A portable, modular parallel wire crane for rescue operations. In: Proceedings - IEEE International Conference on Robotics and Automation (2010). https://doi.org/10.1109/ROBOT.2010.5509299
13. Metillon, M., Cardou, P., Subrin, K., Charron, C., Caro, S.: A cable-driven parallel robot with full-circle end-effector rotations. J. Mech. Robot. (2021). https://doi.org/10.1115/1.4049631
14. Miermeister, P., Kraus, W., Pott, A.: Differential kinematics for calibration, system investigation, and force based forward kinematics of cable-driven parallel robots. In: Mechanisms and Machine Science (2013). https://doi.org/10.1007/978-3-642-31988-4_20
15. Mishra, U.A., Chawla, I., Pathak, P.M.: On determining shortest path in joint space of a cable-driven parallel robot for point-to-point motion. In: 2020 28th Mediterranean Conference on Control and Automation, MED 2020 (2020). https://doi.org/10.1109/MED48518.2020.9183198
16. Notash, L.: Artificial neural network prediction of deflection maps for cable-driven robots. In: Proceedings of the ASME Design Engineering Technical Conference (2020). https://doi.org/10.1115/DETC2020-22513
17. Pott, A.: Influence of pulley kinematics on cable-driven parallel robots. In: Latest Advances in Robot Kinematics (2012). https://doi.org/10.1007/978-94-007-4620-6_25
18. Pott, A., Schmidt, V.: On the forward kinematics of cable-driven parallel robots. In: IEEE International Conference on Intelligent Robots and Systems (2015). https://doi.org/10.1109/IROS.2015.7353818

19. Rasheed, T., Long, P., Caro, S.: Wrench-feasible workspace of mobile cable-driven parallel robots. J. Mech. Robot. **12**(3) (2020). https://doi.org/10.1115/1.4045423. https://hal.archives-ouvertes.fr/hal-02379201
20. Ruiz, A.L.C., Caro, S., Cardou, P., Guay, F.: ARACHNIS: analysis of robots actuated by cables with handy and neat interface software. In: Mechanisms and Machine Science (2015). https://doi.org/10.1007/978-3-319-09489-2_21
21. Santos, J.C., Gouttefarde, M.: A Real-Time Capable Forward Kinematics Algorithm for Cable-Driven Parallel Robots Considering Pulley Kinematics (2021). https://doi.org/10.1007/978-3-030-50975-0_25
22. Su, Y.X., Duan, B.Y., Nan, R.D., Peng, B.: Development of a large parallel-cable manipulator for the feed-supporting system of a next-generation large radio telescope. J. Robot. Syst. (2001). https://doi.org/10.1002/rob.8102
23. Tang, H., Notash, L.: Neural network based transfer learning of manipulator inverse displacement analysis. In: Proceedings of the ASME Design Engineering Technical Conference (2020). https://doi.org/10.1115/DETC2020-22508

Statics and Force Distribution

Maximal Cable Tensions of a N-1 Cable-Driven Parallel Robot with Elastic or Ideal Cables

Jean-Pierre Merlet(✉)

HEPHAISTOS Project, Inria Sophia-Antipolis, Sophia-Antipolis, France
Jean-Pierre.Merlet@inria.fr

Abstract. Determining what will be the maximal cable tensions of a cable-driven parallel robot (CDPR) when it moves over a given workspace is an important step in the design phase as it will allow to choose the cable diameter and to provide a requested information for tuning the CDPR actuation. In this paper we consider a suspended N-1 CDPR with n cables where all cables are attached at the same point, which leads to a 3-dof robot. We assume a quasi-static behavior of the robot and assume that the cable are either ideal or elastic so that we neglect the sagging effect. Under these assumption we show that the maximum of the cable tensions may be determined in a very fast way by solving a set of second-order polynomials which will lead to the poses at which the maximum of each cable tension will occur. For example for a four-cables CDPR determining the maximal cable tension requires to solve at most 149 s order polynomials.

Keywords: Cable-driven parallel robot · Static · Cable tension

1 Introduction

Determining the largest cable tensions over a given workspace is an essential part of the design of a CDPR. This problem has been addressed for classical parallel robots [1,2] and has been used for their design [3]. However this problem is somewhat different for CDPR because of their redundancy, configuration changes, cable slackness and influence of the discrete time control [4] that may lead to different tensions for a given pose of the CDPR or on a given trajectory. Most works on static analysis address only indirectly this issue either by computing different types of workspace under the constraint that the tensions lie within given bounds [5–8] or by trying to impose by control an optimal force distribution [9–13]. In the later case the force distribution is difficult to ensure for fully constrained CDPRs because of the difficulty to reliably measure the cable tensions, while for suspended CDPRs this scheme can only be adopted if the cable are elastic (for non-elastic cables executing a given tension control scheme at a given pose requires an **exact** control of the cable lengths, an assumption

M. Gouttefarde et al. (Eds.): CableCon 2021, MMS 104, pp. 79–89, 2021.
https://doi.org/10.1007/978-3-030-75789-2_7

that is clearly unrealistic). However for safety reasons it is of interest to determine the largest cable tensions in the worst case (i.e. with no assumption on the control) for a given CDPR when the platform moves in a given workspace. Clearly dynamics will play a role on the tensions [14–16] but however a first step is to consider only quasi-static motion.

To the best of the author knowledge the only work having addressed the determination of the largest cable tensions has been presented by Riehl [17] for a 3-1 CDPR with sagging cables. For a given pose the cable tensions are determined assuming ideal cables and then a numerical procedure is used to derive the tensions with sagging cables. Then the given workspace is sampled (i.e. the workspace is approximated by a finite set of poses, called *node poses*) and the maximum cable tensions is derived from the calculation at the node poses. The first limitation of this work is that the inverse kinematics (IK) of the 3-1 CDPR with sagging cable has most probably a single solution while for a N-1 CDPR with $N > 3$ the IK may have multiple solutions so that all solutions of the IK have to be determined in order to determine the largest cable tensions in the worst case. The second limitation is related to the sampling of the workspace which leads to a large computation time for determining an approximation of the largest tensions as the real maximum may possibly not occur at a node pose. For example for a $5 \times 5 \times 5$ m workspace that is sampled with a 5 cm step size with a computation time for calculating the cable tensions at a node of 5 ms we will get a computation time of 10 mn 25 s for determining the maximal cable tensions, which is problematic if this calculation is used in a design process that require a large number of such a calculation. The objective of this paper is to show that the pose at which the maximal tension will occur can be calculated exactly in a very short time.

In this paper we will consider a suspended 3 dof N-1 CDPR with $N \geq 4$ cables attached to the same point on the platform so that this robot allows only for translational motions (Fig. 1). Furthermore we will assume a geometry that is respected for all the existing prototypes: the exit point A_i have the same height z_a while the convex hull of the projection of these points on the ground is a rectangle $\mathscr{R}$. A consequence is that the available workspace for this CDPR in terms of horizontal motions is $\mathscr{R}$. Regarding cable model we will assume ideal cables(no mass, no elasticity). Note that the results we will obtain also cover the case of elastic cables with no mass, whatever the elasticity model is, as elasticity just play a role for determining the cable lengths to reach a given pose but does not influence the cable tensions at this pose.

As mentioned previously the exit point of the cables will be denoted A_i while the common point of the cables will be B with coordinates (x_b, y_b, z_b). We number the A_i point in such a way that the points A_1, A_2, A_3 are three corners of the rectangle $\mathscr{R}$ and we define a reference frame whose origin will be A_1 (whose coordinates are therefore $(0, 0, z_a)$) and A_2, A_3 will have as coordinates ($x_{a_2} > 0$,0,z_a), ($x_{a_3} \geq 0, ya_3 > 0, 0$), while the others A_i coordinates are ($0 \leq x_{a_i} \leq x_{a_2}$,$0 \leq ya_i \leq y_{a_3}, z_a$). We will assume that $x_{a_i} = x_{a_2}$ or $y_{a_i} = y_{a_3}$ so that any A_i is located on one edge of the rectangle $\mathscr{R}$. The point on the ground

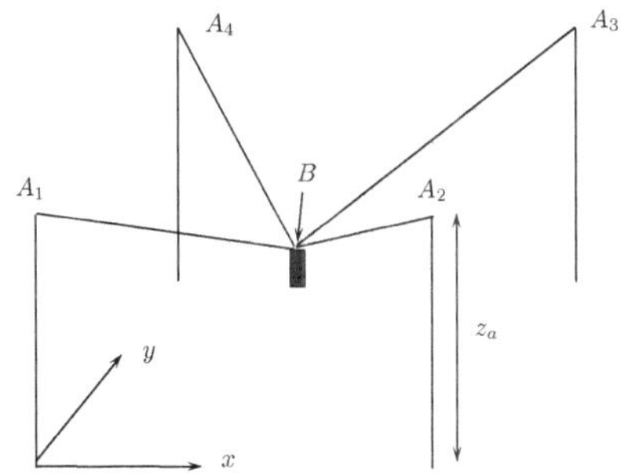

Fig. 1. A example of a $N-1$ CDPR

that results from a vertical projection of a point M will be denoted M^p. We will also assume that the CDPR platform is only submitted to gravity and that its mass is m. In a quasi-static case the center of mass of the platform is located below B so that no torque is exerted at B. We will also assume that the set of poses of the CDPR for which we will calculate the maximal cable tensions (the *desired workspace*) is a rectangle included in $\mathscr{R}$ that is defined by the constraints $0 \leq x_r^1 \leq x \leq x_r^2 \leq x_{a_2}$, $0 \leq y_r^1 \leq y \leq y_r^2 \leq y_{a_3}$. Clearly the maximal height of the desired workspace should be lower than z_a.

2 Calculation of the Maximal Cable Tensions

For ideal cable a CDPR property is that most of the time only 3 of the cables will be under tension. Indeed when 3 cables are under tension their lengths leads to a unique possible pose B_a for B: hence to have any other cable j under tension at B_a may occurs only when its length ρ_j is exactly $||\mathbf{A_j B_a}||$ as if $\rho_j > ||\mathbf{A_j B_a}||$, then cable j is slack and if $\rho_j < ||\mathbf{A_j B_a}||$, then B will move to a pose different from B_a. A consequence is that trying to create by control a configuration in which 4 cables are under tension will require an infinite accuracy on the measurement of the cable lengths together with a exact knowledge of the location of the A_i points. However having more than 3 cables under tension may occur occasionally for example when we initially set the lengths of cable 1, 2, 3 in such a way that B_a^p lies in the triangle $A_1^p A_2^p A_3^p$ while cable 4 is slack and then coil cable 4 until its lengths is largely lower than $||\mathbf{A_j B_a}||$. Hence for computing exactly the maximum cable tensions we have also to consider that more than 3 cables are under tension as this temporary configuration may occur on any trajectory whatever the control law is and although we are not able to determine when this configuration will occur.

In the next section we will first start to investigate the case where only 3 cables are under tension, the others being slack.

2.1 Maximal Tension with only 3 Cables Under Tension

For determining the maximal tension in all the n cables of the CDPR under the assumption that only 3 of them are under tension we will have to consider all the combinations of 3 cables among the n cable i.e. $n(n-1)(n-2)/6$ cases. However the treatment of all these cases is basically the same so that we may consider the example case of cables 1, 2, 3 being under tension.

If we assume that cable 1, 2 and 3 are under tension, while the other cables are slack and the CDPR is in mechanical equilibrium, then the point B_a^p should lie in the triangle $A_1^p A_2^p A_3^p$. If τ_i denotes the tension of cable i and ρ_i its length, then the mechanical equilibrium condition is a linear system of 3 equations in the τ_1, τ_3, τ_3 unknowns so that we get

$$\tau_1 = \frac{mg\,\rho_1\,(xy_{a_3} + x_{a_2}\,y - x_{a_2}\,y_{a_3} - x_{a_3}\,y)}{x_{a_2}\,y_{a_3}\,(z - z_a)}$$

$$\tau_2 = -\frac{mg\,\rho_2\,(xy_{a_3} - x_{a_3}\,y)}{x_{a_2}\,y_{a_3}\,(z - z_a)} \qquad \tau_3 = -\frac{mg\,\rho_3\,y}{y_{a_3}\,(z - z_a)}$$

where $z - z_a$ is negative. We notice that τ_i may be written as $\tau_i = mg\rho_i U_i(x, y)/V_i(z)$ where U_i is linear in x, y, while V_i is always negative.

Regarding τ_1, the term $U_1 = xy_{a_3} + x_{a_2}\,y - x_{a_2}\,y_{a_3} - x_{a_3}\,y$ is the signed distance of B to the line 23 which is negative as soon as B^p lies in the triangle $A_1^p A_2^p A_3^p$, so that $\tau_1 > 0$. Regarding τ_2, the term $xy_{a_3} - x_{a_3}\,y$ is positive as soon as B^p is inside the triangle $A_1^p A_2^p A_3^p$ so that τ_2 is positive in that case, while τ_3 is positive as soon as $y > 0$.

We may now look at the influence of the altitude z on the value of τ_i: we notice that this influence is related to the variation of the term $Z = \rho_i/(z - z_a)$ as each τ_i is established as $K_i\rho_i/(z - z_a)$ where the $K_i < 0$ are not dependent upon z. As ρ_i may be written as $\sqrt{(x - x_{a_i})^2 + (y - ya_i)^2 + (z - z_a)^2}$ the derivative of Z with respect to z is always negative so that we may state a rather trivial and expected result:

Theorem 1. if only 3 cables are under tension the largest cable tension will be obtained for the highest altitude of the platform

Note that if the A_i are not lying in a horizontal plane the above theorem will still hold although the tension will increase when the distance between B and the plane $A_1 A_2 A_3$ decreases (implying that it will also hold if pulleys are used at the A_i points).

Now let us look at the derivatives of the τ_i with respect to x, y. There derivatives may be written as

$$\frac{\partial \tau_i}{\partial x} = \frac{\rho_i^2 \partial U_i/\partial x + U_i(x - x_{a_i})}{W_i} \qquad \frac{\partial \tau_i}{\partial y} = \frac{\rho_i^2 \partial U_i/\partial y + U_i(y - ya_i)}{W_i}$$

where W_i has a constant sign. Looking at the numerator of these derivatives we note that because of the linearity of U_i the term $\partial U_i/\partial x, y$ is a constant so that

$\rho_i^2 \partial U_i/\partial x, y$ is a second order function in x, y and so are also the terms $U_i(x - xa_{a_i}), U_i(y - y_{a_i})$. Hence the numerators of $\partial\tau_i/\partial x, \partial\tau_i/\partial y$ are two quadratic equations $\mathscr{C}_i^x, \mathscr{C}_i^y$. Note that we get 2 such equations for τ_1, τ_2 but only one for τ_3 as $\partial\tau_3/\partial x$ cancel only for $y = 0$ or $x = x_{a_3}$: hence for the whole set of cables we get a total of 5 quadratic equations. The conics $\mathscr{C}_i^x = 0, \mathscr{C}_i^y = 0$ split the desired workspace in different components in which the derivatives of τ_i with respect to x, y have a constant sign. A consequence is that the extremum of a cable tension may be obtained at:

- the set of poses S_C which satisfy $\mathscr{C}_i^x = \mathscr{C}_i^y = 0$: the resultant with respect to x of these two equations always factor out in either one quadratic equation in y or x (for τ_3) or in two quadratic equations (for τ_1, τ_2). Determining the intersection of the conics leads to up to 4 poses for each τ_i that are obtained by solving two quadratic polynomials so that we have to solve a total of 6 quadratic polynomials
- the set S_E consisting of the three corners of the triangle $A_1^p A_2^p A_3^p$ that are projected at the maximum altitude
- the set S_B for of the poses for which either x or y is extremal (i.e. being equal to $x_r^1, x_r^2, y_r^1, y_r^2$ and $\mathscr{C}_i^x = 0$ o r$\mathscr{C}_i^y = 0$. As we have 5 conics and four extremal coordinates determining this set require to solve 20 quadratic polynomials
- the set S_U of poses such that $U_j = 0, j \in [2, 3]$ (which defines a line in the $x - y$ plane) and $\mathscr{C}_i^x = 0$ or $\mathscr{C}_i^y = 0$ with $i \neq j$. In that case we have either $\tau_2 = 0$ or $\tau_3 = 0$ (only 2 cables are under tension) and S_U is obtained by solving 7 quadratic polynomials

Note that we retain as member of the sets only the one that are inside the desired workspace. Obtaining the sets S_B, S_C, S_E, S_U therefore require to solve a total of 33 s order polynomials. As we have to repeat the process for each triple of cables we have to solve $33n(n-1)(n-2)/6$ quadratic polynomials. For example if we have a 4-cables CDPR we will have to solve 132 s order polynomials.

Examples: The length unit in this paper is the centimeter. For the first example we choose $x_{a_2} = 500$, $x_{a_3} = 50$, $y_{a_3} = 500$, the workspace being restricted to $xi^1 = 10, xr^2 = 490, yr^1 = 10, yr^2 = 490$ and we consider the cable 1, 2 and a cable attached at (50,500). The value of z_a is set to 450, the load weight is 100N, while the height of the workspace is limited to 300. A second example just differs from the first one by having $x_{a_3} = 0$. In Fig. 2 we present the various geometrical elements that are considered for the calculation are shown, the points that are used to compute the maximal tensions (in solid circles) while Table 1 presents the maximal tensions.

Table 1. Maximal cable tension for the 2 examples

	x	y	τ max
Cable 1 (1st example)	21.184	211.844	100.06
Cable 2 (1st example)	309.962	10	99.82
Cable 3 (1st example)	31.02	310.23	100.36
Cable 1 (2nd example)	183.379	10	96.94
Cable 2 (2nd example)	309.23	10	100.143
Cable 3 (2nd example)	10	309.23	100.143

2.2 Maximal Tensions with 4 Cables Under Tension

As mentioned previously we cannot by control put a CDPR in a configuration where 4 cables (or more) are under tension and stop the CDPR in that configuration. But during the execution of a trajectory such a configuration may temporarily occur after which the CDPR will move to a configuration with three or two cables only are under tension. Intuitively it may be thought that at a given pose when having four cables under tension instead of three will lead to a decrease of the cable tensions but unfortunately intuition is wrong. Indeed as seen in the 3-cables case the maximum of τ_1 is obtained for a pose that is usually on the line joining A_1 to another of the A_i points. Now if we introduce a fourth cable we have to consider the τ_1 that will be obtained for a pose lying on the line A_1A_4 and it may perfectly occurs that there is pose on that lien such that τ_1 is larger than the maximum obtained in the 3-cables case.

Consider for example a four cables CDPR with A_i coordinates $(0, 0, z_a), (x_{a_2} > 0, 0, z_a), (0, ya_3 > 0, z_a)$ while the A coordinates of a fourth cable are $(0 < x_{a_4} \le x_{a_2}, y_{a_3}, z_a)$. Using the static equilibrium linear equations we may establish for a given pose $\mathbf{X}$ the tensions in the cables 1, 2, 3 as a function of the tension τ_4 in the fourth cable. These tensions may be written as

$$\tau_i = \tau_i^0 + w_i\tau_4, i \in [1, 3] \tag{1}$$

where τ_i^0 is the tension when only the 3 cables 1, 2, 3 are under tension (which is positive if $\mathbf{X}^p$ lies in the triangle $A_1^pA_2^pA_3^p$) and the w_i coefficients are obtained as:

$$w_1 = \frac{x_{a_4}\rho_1}{\rho_4 x_{a_2}} \quad w_2 = -\frac{x_{a_4}\rho_2}{\rho_4 x_{a_2}} \quad w_3 = -\frac{\rho_3}{\rho_4}$$

so that $w_1 > 0, w_2 < 0, w_3 < 0$. Hence if indeed τ_2, τ_3 decrease when a positive tension is applied on cable 4, at the opposite τ_1 will increase. As we cannot master the tension τ_4 it is therefore necessary to investigate its increasing effect on τ_1. However there is an upper limit for the value of τ_4 which is obtained when either τ_2 or τ_3 is equal to 0 as this implies a change in the CDPR configuration. Note also that the maximum for τ_1 may be obtained at a pose that is different from the one we have obtained when calculating the maximum for the 3-cables case. Indeed the increase of τ_1 due to the presence of τ_4 is dependent upon the

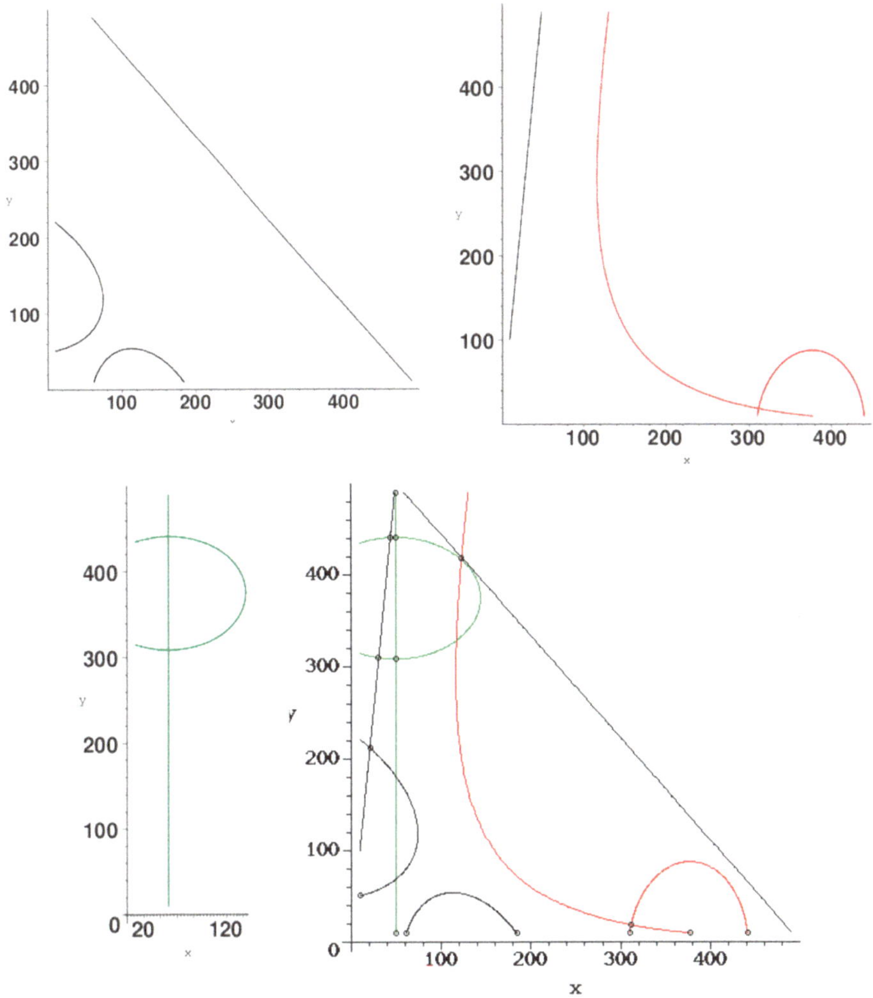

Fig. 2. The elements considered for the calculation of the maximal tensions. Top left we have the 2 conics and line $U_1 = 0$ for cable 1, top right the same elements for cable 2, bottom left for cable 3 (which has a single conic). At the bottom right we have the same elements in a single drawing and In solid circle we have the poses that are considered for the calculation.

pose so that there is no reason that $\tau_1^0 + w_i\tau_4$ may not be larger than the same quantity obtained at the pose where τ_1 has been the maximum for the 3-cables case.

The maximum of τ_1 is obtained at a pose such that either τ_2 or τ_3 (or both) cancels or the derivatives of τ_1 with respect to x, y cancel. The later case is not possible as the numerator of both derivatives is equal to $\rho_1 mg$. We will thus determine τ_1 if $\tau_2 = 0$, $\tau_3 = 0$, $\tau_2 = \tau_3 = 0$ and we will retain the one leading to the largest τ_1.

Case with $\tau_2 = 0$. First we determine the values of τ_4 that cancel τ_2 and report these values in τ_1 so that τ_1 is now a function of only x, y, z. We then consider the derivatives of τ_1 with respect to x, y and determine the constraints that cancel these terms. The derivative of τ_1 with respect to x cancel only for $x = 0$ or $y = y_{a_3}$ while the derivative of τ_1 with respect to y leads to a conic $C_{y_2} = 0$.

Case with $\tau_3 = 0$. First we determine the values of τ_4 that cancel τ_2 and report these values in τ_1 so that τ_1 is now a function of only x, y, z. The derivatives of τ_1 with respect to x, y leads to two conics $C_{x_3} = 0, C_{y_3} = 0$.

Case with $\tau_2 = \tau_3 = 0$. We have also to consider the case where τ_4 cancel both τ_2 and τ_3: this will occur if $x = x_{a_4} y / y a_{a_3}$. We report x and the corresponding value of τ_4 into τ_1 and then calculate the derivative of τ_1 with respect to y leads to a second order polynomial P_y in y.

Hence the maximum of τ_1 will be obtained at a pose such that:

- x or y have one of the extremal value $x_r^1, x_r^2, y_r^1, y_r^2$ and C_{x_3} or C_{y_3} or C_{y_2} cancel. As x or y are fixed the conic leads to a quadratic polynomial in the remaining variable. As we have 3 conics and four extremal values we have to solve 12 quadratic polynomials.
- $C_{x_3} = C_{y_3} = 0$: the resultant of C_{x_3} and C_{y_3} in x factors out in 2 quadratic polynomials. Therefore to deal with this case we have to solve 4 quadratic polynomials
- the roots of P_y

Overall getting the maximum for the 4-cables cases amount to solve 17 quadratic polynomials. In all this cases we get a set of possible poses from which we eliminate the one that are outside the desired workspace. For each of the remaining poses we calculate the value of τ_1 and retain the maximal value. We have to repeat this process by considering all $A = \prod_{j=0}^{n-2}(n - j)$ quadruples of cables among the n cables, thereby having to solve $17A$ quadratics.

Note that in this section we have considered the case where both τ_2 and τ_3 cancel, so that we also consider the case where only $n - 2$ cables are under tension.

In summary for a 4-cables CDPR determining the maximal tension for all cables requires at most to solve $17 + 132 = 149$ s order polynomials, thereby leading to a very fast algorithm.

Example: using the same data than for the second example of the previous section with $x_{a_4} = 250$, $y_{a_4} = 500$ leads to a maximum of τ_1 of 107.743 at the pose $x = 103.197$, $y = 206.394$. We note that this value is approximately 10% higher than the value we have obtained for three cables. The pose at which this maximum is obtained is also different from the pose obtained in the three-cables case.

3 Extension to $N > 4$ Cables

We may generalize the result obtained in the previous section to a CDPR with an arbitrary number of cables. Again we consider the pose $\mathbf{X}$ whose $\mathbf{X}^p$ is included in the triangle $A_1^pA_2^pA_3^p$ where the coordinates of A_1, A_2, A_3 are $(0,0,z_a), (x_{a_2} > 0, 0, z_a), (0, y_{a_3>0}, z_a)$. We will assume that the other cables have as coordinates either (x_{a_2}, y_{a_j}, z_a) with $0 \leq y_{a_j} \leq y_{a_3}$ or (x_{a_j}, y_{a_3}, z_a) with $0 \leq x_{a_j} \leq x_{a_2}$ i.e. A_j lies on an edge of the rectangle $\mathscr{R}$. The static equilibrium condition may be written as

$$\mathbf{J_3}(\tau_1,\tau_2,\tau_3)^T = (\sum_{k=4}^{k=n} -\frac{(x-x_{a_k})\tau_k}{\rho_k}, \sum_{k=4}^{k=n} -\frac{(y-y_{a_k})\tau_k}{\rho_k}, -mg - \sum_{k=4}^{k=n} \frac{(z-z_a)\tau_k}{\rho_k})^T \tag{2}$$

where $\mathbf{J_3}$ is the transpose of the inverse kinematic jacobian restricted to cable 1, 2, 3. Let us define $\mathbf{J_2}$ as the matrix derived from $\mathbf{J_3}$ by substituting the second column of $\mathbf{J_3}$ by (u_1, u_2, u_3). The determinant of $\mathbf{J_2}, \mathbf{J_3}$ are obtained as

$$|\mathbf{J_2}| = -\frac{y_{a_3}(u_1(z-z_a) - u_3 x)}{\rho_1\rho_3} \qquad |\mathbf{J_3}| = \frac{(z-z_a)x_{a_2}y_{a_3}}{\rho_1\rho_2\rho_3} \tag{3}$$

so that $-|\mathbf{J_2}|/|\mathbf{J_3}|$ is $(u_1(z-z_a)-u_3x)\rho_2/((z-z_a)x_{a_2})$. If we set u_1, u_2, u_3 to the components of the right-hand side of Eq. (2) this ratio will be equal to τ_2. Consequently τ_2 may be written as

$$\tau_2 = a_0 + \sum_{k=4}^{k=n} a_k\tau_k$$

where the a_k are constants that depend only upon the CDPR pose and geometry. The coefficient a_k of τ_k in τ_2 is

$$a_k = \frac{((z-z_a)\frac{(x-x_{a_k})}{\rho_k} - x\frac{(z-z_a)}{\rho_k})\rho_2}{(z-z_a)x_{a_2}} = -\frac{x_{a_k}\rho_2}{x_{a_2}\rho_k}$$

which is always negative. Thus any additional cable having a positive tension will reduce the value of τ_2. The same procedure may be used to determine the coefficient of τ_k in τ_3 whose value is $-y_{a_k}\rho_3/(y_{a_3}\rho_k)$ whose value is also always negative so that τ_3 decreases when additional cables are used. On the other hand the coefficient c_{k1} of τ_k in τ_1 is

$$c_{k1} = -\frac{\rho_1(y_{a_3}(x_{a_2}-x_{a_k}) - x_{a_2}y_{a_k})}{\rho_k y_{a_3} x_{a_2}}$$

as we have either $x_{a_k} = x_{a_2}$ or $y_{a_k} = y_{a_3}$, then we get

$$c_{k1} = \frac{\rho_1 y_{a_k}}{y_{a_3}\rho_k} \qquad \text{or} \qquad c_{k1} = \frac{\rho_1 x_{a_k}}{x_{a_2}\rho_k}$$

so that c_{k1} is always positive. Hence τ_1 increases whenever a cable with a positive tension is added.

Theorem 2. if we consider a triplet of cables that represent three successive corners i, j, k and two edges of the desired workspace, then for any number of additional cables l whose A lies on one of the two other edges of the workspace a positive tension leads to an increase of one element in the set $\{\tau_i, \tau_j, \tau_k\}$ and to a decrease for the other elements of the set.

Let us assume that τ_i is the cable tension that increases with τ_l. The derivatives of c_{k1} with respect to x, y is always a quadratic equation and consequently determining the value of τ_l that leads to the largest τ_i under the constraint $\tau_j \geq 0, \tau_k \geq 0$ follows the same process than for the 4-cables case.

4 Conclusion

We have proposed in this paper a fast algorithm for computing the maximal cable tensions of a N-1 CDPR over a given workspace that provides exactly the result by solving only a limited set of 2nd order equations, without having to sample the workspace. Furthermore as all these equations are independent the algorithm may be implemented on a multi-core processor so that the computation time will be extremely low. A first possible extension will be to deal with an arbitrary disposition of the winch points together with a more general way to describe the workspace: we believe that as soon as this description involves only linear elements or is parametric this extension will not be difficult. A second possible extension will be to consider an arbitrary number of cables: here again although we will have to manage the combinatorial aspect the principle of the proposed algorithm will remain the same. Finally we may consider uncertainties in the design: as the calculation relies only on simple, low-degree, algebraic equations it will be relatively easy to determine the worst case maximal tension for example by using interval analysis.

Regarding open issues we may consider taking into account the cables sagging: here the open question for the N-1 case is if the maximal cable tensions with sagging will be obtained at the same pose than when considering ideal cables (in which case taking sagging into account will not drastically change the algorithm, including taking into account the uncertainties on the robot geometry) ? We may have also to consider the case where the platform exhibits limited oscillations so that the force exerted at B lie in a cone.

Finally extending this work to 6 dof CDPRs will be quite complex as cable/cable and platform/cable interference and limitation on the cable lengths will have to be taken into account.

Extension to a dynamics analysis for a given trajectory appears to be possible as the platform dynamics may be solved to determine the forces exerted at B but will be more complex as time will be an additional variable so that a temporal-geometrical analysis will have to be performed.

References

1. Merlet, J.-P.: Efficient estimation of the extremal articular forces of a parallel manipulator in a translation workspace. In: IEEE International Conference on Robotics and Automation, Louvain, 18–20 May 1998, pp. 1982–1987 (1998)
2. Flores, F.G., Kecskeméthy, A., Pöttker, A.: Workspace analysis and maximal force calculation of a face-shovel excavator using kinematical transformers. In: 12th IFToMM World Congress on the Theory of Machines and Mechanisms, Besancon, 18–21 June 2007 (2007)
3. Ur-Rehman, R., Caro, S., Chablat, D., Wenger, P.: Multi-objective path placement of parallel kinematics machines based on energy consumption, shaking forces and maximum actuator torques: application to the Orthoglide. Mech. Mach. Theory **45**(8), 1125–1141 (2010)
4. Merlet, J.-P.: Simulation of discrete-time controlled cable-driven parallel robots on a trajectory. IEEE Trans. Robot. **33**(3), 675–688 (2017)
5. Gouttefarde, M., Gosselin, C.M.: Analysis of the wrench-closure workspace of planar parallel cable-driven mechanisms. IEEE Trans. Robot. **22**(3), 434–445 (2006)
6. Lau, D., Oetomo, D., Halgamuge, S.K.: Wrench-closure workspace generation for cable driven parallel manipulators using a hybrid analytical-numerical approach. ASME J. Mech. Des. **133**(7), 071004 (2011)
7. Ouyang, B., Shang, W.-W.: Wrench-feasible workspace based optimization of the fixed and moving platforms for cable-driven parallel manipulators. Robot. Comput.-Integr. Manufact. **30**(6), 629–635 (2014)
8. Pott, A., Kraus, W.: Determination of the wrench-closure translational workspace in closed-form for cable-driven parallel robots. In: IEEE International Conference on Robotics and Automation, Stockholm, 16–20 May 2016 (2016)
9. Bury, D., et al.: Continuous tension validation for cable-driven parallel robots. In: IEEE International Conference on Intelligent Robots and Systems (IROS), Las Vegas, 25–29 October 2020 (2020)
10. Borgstrom, P.H., et al.: Rapid computation of optimally safe tension distributions for parallel cable-driven robots. IEEE Trans. Robot. **25**(6), 1271–1281 (2009)
11. Bruckmann, T., Pott, A., Hiller, M.: Calculating force distributions for redundantly actuated tendon-based Stewart platforms. In: ARK, Ljubljana, 26–29 June 2006, pp. 403–412 (2006)
12. Fortin-Côté, A., Cardou, P., Gosselin, C.: A tension distribution algorithm for cable-driven parallel robots operating beyond their wrench-feasible workspace. In: 6th International Conference on Control, Automation and Systems (ICCAS), Gyeongju, 16–19 October 2016 (2016)
13. Gouttefarde, M., et al.: A versatile tension distribution algorithm for n-DOF parallel robots driven by n + 2 cables. IEEE Trans. Robot. **31**(6), 1444–1457 (2015)
14. Barrette, G., Gosselin, C.: Determination of dynamic workspace of cable-driven planar parallel mechanisms. ASME J. Mech. Des. **127**(2), 242–248 (2005)
15. Korayem, M.H., Bamdad, M.: Dynamic load-carrying capacity of cable-suspended parallel manipulators. Int. J. Adv. Manuf. Technol. **44**(7–8), 829–840 (2009)
16. Kraus, W., Pott, A.: Scenario-based dimensionning of the actuator of parallel cable-driven robots. In: 4th European Conference on Mechanism Science (Eucomes), Santander, 19–21 September 2012, pp. 131–139 (2012)
17. Riehl, N., et al.: On the determination of cable characteristics for large dimension cable-driven parallel mechanisms. In: IEEE International Conference on Robotics and Automation, Anchorage, 3–8 May 2010, pp. 4709–4714 (2010)

Wrench Analysis of Kinematically Redundant Planar CDPRs

Adhiti Raman(✉), Matthias Schmid, and Venkat N. Krovi

Clemson University, Greenville, SC 29607, USA
{adhitir,schmidm,vkrovi}@clemson.edu

Abstract. A fully-constrained $n - DOF$ cable-driven parallel robot (CDPR) has wrench closure if there are $n + 1$ cables exerting positive tensions spanning the wrench space. However, the quality of wrench closure is often dependent on the geometric configuration of the supporting in-parallel chains of the CDPR. The reconfigurability endowed by adding in-chain kinematic and/or actuation redundancy to a conventional cable robot could greatly improve quality of the workspace. However, the status of various joints (active, passive or locked) affect the complexity of the systematic formulation and ultimate wrench-based analysis. Past efforts have tended to equilibrate the forces in these systems in such a way as to avoid kinematic redundancies. To this end, we formulate the kinematics of the redundant reconfigurable CDPR using matrix Lie group formulation (to allow ease of formulation and subsequent generalizability). Reciprocity (and selective reciprocity) permits the development of wrench analyses including the partitioning of actuation vs structural equilibration components. The total wrench set is greatly expanded both by the addition of kinematic redundancy and selective actuation/locking of the joints. The approach adopted facilitates the holistic determination of the true wrench polytope which accounts for the wrench contributions from all actuation sources. All these aspects are examined with variants of a 4-PRPR planar cable driven parallel manipulator (with varied active/passive/locked joints).

Keywords: Wrench analysis · Reconfigurable cable robot · Reciprocal screws

1 Introduction

Traditional cable-driven parallel robots (CDPRs) are parallel manipulators whose individual chains may not feature kinematic redundancy (e.g. R$\underline{P}$R in planar, S$\underline{P}$S/S$\underline{P}$U in spatial). However, due to the unilateral cable tension conditions, they require at minimum $n+1$ cables for an $n - DOF$ task space which creates actuation redundancy.

Supported by NSF Award #IIS-1924721.

M. Gouttefarde et al. (Eds.): CableCon 2021, MMS 104, pp. 90–104, 2021.
https://doi.org/10.1007/978-3-030-75789-2_8

The placement of the cable winching mechanism on a mobile platform allows the ability to change the geometry of the attachment points of the cables creating a reconfigurable CDPR (r-CDPR). Such additional kinematic and/or actuation redundancy creates potential for enhancing the performance of conventional cable robots. For example, the added mobility offers advantages such as the ability to increase the wrench feasible workspace (WFW) as proposed by Rosati et al. [20] or improve the quality of the WFW for a span of end-effector poses. This mobility can be incorporated by housing the winching mechanism on mobile differential drive robots that are independently steered, as seen i n [17], or looping the cable through a pulley on linear guides as modeled and implemented in [24]. While early efforts [4] speculated on the value of adding reconfigurability to CDPRs, specifically the location of the control and actuator packages, this prospect was not fully analyzed until later. Zhou [23] used the reconfigurability offered by an r-CDPR with mobile attachment points to adjust cable tensions to suit the task and modulate control the stiffness at the end-effector. Nyugen et al. [16] proposed an r-CDPR alternative to maintenance lifts in aircraft hangers and resolved the redundancy by optimizing for stiffness and power consumption. Rasheed et al. [19] proposed a spatial r-CDPR wherein the columns supporting the fully-constrained CDPR are independently movable on differential drive robots (forerunner to [17]). Gagliardini et al. [11] presented a spatial reconfigurable cable robot for sandblasting and painting of large structures and proposed an optimization algorithm for planning the reconfiguration. Erskine et al. [8] performed the wrench analysis on a reconfigurable robot where the cable lengths are kept constant and the cable attachment points are connected to drones. In our own past work, we have used base mobility in combination with the tension isolation module developed by Zhou [23] on a 3-cable planar r-CDPR to modulate the stiffness at the end-effector [18].

The modeling and formulation for CDPRs have largely relied on vector loop closure. In some papers, this relationship has also been easily derived through screw theoretic analysis as described in Bosscher et al. [2]. For a fully-constrained CDPR, the force and moment applied on the payload is related to the cable tension directly by J^T (which is composed of the wrenches along the cables). Traditionally, cables have been modeled as unilateral prismatic joints surrounded by two passive revolute joints, i.e. they can exert forces along an axis, in a certain direction. Such a formulation treats the cable and attachment points as a "proper" serial chain (3 joints for 3 degrees of freedom at the end-effector). In effect the cable is treated as a two force member (in the planar case) and defines the direction of the line of action of a pure force. If the attachment points of the cables are known, one can write down the equations of static equilibrium in a matrix form where the columns of the subsequent Jacobian form the pure force direction and the magnitudes of the forces are unknown. Such a formulation makes it amenable for subsequent linear matrix analysis. Although this has proven adequate for CDPRs, the formulation and wrench analyses r-CDPRs has not yet been undertaken without decoupling the mobile base attachment points from the CDPR. In this paper we systematically formulate and analyze a planar r-CDPRs with multiple cases of in-chain kinematic and actuation redundancy by varying status of joints (active/passive/locked). We formulate the kinematics using reciprocal screw theory [1] applied using matrix Lie group formulation following the formulation style and notation in Murray et al. [15]. The use of reciprocity, selective reciprocity with matrix Lie groups permits the systematic, analytical and computationally-efficient treatment of r-CDPRs as a sub-class of articulated multi-body systems as well as ease of generalization to the fully-spatial case.

Further, while there have been many studies encompassing wrench feasibility analyses of CDPRs [3,5,12,13], these do not consider the effect of kinematically-redundant supporting chains. Even in the studies by Erskine [8] and Rasheed [19] that deal specifically with the wrench analysis of reconfigurable CDPRs, the forces at the payload and mobile bases are equilibrated separately. Therefore, the determined wrench set available at the end-effector does not directly take into consideration the wrench contributions at the mobile joints. The separation of the force equilibration also necessitates the redundancy resolutions to take cascaded forms where the actuation and kinematic redundancies are resolved asynchronously. In contrast, we have reformulated the kinetostatics to consider these forces within the same wrench Jacobian and also determine the true wrench set available at the end-effector.

The wrench set determination was carried out with the convex hull method described by Bouchard [5]. Studies by Bosscher [3], Firmani [9,10] and Boudreu [6] all consider the convex hull determination for examining the available wrench set for parallel structures. Bouderu specifically tackles the problem of a kinematically redundant (rigid-link) planar parallel manipulator and shows that the convex hull, sliced at $m_z = 0$ for all possible kinematic configurations of the redundant manipulator presents the same results as the optimization schema to produce maximum forces at the end-effector for a kinematically redundant 3-RPRP manipulator by Weihmann [21].

In this work, after a brief introduction of the mathematical background (in Sect. 2), we represent the formulation of the r-CDPR using Lie groups (in Sect. 3.1) by following the procedure and notation established in [15]. In subsequent sections of this paper we model the reciprocal wrenches that span the wrench space of the r-CDPR (Sect. 3.2) and discuss how the in-chain redundancies improve the dexterity and wrench set of the manipulator. Finally, a study of wrench feasibility shows the value added by the redundancy and is demonstrated in the paper with the analysis and visualization of available wrench sets (and how they change shape) at the end-effector through graphical representations of its wrench polytope (Sect. 4).

2 Mathematical Methods

2.1 Lie Group Formulations

Lie groups are a group of symmetry transformations that form smooth differentiable manifolds. Lie groups include rotations, translations, scaling, and other geometric transformations. Despite the many parallels existing between the unilateral tension requirements and unidirectional normal-force constraints arising in multi-fingered hands, multi-legged walkers and multi-arm cooperative-manipulation, efforts to relate this wealth of literature to cable robots have been very limited. Such an approach offers us an opportunity to pursue a coordinate-free geometric formulation as well as connect this formulation to the rich literature of geometric treatment of articulated in-parallel multibody systems.

The notation for the Lie group formulation is largely based on the notation in [15], where $\{s\}$ forms the inertially-fixed spatial frame and $\{b\}$ forms the body-fixed or moving frame and homogeneous transformation of the body-fixed frame with respect to the spatial frame is given by $g_{sb} \in SE(3)$. In vector space, the body twist is ${}^b t_{sb} = \begin{bmatrix} v \, \omega \end{bmatrix}^T$ and is related to the spatial twist by an adjoint transformation, ${}^s t_{sb} = Ad_{g_{sb}} {}^b t_{sb}$ where $Ad_{g_{sb}} = \begin{bmatrix} R_{sb} & \hat{p}_{sb} R_{sb} \\ 0 & R_{sb} \end{bmatrix}$ (in planar: $Ad_{g_{sb}} = \begin{bmatrix} R_{sb} & \begin{bmatrix} p_y \\ -p_x \end{bmatrix} \\ 0 & 1 \end{bmatrix}$).

2.2 Reciprocal Wrenches

There have been many successful attempts to unify the terminology and formulation surrounding the design of parallel robots. In [7], Bruyninckx proposes a unified method of wrench formulation using screws that is equally applicable to the formulation of serial-chain as well as in-parallel manipulators.

For a given serial chain, each column of its Jacobian, J, represents the partial twist at the end-effector which is contributed by that joint. In other words, it is the instantaneous motion of the end-effector caused by the ith joint when all but the ith joint is locked [14]. The span of the all the twists that make up a Jacobian is called the twist space of the robot.

A *reciprocal wrench* is then any wrench that is reciprocal to a twist in the span of J such that, $w_r^T t = 0, \forall t \in span(J)$. In other words, given a serial chain the reciprocal wrench to the twists that form the Jacobian will be that set of wrenches that does zero work on $t \in span(J)$. The wrench will then belong to the null space of the Jacobian. The number of reciprocal wrenches that exist will be $(6 - n)$ for spatial and $(3 - n)$ for planar, where $n = rank(J)$.

A *selectively non-reciprocal wrench* (SNRW) is a special case of reciprocal wrenches. It is a wrench that is reciprocal only to the ith twist and non-reciprocal to all other joints.

$$w_r^T t = \delta_i \text{ with } \begin{cases} (\delta_i)_j = 1, & \text{if } i = j \\ (\delta_i)_j = 0, & \text{if } i \neq j \end{cases} \tag{1}$$

The SNRW of a joint is the partial wrench at the end-effector contributed by joint i. In this sense it is analogous to the assembly of partial twists forming the twist Jacobian. The assembly of the SNRW for each of the joints forms the wrench Jacobian, J_w. For some unique cases, (e.g. a serial robot) this Jacobian is the same as the transpose of the twist Jacobian, J. The selectively non-reciprocal wrench (or screw) described here is the same as the special case of reciprocal screws described by Agrawal [1].

2.3 Available Wrench Set

The Available Wrench Set (AWS) is the set of all forces at the end-effector that are achievable without violating the force limits at the actuated joints. These limits are determined by the setting the joint torques to their extreme values and plotting the limits of wrenches they cause at the end-effector. Bouchard et al. [5] determined that the zonotope generated at the end-effector for an equilibrated system is a convex polytope called the wrench polytope. The wrench limits can also be determined as ellipsoids, tangential to the manipulability ellipsoids introduced by Yoshikawa [22]. However, as the ellipsoids an approximation of the wrench polytope, they do not represent the AWS in its entirety and thus not considered here.

2.4 Wrench Polytope

The wrench polytope is a region of end-effector wrenches determined by the bounds on the range space of joint forces. For the planar case, this polytope is graphically represented by a three dimensional shape in the task space. Given the current pose of the end-effector, the vertices of the polytope are determined by mapping the extreme values of the joint torques ($f_i \in \{f_{max}, f_{min}\}$) to task space wrenches using the wrench Jacobian. If the number of actuated joints in a parallel robot is given by n, then the resultant number of vertices forming the polytope is given by 2^n,

$$F_{O_{ext}} = J_w f_{C_{ext}}; f_{C_{ext}} = \begin{bmatrix} f_{1_{ext}} & f_{2_{ext}} & \dots & f_{n_{ext}} \end{bmatrix}^T$$

where n is the number of actuators and $f_{n_{ext}}$ represents an extreme value of f_i.

For the redundant manipulator, a number of these vertices thus determined lie inside the overall convex hull. However, the internal facets determined by connecting all the (internal and external) vertex points also represent $n-2$ actuators working at the extreme limits.

3 r-CDPR Formulation

All these aspects are examined with variants of a 4-PRPR planar cable-driven parallel manipulator (with varied active/passive/locked joints) to serve as an exemplar of the reconfigurable CDPRs (r-CDPR) under study. In the rest of the paper, we will use the following notation for the planar 4-PRPR with frames as shown in Fig. 1.

We define the payload or end-effector as a rigid body connected via cables to the main frame or to sliders that move along the main frame. The cables

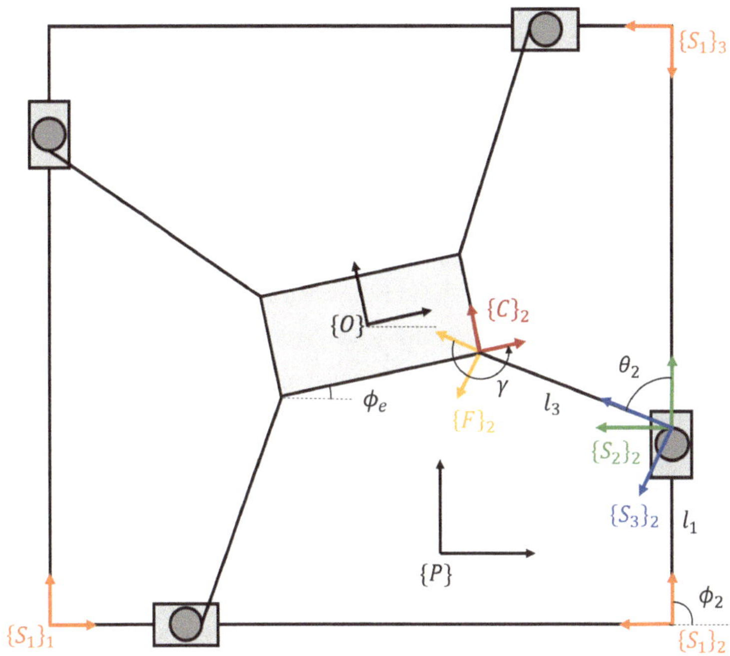

Fig. 1. Planar 4-PRPR CDPR (Planar r-CDPR)

are considered to have no mass and no slack, and the sliders moving along the frame are not inhibited by friction. The world frame and end-effector frame are denoted by $\{P\}$ and $\{O\}$ respectively. The frame corresponding to the geometry of the CDPR are given by $\{S_1\}_i$ where i corresponds to the ith chain. Every subsequent frame in the chain is given by $\{S_j\}_i$ for j joints in each chain. The final frame in the chain, the frame that shares and origin with the contact point on the surface, is given by $\{F\}_i$.The contact point of each cable with the end-effector is given by $\{C\}_i$, and it is assumed that that the orientation of $\{C\}_i$ is fixed with respect to $\{O\}$. The angle between $\{F\}_i$ and $\{C\}_i$ is γ_i.

3.1 Twist Assembled Jacobian

For the ith serial chain, the twist from it's base frame $\{S_1\}_i$ to final frame $\{F\}_i$ is given by $t^{s_{1_i}}_{s_{1_i} f_i}$. Dropping subscript i,

$$t^{s_1}_{s_1 f} = \begin{bmatrix} t^{s_1}_{s_1 s_2} & t^{s_1}_{s_2 s_3} & \dots & t^{s_1}_{s_n f} \end{bmatrix}$$

$$t^{s_1}_{s_1 f} = \begin{bmatrix} \hat{t}^{s_1}_{s_1 s_2} & Ad_{g_{s_1 s_2}} \hat{t}^{s_2}_{s_2 s_3} & \dots & Ad_{g_{s_1 s_n}} \hat{t}^{s_n}_{s_n f} \end{bmatrix} \begin{bmatrix} \dot{q}_{01} \\ \dot{q}_{12} \\ \vdots \\ \dot{q}_{n-1,n} \end{bmatrix}$$

$$t^{s_1}_{s_1 f} = J^{s_1}_{s_1 f}(q)_i \dot{q}_i = J_s(q)_i \dot{q}_i$$

Each column of the twist assembled Jacobian or the structure Jacobian, $J_s(q)_i$, forms the twist between two consecutive frames, $k-1$ and k, expressed in the spatial frame $\{S_1\}$ by adjoint transformations, ${}^F t_{k-1,k} = Ad_{g_{F,k-1}} {}^{k-1}t_{k-1,k}$.

The attachment point of a cable on the payload the cable robot is analogous to a the modelling of a finger contact on a grasped object by Murray, Li and Sastry [15]. The wrench contribution from tension in the cable is along the direction of $\{F\}_i$ applied on $\{C_i\}$. This wrench is given by F_{c_i},

$$F_{c_i} = \begin{bmatrix} -\cos(\gamma) \\ -\sin(\gamma) \\ 0 \end{bmatrix} f_i, \forall f_i \geq 0$$

where f_i is the magnitude of the tension. Transforming this force into the object frame $\{O\}$,

$$F_{O_i} = Ad^T_{g^{-1}_{oc_i}} F_{c_i} = Ad^T_{g^{-1}_{oc_i}} N_{c_i} f_i = P_i f_i$$

where N_{c_i} is the direction vector of the ith cable with respect to the contact frame $\{C_i\}$, and P_i is the ith column of the pulling map. The total wrench is the summation of the individual i^{th} contributions. As the contact point is fixed to the object, the relative velocity between the contact frame and the cable frame is zero in x, y and z. Thus the relationship $N^T_{c_i} t^{c_i}_{f_i c_i}$ equals zero. With this relationship follows the computation of the relationship between the active and passive joint twists of serial chain and twist at the end-effector. The complete derivation is given in [15] for the analogous case of multi-fingered grasping.

The forward-kinematic relationship (joint- to task-space) for the manipulator is given by,

$$J_c = \begin{bmatrix} N^T_{c_1} \mathrm{Ad}^{-1}_{g_{s_1 c_1}} J^{s_1}_{s_1 f_1}(q_{f_1}) & & 0 \\ & \ddots & \\ 0 & & N^T_{c_k} \mathrm{Ad}^{-1}_{g_{s_k c_k}} J^{s_k}_{s_k f_k}(q_{f_k}) \end{bmatrix}$$

$$J_c \dot{q} = P^T t^o_{po}$$

where P is the pulling map and J_c is the cable Jacobian, analogous to the grasp map and hand Jacobian. For the fixed base standard case, the cable Jacobian is an identity matrix.

3.2 Wrench Assembly

The body twist at $\{O\}$ for a single chain is given by,

$$t^o_{s_1 o} = Ad^{-1}_{g_{s_1 o}} J^{s_1}_{s_1 o}(q)_i \dot{q}_i$$

For the 4-cable planar r-CDPR, this is

$$\begin{bmatrix} \dot{x} \\ \dot{y} \\ \dot{\phi} \end{bmatrix} = \begin{bmatrix} \cos(\gamma+\theta_2) & y_{oc}+l_3\sin\gamma & \cos\gamma & y_{oc} \\ -\sin(\gamma+\theta_2) & l_3\cos\gamma - x_{oc} & -\sin\gamma & -x_{oc} \\ 0 & 1 & 0 & 1 \end{bmatrix} \begin{bmatrix} \dot{l}_1 \\ \dot{\theta}_2 \\ \dot{l}_3 \\ \dot{\gamma} \end{bmatrix}$$

where l_1 and l_3 at the linear translations, which can be locked, passive or active and θ_2 and γ are the passive rotations.

Case 1: Passive Motion Accommodation (4-PRPR): If the entire system is passive, the system should be able to accommodate any arbitrary twist applied at the end-effector. The total rank of the assembled Jacobian for a singular chain is 3, while the dimensions of its columns is four. Thus, there is a linear dependency that must exist between the columns. However, we see that at any time, at least 3 of the four columns are independent and is thus able to passively accommodate any twist at the end-effector except when $l_3 = 0$ or when $\theta_2 = 0$. For these conditions, another set of columns becomes linearly dependent thus reducing the rank to 2.

Case 2: Standard, Fully-Constrained CDPR (4-R$\underline{\text{P}}$R): With the mobile bases locked and fixed, each individual chain forms a planar R$\underline{\text{P}}$R serial chain attached at the end-effector and twists associated with the locked joint drop out of consideration. Thus each chain forms a class-1 type "proper" serial chain, wheres the $DOFs$ of the joint equal the $DOFs$ at the end-effector. The wrench at the end-effector is contributed to only by the actuated joint. The SNRW at this joint is given by,

$$\begin{bmatrix} t_{\theta_2} & t_\gamma \end{bmatrix}^T w_{l_3} = 0$$

$$w_{l_3} = \begin{bmatrix} -\cos\gamma \\ \sin\gamma \\ x_{oc}\sin\gamma + y_{oc}\cos\gamma \end{bmatrix}$$

Assembling the wrenches for all the four chains, we get,

$$\begin{bmatrix} w_{l_{3_1}} & w_{l_{3_2}} & w_{l_{3_3}} & w_{l_{3_4}} \end{bmatrix} \begin{bmatrix} f_{c_1} \\ f_{c_2} \\ f_{c_3} \\ f_{c_4} \end{bmatrix} = F_o$$

The assembled wrenches form the wrench Jacobian $\mathbf{J}_w$ which takes the same form as the pulling map, P.

Case 3: Reconfigurable CDPR (4-$\underline{\text{P}}$R$\underline{\text{P}}$R): Each chain connecting to payload is a $4-DOF$ $\underline{\text{P}}$R$\underline{\text{P}}$R mechanism. Thus, each chain forms a class-3 type redundant serial chain. Any wrench at the end-effector will have a non-unique and possibly undefinable contribution at the joint. Thus, we break each chain off after the first 3 (PRP) joints to form a "proper" serial chain. Any wrench at frame $\{F\}$ will have a unique solution at the joints. The forces on the payload from these wrenches are then summed and translated into the body frame at $\{O\}$.

$$\begin{bmatrix} t_{\theta_2} & t_{l_3} \end{bmatrix}^T w_{l_1} = 0$$

$$\begin{bmatrix} t_{\theta_2} & t_{l_1} \end{bmatrix}^T w_{l_3} = 0$$

$$w^f_{l_1} = \begin{bmatrix} 0 \\ -1 \\ l_3 \end{bmatrix}$$

$$w^f_{l_3} = \begin{bmatrix} -\sin(\theta_2) \\ -\cos(\theta_2) \\ l_3\cos(\theta_2) \end{bmatrix}$$

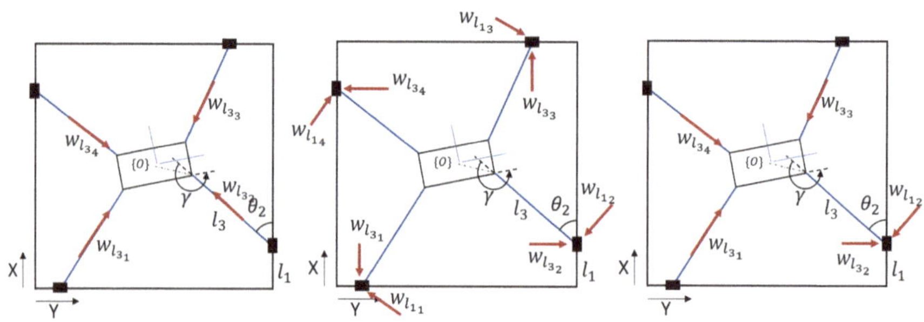

Fig. 2. Lines of action of reciprocal wrenches on a $3-DOF$ planar model for (a) Case 2, (b) Case 3 and (c) Case 4

These wrenches are then transformed to the body-frame $\{O\}$.

$$w_{l_1} = Ad^T_{g^{-1}_{of}} w^f_{l_1} = \begin{bmatrix} -\sin(\gamma) \\ -\cos(\gamma) \\ l_3 - x_{oc}\cos(\gamma) + y_{oc}\sin(\gamma)) \end{bmatrix}$$

$$w_{l_3} = Ad^T_{g^{-1}_{of}} w^f_{l_3} = \begin{bmatrix} -\sin(\gamma+\theta_2) \\ -\cos(\gamma+\theta_2) \\ l_3\cos(\theta_2) - x_{oc}\cos(\gamma+\theta_2) + y_{oc}\sin(\gamma+\theta_2) \end{bmatrix}$$

The collected wrench Jacobian is an assembly of individual wrenches. The lines of actions of these forces are given in Fig. 2. For a CDPR, the direction of the forces should be such that they always provide positive tensions at the cables.

Case 4: Single Mobile Base (3-R$\underline{P}$R, 1-$\underline{P}$R$\underline{P}$R): We break each chain at the point connecting to the payload $\{C_i\}$ and the forces at $\{C_i\}$ are translated into the $\{O\}$ frame. The 3 R$\underline{P}$R chains form a $2 - DOFR\underline{P}$ "deficient" serial chain. Any wrench at frame $\{F\}$ will have non-unique solutions at the joints. Given a reciprocal wrench for the entire chain, the SNRWs will be the set of reciprocal wrench minus the wrench reciprocal to the entire chain. The forces on the payload from these wrenches are then summed and translated into the body frame. The reciprocal wrench for the entire chain given by,

$$\begin{bmatrix} t_{\theta_2} & t_{l_3} \end{bmatrix}^T w_r = 0; w_r = \begin{bmatrix} 0 \\ -1 \\ l_3 \end{bmatrix}$$

For each joint there exist two wrenches spanned by the null space. For the active joint these are,

$$W_{l_3} = \begin{bmatrix} 1 & 0 \\ 0 & -1 \\ 0 & l_3 \end{bmatrix}$$

As one of these screws corresponds to the reciprocal wrench for the entire chain, it cannot be unique, and thus the true selective reciprocal wrench is given by the first column. The SNRWs for the remaining chain are described in Case 3. The wrenches are transformed from frame $\{F\}$ to frame $\{O\}$ as before and we see that the wrench contribution of the individual cables takes a form similar to Case 2, as expected.

$$W_{O_i} = \begin{bmatrix} -\cos(\gamma) \\ \sin(\gamma) \\ x_{oc}\sin(\gamma) + y_{oc}\cos(\gamma) \end{bmatrix}$$

Assembling the wrenches, we extract a wrench Jacobian of size 3 by 5 where 5 is the number of actuated joints.

4 Analysis of the Available Wrench Set

The following section contains a study on the determination of the wrench set available at the end-effector for two of the cases described above. This is done by investigating the convex polytopes generated by joint actuators at their extreme values. Any wrench at the end-effector that can be sustained by the joints must lie within the boundaries of that polytope.

The wrench-feasible workspace (WFW) of a parallel cable-driven mechanism is the set of poses of its mobile platform for which the cables can balance any wrench in a specified set of wrenches, such that the tension in each cable (as well as the forces at the mobile bases for an r-CDPR) remains within a prescribed range. The boundaries of the WFW are determined by the geometric properties of the Available Wrench Set [3].

Traditionally, the AWS is determined only by the tension contribution of the cables. Even in studies of reconfigurable robots, the wrench contributions of the mobile bases are analysed independently. In this paper, we will make use of the formulations in Sect. 3 to visualize the combined wrench set.

We assume that the cables contain negligible mass and have no slack even at minimum tensions. The end-effector is a rigid body and the dynamic effect of gravity is considered negligible as well. The frame of the planar manipulator is 1 m × 1 m, and the platform is 0.39 m × 0.16 m. The minimum and maximum forces at the cables ranges from 0 N to 1 N and at the mobile base prismatic joints from −0.1 N to 0.1 N.

The quality o f the Available Wrench Set provides an indication of the quality of the wrench feasible workspace at that point, but does not give an indication of the overall quality of the workspace over a span of poses. Future efforts will involve determining the quality of the wrench feasible workspace around the locii of the current pose to update the limits of the available end-effector wrenches and twists for input into r-CDPR control.

4.1 Available Wrench Set for the 4-R$\underline{\text{P}}$R CDPR

The wrench polytope of the 4-R$\underline{\text{P}}$R CDPR is shown in Fig. 3. As expected, the number of vertices for the wrench polytope are 16, as there are 16 combinations of maximum torques at the end-effector. The red hypercube in Fig. 3 encapsulates the set of wrenches for which the magnitude of tension in cable 1 is maximum and the green hypercube encapsulates the set of wrenches for which the magnitude of the tension in cable 1 is at its minimum. The regions between them and within the polytope shows the region in which this actuator works under its torque capabilities, but another set of actuators sustains extreme values. The polytope generated at $\phi = 0$ is symmetric along all three axis. It contains 14 external vertices and two internal vertices with are coincident. This marks points where all of the actuators have either their maximum or minimum values. This is aligned with intuition. To achieve zero wrench at the end-effector all of the forces at the joints should either be fighting against each other or be zero. The closer the pose moves towards the extreme value of orientation ($\phi = -0.4$ or 0.4), the less it is able to sustain a moment in that same direction. This is seen by the polytope moving further along the positive m_z axis as the pose moves to $\phi = -0.35$ and along the negative axis when the pose moves to $\phi = 0.35$. At these positions, the cable actuators are less able to sustain transition actuator values, as seen by either disappearing hypercubes for cable 1 in Fig. 3 (b) or fully aligned hypercubes in Fig. 3(c). The polytopes become truly cuboidal at $\phi = \pm 0.4$.

As the number of actuators in the chain are increased, so do the number of vertices for the wrench polytope. For a 4 cable r-CDPR with 8 actuators, the corresponding wrench polytope contains $2^8 = 256$ vertices. With increasing number of vertices, the creation of polytopes becomes increasingly complex to visualize. Thus for the next case, we will study the wrench polytope of the case with a single mobile base.

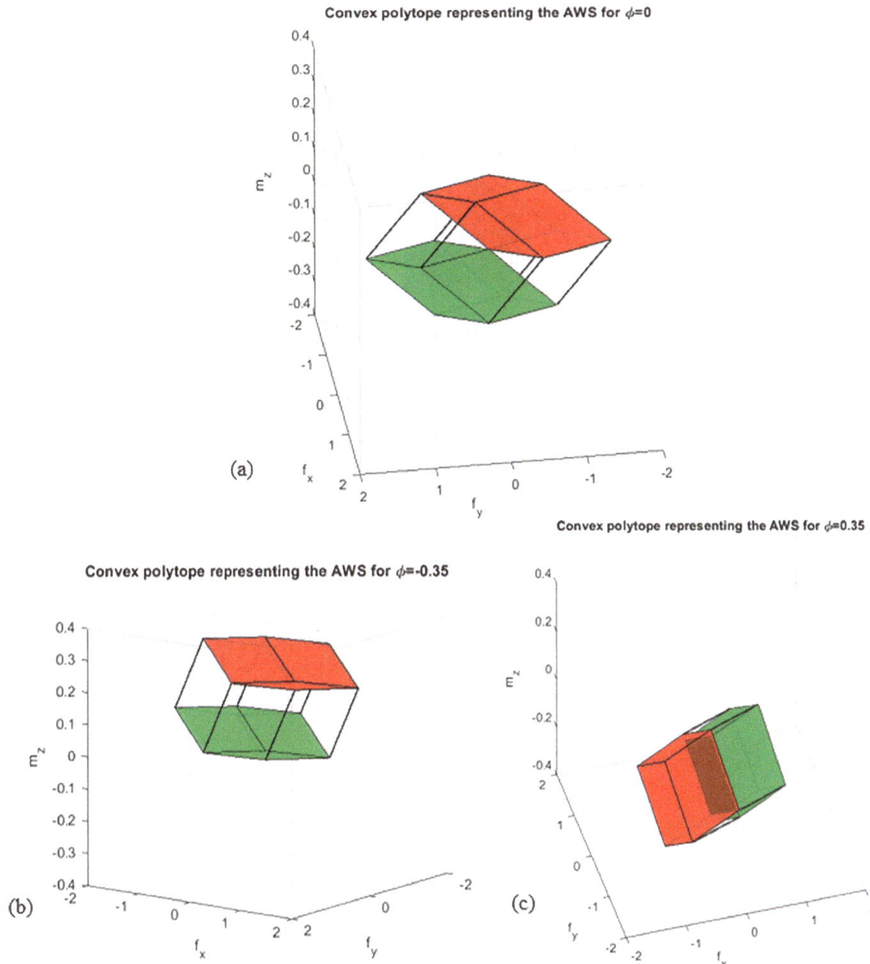

Fig. 3. Convex polytope representing AWS for the 4-R$\underline{P}$R CDPR at (a) $\phi_e = 0$ (b) $\phi_e = -0.35$ and (c) $\phi_e = 0.35$

4.2 Available Wrench Set for the 3-R$\underline{P}$R, 1-$\underline{P}$R$\underline{P}$R r-CDPR

The wrench polytope for this case is represented in Fig. 4. The number of vertices for this structure is 16 (2^5), and similar to the previous figure, red denotes the set of vertex points encapsulating a volume where forces at the slider are maximum and green denotes the set of points encapsulating a volume where forces at the slider are minimum. The individual polytopes themselves are reminiscent of the prior case with 4 cable actuators. The individual hypercubes that comprise this polytope similarly represent the volumes enclosed by cable 1 at its maximum and minimum limit respectively.

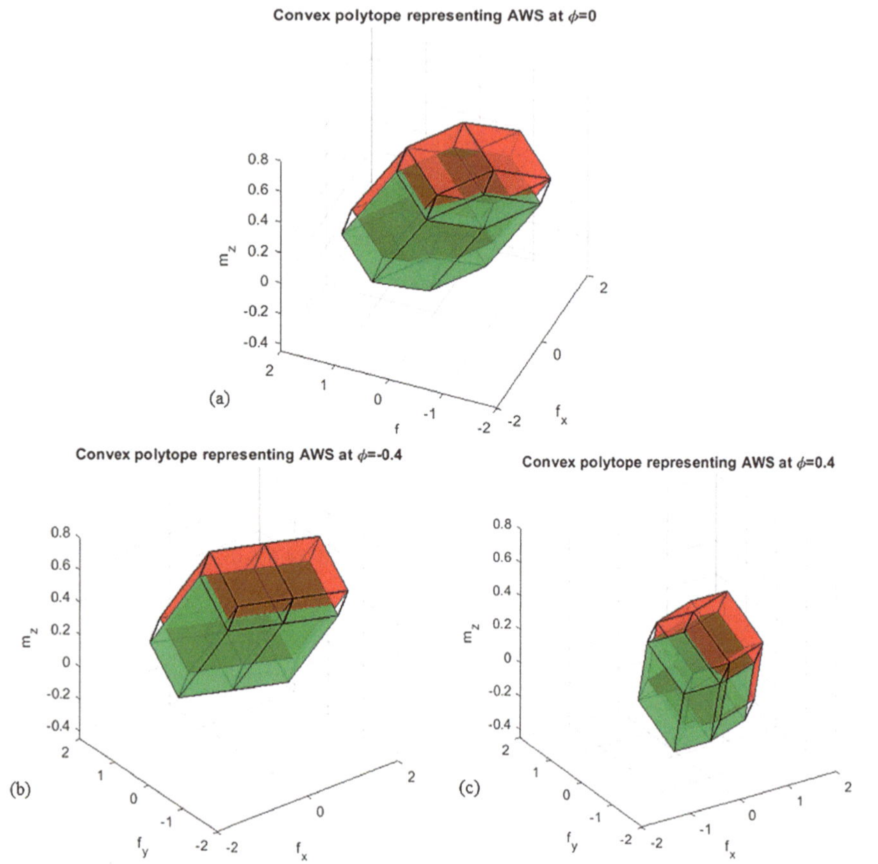

Fig. 4. Convex polytope representing AWS for the 3RPR, 1-PRPR r-CDPR at (a) $\phi_e = 0$ (b) $\phi_e = -0.4$ and (c) $\phi_e = 0.4$

In part (b) we notice that as ϕ approaches an extreme limit (-0.4), the forces at the cables reach their limit conditions and can no longer generate a pose with $\phi < -0.4$. Further, the lowest vertex of the polytope lies just above the $m_z = 0$ plane, which indicates that a force with further negative moment can no longer be applied or sustained. However, this is not the case in (c) when ϕ reaches its symmetrically opposite limit of 0.4. These angles are physically realizable due to the additional effect of the mobile slider. The end-effector is capable of sustaining further positive moments at this position indicating that Available Wrench Set at the end-effector is greatly increased simply with the addition of a single mobile base. The true performance and quality of these wrench polytopes over the span of the entire workspace can be evaluated on the basis of better metrics than the pure interpretation of graphical representations and will be part of future work.

5 Conclusion

The wrench formulation through Lie group analysis and subsequent analysis of the wrench polytopes is presented for the fixed fully-constrained planar CDPR and for a reconfigurable CDPR with mobile cable attachment points. Unlike previous established efforts, this approach of formulation does not require the mobile bases and the CDPR to be equilibrated separately. The use of Lie groups was favored here as it aids the formulation (over traditional line-geometry based-screws or vector loop-closure based kinetostatics) because: (a) the modelling lends itself well to generalizations in higher dimensions; and (b) allows direct exploitation of computationally-assisted matrix-vector formulation approaches. The assembly of the wrench Jacobian is based on the determination of reciprocal wrenches (reciprocal screws) of the individual proper- and redundant- chains. Both parallel mechanisms are redundant, but while a fixed base contains only actuation redundancy, the r-CDPR contains kinematic and further actuation redundancy. The wrench Jacobian can be directly used to graphically represent the true Available Wrench Set at the end-effector that fully represents the tensions at the cable as well as the forces at the mobile bases. We show that with actuation and/or kinematic redundancy at the level of individual chains, the resulting set of wrenches that can be applied and sustained at those poses is significantly increased.

References

1. Agrawal, S.K.: A study of in-parallel manipulators (1991)
2. Bosscher, P., Ebert-Uphoff, I.: Wrench-based analysis of cable-driven robots. In: Proceedings of IEEE International Conference on Robotics and Automation, ICRA 2004, vol. 5, pp. 4950–4955. IEEE (2004)
3. Bosscher, P., Riechel, A.T., Ebert-Uphoff, I.: Wrench-feasible workspace generation for cable-driven robots. IEEE Trans. Rob. **22**(5), 890–902 (2006)
4. Bostelman, R.V., Jacoff, A.S., Proctor, F.M., Kramer, T.R., Wavering, A.J.: Cable-based reconfigurable machines for large scale manufacturing (2000)
5. Bouchard, S., Gosselin, C., Moore, B.: On the ability of a cable-driven robot to generate a prescribed set of wrenches (2010)
6. Boudreau, R., Nokleby, S., Gallant, M.: Wrench capabilities of a kinematically redundant planar parallel manipulator. Robotica 1–16 (2021)
7. Bruyninckx, H., De Schutter, J.: Unified kinetostatics for serial, parallel and mobile robots. In: Lenarčič, J., Husty, M.L. (eds.) Advances in Robot Kinematics: Analysis and Control, pp. 343–352. Springer, Dordrecht (1998)
8. Erskine, J., Chriette, A., Caro, S.: Wrench analysis of cable-suspended parallel robots actuated by quadrotor unmanned aerial vehicles. J. Mech. Robot. **11**(2) (2019)
9. Firmani, F., Zibil, A., Nokleby, S.B., Podhorodeski, R.P.: Wrench capabilities of planar parallel manipulators. part i: wrench polytopes and performance indices. Robotica **26**(6), 791 (2008)
10. Firmani, F., Zibil, A., Nokleby, S.B., Podhorodeski, R.P.: Wrench capabilities of planar parallel manipulators. part ii: redundancy and wrench workspace analysis. Robotica **26**(6), 803 (2008)

11. Gagliardini, L., Caro, S., Gouttefarde, M., Girin, A.: Discrete reconfiguration planning for cable-driven parallel robots. Mech. Mach. Theory **100**, 313–337 (2016)
12. Gouttefarde, M., Collard, J.F., Riehl, N., Baradat, C.: Geometry selection of a redundantly actuated cable-suspended parallel robot. IEEE Trans. Rob. **31**(2), 501–510 (2015)
13. Kamali, K., Joubair, A., Bonev, I.A.: Optimizing cable arrangement in cable-driven parallel robots to improve the range of available wrenches (2018)
14. Mohamed, M., Duffy, J.: A direct determination of the instantaneous kinematics of fully parallel robot manipulators (1985)
15. Murray, R.M., Li, Z., Sastry, S.S., Sastry, S.S.: A Mathematical Introduction to Robotic Manipulation. CRC Press, Boca Raton (1994)
16. Nguyen, D.Q., Gouttefarde, M.: Study of reconfigurable suspended cable-driven parallel robots for airplane maintenance. In: 2014 IEEE/RSJ International Conference on Intelligent Robots and Systems, pp. 1682–1689. IEEE (2014)
17. Pedemonte, N., et al.: Fastkit: a mobile cable-driven parallel robot for logistics. In: Grau, A., Morel, Y., Puig-Pey, A., Cecchi, F. (eds.) Advances in Robotics Research: From Lab to Market, pp. 141–163. Springer, Cham (2020)
18. Raman, A., Schmid, M., Krovi, V.: Stiffness modulation for a planar mobile cable-driven parallel manipulators via structural reconfiguration (2020)
19. Rasheed, T., Long, P., Marquez-Gamez, D., Caro, S.: Available wrench set for planar mobile cable-driven parallel robots. In: 2018 IEEE International Conference on Robotics and Automation (ICRA), pp. 962–967. IEEE (2018)
20. Rosati, G., Zanotto, D.: A novel perspective in the design of cable-driven systems. In: ASME International Mechanical Engineering Congress and Exposition, vol. 48739, pp. 617–625 (2008)
21. Weihmann, L., Martins, D., Coelho, L.: Force capabilities of kinematically redundant planar parallel manipulators (2011)
22. Yoshikawa, T.: Manipulability of robotic mechanisms. Int. J. Robot. Res. **4**(2), 3–9 (1985)
23. Zhou, X., Jun, S.K., Krovi, V.: Wrench reconfigurability via attachment point design in mobile cable robots. In: International Design Engineering Technical Conferences and Computers and Information in Engineering Conference, vol. 55935, p. V06AT07A072. American Society of Mechanical Engineers (2013)
24. Zhou, X., Tang, C.P., Krovi, V.: Analysis framework for cooperating mobile cable robots. In: 2012 IEEE International Conference on Robotics and Automation, pp. 3128–3133. IEEE (2012)

Nonlinear Observer-Based Tension Distribution for Cable-Driven Parallel Robots

Adel Ameri[1], Amir Molaei[2], and Mohammad A. Khosravi[1](✉)

[1] Department of Electrical Engineering, Amirkabir University of Technology (Tehran Polytechnic), Tehran, Iran
{adelameri, m.a.khosravi}@aut.ac.ir

[2] Department of Mechanical Engineering, Concordia University, Montreal, Quebec, Canada
a_molaei@encs.concordia.ca

Abstract. This paper addresses a new methodology for positive tension distribution in cable-driven parallel robots (CDPRs) employing a nonlinear disturbance observer. Cables can merely apply positive tensile force; thus, a CDPR's control scheme should keep the cable tension positive. This is generally provided with a redundancy resolution (RR) scheme, which maps the control effort into a positive range without changing the motion characteristics of the robot. RR methods are Jacobian-dependent and prone to kinematic uncertainties. Thus, such uncertainties induce extra errors in the computation of the cable tension, apart from the control scheme sensitivity to dynamic uncertainties and external disturbances. Additionally, RR demands high-performance computing due to the optimization-based nature of the approach. The proposed control strategy eliminates the complexity of having RR in the control architecture by incorporating a positive saturation-type function to map the control effort into a positive range and a nonlinear disturbance observer to compensate for the saturation effects. The stability of the proposed control scheme is investigated through Lyapunov's second method in detail. Furthermore, its performance is compared with a conventional RR-based method for a spatial six-degree of freedom CDPR, considering uncertainties.

Keywords: Cable-driven parallel robots · Cable positive tension · Disturbance compensation · Sliding mode control · Robust control · Lyapunov analysis

1 Introduction

In cable-driven parallel robots (CDPRs), the motion of the end-effector (EE) is provided through a cable system in a closed kinematic chain [1]. CDPRs have considerably lower inertial properties, compared to their rigid counterparts, making them a unique candidate for high-speed manipulation [2]. The

M. Gouttefarde et al. (Eds.): CableCon 2021, MMS 104, pp. 105–116, 2021.
https://doi.org/10.1007/978-3-030-75789-2_9

flexibility of the cable allows modularity and reconfigurability in design [3]. On the other hand, as cables merely apply tensile force, the control input should be positive, bringing complexity in the control system design. In conventional methods for CDPR's control, the controller is first designed without considering the constraints on the bounds of the controller. Then, a redundancy resolution (RR) method is utilized to map the control effort into a positive range, without changing the motion characteristics of the robot. The RR method is based on the Jacobian matrix, and in the presence of the kinematic uncertainties, extra errors are induced in the positive tension computation. As RR is generally formulated as an optimization problem, the computation time of the algorithm plays an essential role in the real-time application. RR method has been studied along with various control methods for positive tension distribution in cable robots [4–8]. In [9], the linear and quadratic programming is used to bias cable tensions in a positive range using the Jacobian null space. In [10], the computation of the cable tension distribution is represented with the 2-norm of the actuator forces as a constrained optimization problem with no upper bound limit for the tensions. In another study, Hassan and Khajepour [11] have studied force distribution in a CDPR as a projection problem of convex sets. Authors in [12] have proposed a geometric optimization algorithm as a convex optimization problem that offers considerable computation time reduction. Although, there has been great progress for the real-time computation of the positive tension in cable robot, due to the optimization-based nature of RR, the proposed methods suffer from the curse of dimensionality in higher degrees of redundancy and do not represent the solution in terms of an explicit function in tension set [13]. Accordingly, most methods as in [9,12,14–17] consider at most two degrees of redundancy for the cable robot which cannot be extended for higher degrees of redundancy.

As mentioned earlier, to the authors' best knowledge, the available methods for positive tension distribution of the cables in CDPRs are formulated using RR schemes, which is an optimization method based on the null space of the Jacobian. As most of the optimization algorithms are iterative, they do not satisfy the real-time computing requirements. In addition, the kinematic uncertainties effect in the Jacobian result in extra errors in the calculation of positive tension distribution. This paper deals with the tension positiveness in CDPRS as a bounded input control problem with no RR in the control scheme. To this end, first, we design a sliding mode controller (SMC), and a positive saturation function is introduced in the control architecture to map the control effort into a positive range. Then, a nonlinear observer (NDO) is proposed to compensate for the effects of the saturation function. The structure of the proposed control scheme is shown in Fig. 1. One of the key aspects of the proposed method is the reduced impact of uncertainties on the performance of the controller owing to its independency with respect to the Jacobian. Additionally, the complexity of the optimization procedure is eliminated from the control system, which makes it suitable for real-time applications.

The rest of the paper is as follows. In Sect. 2, the dynamic model and the controller design are presented. The stability analysis of the controller is covered in Sect. 3. Section 4 validates the proposed controller efficacy in detailed performance analysis. Conclusions are discussed in Sect. 5.

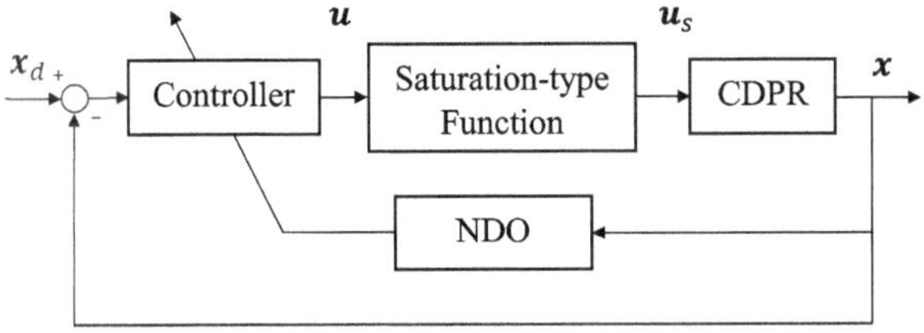

Fig. 1. Block diagram of the proposed control architecture.

2 Dynamic Modeling and Control

The dynamic model of an n degrees of freedom (DOF) CDPRs with ideal cables can be written in a closed form as the following:

$$\boldsymbol{M}(\boldsymbol{x})\ddot{\boldsymbol{x}} + \boldsymbol{C}(\boldsymbol{x}, \dot{\boldsymbol{x}})\dot{\boldsymbol{x}} + \boldsymbol{G}(\boldsymbol{x}) = -\boldsymbol{J}^T(\boldsymbol{x})\boldsymbol{u}, \quad \boldsymbol{u} > \boldsymbol{0} \tag{1}$$

in which $\boldsymbol{x} \in \mathbb{R}^n$ is the generalized coordinates vector, $\boldsymbol{u} = [u_1, u_2, ..., u_m]^T \in \mathbb{R}^m$ is the cable forces, $\boldsymbol{M}(\boldsymbol{x})$ is the mass matrix, $\boldsymbol{C}(\boldsymbol{x}, \dot{\boldsymbol{x}})$ represents the Coriolis/centripetal forces, $\boldsymbol{G}(\boldsymbol{x})$ denotes the gravity force, and $\boldsymbol{J}$ is Jacobian matrix of the robot. As CDPR has a redundant kinematic structure, the Jacobian matrix is non-square. The task space force of the EE, $\boldsymbol{F}$, can be mapped to the cables tension vector using the Jacobian matrix as follows:

$$\boldsymbol{F} = -\boldsymbol{J}^T(\boldsymbol{x})\boldsymbol{u} \tag{2}$$

If $\boldsymbol{J}^T\boldsymbol{J}$ is invertible, (1) is an under-determined set of equations and there would be infinite values for $\boldsymbol{u}$ satisfying the Eq. (1). In the well-known so-called internal force concept the general solution of (1) is as follow:

$$\boldsymbol{u} = \bar{\boldsymbol{u}} + \boldsymbol{Q} \tag{3}$$

where $\bar{\boldsymbol{u}} = -\boldsymbol{J}^\dagger\boldsymbol{F}$ is the minimum norm solution of (2) without considering the positivity constraint of the cable tension and $\boldsymbol{Q}$ lies in the null space of $\boldsymbol{J}^T$ guaranteeing the positiveness of the cable tension. $\boldsymbol{Q}$ can be physically interpreted as the internal forces and may increase the stiffness of the cable robot [18]. Due to the uncertainty in the kinematics and the Jacobian, finding internal force $\boldsymbol{Q}$ using Jacobian results an additional error in the computation of the cable tension $\boldsymbol{u}$. In other words, $\boldsymbol{Q}$ has undesirable effects in the tracking performance and leads inevitable errors. For an n DOFs CDPR with mass-less non-flexible cables, the governing equation of motion can be rewritten as:

$$\boldsymbol{M}(\boldsymbol{x})\ddot{\boldsymbol{x}} + \boldsymbol{C}(\boldsymbol{x}, \dot{\boldsymbol{x}})\dot{\boldsymbol{x}} + \boldsymbol{G}(\boldsymbol{x}) + \boldsymbol{u}_d = -\boldsymbol{J}^T(\boldsymbol{x})\boldsymbol{u}_s \tag{4}$$

where $\boldsymbol{u}_d$ is the external disturbances on the EE. In addition, $\boldsymbol{u}_s = [u_{s_1}, ..., u_{s_m}]^T$ is the saturated control law with a hyperbolic tangent based function with positive lower and upper bounds as $0 < u_{min} < u_{max}$.

$$u_{s_i} = u_q \tanh\left(\frac{u_i - u_p}{u_q}\right) + u_p \tag{5}$$

where $\boldsymbol{u} = [u_1, ..., u_m]^T$ is ideal control input, $u_q = (u_{max} - u_{min})/2$ and $u_p = (u_{max} + u_{min})/2$.

Assume that the kinematic and dynamic modelling uncertainties are considered as the following form:

$$\begin{aligned} \hat{\boldsymbol{M}}(\boldsymbol{x}) &= \boldsymbol{M}(\boldsymbol{x}) + \Delta\boldsymbol{M}, \quad \hat{\boldsymbol{J}}(\boldsymbol{x}) = \boldsymbol{J}(\boldsymbol{x}) + \Delta\boldsymbol{J} \\ \hat{\boldsymbol{C}}(\boldsymbol{x}, \dot{\boldsymbol{x}}) &= \boldsymbol{C}(\boldsymbol{x}, \dot{\boldsymbol{x}}) + \Delta\boldsymbol{C}, \quad \hat{\boldsymbol{G}}(\boldsymbol{x}) = \boldsymbol{G}(\boldsymbol{x}) + \Delta\boldsymbol{G} \end{aligned} \tag{6}$$

in which $\hat{\boldsymbol{M}}(\boldsymbol{x})$, $\hat{\boldsymbol{C}}(\boldsymbol{x}, \dot{\boldsymbol{x}})$ and $\hat{\boldsymbol{G}}(\boldsymbol{x})$ are the nominal dynamic matrices and $\hat{\boldsymbol{J}}^T(\boldsymbol{x})$ is nominal Jacobian, in which $\hat{}$ represents uncertainty.
Substituting (6) in (4), yields,

$$\hat{\boldsymbol{M}}(\boldsymbol{x})\ddot{\boldsymbol{x}} + \hat{\boldsymbol{C}}(\boldsymbol{x}, \dot{\boldsymbol{x}})\dot{\boldsymbol{x}} + \hat{\boldsymbol{G}}(\boldsymbol{x}) = -\hat{\boldsymbol{J}}^T(\boldsymbol{x})\boldsymbol{u}_s + \boldsymbol{d} \tag{7}$$

where

$$\boldsymbol{d} = \Delta\boldsymbol{M}\ddot{\boldsymbol{x}} + \Delta\boldsymbol{C}\dot{\boldsymbol{x}} + \Delta\boldsymbol{G} + \Delta\boldsymbol{J}\boldsymbol{u}_s - \boldsymbol{u}_d. \tag{8}$$

In our methodology, we divide the control law as the following:

$$\boldsymbol{u}_s = \boldsymbol{u} + \Delta\boldsymbol{u} \tag{9}$$

in which $\boldsymbol{u}_s$ is the constrained input to the cables, and $\boldsymbol{u}$ is the ideal control law. The ideal control law can be obtained using any type of controller without considering the positiveness constraint of the cable force. $\Delta\boldsymbol{u}$ is the difference between the constrained positive input and the ideal control law. $\boldsymbol{u}_s$ is obtained by passing $\boldsymbol{u}$ from the unidirectional hyperbolic tangent function in the control architecture. In the proposed approach, the effect of saturation, $\Delta\boldsymbol{u}$, is considered as a disturbance and we estimate and compensate undesirable effects of the saturation function.
Substituting (9) in (7) results:

$$\hat{\boldsymbol{M}}(\boldsymbol{x})\ddot{\boldsymbol{x}} + \hat{\boldsymbol{C}}(\boldsymbol{x}, \dot{\boldsymbol{x}})\dot{\boldsymbol{x}} + \hat{\boldsymbol{G}}(\boldsymbol{x}) = -\hat{\boldsymbol{J}}^T(\boldsymbol{x})\boldsymbol{u} - \hat{\boldsymbol{J}}^T(\boldsymbol{x})\Delta\boldsymbol{u} + \boldsymbol{d} \tag{10}$$

which can be rewritten in the state space as:

$$\begin{cases} \dot{\boldsymbol{x}}_1 = \boldsymbol{x}_2 \\ \dot{\boldsymbol{x}}_2 = \hat{\boldsymbol{g}}(\boldsymbol{x}_1)(\boldsymbol{u} + \Delta\boldsymbol{u}) + \hat{\boldsymbol{f}}(\boldsymbol{x}_1, \boldsymbol{x}_2) + \hat{\boldsymbol{M}}^{-1}(\boldsymbol{x}_1)\boldsymbol{d} \end{cases} \tag{11}$$

where $\boldsymbol{x}_1 = \boldsymbol{x} \in \mathbb{R}^n$ and $\boldsymbol{x}_2 = \dot{\boldsymbol{x}} \in \mathbb{R}^n$. Furthermore, the nonlinear function $\hat{\boldsymbol{g}}$ and $\hat{\boldsymbol{f}}$ in (11) are defined as:

$$\begin{cases} \hat{\boldsymbol{g}} = -\hat{\boldsymbol{M}}^{-1}(\boldsymbol{x}_1)\hat{\boldsymbol{J}}^T(\boldsymbol{x}_1) \\ \hat{\boldsymbol{f}} = -\hat{\boldsymbol{M}}^{-1}(\boldsymbol{x}_1)(\hat{\boldsymbol{C}}(\boldsymbol{x}_1, \boldsymbol{x}_2)\boldsymbol{x}_2 + \hat{\boldsymbol{G}}(\boldsymbol{x}_1)). \end{cases} \tag{12}$$

Thus, the dynamic equation (11) can be rewritten as:

$$\begin{cases} \dot{\boldsymbol{x}}_1 = \boldsymbol{x}_2 \\ \dot{\boldsymbol{x}}_2 = \hat{\boldsymbol{g}}(\boldsymbol{x}_1)\boldsymbol{u} + \hat{\boldsymbol{f}}(\boldsymbol{x}_1, \boldsymbol{x}_2) + \boldsymbol{d}_{total} \end{cases} \tag{13}$$

where

$$\boldsymbol{d}_{total} = \boldsymbol{M}^{-1}(\boldsymbol{x}_1)\left(-\boldsymbol{J}^T(\boldsymbol{x}_1)\Delta\boldsymbol{u} + \boldsymbol{d}\right) \tag{14}$$

The total disturbance $\boldsymbol{d}_{total}$ is to be estimated and its effect should be compensated with a feed-forward term in control law. to this aim, a nonlinear disturbance observer is designed as:

$$\begin{cases} \dot{\hat{\boldsymbol{x}}}_1 = \boldsymbol{x}_2 \\ \dot{\hat{\boldsymbol{x}}}_2 = \hat{\boldsymbol{g}}(\boldsymbol{x}_1)\boldsymbol{u} + \hat{\boldsymbol{f}}(\boldsymbol{x}_1, \boldsymbol{x}_2) + \hat{\boldsymbol{d}}_{total} \\ \dot{\hat{\boldsymbol{d}}}_{total} = \boldsymbol{R}^2\left(-a_1\tan^{-1}\left(b_1\left(\hat{\boldsymbol{x}}_2 - \boldsymbol{x}_2\right)\right) - a_2\tan^{-1}(b_2\boldsymbol{R}^{-1}\hat{\boldsymbol{d}}_{total})\right). \end{cases} \tag{15}$$

in which $\hat{\boldsymbol{x}}_1 \in \mathbb{R}^n$, $\hat{\boldsymbol{x}}_2 \in \mathbb{R}^n$ and $\hat{\boldsymbol{d}}_{total} \in \mathbb{R}^n$ are the estimations of $\boldsymbol{x}_1$, $\boldsymbol{x}_2$ and $\boldsymbol{d}_{total}$, respectively. $\boldsymbol{R} = diag(r_1, ..., r_n)$ is a diagonal parameter matrix with positive values, which highly effects the convergence rate of NDO and a_1, a_2, b_1 and b_2 are positive parameters. The stability and convergence of (15) is proved in [19].

Remark 1: The total disturbance of the system $\boldsymbol{d}_{total}(\boldsymbol{x}_1)$ is bounded for all the possible configuration of the robot in the wrench feasible workspace as discussed in [20]. Thus, there exist a positive constant d_{max} where $\|\boldsymbol{d}_{total}\| \leq d_{max}$.

We define the tracking errors as $\boldsymbol{e}_{x_1} = \boldsymbol{x}_{1d} - \boldsymbol{x}_1$ and $\dot{\boldsymbol{e}}_{x_1} = \boldsymbol{e}_{x_2} = \boldsymbol{x}_{2d} - \boldsymbol{x}_2$, in which $\boldsymbol{x}_{1d}$ and $\boldsymbol{x}_{2d}$ are desired trajectories. The sliding surface for the controller is defined as:

$$\boldsymbol{s} = \boldsymbol{e}_{x_2} + \boldsymbol{K}\boldsymbol{e}_{x_1}. \tag{16}$$

In the above equation $\boldsymbol{K}$ is a $n \times n$ diagonal positive definite matrix. Thus the ideal designed control law is proposed as:

$$\boldsymbol{u} = \hat{\boldsymbol{g}}^{\dagger}\left(-\hat{\boldsymbol{f}}(\boldsymbol{x}_1, \boldsymbol{x}_2) + \dot{\boldsymbol{x}}_{2d} + \boldsymbol{K}\boldsymbol{e}_{x_2} + \boldsymbol{\Gamma}\boldsymbol{s} + \boldsymbol{\rho}\tanh\left(\boldsymbol{\epsilon}^{-1}\boldsymbol{s}\right) - \hat{\boldsymbol{d}}_{total}\right) \tag{17}$$

in which $\boldsymbol{\Gamma}$ and $\boldsymbol{\rho}$ are $n \times n$ diagonal positive definite matrices and $\hat{\boldsymbol{g}}^{\dagger}$ is the pseudo inverse of $\hat{\boldsymbol{g}}$. Furthermore, $\boldsymbol{\epsilon} = diag(\epsilon_1, ..., \epsilon_n)$ is the threshold width on the boundary layer of $\boldsymbol{s}$, $tanh(\boldsymbol{s}) = [tanh(s_1), tanh(s_2), ..., tanh(s_n)]^T$ is the hyperbolic tangent function. It is noted that $\boldsymbol{\Gamma}$ and $\boldsymbol{\rho}$ in (17) are the design parameters to be determined for guaranteeing the stability.

3 Stability Analysis

In this section we analyze the stability of the system (13) with the control law in (17) using lyapunov method. The Lyapunov candidate function is chosen as follows:

$$\boldsymbol{V} = \frac{1}{2}\boldsymbol{s}^T\boldsymbol{s} + \frac{1}{2}\tilde{\boldsymbol{d}}^T\tilde{\boldsymbol{d}} \tag{18}$$

where

$$\tilde{\boldsymbol{d}} = \boldsymbol{d}_{total} - \hat{\boldsymbol{d}}_{total} \tag{19}$$

is the disturbance estimation error and its boundedness is guaranteed in [19]. Taking the time derivative of the Lyapunov function (18) yields:

$$\dot{\boldsymbol{V}} = \boldsymbol{s}^T \dot{\boldsymbol{s}} + \tilde{\boldsymbol{d}}^T \dot{\tilde{\boldsymbol{d}}}. \tag{20}$$

The time derivative of the sliding surface is also as:

$$\dot{\boldsymbol{s}} = \boldsymbol{k}\dot{\boldsymbol{e}}_{x_1} + \dot{\boldsymbol{e}}_{x_2} = \boldsymbol{K}\boldsymbol{e}_{x_2} + \dot{\boldsymbol{e}}_{x_2}. \tag{21}$$

Substituting (17) into (13) yields:

$$\dot{\boldsymbol{e}}_{x_2} = -(\boldsymbol{\Gamma}\boldsymbol{s} + \rho tanh(\epsilon^{-1}\boldsymbol{s}) + \tilde{\boldsymbol{d}} + \boldsymbol{K}\boldsymbol{e}_{x_2}) \tag{22}$$

and substituting (22) and (21) into (20) results:

$$\dot{\boldsymbol{V}} = \boldsymbol{s}^T(\boldsymbol{K}\boldsymbol{e}_{x_2} - (\boldsymbol{\Gamma}\boldsymbol{s} + \rho tanh\left(\epsilon^{-1}\boldsymbol{s}\right) + \tilde{\boldsymbol{d}} + \boldsymbol{K}\boldsymbol{e}_{x_2})) + \tilde{\boldsymbol{d}}^T \dot{\tilde{\boldsymbol{d}}}. \tag{23}$$

By simplifying and sorting (23) we have:

$$\dot{\boldsymbol{V}} = -\boldsymbol{s}^T\boldsymbol{\Gamma}\boldsymbol{s} + \tilde{\boldsymbol{d}}^T \dot{\tilde{\boldsymbol{d}}} - \boldsymbol{s}^T \rho tanh\left(\epsilon^{-1}\boldsymbol{s}\right) - \boldsymbol{s}^T\tilde{\boldsymbol{d}}. \tag{24}$$

By choosing $\boldsymbol{\rho} > \left|\tilde{\boldsymbol{d}}\right|$ and assuming $\boldsymbol{\delta}_d$ as the upper bound of $\left|\tilde{\boldsymbol{d}}^T \dot{\tilde{\boldsymbol{d}}}\right|$, Eq. (24) becomes:

$$\dot{\boldsymbol{V}} < -\boldsymbol{s}^T\boldsymbol{\Gamma}\boldsymbol{s} + \boldsymbol{\delta}_d. \tag{25}$$

Choosing a proper $\boldsymbol{\Gamma}$, the derivative of lyapunov function with respect to the time would be negative $\dot{\boldsymbol{V}} < 0$. Hence the stability of the tracking errors of $\boldsymbol{e}_{x_1}$ and $\boldsymbol{e}_{x_2}$ are guaranteed.

4 Performance Analysis

To verify the effectiveness of the proposed control algorithm, a simulation study has been performed on a six DOF spatial cable robot. The performance of the control architecture is compared with that of the null space projection method for a given trajectory in the wrench feasible workspace of the robot. Additionally, a sinusoidal trajectory is also investigated for further performance analysis.

4.1 Simulation Setup

For the spatial case of the CDPR, a 6 DOF manipulator with two degrees of redundancy as described in [21] is chosen. The configuration of this CDPR is shown in Fig. 2.

Denoting the generalized coordinate vector as $\boldsymbol{x} = [x_p, y_p, z_p, \alpha, \beta, \gamma]^T$ with $\boldsymbol{\theta} = [\alpha, \beta, \gamma]^T$ as Pitch–Roll–Yaw angles, the angular velocity of the EE can be written in the following form:

$$\boldsymbol{\omega} = \boldsymbol{E}\dot{\boldsymbol{\theta}}, \qquad \dot{\boldsymbol{\theta}} = [\dot{\alpha}, \dot{\beta}, \dot{\gamma}]^T \tag{26}$$

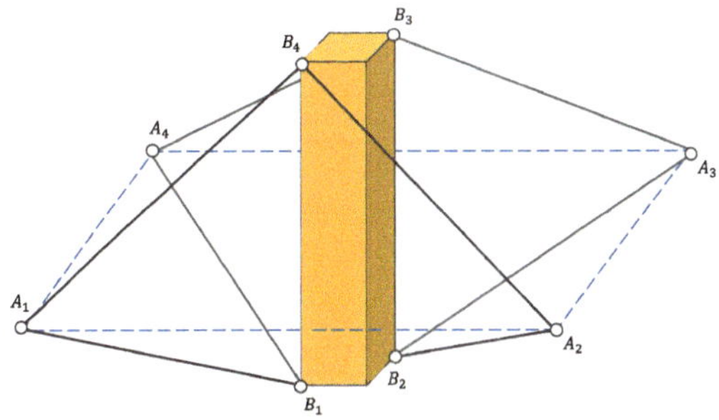

Fig. 2. Configuration of the six DOF spatial cable robot.

in which

$$\boldsymbol{E} = \begin{bmatrix} \cos(\beta)\cos(\gamma) & -\sin(\gamma) & 0 \\ \cos(\beta)\sin(\gamma) & \cos(\gamma) & 0 \\ -\sin(\beta) & 0 & 1 \end{bmatrix} \tag{27}$$

Considering the general dynamic model (1), the corresponding $\boldsymbol{M}$, $\boldsymbol{C}$ and $\boldsymbol{G}$ matrices of the dynamic model for the 6 DOF CDPR may be derived based on [21] as:

$$\boldsymbol{M}(\boldsymbol{x}) = \begin{bmatrix} m\boldsymbol{I}_{3\times3} & \boldsymbol{0}_{3\times3} \\ \boldsymbol{0}_{3\times3} & \boldsymbol{E}^T\boldsymbol{I}_P\boldsymbol{E} \end{bmatrix} \quad , \quad \boldsymbol{G}(\boldsymbol{x}) = \begin{bmatrix} -m\boldsymbol{g} \\ \boldsymbol{0}_{3\times1} \end{bmatrix}$$

$$\boldsymbol{C}(\boldsymbol{x}, \dot{\boldsymbol{x}})\dot{\boldsymbol{x}} = \begin{bmatrix} \boldsymbol{0}_{3\times1} \\ \boldsymbol{E}^T(\boldsymbol{I}_P\dot{\boldsymbol{E}}\dot{\boldsymbol{\theta}} + (\boldsymbol{E}\dot{\boldsymbol{\theta}}) \times \boldsymbol{I}_P(\boldsymbol{E}\dot{\boldsymbol{\theta}})) \end{bmatrix}$$

In these equations, m denotes the mass of the EE and $\boldsymbol{I}_P$ is its inertia tensor; $\boldsymbol{I}_{3\times3}$ is an identity matrix, and $\boldsymbol{g}$ accounts for the gravity acceleration vector. In this study $\boldsymbol{g} = [0, 0, 9.8]^T$, $m = 2.5$, and $\boldsymbol{I}_P = diag[0.212, 0.225, 0.03]$ all in *SI* units are chosen for the simulation purpose.

4.2 Simulation Results

To evaluate the tracking performance of the proposed control scheme, suppose the robot is initially at the origin and has to track the following trajectory in the wrench feasible workspace of the robot.

$$\begin{cases} x_d = y_d = 0.3 + 1.5e^{-t} - 1.8e^{-t/1.2} \\ z_d = 0.2 + e^{-t} - 1.2e^{-t/1.2} \\ \alpha_d = \beta_d = 0 \\ \gamma_d = 0.2 + e^{-t} - 1.2e^{-t/1.2} \end{cases} \tag{28}$$

To investigate the internal force concept method, we employ an sliding mode controller with the null space projection of the cable tension to ensure the positive tension of the cables. The gain matrices of the controller are chosen as $\epsilon = 0.01\boldsymbol{I}_{6\times6}$, $\boldsymbol{K} = 5\boldsymbol{I}_{6\times6}$, $\boldsymbol{\rho} = 1\boldsymbol{I}_{6\times6}$ and $\boldsymbol{\Gamma} = 5\boldsymbol{I}_{6\times6}$ to satisfy the stability conditions. The tracking performance of the controller for the ideal condition and with uncertainty is compared in Fig. 3. The uncertainty is considered to be 10% for both the kinematic and dynamic parameters. As it is clear from the error graph, the controller has a pretty good performance in the absence of uncertainty, while by introducing kinematic uncertainty to the model, there are substantial tracking errors.

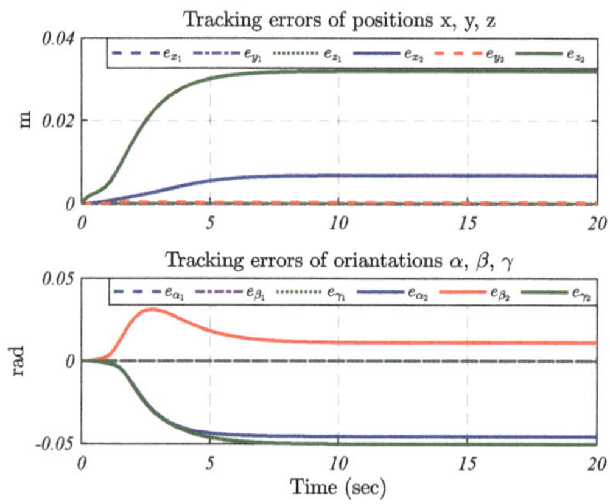

Fig. 3. Trajectory tracking errors of the sliding mode controller along with internal force concept method in ideal conditions (index 1) and with kinematic and dynamic uncertainties (index 2).

The idea behind our proposed method is to make use of the ideal controller for the cable robot without changing its parameters. Having said this, the controller is kept to be the same, with the saturation-type function as described in (5) and the disturbance compensator as defined in (15). The upper and the lower bounds of the saturation-type function are chosen as $u_{max} = 200$, $u_{min} = 1$. Furthermore, the gains of NDO are chosen as $\boldsymbol{R} = 3000\boldsymbol{I}_{6\times 6}$, $a_1 = 2$,$a_2 = 1$, $b_1 = 10$, and $b_2 = 0.5$. Considering the same uncertainties as above, the tracking performance of the proposed control architecture is shown in Figs. 4 and 5. As it is clear by adding uncertainties to the model, the controller follows the given trajectory with the error value in the order of 10^{-5} for the $[x, y, z]$ and 10^{-4} for the $[\alpha, \beta, \gamma]$. A comparison of the root mean square error of the tracking for both of the studies is summarized in Table 1.

For further investigation of the performance of the two methodologies, we also make a comparison for the cable tension distributions, Fig. 6. The average value of the cable tensions within the whole trajectory for the two cases is summarized in Table 2. By referring to this table, the average value of the total average tensions is 39.34 and 50.94 for the two cases. Having less average tension in our proposed methodology has advantages in terms of cost, maintenance, and reliability. Additionally, with lower average cable tension, the robot can track the trajectory with less energy. The total disturbances for the generalized coordinate acceleration and the observed value of the disturbance are also represented

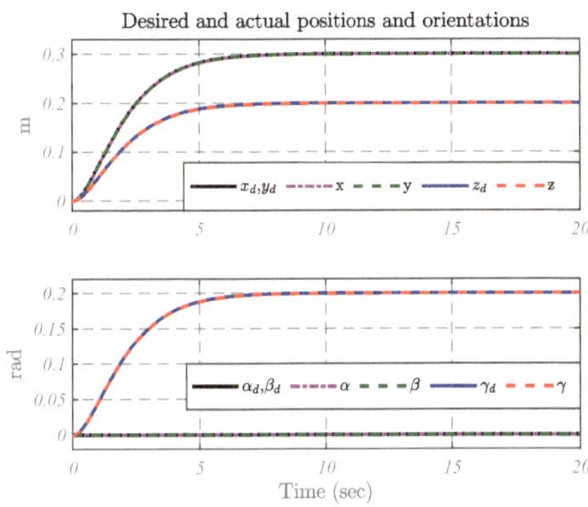

Fig. 4. The desired and actual trajectory of the EE for the proposed control algorithm in the presence of uncertainties.

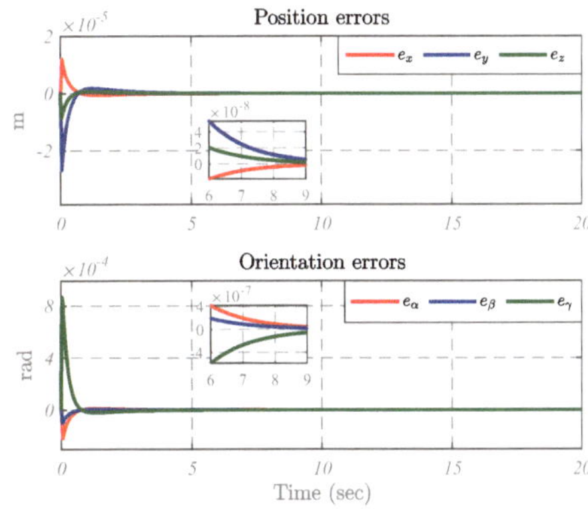

Fig. 5. The tracking errors of the EE pose for the proposed control algorithm in the presence of the uncertainties.

Table 1. Root mean square of the tracking errors

	Proposed approach	Internal force approach
e_x [m]	9.9×10^{-7}	5.9×10^{-3}
e_y [m]	2.1×10^{-6}	1.7×10^{-4}
e_z [m]	7×10^{-7}	2.9×10^{-2}
e_α [rad]	1.7×10^{-5}	4.1×10^{-2}
e_β [rad]	7.6×10^{-6}	1.4×10^{-2}
e_γ [rad]	7.2×10^{-5}	4.5×10^{-2}

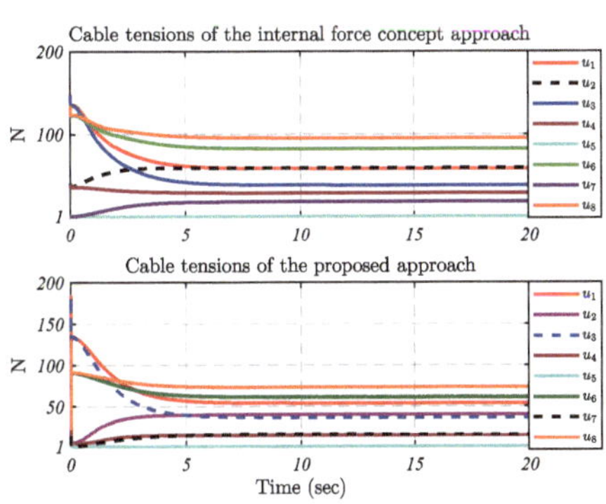

Fig. 6. The tension distribution of the cable robot in the presence of uncertainties (internal force concept and proposed approaches).

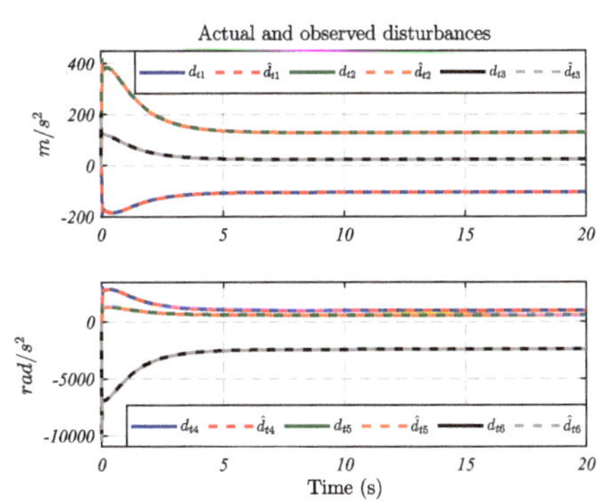

Fig. 7. The actual and estimated disturbances using the NDO.

Table 2. Average values of the tensions [N]

	Proposed approach	Internal force approach
u_1	62.12	67.08
u_2	37.73	57.37
u_3	45.78	48.83
u_4	13.95	30.22
u_5	1.00	1.00
u_6	64.90	87.42
u_7	13.61	16.61
u_8	75.61	98.96

in Fig. 7. The total disturbance signal includes the saturation disturbance and uncertainty in the kinematics and dynamics. As seen in this graph, the NDO has a desirable performance and smoothly tracks the disturbance signal.

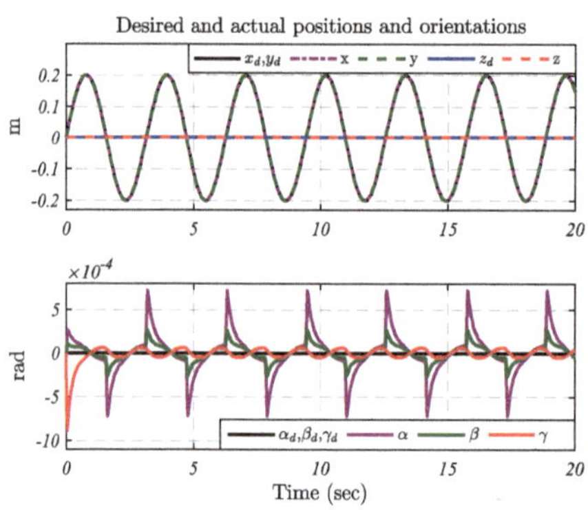

Fig. 8. The actual and desired positions and orientations for the sinusoidal reference trajectory.

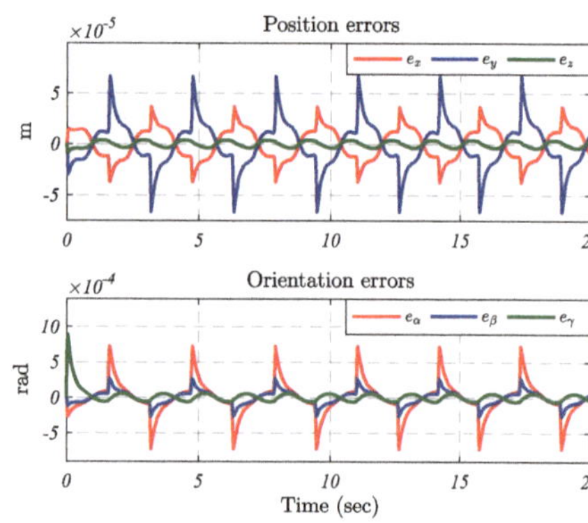

Fig. 9. The tracking errors of the EE pose for the sinusoidal reference trajectory.

To verify the reachable bandwidth of the closed-loop system a sinusoidal trajectory is chosen for x and y while keeping other coordinates at zero.

$$\begin{aligned} x_d &= y_d = 0.2\sin(2t) \\ z_d &= \alpha_d = \beta_d = \gamma_d = 0 \end{aligned} \tag{29}$$

As is shown in Figs. 8 and 9, the proposed control scheme is capable of performing such a maneuver, while the positioning errors remain small. Moreover, as it is depicted in Fig. 10, it is observed that all the cables are in tension throughout the given trajectory.

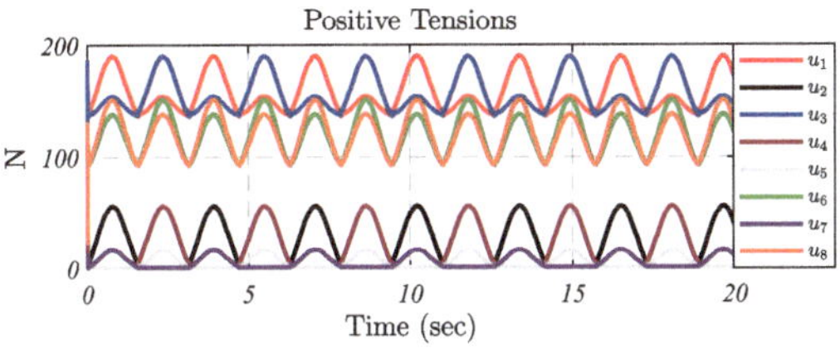

Fig. 10. The tension distribution for the sinusoidal trajectories.

5 Conclusion

In this paper, the positive tension distribution of cable-driven parallel robots is investigated with a control-based viewpoint. The conventional approaches in literature have dealt with this problem using a redundancy resolution (RR) scheme. First, a controller is designed regardless of the limits on the actuator effort, then RR is used to map the control effort in a positive range. In our proposed approach, we introduce a saturation function to map the control effort in the positive range, and the effects of saturation are then compensated using a nonlinear disturbance observer. The results of our study show that the proposed architecture can suitably estimate the induced disturbance of the saturation function along with external ones. To evaluate the effectiveness of the proposed method, a simulation study is also done comparing the proposed method with RR schemes. Comparing the tracking performance of the proposed control method with the conventional ones shows superior trajectory tracking of our method in the presence of uncertainties. Additionally, as RR methods are based on numerical optimization, there are concerns related to the real-time application of it. In the general case of an n DOF CDPR, the worst-case computation time of the RR scheme is unpredictable due to the iterative nature of the optimization methods, which is removed in the proposed approach.

References

1. Taghirad, H.D.: Parallel Robots: Mechanics and Control. CRC Press, Boca Raton (2013)
2. Rushton, M., Jamshidifar, H., Khajepour, A.: Multiaxis reaction system (MARS) for vibration control of planar cable-driven parallel robots. IEEE Trans. Robot. **35**(4), 1039–1046 (2019)
3. Abbasnejad, G., Eden, J., Lau, D.: Generalized ray-based lattice generation and graph representation of wrench-closure workspace for arbitrary cable-driven robots. IEEE Trans. Robot. **35**(1), 147–161 (2018)
4. Khosravi, M.A., Taghirad, H.D.: Robust PID control of fully-constrained cable driven parallel robots. Mechatronics **24**(2), 87–97 (2014)
5. Oh, S.R., Agrawal, S.K.: Generation of feasible set points and control of a cable robot. IEEE Trans. Robot. **22**(3), 551–558 (2006)

6. Qi, R., Rushton, M., Khajepour, A., Melek, W.W.: Decoupled modeling and model predictive control of a hybrid cable-driven robot (HCDR). Robot. Auton. Syst. **118**, 1–12 (2019)
7. Babaghasabha, R., Khosravi, M.A., Taghirad, H.D.: Adaptive robust control of fully-constrained cable driven parallel robots. Mechatronics **25**, 27–36 (2015)
8. Taghirad, H.D., Bedoustani, Y.B.: An analytic-iterative redundancy resolution scheme for cable-driven redundant parallel manipulators. IEEE Trans. Robot. **27**(6), 1137–1143 (2011)
9. Oh, S.R., Agrawal, S.K.: Cable suspended planar robots with redundant cables: controllers with positive tensions. IEEE Trans. Robot. **21**(3), 457–465 (2005)
10. Agahi, M., Notash, L.: Redundancy resolution of wire-actuated parallel manipulators. Trans. Can. Soc. Mech. Eng. **33**(4), 561–573 (2009)
11. Hassan, M., Khajepour, A.: Optimization of actuator forces in cable-based parallel manipulators using convex analysis. IEEE Trans. Robot. **24**(3), 736–740 (2008)
12. Gouttefarde, M., Lamaury, J., Reichert, C., Bruckmann, T.: A versatile tension distribution algorithm for n-DOF parallel robots driven by n+2 cables. IEEE Trans. Robot. **31**(6), 1444–1457 (2015)
13. Geng, X., Li, M., Liu, Y., Li, Y., Zheng, W., Li, Z.: Analytical tension-distribution computation for cable-driven parallel robots using hypersphere mapping algorithm. Mech. Mach. Theory **145**, 103692 (2020)
14. Gouttefarde, M., Collard, J.F., Riehl, N., Baradat, C.: Geometry selection of a redundantly actuated cable-suspended parallel robot. IEEE Trans. Robot. **31**(2), 501–510 (2015)
15. Lamaury, J., Gouttefarde, M.: Control of a large redundantly actuated cable-suspended parallel robot. In: 2013 IEEE International Conference on Robotics and Automation, pp. 4659–4664. IEEE (2013)
16. Pott, A., Mütherich, H., Kraus, W., Schmidt, V., Miermeister, P., Verl, A.: IPAnema: a family of cable-driven parallel robots for industrial applications. In: Cable-Driven Parallel Robots, pp. 119–134. Springer (2013)
17. Lim, W.B., Yeo, S.H., Yang, G.: Optimization of tension distribution for cable-driven manipulators using tension-level index. IEEE/ASME Trans. Mechatron. **19**(2), 676–683 (2013)
18. Behzadipour, S., Khajepour, A.: Stiffness of cable-based parallel manipulators with application to stability analysis. J. Mech. Des. **128**(1), 303–310 (2006)
19. Bu, X.W., Wu, X.Y., Chen, Y.X., Bai, R.Y.: Design of a class of new nonlinear disturbance observers based on tracking differentiators for uncertain dynamic systems. Int. J. Control Autom. Syst. **13**(3), 595–602 (2015)
20. Xiao, B., Yin, S., Kaynak, O.: Tracking control of robotic manipulators with uncertain kinematics and dynamics. IEEE Trans. Ind. Electron. **63**(10), 6439–6449 (2016)
21. Khosravi, M.A., Taghirad, H.D.: Dynamic modeling and control of parallel robots with elastic cables: singular perturbation approach. IEEE Trans. Robot. **30**(3), 694–704 (2014)

A Practical Force Correction Method for Over-Constrained Cable-Driven Parallel Robots

Marc Fabritius[1(✉)], Christoph Martin[1], Guillermo Rubio Gomez[2], Werner Kraus[1], and Andreas Pott[3]

[1] Fraunhofer Institute for Manufacturing, Engineering and Automation IPA, 70569 Stuttgart, Germany
marc.fabritius@ipa.fraunhofer.de

[2] School of Industrial and Aerospace Engineering, Av. Carlos III, 45071 Toledo, Spain

[3] University of Stuttgart, 70174 Stuttgart, Germany

Abstract. Cable-driven parallel robots can have a much larger workspace than other parallel robots with rigid links or conventional serial robots. This property comes at the cost of more complex workspace calculations and control schemes that are necessary to account for the elasticity and unilateral force transmission of their parallel cable links. In practice, most cable-driven parallel robots cannot achieve the full workspace that is predicted by theoretical models. This is due to calibration errors and simplified modelling assumptions in the control schemes.

While most previous works on this subject have focused on creating more accurate and complex models, the goal of this work is to increase the workspace volume that cable-driven parallel robots can realize in practice by using a simple model coupled with a new force correction method that is robust to modelling errors and uncertainties. The new method applies force corrections within the nullspace of the structure matrix in order to keep the cable forces within their limits. Experiments show that the new method can significantly increase the workspace when used in addition to basic kinematic codes. In simulations this combination achieves the same workspace as complex controllers that require the precise knowledge of many additional model parameters.

Keywords: Cable-driven parallel robots · Force distribution method · Workspace analysis · Simulation

1 Introduction

Cable-Driven Parallel Robots (CDPRs) are a class of parallel manipulators in which rigid links are replaced by actuated cables. Controlling the cables' lengths allows control of the platform position and orientation [15]. The large workspace of CDPRs is one of their most important features and distinguishes them from other types of parallel or serial robots [16]. The wrench-feasible workspace of a

M. Gouttefarde et al. (Eds.): CableCon 2021, MMS 104, pp. 117–128, 2021.
https://doi.org/10.1007/978-3-030-75789-2_10

CDPR is defined as the set of platform poses where a cable force distribution exists that simultaneously satisfies the static equilibrium of the robot for a given set of external wrenches and lies within the feasible cable force limits [15]. For over-constrained CDPRs, there can be infinitely many solutions to this problem. Several approaches have been proposed to calculate solutions: the problem can be treated as a constrained optimization problem [14,15], by using the Dykstra algorithm [7], or with non-iterative methods [1].
In practice, a common approach is to control CDPRs using only a kinematic model. While this is favorable due to its simplicity, such controllers can often only realize a subset of the theoretical wrench-feasible workspace of a CDPR.

Several reasons, such as inaccuracies in the calibration and nonlinear effects in the cables, which are often neglected in kinematic models, contribute to this phenomenon [8,13]. Moreover, CDPRs are stiff systems, i.e. small cable length variations can yield to large force variation. Therefore small calibration errors can have a large influence on the cable forces and the workspace boundaries. Additionally, it has been shown that cable pretension plays a key role for the size of the workspace that can be realized in practice when the CDPR is only controlled by kinematic means [8, see p. 91].

One possible approach to compensate cable effects such as elasticity [10], sagging [4,5], pulley friction [9] or hysteresis [2], is to use more complex CDPR models in the controller. This approach has several drawbacks: complex models contain nonlinear equations, which are difficult to implement within the constraints of a real-time capable system. Furthermore, these models depend on many parameters that are difficult to determine or measure in an accurate and reliable manner. For example, the Young modulus and hysteresis parameters of Dyneema cables are known to depend on environmental influences like temperature and humidity [6].

This paper presents a novel control approach which aims at increasing the portion of the workspace that can be effectively realized by CDPRs in practice. Based on a simple CDPR model that considers the pulley geometry and can be implemented in real-time [11], a new algorithm computes force corrections that keep the cable forces within their limits. These corrections are applied to the CDPR by means of cable length offsets. Since the algorithm does not require the precise knowledge of any cable or platform parameters, it is robust to uncertainty or changes in the CDPR's model parameters. Hence, it is well suited for applications such as pick-and-place, where the external wrench of the platform changes over time.

The following work is organized as follows: Sect. 2 lays out the mathematical foundations and the nomenclature used in this work. The new approach is presented in Sect. 3. In Sect. 4, it is then evaluated in experiments and simulations. Finally, the conclusion and outlook are given in Sect. 5.

2 Notation

CDPRs are parallel robots that use cables to manipulate a mobile platform. They are classified based on the number of their cables $m \in \mathbb{N}$ and degrees of freedom o f their platform $n \in \mathbb{N}$. This work focuses on over-constrained CDPRs with $m = 8$ and $n = 6$. The degrees of freedom of such a CDPR consist of the position vector $\boldsymbol{r} \in \mathbb{R}^3$ and orientation matrix $\boldsymbol{R} \in \mathrm{SO}(3)$ of its platform. Together, they form the pose $(\boldsymbol{r}, \boldsymbol{R}) \in \mathrm{SE}(3)$, which determines the transformation from the fixed CDPR coordinate system $\mathcal{K}_0$ to the mobile platform coordinate system $\mathcal{K}_p$. Each cable $i \in \{1, ..., m\}$ of a CDPR is described by its fixed origin $\boldsymbol{a}_i \in \mathbb{R}^3$ in $\mathcal{K}_0$ and its attachment point on the platform $\boldsymbol{b}_i \in \mathbb{R}^3$ in $\mathcal{K}_p$. Using the platform pose, the attachment point on the platform can also be expressed in $\mathcal{K}_0$ as $\boldsymbol{r}+\boldsymbol{R}\boldsymbol{b}_i$. For the behavior of the cables between their endpoints $\boldsymbol{a}_i$ and $\boldsymbol{b}_i$, various models have been proposed in the literature that consider different effects, such as pulleys [11] or the cables' forces, elasticity and weight [3].

Regardless of the cable model, the equilibrium of forces and moments acting on the platform has to be satisfied for all static platform poses $(\boldsymbol{r}, \boldsymbol{R})$. This equilibrium is calculated based on the structure matrix $\boldsymbol{A}^\mathsf{T} \in \mathbb{R}^{n\times m}$ that contains the directions in which the cables can exert forces on the platform. Its i-th column $\boldsymbol{A}_i^\mathsf{T}$ corresponds to the force and moment exerted by the i-th cable

$$\boldsymbol{A}_i^\mathsf{T} = \begin{bmatrix} \boldsymbol{u}_i \\ \boldsymbol{R}\boldsymbol{b}_i \times \boldsymbol{u}_i \end{bmatrix} \in \mathbb{R}^6 \,. \tag{1}$$

Hereby $\boldsymbol{u}_i \in \mathbb{R}^3$ with $\|\boldsymbol{u}_i\| = 1$ denotes the direction in which the i-th cable leaves the platform at the point $\boldsymbol{b}_i$. Considering the cable forces $\boldsymbol{f} \in \mathbb{R}^m$ and an external wrench $\boldsymbol{w}_\text{ext} \in \mathbb{R}^n$ acting on the platform, the wrench equilibrium takes the form

$$\boldsymbol{A}^\mathsf{T}\boldsymbol{f} + \boldsymbol{w}_\text{ext} = \boldsymbol{0} \,. \tag{2}$$

A CDPR pose $(\boldsymbol{r}, \boldsymbol{R})$ is called feasible if Eq. (2) is satisfied and the applied cable forces $\boldsymbol{f}$ are within the limits

$$f_\text{min} \leq \boldsymbol{f}_i \leq f_\text{max} \qquad \text{for all } i = 1, ..., m \,. \tag{3}$$

Over-constrained CDPRs are usually designed in such a way that for platform positions $\boldsymbol{r}$ within the envelope of the cable origin points $\{\boldsymbol{a}_1, ..., \boldsymbol{a}_m\}$, the structure matrix $\boldsymbol{A}^\mathsf{T}$, that is calculated by the kinematic model from [11], has full rank. This means that for these poses there exists a $(m-n)$ dimensional nullspace of $\boldsymbol{A}^\mathsf{T}$ in $\mathbb{R}^m$. The cable forces in this subspace can be expressed in terms of a nullspace basis $\boldsymbol{N} \in \mathbb{R}^{m\times(m-n)}$ and a corresponding coordinate vector $\boldsymbol{\alpha} \in \mathbb{R}^{(m-n)}$. By definition, these forces satisfy

$$\boldsymbol{A}^\mathsf{T}\boldsymbol{N}\boldsymbol{\alpha} = \boldsymbol{0} \qquad \text{for all } \boldsymbol{\alpha} \in \mathbb{R}^{(m-n)}. \tag{4}$$

The force correction method presented in the next section uses this property to ensure that the cable forces stay within the limits from Eq. (3) without altering the platform pose by violating the wrench equilibrium.

3 A Force Correction Method Considering the Nullspace

The force correction method in Algorithm 1 is designed to apply cable lengths offsets $\boldsymbol{\Delta l}$ to the CDPR, that keep its cable forces $\boldsymbol{f}$ within their limits and do not change the platform pose $(\boldsymbol{r}, \boldsymbol{R})$. To achieve this goal, the algorithm only considers force corrections within the nullspace of the structure matrix $\boldsymbol{A}^\mathsf{T}$, which can be expressed as a vector $\boldsymbol{\alpha}$ in the nullspace basis $\boldsymbol{N}$.

Algorithm 1: Force Correction Method

Input: cable forces $\boldsymbol{f} \in \mathbb{R}^m$, cable lengths $\boldsymbol{l} \in \mathbb{R}^m$, nullspace basis $\boldsymbol{N} \in \mathbb{R}^{m \times (m-n)}$

1 $v_{\text{norm}} = 0$
2 $\boldsymbol{\alpha} = \boldsymbol{0} \in \mathbb{R}^{(m-n)}$
for $i = 1, ..., m$ **do**
    $v = 0$
3   **if** $\boldsymbol{f}_i - f_{\max} > 0$ **then** $v = \boldsymbol{f}_i - f_{\max}$
4   **if** $\boldsymbol{f}_i - f_{\min} < 0$ **then** $v = \boldsymbol{f}_i - f_{\min}$
5   $v_{\text{norm}} = v_{\text{norm}} + |v|$
6   $\boldsymbol{\alpha} = \boldsymbol{\alpha} - v \cdot |v| \cdot \frac{\boldsymbol{N}_i}{\|\boldsymbol{N}_i\|^2}$
end
if $v_{\text{norm}} > 0$ **then**
7   $\boldsymbol{\alpha} = \frac{\boldsymbol{\alpha}}{v_{\text{norm}}}$
end
8 $\boldsymbol{\beta} = \text{PI}(\boldsymbol{\alpha})$
9 $\boldsymbol{\Delta l} = \text{diag}(\boldsymbol{l})\, \boldsymbol{N} \boldsymbol{\beta}$

Output: cable lengths offsets $\boldsymbol{\Delta l}$

Algorithm 1 calculates the force limit violation v of each cable i in Lines 3 and 4.

The sum of the absolute values of these violations v_{norm} is computed in Line 5 to establish a cumulative measure for the force limit violations of all cables.

If only one cable i violates the force limits, the correction in the nullspace $\boldsymbol{\alpha}$ that eliminates its violation has to satisfy

$$\boldsymbol{N}_i^T \boldsymbol{\alpha} = -v \,. \tag{5}$$

Hereby $\boldsymbol{N}_i \in \mathbb{R}^{(m-n)}$ denotes the i-th row of the nullspace basis. This condition is satisfied by $\boldsymbol{\alpha}_i = -v \cdot \frac{\boldsymbol{N}_i}{\|\boldsymbol{N}_i\|^2}$. If multiple cables violate the force limits, a weighted sum of these directions in the nullspace is calculated in Line 6.

The correction direction $\boldsymbol{\alpha}_i$ for each cable i is weighted with the absolute value of its violation $|v|$. This design choice is made to prioritize the correction of larger violations. In Line 7, $\boldsymbol{\alpha}$ is normalized if it is not equal to zero, which is implied by the condition $v_{\text{norm}} > 0$.

If $\boldsymbol{\alpha} = \mathbf{0}$, then either all cable forces are within their limits or the weighted directions in the nullspace that correct the violations of individual cables negate each other in Line 6. Therefore, $\boldsymbol{\alpha} = \mathbf{0}$ is considered to be the optimal state that the algorithm aims to achieve through force corrections in the nullspace. To move the system to this state, $\boldsymbol{\alpha}$ is passed as an input to a proportional–integral–controller (PI) in Line 8. The controller also filters out noise and jumps from the cable forces $\boldsymbol{f}$, such that the output of the algorithm $\Delta \boldsymbol{l}$ is smooth. Note that a PI-controller is chosen for simplicity and could be replaced by any other controller. The controller output $\boldsymbol{\beta} \in \mathbb{R}^{(m-n)}$ are also coordinates in the nullspace basis $\boldsymbol{N}$.

To translate the force corrections $\boldsymbol{N}\boldsymbol{\beta}$ into cable lengths offsets $\Delta \boldsymbol{l}$ in Line 9, it is assumed that the cables behave similar to linear springs whose stiffness depends on the lengths of the cables $\boldsymbol{l}$. Therefore, the cable force offsets are multiplied with the diagonal matrix $\operatorname{diag}(\boldsymbol{l})$ whose entries are the components of $\boldsymbol{l}$. The output $\Delta \boldsymbol{l}$ of the algorithm is applied to the CDPR by adding it to the cable lengths $\boldsymbol{l}$ that are commanded to the winches. Afterwards, the resulting cable forces $\boldsymbol{f}$ are measured and Algorithm 1 is executed again. This cycle is repeated indefinitely when the force correction method is activated on a CDPR.

4 Evaluation Through Experiments and Simulation

The force correction method presented in the last section is evaluated in experiments on the IPAnema 3 [12] which is an over-constrained CDPR with 8 cables. Furthermore, the behavior of the force correction method is simulated, using the open-source CDPR simulation software WireX [1].

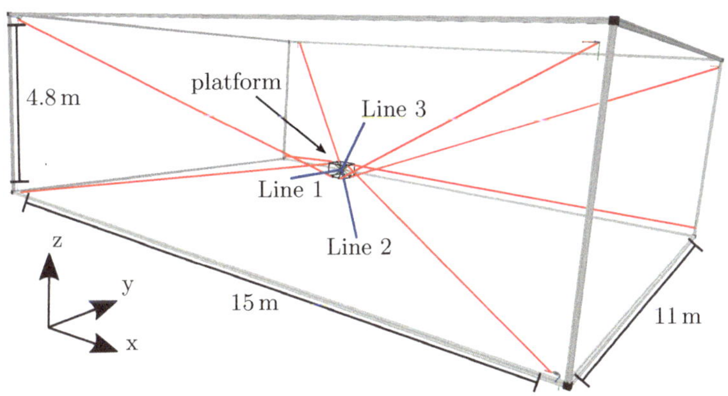

Fig. 1. Visualization of the IPAnema 3s

Figure 1 shows a visualization of the IPAnema 3. The platform is depicted in the center of the robot frame (grey) to which it is connected via its cables (red).

[1] https://gitlab.cc-asp.fraunhofer.de/wek/wirex.git, [accessed: 04-February-2021].

Additionally, three lines are displayed (blue) on which experiments and simulations are conducted. The fixed frame, that holds the cables' origin points $\boldsymbol{a}_i$, has the dimensions of $15 \times 11 \times 4.8$ meters. The mobile platform's body, on which the cables' endpoints $\boldsymbol{b}_i$ are located, has the size: $0.5 \times 0.5 \times 0.35$ meters. The weight of the platform is 30 kg, and for the purpose of this work, the cable force limits are set to

$$f_{\min} = 300\,\mathrm{N} \quad \text{and} \quad f_{\max} = 900\,\mathrm{N}\,. \tag{6}$$

While the IPAnema 3 is designed for cable forces of up to 3 000 N, these narrow limits are chosen such that the three lines from Fig. 1 extend beyond the boundaries of the resulting wrench-feasible workspace.

The controller of the IPAnema 3 is based on the inverse kinematics from [11] which does not consider the elasticity or weight of the cables. The performance of this controller (abbreviated as IK) is compared to that obtained with the new force correction method (abbreviated as FC) in both experiments and simulations. Additionally, only in simulation, the catenary-pulley inverse kinematics from [4] (abbreviated as CP), which accounts for the pulleys, elasticity and weight of the cables is also compared. This complex kinematic code requires the knowledge of more parameters and serves as a benchmark for the maximal workspace volume possible in theory. The comparison of IK, FC and CP methods is done along the three lines that are shown in blue in Fig. 1. They extend from the center of the workspace $\boldsymbol{r}_0 = \boldsymbol{0}$ in different directions towards the points

$$\boldsymbol{r}_1 = \begin{bmatrix} 0\,\mathrm{m} \\ -2\,\mathrm{m} \\ 0\,\mathrm{m} \end{bmatrix}, \quad \boldsymbol{r}_2 = \begin{bmatrix} 2\,\mathrm{m} \\ -2\,\mathrm{m} \\ -1\,\mathrm{m} \end{bmatrix}, \quad \boldsymbol{r}_3 = \begin{bmatrix} 2\,\mathrm{m} \\ -2\,\mathrm{m} \\ 1.5\,\mathrm{m} \end{bmatrix}. \tag{7}$$

The platform orientation $\boldsymbol{R} = \boldsymbol{I}_{3\times3}$ remains unchanged along the lines. Each line is discretized in 50 equally spaced points, where the measurements are conducted.

In the experiments on the IPAnema 3, the cable forces $\boldsymbol{f}$ are measured using force sensors in the winches. The platform pose $(\boldsymbol{r}, \boldsymbol{R})$ is measured with a Leica AT960 laser tracker and a 6-dof T-Mac[2]. The T-Mac is mounted on the platform and allows a direct measurement of its pose with an accuracy of $\pm\left(15\,\mu m + 6\frac{\mu m}{\mathrm{m}}\right)$ for its position and $\pm 0.01°$ for orientation. The position accuracy decreases by $6\,\mu m$ per meter distance and the maximal range for the 6-dof measurement of the T-Mac is 10 meters.

Algorithm 1 is implemented in the PLC of the IPAnema 3 and executes every one-millisecond-cycle when activated. The calculation of the nullspace basis $\boldsymbol{N}$ is performed in Python using a singular value decomposition of the structure matrix $\boldsymbol{A}^\mathsf{T}$ that is supplied by the kinematic model in [11]. The PI-controller from Line 8 of Algorithm 1 is implemented with the diagonal gain matrices $K_p = 2.4 \cdot 10^{-7} \cdot \boldsymbol{I}_{2\times2}$ and $T_i = 50 \cdot \boldsymbol{I}_{2\times2}$ in standard form. For all platform positions in the experiments, these gains lead to a similar behavior as the system converges to a stationary state with $\boldsymbol{\alpha} = \boldsymbol{0}$ and constant cable forces $\boldsymbol{f}$.

[2] https://www.hexagonmi.com/de-de/products/laser-tracker-systems/leica-absolute-tracker-at960, [accessed: 04-February-2021].

In the simulation within WireX, the behavior of the CDPR is modeled using the forward kinematics of the catenary-pulley model from [3], which accounts for the elasticity and weight of the cables. The simulation model uses the same geometry parameters as the IPAnema 3 controller and the following additional parameters for its cables. The Dyneema cables of the IPAnema 3 have a diameter of 6 mm and a weight of 23 g per meter of unstretched cable. The product of their Young modulus E and cross sectional area A is set to $EA = 3 \cdot 10^6$ N.

For the simulations, the gains of the PI-controller in Algorithm 1 are set to $K_p = \frac{1}{EA} \cdot \boldsymbol{I}_{2\times 2}$ and $T_i = \boldsymbol{I}_{2\times 2}$. Like in the experiments, Algorithm 1 converges to a stationary state for all positions on the three lines specified in Fig. 1 after a few iteration cycles.

Figures 2, 3, 4 and 5 compare the previously described implementations of Algorithm 1 (FC) with a simple (IK) and a complex (CP) kinematic code from [11] and [4] in terms of their cable forces, nullspace coordinates and accuracy. The displayed values are obtained after the robot reaches a stationary state at each measurement position.

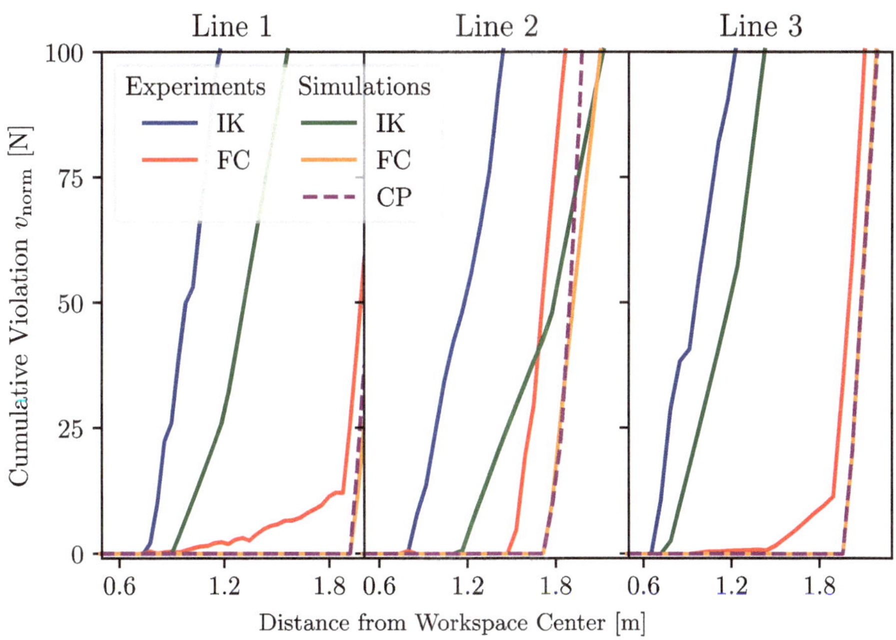

Fig. 2. Cumulative violation measure v_{norm} for different methods.

Figure 2 shows how the cumulative force limit violation measure v_{norm} defined in Algorithm 1 evolves along the three lines. In the experiments, the FC method keeps v_{norm} close to zero for additional distances of 0.734 m, 1.102 m and 1.176 m on the Lines 1–3, compared to the IK method. These are percentual increases of about 250% on each line. The simulations show a similar behavior with extensions of the realizable workspace of 0.693 m, 1.040 m and 1.110 m on each line, respectively. This represents a percentual increases of about 221% on all three lines. The simulated FC method achieves the same workspace ($v_{\text{norm}} = \mathbf{0}$) as the benchmark CP method.

To investigate the cable forces based on which v_{norm} is calculated, Fig. 3 exemplarily depicts them along Line 1. In both experiments and simulations, the cable forces of the FC method stick to force limits when the corresponding forces of the IK method move past the limits. The general behavior and magnitude of the cable forces are the same in experiments and simulations. A noticeable difference is that in the experiments a random noise can be observed in the cable forces, while in simulation, they are smooth.

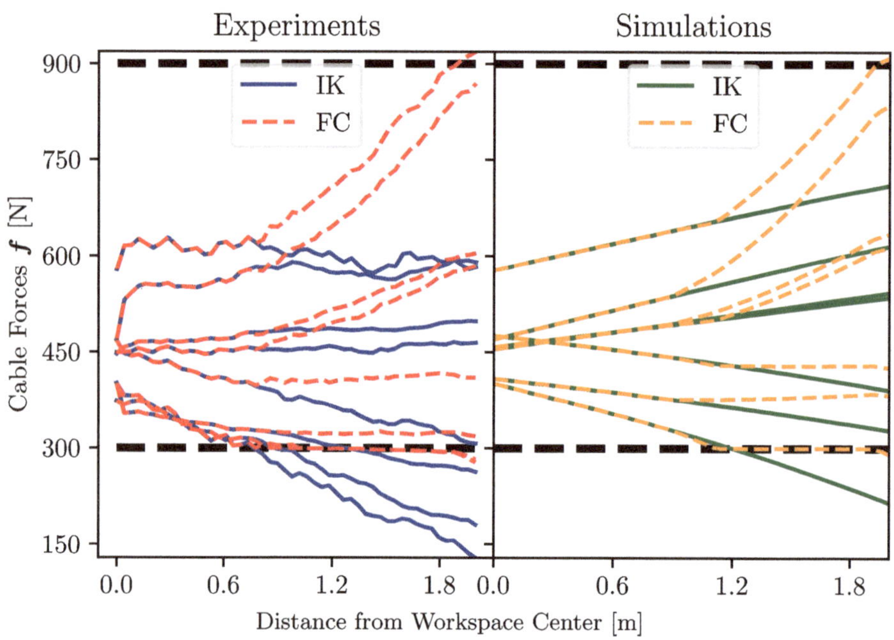

Fig. 3. Cable forces $\boldsymbol{f}$ of IK and FC methods along Line 1

Figure 4 shows the outputs $\boldsymbol{\beta} \in \mathbb{R}^2$ of the PI-controller from Algorithm 1 in experiments and simulations. In the simulations, the general behavior of $\boldsymbol{\beta}$ is the same along all three lines: first $\boldsymbol{\beta}$ is zero until the point where the v_{norm} of the simulated IK method departs from zero in Fig. 2. Then, both components increase linearly with the same slope. For one of the components this increase is slightly delayed in terms of the distance from the workspace center. At the point where the v_{norm} of the simulated FC method departs from zero in Fig. 2, the linear increase of $\boldsymbol{\beta}$ is stopped and its components remain at their current level

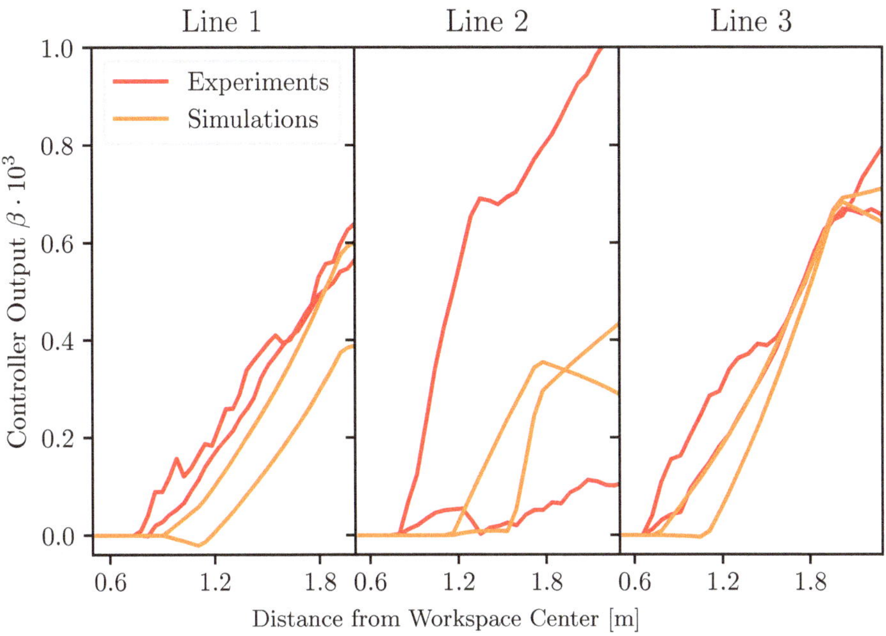

Fig. 4. Controller output β of the FC method

or decrease slightly. On Lines 1 and 3, the same behavior can also be observed in the experiments.

On Line 2, this is not the case. In the experiment, one component of $\boldsymbol{\beta}$ becomes very large while the other remains close to zero. This substantially different behavior could be caused by unwanted elongations of the Dyneema cables due to hysteresis in between the experiments. Interestingly, this does not seem to have a negative effect on the workspace extension achieved by the FC method along Line 2 in Fig. 2.

Figure 5 investigates the influence of the FC method on the accuracy of the robot compared to the IK method. For both methods, the position and orientation errors increase as a function of the distance from the center of the workspace. In the experiments, the scale of the errors is much larger than in the simulations. This is likely due to calibration errors and robot behavior, such as creep in the cables, that is not captured by the simulation model. On Lines 1 and 2, the errors of the FC method are slightly larger than those of the IK method, while on Line 3 they are lower. In both experiments and simulations, the accuracy of the CDPR is not significantly changed by the FC method in comparison to the IK method.

From the analysis of Figs. 2, 3, 4 and 5 we can conclude that the new FC method can significantly increase the portion of the workspace, that can be realized by the controller, when compared with the IK method. The impact of the FC method on the robot's accuracy is negligible.

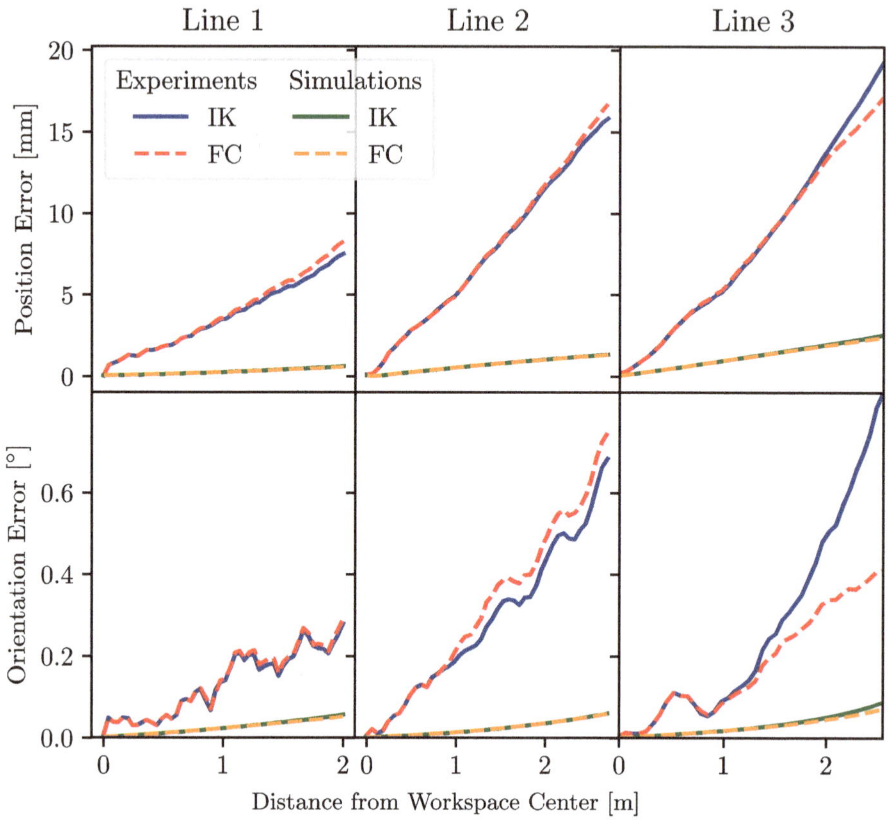

Fig. 5. Position and orientation errors of the IK and FC methods.

Due to the qualitatively similar behavior of simulations and experiments in terms of their achieved workspaces along the three lines shown in Fig. 2, a further study of the realized workspace volumes in simulation may also be indicative for the behavior of the CDPR in practice. Therefore, this comparison is carried out in WireX and its results are presented in the following.

Figure 6 shows three slices of the simulated realized workspaces along the principal axes in the center of the CDPR frame. In all three planes, the FC method (yellow) achieves a significant extension of the realized workspace when compared to the IK method (green). The workspace volume of the FC method is $77.3\,\mathrm{m}^3$, which is three times larger than the $25.9\,\mathrm{m}^3$ volume achieved by the IK method. As observed in Fig. 2, the FC method achieves exactly the same workspace volume as the CP method, which is therefore not shown in Fig. 6. This means that the FC method achieves the maximal possible workspace in simulation for the given CDPR model parameters.

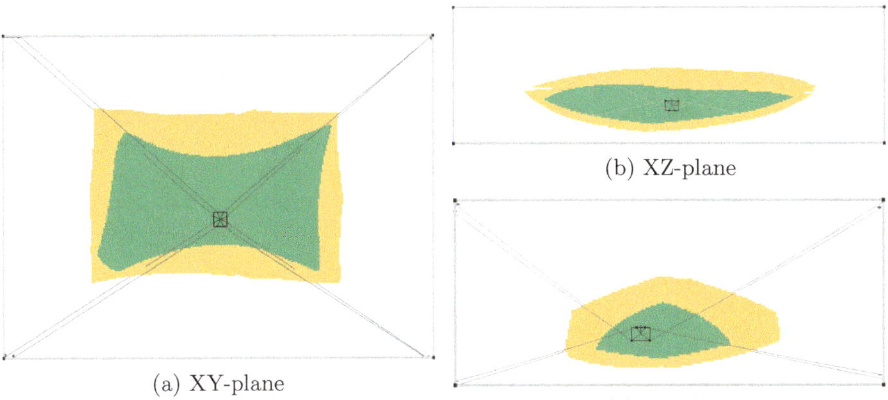

(a) XY-plane

(b) XZ-plane

(c) YZ-plane

Fig. 6. Realized workspace of FC (yellow) and IK (green) methods in simulation

5 Conclusion and Outlook

This work introduces a new force correction method for CDPRs which utilizes a simple cable model and applies a cable force correction if the forces exceed their limits. The correction achieves a significant increase in the realized workspace volume of up to 300% as measured in experiments and simulations, compared to a simple controller based only on a kinematic CDPR model.

Compared to other methods based on more complex models, it achieves the same workspace volume in simulations, but offers the practical benefit of not requiring the knowledge of parameters that are difficult to measure or may change over time like the cables' Young modulus or the platforms' weight distribution. These benefits over both simple kinematic codes and complex models, make the new method well suited to be applied on CDPRs in practice.

Future research should investigate the behavior of the proposed method on dynamic trajectories and different CDPRs. Furthermore, the idea of force corrections within the nullspace could also be used to keep the cable forces close to a desired level.

Acknowledgment. The authors wish to thank the university of Castilla-La Mancha and the European Social Fund for their financial support through the pre-doctoral funding [2019/451] and mobility grant for doctoral students 2020.

References

1. Bouchard, S., Gosselin, C., Moore, B.: On the ability of a cable-driven robot to generate a prescribed set of wrenches. J. Mech. Robot. **2**(1) (2010)
2. Do, T., Tjahjowidodo, T., Lau, M., Yamamoto, T., Phee, S.: Hysteresis modeling and position control of tendon-sheath mechanism in flexible endoscopic systems. Mechatronics **24**(1), 12–22 (2014)

3. Fabritius, M., Pott, A.: A forward kinematic code for cable-driven parallel robots considering cable sagging and pulleys. In: International Symposium on Advances in Robot Kinematics, pp. 218–225. Springer (2020)
4. Fabritius, M., Pott, A.: An inverse kinematic code for cable-driven parallel robots considering cable sagging and pulleys. In: European Conference on Mechanism Science, pp. 423–431. Springer (2020)
5. Gouttefarde, M., Nguyen, D.Q., Baradat, C.: Kinetostatic analysis of cable-driven parallel robots with consideration of sagging and pulleys. In: Advances in Robot Kinematics, pp. 213–221. Springer (2014)
6. Govaert, L., Peijs, T.: Tensile strength and work of fracture of oriented polyethylene fibre. Polymer **36**(23), 4425–4431 (1995)
7. Hassan, M., Khajepour, A.: Analysis of bounded cable tensions in cable-actuated parallel manipulators. IEEE Trans. Robot. **27**(5), 891–900 (2011)
8. Kraus, W.: Force control of cable-driven parallel robots. Ph.D. thesis, University of Stuttgart (2016)
9. Kraus, W., Kessler, M., Pott, A.: Pulley friction compensation for winch-integrated cable force measurement and verification on a cable-driven parallel robot. In: IEEE International Conference on Robotics and Automation (ICRA), pp. 1627–1632. IEEE (2015)
10. Merlet, J.P.: Analysis of wire elasticity for wire-driven parallel robots. In: Proceedings of EUCOMES 08, pp. 471–478. Springer (2009)
11. Pott, A.: Influence of pulley kinematics on cable-driven parallel robots. In: Latest Advances in Robot Kinematics, pp. 197–204. Springer (2012)
12. Pott, A., Mütherich, H., Kraus, W., Schmidt, V., Miermeister, P., Verl, A.: IPAnema: a family of cable-driven parallel robots for industrial applications. In: Bruckmann, T., Pott, A. (eds.) Cable-Driven Parallel Robots, pp. 119–134. Springer (2013)
13. Riehl, N., Gouttefarde, M., Krut, S., Baradat, C., Pierrot, F.: Effects of non-negligible cable mass on the static behavior of large workspace cable-driven parallel mechanisms. In: 2009 IEEE international conference on robotics and automation, pp. 2193–2198. IEEE (2009)
14. Shiang, W.J., Cannon, D., Gorman, J.: Optimal force distribution applied to a robotic crane with flexible cables. In: Proceedings 2000 ICRA. Millennium Conference. IEEE International Conference on Robotics and Automation. Symposia Proceedings (Cat. No. 00CH37065), vol. 2, pp. 1948–1954. IEEE (2000)
15. Verhoeven, R.: Analysis of the workspace of tendon-based Stewart platforms. Ph.D. thesis, University Duisburg-Essen (2004)
16. Zi, B., Duan, B., Du, J., Bao, H.: Dynamic modeling and active control of a cable-suspended parallel robot. Mechatronics **18**(1), 1–12 (2008)

Force-Distribution Sensitivity to Cable-Tension Errors: A Preliminary Investigation

Valentina Mattioni(✉), Edoardo Idà, and Marco Carricato

Department of Industrial Engineering, University of Bologna, Bologna, Italy
{valentina.mattioni,edoardo.ida2,marco.carricato}@unibo.it
http://www.irmalab.org/

Abstract. In order to control cable-driven parallel robots (*CDPRs*), it is necessary to keep all cable tensions within (positive) known limits during motion. For *CDPRs* having more cables than end-effector degrees of freedom, a feasible force distribution within cables should be computed beforehand. This paper aims at evaluating how a tension error in one cable affects the overall distribution of tensions in the other cables, by focusing on planar overconstrained *CDPRs* with four cables. The cable whose tension error more limitedly impact the force distribution is analytically determined by computing the right nullspace of the manipulator structure matrix. It is then shown how the cable least influencing the force distribution changes throughout the wrench-feasible workspace. Lastly, the results of the proposed analysis are used to perform a motion-control experiment on a prototype, where, for any configuration of the end-effector, the cable least influencing the force distribution is tension-controlled, while the remaining ones are length-controlled.

Keywords: Overconstrained cable-driven parallel robots · Force distribution · Workspace computation

1 Introduction

A parallel robot, whose rigid links are replaced by flexible cables, is known as a Cable-Driven Parallel Robot (*CDPR*). Cables are coiled and uncoiled by motorized winches, usually placed on a fixed base, and drive the end-effector (*EE*) throughout its workspace (*WS*). The use of light cables, instead of rigid links, brings numerous advantages in term of *WS* dimension, dynamic performances, payload capability and robot cost. These features make *CDPRs* suitable for a variety of applications in the field of logistics [1], construction [2,3] and pick and place [4], just to name a few. *CDPRs* control is complicated by the fact that cables can only exert tensile forces, thus, they cannot generate wrenches in every direction [5]. To overcome this issue, usually they include more actuated cables than those that are strictly needed to control the platform pose, with cables pulling one against the other. Overconstrained *CDPRs* (*OCDPRs*)

M. Gouttefarde et al. (Eds.): CableCon 2021, MMS 104, pp. 129–141, 2021.
https://doi.org/10.1007/978-3-030-75789-2_11

require special control techniques, usually based on the closed-loop control of cable tensions between a minimum and a maximum value. Cable-force values can be computed by solving the static equations for a given pose of the *EE*, and employed for the determination of torque set-points for the actuators. However, under the assumption of inextensible cables, the static-equilibrium problem for an overconstrained system is underdetermined, thus admitting an infinite number of solutions for every *EE* pose. This makes the determination of cable forces non-trivial. This problem is known in the literature as Force or Tension Distribution (*FD*) computation. *FD* computation can be formulated as an optimization problem [6], or, if real-time capability is an application requirement [7], methods with less computation complexity, such as geometrical approaches, can be used [8,9].

The aim of this work is to evaluate how *FD* varies with respect to an error in the force of one cable, and to determine which cable-tension error influences the least the *FD*. In other words, the *FD* sensitivity to one cable tension error is evaluated and the cable whose tension error results in the minimum *FD* sensitivity is identified. Then, the wrench-feasible workspace of a planar *OCDPR* with two different cable arrangements is computed, according to a well-known kernel-based approach, and it is shown that the cable least influencing the *FD* varies in the *WS*. The identification of this cable is then used for the preliminary test of a hybrid-input control strategy of a 4-cable 3-*DoF CDPR*. This strategy, introduced in [10] and [11], requires that three cables, equal to the number of platform *DoFs*, are length-controlled, whereas the redundant cable is force-controlled. This simple feedforward control scheme provides a good *EE* motion accuracy, while it succeeds in maintaining all cables taut throughout the *WS*, without requiring more involved cascade closed-loop controllers [12]. References [10] and [11] suggest no procedure for the selection of the force-controlled cable, while, in this paper, it is determined by way of the results of sensitivity analysis. The aim is to ensure that the *FD* error is kept as low as possible, while preserving the *WS* dimensions.

The paper is organised as follows. Section 2 describes the kinematic model of an overconstrained planar *CDPR* with 4 cables. Section 3 provides the *FD* formulation and the derivation of its sensitivity to an error in one cable tension. The *WS* computation is shown in Sect. 4, as an application example. In Sect. 5, preliminary experimental results for the control of a prototype are reported. Finally, results of simulations and experiments are discussed.

2 Kinematic Model

An *overconstrained planar CDPR* consists of a moving platform driven by n actuated cables, with n greater than the *EE DoFs*, i.e. 3. The degree of redundancy can be defined as $\mu = n - 3$. In this paper, $n = 4$ and $\mu = 1$. Moreover, cables are considered massless and inextensible, and modeled as the line segments between the anchor points on the platform and the ones on the fixed base (A_i and B_i respectively, for $i = 1, ..., n$, as in Fig. 1). An inertial frame Oxy and a

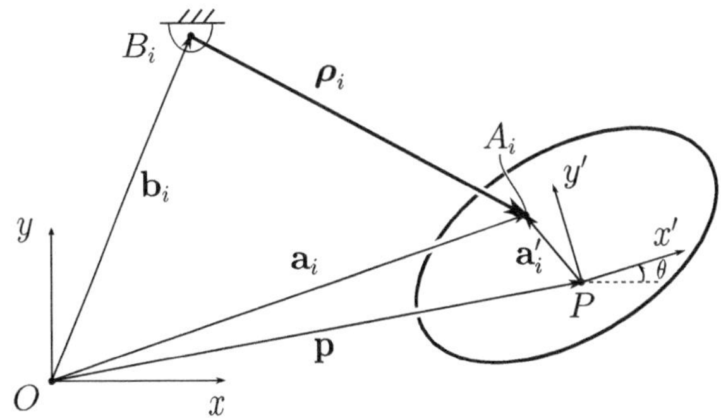

Fig. 1. Geometry of the i-th constraint of a planar *CDPR*.

mobile frame $Px'y'$, attached to the platform center of mass P, are introduced. Points P, A_i and B_i are described by vectors $\mathbf{p}$, $\mathbf{a}_i$ and $\mathbf{b}_i$ with respect to the inertial frame, whereas they are identified by $\mathbf{p}'$, $\mathbf{a}'_i$ and $\mathbf{b}'_i$ in the mobile frame. The orientation of the platform is described by the rotation matrix $\mathbf{R}(\theta)$. The i-th cable vector is:

$$\boldsymbol{\rho}_i = \mathbf{a}_i - \mathbf{b}_i = \mathbf{p} + \mathbf{R}(\theta)\mathbf{a}'_i - \mathbf{b}_i, \quad \mathbf{R}(\theta) \triangleq \begin{bmatrix} cos(\theta) & -sin(\theta) \\ sin(\theta) & cos(\theta) \end{bmatrix} \tag{1}$$

If the pose $[\mathbf{p}^T, \theta]^T$ of the platform is assigned, the i-th cable length l_i can be determined from the constraint imposed by the i-th cable:

$$\boldsymbol{\rho}_i^T \boldsymbol{\rho}_i - l_i^2 = 0 \tag{2}$$

The unit vector of the i-th cable, pointing from the base towards the platform, is:

$$\mathbf{t}_i = \frac{\boldsymbol{\rho}_i}{l_i} \tag{3}$$

3 Force-Distribution Sensitivity

The static equilibrium of the platform can be formulated as:

$$\mathbf{J}^T \boldsymbol{\tau} - \mathbf{W} = \mathbf{0} \tag{4}$$

where $\mathbf{J}^T \in \mathbb{R}^{3\times 4}$, referred to as *structure matrix*, is the transpose of the Jacobian matrix of the inverse kinematics [13], whose i-th row is:

$$\mathbf{J}_i = \left[\mathbf{t}_i{}^T \; -\mathbf{t}_i^T \mathbf{E}\mathbf{R}\mathbf{a}'_i\right], \quad \mathbf{E} \triangleq \begin{bmatrix} 0 & -1 \\ 1 & 0 \end{bmatrix} \tag{5}$$

$\boldsymbol{\tau} \in \mathbb{R}^4$ is the array of cable tensions, and $\mathbf{W} = mg\,\left[0\;1\;0\right]^T \in \mathbb{R}^3$ is the external wrench if only gravity is acting on the system (m is the *EE* mass and $g = -9.81$ m/s^2).

For an *OCDPR*, if cable elasticity is not taken into account, the inverse static problem is underdetermined, due to actuation redundancy [14], thus it admits infinite solutions in the cable tensions. This problem is known in the literature as Force or Tension Distribution (*FD*) computation. The *FD* problem can be solved by different approaches, carefully analysed and compared in [15], depending on application requirements. In general, for cables to remain taut, while the *EE* moves along a prescribed trajectory, an acceptable *FD* must be determined for each pose.

To derive the cable-tension sensitivity index, suitable partitions of the structure matrix and the tension array are introduced into the static equation (4). These partitions are based on the assumption that one cable, whose tension is called τ_c, can be force-controlled, whereas the tensions of the remaining cables, $\boldsymbol{\tau}_d$, depend on τ_c and the external wrench $\mathbf{W}$. If we assume, for the time being, that the 4th cable is force-controlled, the array $\boldsymbol{\tau}$ of cable tensions is:

$$\boldsymbol{\tau} \triangleq \begin{bmatrix} \boldsymbol{\tau}_d \\ \tau_c \end{bmatrix} = \begin{bmatrix} \boldsymbol{\tau}_0 + \boldsymbol{\Delta\tau} \\ \tau_c \end{bmatrix} \tag{6}$$

and the structure matrix can be partitioned as:

$$\mathbf{J}^T = \begin{bmatrix} \mathbf{J}_d \; \mathbf{J}_c \end{bmatrix} \tag{7}$$

where $\mathbf{J}_d \in \mathbb{R}^{3\times 3}$ and $\mathbf{J}_c \in \mathbb{R}^3$. Introducing definitions (6) and (7) in Eq. (4) yields:

$$\mathbf{W} = \mathbf{J}^T\boldsymbol{\tau} = \begin{bmatrix} \mathbf{J}_d \; \mathbf{J}_c \end{bmatrix} \begin{bmatrix} \boldsymbol{\tau}_0 + \boldsymbol{\Delta\tau} \\ \tau_c \end{bmatrix} = \mathbf{J}_d(\boldsymbol{\tau}_0 + \boldsymbol{\Delta\tau}) + \tau_c\mathbf{J}_c \tag{8}$$

If $\mathbf{J}_d$ is invertible (which is true if the *EE* is not in a geometrically singular configuration), $\boldsymbol{\tau}_0$ is defined as the solution of Eq. (8) when $\boldsymbol{\tau}_c = 0$ and $\boldsymbol{\Delta\tau} = \mathbf{0}$:

$$\boldsymbol{\tau}_0 \triangleq \mathbf{J}_d^{-1}\mathbf{W} \tag{9}$$

The expression of $\boldsymbol{\Delta\tau}$ is deduced from Eq. (8), by considering Eq. (9):

$$\boldsymbol{\Delta\tau} = -\tau_c\mathbf{J}_d^{-1}\mathbf{J}_c \tag{10}$$

Following the geometrical approach proposed in [7], the force distribution $\boldsymbol{\tau}$ in Eq. (6) can be divided into two terms: $\boldsymbol{\tau}_{ls}$, obtained from the solution of the linear system (9), and $\boldsymbol{\tau}_{ker}$, which provides a variation of the overall distribution that still satisfies the *EE* equilibrium (4):

$$\boldsymbol{\tau} = \boldsymbol{\tau}_{ls} + \boldsymbol{\tau}_{ker} \tag{11}$$

With the definition given in Eq. (6), the first contribution is expressed as:

$$\boldsymbol{\tau}_{ls} = \begin{bmatrix} \boldsymbol{\tau}_0 \\ 0 \end{bmatrix} = \begin{bmatrix} \mathbf{J}_d^{-1}\mathbf{W} \\ 0 \end{bmatrix} \tag{12}$$

while the contribution $\boldsymbol{\tau}_{ker}$ can be written as:

$$\boldsymbol{\tau}_{ker} = \begin{bmatrix} \boldsymbol{\Delta\tau} \\ \tau_c \end{bmatrix} = \tau_c \begin{bmatrix} -\mathbf{J}_d^{-1}\mathbf{J}_c \\ 1 \end{bmatrix} = \tau_c\mathbf{J}^{\perp}, \quad \mathbf{J}^{\perp} \triangleq \begin{bmatrix} -\mathbf{J}_d^{-1}\mathbf{J}_c \\ 1 \end{bmatrix} \tag{13}$$

$\mathbf{J}^{\perp} \in \mathbb{R}^4$ is the right nullspace of the (3×4) matrix $\mathbf{J}^T$, namely the 4-dimensional vector such that $\mathbf{J}^T\mathbf{J}^{\perp} = \mathbf{0}_{3\times 1}$. Indeed:

$$\mathbf{J}^T\mathbf{J}^{\perp} = \begin{bmatrix}\mathbf{J}_d \ \mathbf{J}_c\end{bmatrix} \begin{bmatrix}-\mathbf{J}_d^{-1}\mathbf{J}_c \\ 1\end{bmatrix} = -\mathbf{J}_c + \mathbf{J}_c = \mathbf{0} \tag{14}$$

The introduction of Eqs. (12) and (13) in Eq. (11) yields:

$$\boldsymbol{\tau} = \begin{bmatrix}\mathbf{J}_d^{-1}\mathbf{W} \\ 0\end{bmatrix} + \tau_c \begin{bmatrix}-\mathbf{J}_d^{-1}\mathbf{J}_c \\ 1\end{bmatrix} = \begin{bmatrix}\mathbf{J}_d^{-1}(\mathbf{W} - \tau_c\mathbf{J}_c) \\ \tau_c\end{bmatrix} \tag{15}$$

So, a *FD* can be determined by choosing a value of $\tau_c \in \mathbb{R}$ and substituting it back in the first three elements at the right-hand side of Eq. (15), in order to compute the remaining three tensions:

$$\boldsymbol{\tau}_d = \boldsymbol{\tau}_0 + \boldsymbol{\Delta\tau} = \mathbf{J}_d^{-1}(\mathbf{W} - \tau_c\mathbf{J}_c) \tag{16}$$

It can be noticed from Eq. (13) that the right nullspace $\mathbf{J}^{\perp}$ of the structure matrix $\mathbf{J}^T$ represents the *force-distribution sensitivity* $\boldsymbol{\sigma}$. Namely, $\boldsymbol{\sigma}$ shows how the overall cable tensions vary due to a unit change in the tension of the force-controlled cable:

$$\boldsymbol{\sigma} = \boldsymbol{\tau}_{ker}\big|_{\tau_c=1} = \begin{bmatrix}\sigma_1 \\ \sigma_2 \\ \sigma_3 \\ \sigma_4\end{bmatrix} \triangleq \tau_c\boldsymbol{J}^{\perp}\big|_{\tau_c=1} = \boldsymbol{J}^{\perp} = \begin{bmatrix}-\boldsymbol{J}_d^{-1}\boldsymbol{J}_c \\ 1\end{bmatrix} \tag{17}$$

A value $|\sigma_i| > 1$ (resp. $|\sigma_i| < 1$) in Eq. (17) means that the tension in cable i varies more (resp. less) than the tension in the 4th cable. Thus, if the cable with the largest $|\sigma_i|$ is (locally) chosen to be force controlled, all other cables are guaranteed to have smaller tension errors than the error committed in the control of that cable. The largest component (in magnitude) of $\boldsymbol{\sigma}$ is denoted $\sigma_{i^\star}$ and the corresponding cable, $i^\star$. As a matter of fact, the preliminary choice of τ_c as the tension of the 4th cable is irrelevant: the choice of another cable would change the numerical value of $\boldsymbol{\sigma}$ but, thanks to the linear nature of the problem, the cable with the highest sensitivity to tension variations would remain the same.

As a result, it is possible to compute $\boldsymbol{\sigma}$ in advance, along with the *WS*, so that, for every *EE* pose, the cable with highest sensitivity can be selected to be force-controlled.

4 Workspace Characterization

As an example of application of the *FD* sensitivity, the constant-orientation wrench-feasible *WS* of a planar *OCDPRs* with two different cable arrangements is presented in this section, and the highest-sensitivity cable $i^\star$ is determined in each *WS* configuration. A kernel-based approach is used for the WS calculation [13], where the nullspace (or kernel) $\mathbf{J}^{\perp}$ is computed as in Eq. (17).

The wrench-feasible *WS* is defined as the set of poses $[\mathbf{p}^T, \theta]^T$ for which:

$$\exists \boldsymbol{\tau} : \quad \mathbf{T}_{min} \leq \boldsymbol{\tau} \leq \mathbf{T}_{max}, \quad \mathbf{J}^T\boldsymbol{\tau} - \mathbf{W} = \mathbf{0} \tag{18}$$

where all elements of the arrays $\mathbf{T}_{min}$ and $\mathbf{T}_{max} \in \mathbb{R}^4$ are equal to the tension limits τ_{min} and τ_{max}, respectively. Considering the partition (6), Eq. (18) can be written as:

$$\tau_{min} \leq \tau_c \leq \tau_{max} \tag{19}$$

$$\boldsymbol{\tau}_{min} \leq \boldsymbol{\tau}_d \leq \boldsymbol{\tau}_{max}, \quad \boldsymbol{\tau}_{min}, \boldsymbol{\tau}_{max} \in \mathbb{R}^3 \tag{20}$$

Due to Eq. (16), Eq. (20) yields:

$$\boldsymbol{\tau}_{min} \leq \mathbf{J}_d^{-1}(\mathbf{W} - \tau_c\mathbf{J}_c) \leq \boldsymbol{\tau}_{max} \tag{21}$$

$$-\boldsymbol{\tau}_{max} + \mathbf{J}_d^{-1}\mathbf{W} \leq \tau_c\mathbf{J}_d^{-1}\mathbf{J}_c \leq -\boldsymbol{\tau}_{min} + \mathbf{J}_d^{-1}\mathbf{W} \tag{22}$$

$$\mathbf{a} \leq \tau_c\mathbf{c} \leq \mathbf{b}, \quad \mathbf{a} \triangleq -\boldsymbol{\tau}_{max} + \mathbf{J}_d^{-1}\mathbf{W}, \quad \mathbf{b} \triangleq -\boldsymbol{\tau}_{min} + \mathbf{J}_d^{-1}\mathbf{W}, \quad \mathbf{c} \triangleq \mathbf{J}_d^{-1}\mathbf{J}_c \tag{23}$$

$$\begin{bmatrix} a_1 \\ a_2 \\ a_3 \end{bmatrix} \leq \tau_c \begin{bmatrix} c_1 \\ c_2 \\ c_3 \end{bmatrix} \leq \begin{bmatrix} b_1 \\ b_2 \\ b_3 \end{bmatrix} \tag{24}$$

Thus, if the intersection between the four inequalities in Eq. (19) and (24) is nonempty, there is at least one value of τ_c for which a feasible *FD* exists, for a given *EE* pose.

By using the above method, the wrench-feasible constant-orientation *WS* of a planar 3-*DoF* 4-cable *CDPR* model may be computed and characterized by the variation of $i^\star$ throughout it. In case of planar *CDPRs*, the constant-orientation *WS* can be easily visualized in 2D, and represented with a regular discrete grid of $N \times M$ points. In the following example, $N = M = 100$. Moreover, $m = 2.5$ kg, $\tau_{min} = 10$ N, and $\tau_{max} = 80$ N.

The *CDPR* at hand (Fig. 2) has rectangular base (0.875 m × 0.700 m) and mobile platform (0.080 m × 0.100 m). The inertial frame Oxy is located in center of the base and the moving frame $Px'y'$ at the center of the *EE*, coinciding with its center of mass. Two different cable arrangements are considered on the same model: a *standard* layout (Fig. 2a), where each cable goes from one vertex of the rectangular base to the corresponding vertex of the *EE*, and a *crossed* layout (Fig. 2b), where cables cross under and above the platform.

The wrench-feasible *WS* with constant orientation $\theta = 0°$ is shown in Fig. 3 for the *standard* (Fig. 3a) and the *crossed* layout (Fig. 3b). Four distinct constant-$i^\star$ areas emerge from the *FD* sensitivity analysis: the change in $i^\star$ occurs exactly when the *EE* position crosses one coordinated axis of the inertial frame Oxy. For the *standard* layout, only the case with $\theta = 0°$ is analysed, since this cable configuration has a very limited orientation capability, as reported in [16]. For the *crossed* layout, four other different cases are investigated. The *WS* for constant

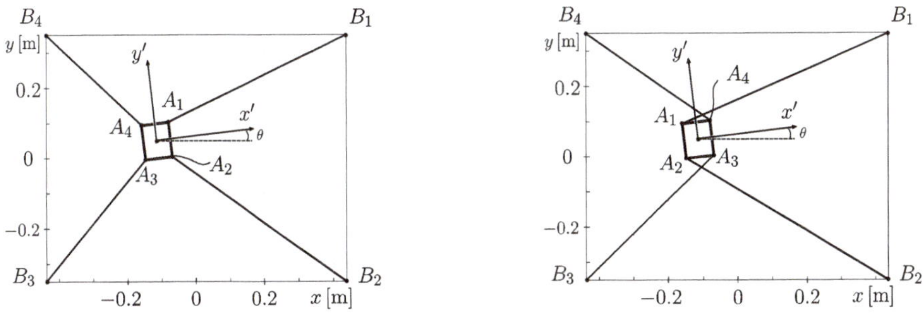

(a) *Standard layout* (b) *Crossed layout*

Fig. 2. Cable arrangements of the planar 4-cable *OCDPR*.

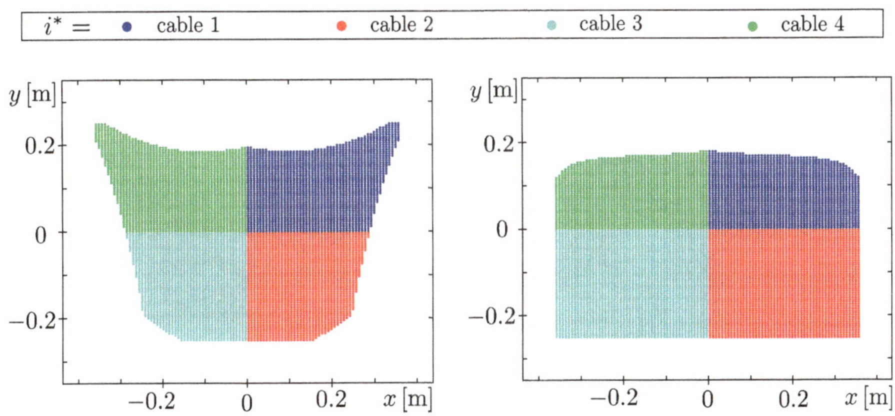

(a) *Standard layout* (b) *Crossed layout*

Fig. 3. Highest-sensitivity cable ($i^{\star}$) on the wrench-feasible *WS* of the planar 4-cable *OCDPR* with constant orientation $\theta = 0°$.

orientations $\theta = \pm 10°$ is shown in Figs. 4a and 4b. It can be noticed that, one sensitivity area, respectively for $i^{\star} = 3$ when $\theta = 10°$ (Fig. 4b), and $i^{\star} = 2$ when $\theta = -10°$ (Fig. 4b), is significantly larger, compared to the case $\theta = 0°$ in Fig. 3b. Also, the dimensions of this area increase for larger values of θ, as shown in Figs. 4c and 4d.

Due to space limitations, the detailed analysis of *FD* sensitivity values σ_i throughout the *WS* is not reported. However, simulation results show that, for all configurations near transition borders, the values of σ_i in contiguous areas are very similar. On the other hand, at the edges of the *WS*, far from transitions, those values significantly differ. For example, in the bottom left corner of the crossed-layout *WS* in Fig. 3b, $|\sigma_3|$ is ten times greater than $|\sigma_4|$. As a consequence, for close-to-transition configurations, it may not be worth changing the force-controlled cable, but it may be beneficial near *WS* edges.

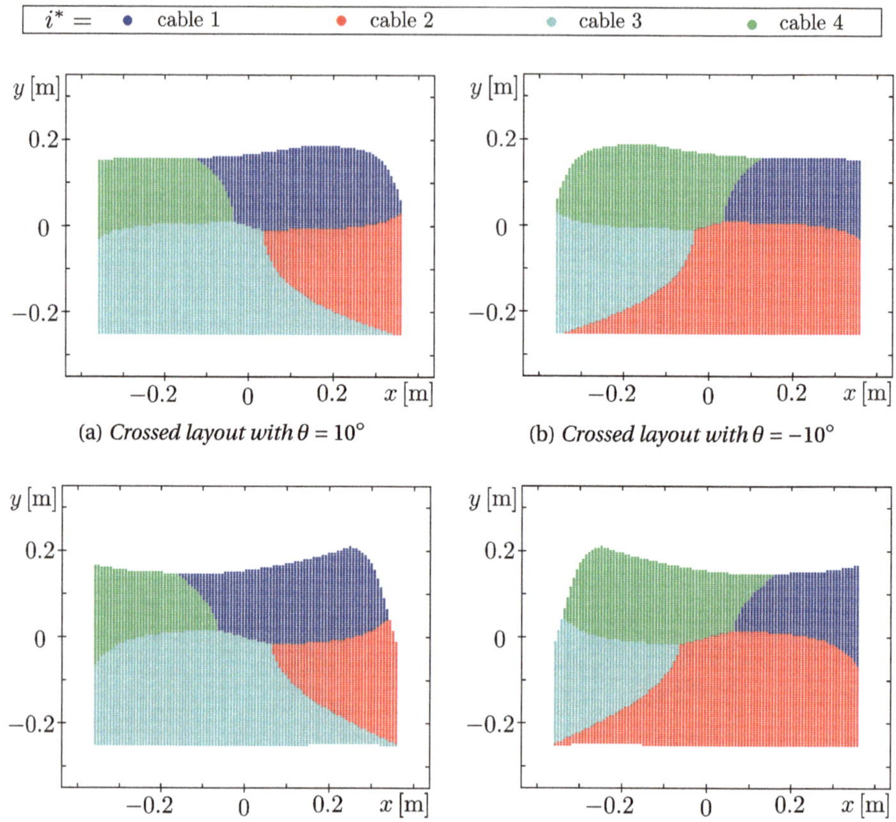

(a) *Crossed layout with* $\theta = 10°$

(b) *Crossed layout with* $\theta = -10°$

(c) *Crossed layout with* $\theta = 20°$

(d) *Crossed layout with* $\theta = -20°$

Fig. 4. Highest-sensitivity cable ($i^\star$) on the wrench-feasible *WS* of the planar 4-cable *OCDPR* with variable constant orientation $\theta = 10°, -10°, 20°$ and $-20°$.

Two facts are noteworthy: (i) the sensitivity areas for negative angles are symmetrical with respect to those for positive angles, and (ii) when $\theta \neq 0°$ the boundaries of the regions where $i^\star$ is constant are not straight lines. These observations are relevant if a hybrid-input control strategy is implemented on the manipulator, as proposed in [10,11]. According to this strategy, three cables are length-controlled, while the remaining one is force-controlled. As introduced in Sect. 3, cable $i^\star$ can be chosen to be force-controlled, since its tension-control error causes the smallest tension errors in the other cables in a given configuration. On the other hand, while the *EE* performs a prescribed trajectory inside the *WS*, the force-controlled cable should change when crossing from one sensitivity area to another. Changing the control input type (length or force) could cause a discontinuity of the control action, since the error tracked by the controller changes. Thus, it can be beneficial to reduce the number of constant-$i^\star$ boundary transitions during motion. To do so, an area with constant $i^\star$ can be

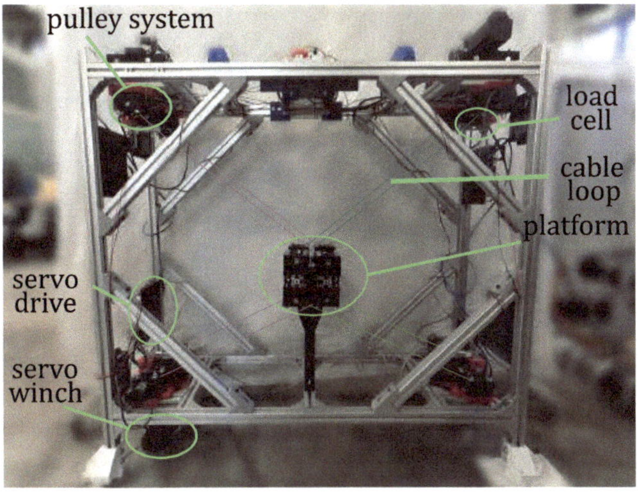

Fig. 5. Planar 3-*DoF* 4-cable *OCDPR* prototype developed at *IRMA L@B*.

enlarged, in order to perform an entire trajectory without crossing such boundaries. For this purpose:

1. for a given robot design, the platform orientation can be varied, unless $\theta = const$ is prescribed by the application;
2. the positions of cable anchor points on the *EE* can be optimized in order to maximize one sensitivity area so that it includes the entire *WS* needed by the application at hand.

5 Experiments on Hybrid-Input Control

In this section, a hybrid-input control strategy is preliminarily investigated, and applied to a planar 3-*DoF* 4-cable *OCDPR* prototype developed at *IRMA L@B* (Fig. 5): this prototype has to perform quasi-static non-contact tasks, such as laser engraving or laser cutting. The prototype presents a *crossed* cable layout, as in Fig. 2b, and its geometrical parameters are those given in Sect. 4. Four self-manufactured servo winches are fixed to the corners of an aluminium-profile base and each one of them drives a cable coiled in a loop onto the pulley system. The high-level robot controller runs on a *Real-Time Linux* PC at 1 kHz rate. When a trajectory in the Cartesian space is assigned, the controller computes the feedforward hybrid inputs for the robot actuators. The servo-winch control schemes are also developed in house and run on a *ST Nucleo-H743ZI* development board at 10 kHz. In addition to controlling the motor angle, this allows us to directly feed a cable-tension command to the drive, to compare said command with the signal of a load-cell embedded in the cable transmission, and regulate the tension by a *PID* controller at drive level.

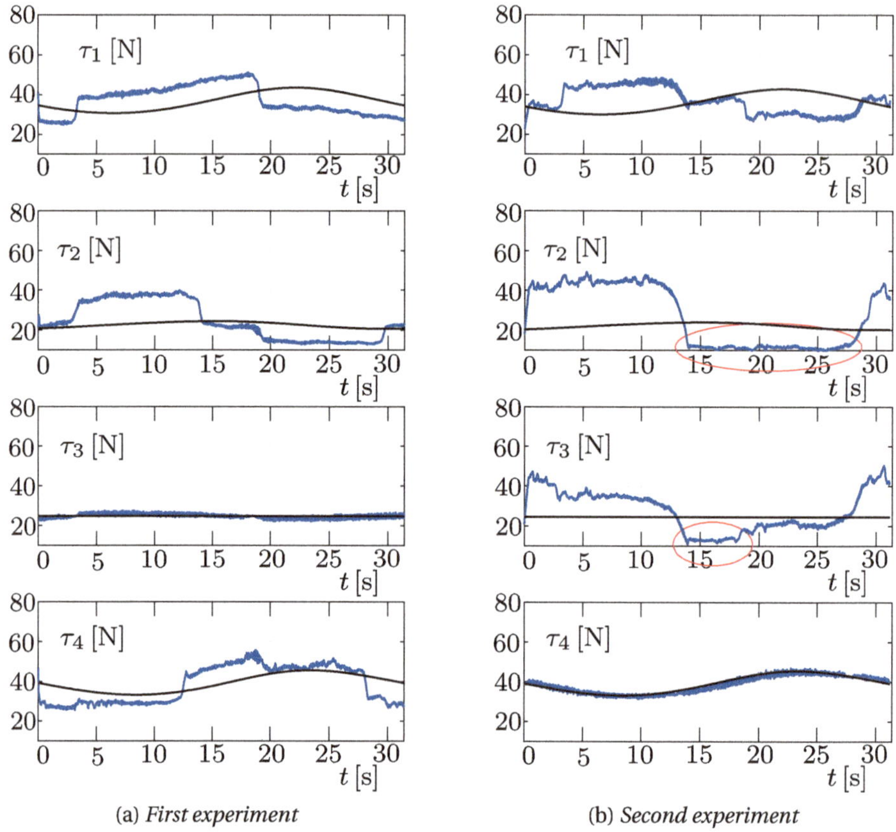

Fig. 6. Real (blue) and computed (black) *FD* during experiments.

The hybrid-input control strategy requires that three cables, equal to the number of platform *DoFs*, are length-controlled while the redundant cable is force-controlled. The rationale behind this choice is that three length-controlled cables are sufficient for precisely controlling the pose of the planar *EE*, whereas the 4th cable is needed to keep the remaining cables under tension. Cable $i^\star$ is chosen to be force-controlled, since errors in its tension estimation (or low-level feedback control) limitedly impact the overall *FD*.

Two experiments are conducted, while the *EE* follows the same circular trajectory with a constant linear speed of 1 cm/s, with $\theta = 0°$, that is:

$$\mathbf{p}(t) = \begin{bmatrix} x \\ y \end{bmatrix} = \begin{bmatrix} P_{0x} + r\cos\left(\frac{2\pi t}{T}\right) \\ P_{0y} + r\sin\left(\frac{2\pi t}{T}\right) \end{bmatrix}, \quad \begin{aligned} P_{0x} = -0.06\text{ m},\ P_{0y} = -0.06\text{ m} \\ r = 0.05\,\text{m},\ T = 31.4\,\text{s} \end{aligned} \tag{25}$$

The value of the constant speed is selected so that a piece of paper may be cut with a 2.5 W laser-diode mounted onto the *EE*. The whole trajectory lies in the area where $i^\star = 3$ in Fig. 3b. The *FD* over the whole trajectory is calculated, by choosing $\tau_c = 25$ N as a constant value and considering gravity as

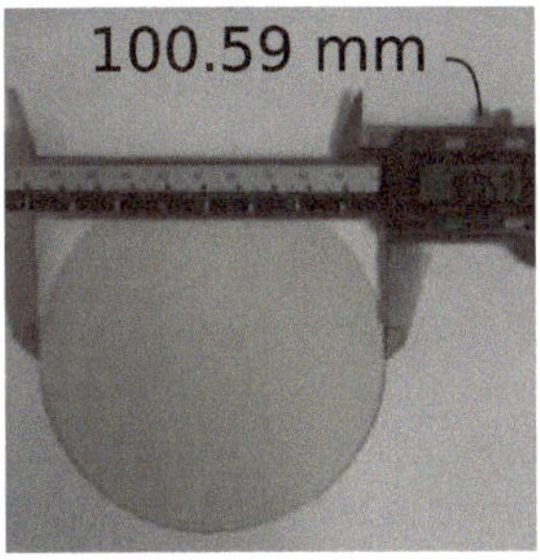

(a) *First experiment*

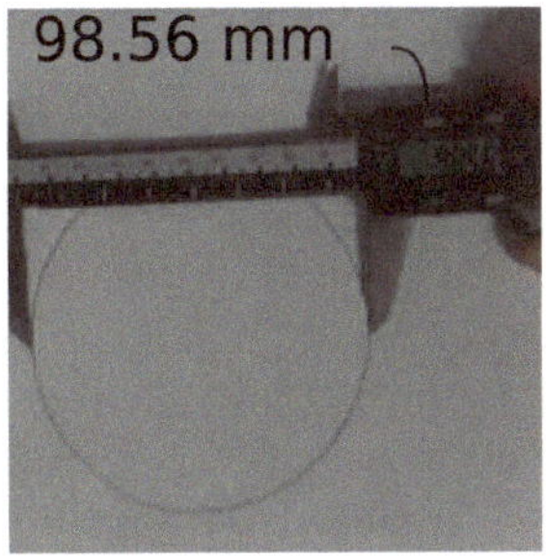

(b) *Second experiment*

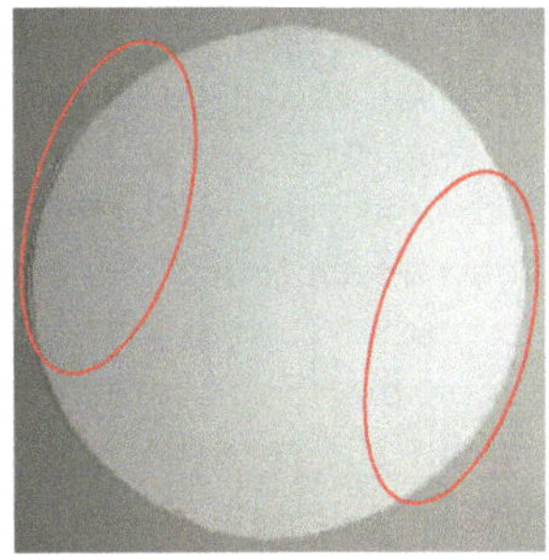

(c) *Superimposed figures*

Fig. 7. Results of the laser-cutting operation.

the only external wrench. In the first experiment, the 3rd cable, whose tension errors influence the *FD* as little as possible, is force controlled, so that $\tau_3 = \tau_c$. In the second experiment, the 4th cable, whose tension errors more influence the *FD*, is force controlled. The tension set-points of the tension-controlled cable are chosen so that the same theoretical *FD* is generated in both experiments.

Experimental results on the actual *FD* attained during the trajectory execution are shown in Fig. 6. It can be noticed that the low-level force controller is effective in maintaining the controlled cable tension very close to the assigned value, but the other cable tensions are quite distant from their theoretical value, which is compatible with the oversimplified model that was adopted and the cheap robot components (most of them are plastic made). However, when the 3rd cable is force-controlled during the first experiment, the other tensions, as measured by the load cells, never fall below the lower cable tension limit $\tau_{min} = 10$ N. Conversely, during the second experiment, when the 4th cable is force-controlled, there is a prolonged period for which the second and third cables are practically slack (see the red-circled areas in Fig. 6b). In practice, by force-controlling the highest-sensitivity cable ($i^\star$), the *FD* errors with respect to the theoretical ones are minimized.

The paper-cutting operation confirms that the first experiment produced good results also in practice, while the results of the second experiments are poorer (see Fig. 7). Figure 7a shows a well-cut circle executed with millimetric accuracy, whereas the second experiment produced a more elliptic shape (without managing to complete the cut), as shown in Fig. 7b. The hybrid-input control strategy, where cable $i^\star$ is force-controlled, produced good preliminary results, despite the hardware limitations.

6 Conclusions

In this paper, the *FD* sensitivity to tension errors in one cable was investigated, and the cable whose tension error plays the least influence was analytically derived. The wrench-feasible *WS* of a 3-*DoF* planar *OCDPR* with 4-cables was calculated with two different cable arrangements, and the areas where different cables have the highest sensitivity were displayed. A hybrid-input control strategy was preliminarily investigated on a prototype, with one cable being force-controlled and the others being length-controlled. The force control of the highest-sensitivity cable showed promising results. This strategy is a simpler alternative to well known cascade position/force control strategies, and provides a good motion accuracy while maintaining all cables taut. In the future, we will expand the concept of *FD* sensitivity to *OCDPRs* with degree of redundancy greater than one, and we will characterize the stability of the hybrid-input controller in case the force controlled cable is changed during a trajectory.

References

1. Bruckmann, T., Lalo, W., Nguyen, K., Salah, B.: Development of a storage retrieval machine for high racks using a wire robot. In: ASME 2012 International Design Engineering Technical Conferences, Chicago, USA, pp. 771–780 (2012)
2. Bruckmann, T., Mattern, H., Spengler, A., Reichert, C., Malkwitz, A., König, M.: Automated construction of masonry buildings using cable-driven parallel robots. In: 33rd International Symposium on Automation and Robotics in Construction, Auburn, USA, pp. 332–340 (2016)
3. Izard, J.-B., Dubor, A., Hervé, P.-E., Cabay, E., Culla, D., Rodriguez, M., Barrado, M.: Large-scale 3D printing with cable-driven parallel robots. Constr. Robot. **1**(1), 69–76 (2017)
4. Kawamura, S., Choe, W., Tanaka, S., Kino, H.: Development of an ultrahigh speed robot FALCON using parallel wire drive systems. J. Robot. Soc. Japan **15**(1), 82–89 (1997)
5. Bouchard, S., Gosselin, C., Moore, B.: On the ability of a cable-driven robot to generate a prescribed set of wrenches. J. Mech. Robot. **2**(1) (2010)
6. Gosselin, C., Grenier, M.: On the determination of the force distribution in overconstrained cable-driven parallel mechanisms. Meccanica **46**(1), 3–15 (2011)
7. Bruckmann, T., Pott, A., Hiller, M.: Calculating force distributions for redundantly actuated tendon-based Stewart platforms. In: Lennarčič, J., Roth, B. (eds.) Advances in Robot Kinematics, pp. 403–412. Springer (2006)
8. Lamaury, J., Gouttefarde, M.: A tension distribution method with improved computational efficiency. In: Bruckmann, T., Pott, A., (eds.) Cable-Driven Parallel Robots, pp. 71–85. Springer (2013)
9. Pott, A.: An improved force distribution algorithm for over-constrained cable-driven parallel robots. In: Thomas, F., Perez Gracia, A. (eds.) Computational Kinematics, pp. 139–146. Springer (2014)
10. Bouchard, S., Gosselin, C.: A simple control strategy for overconstrained parallel cable mechanisms. In: Proceedings of the 20th Canadian Congress of Applied Mechanics (CANCAM: Montreal, p. 2005. Quebec, Canada (2005)
11. Bruckmann, T., Mikelsons, L., Hiller, M., Schramm, D.: A new force calculation algorithm for tendon-based parallel manipulators. In: IEEE/ASME International Conference on Advanced Intelligent Mechatronics, Zurich, Switzerland, pp. 1–6 (2007)

12. Kraus, W., Schmidt, V., Rajendra, P., Pott, A.: System identification and cable force control for a cable-driven parallel robot with industrial servo drives. In: 2014 IEEE International Conference on Robotics and Automation (ICRA), Hong Kong, China, pp. 5921–5926 (2014)
13. Verhoeven, R., Hiller, M., Tadokoro, S.: Workspace, stiffness, singularities and classification of tendon-driven Stewart platforms. In: Lenarčič, J., Husty, M.L. (eds.) Advances in Robot Kinematics: Analysis and Control, pp. 105–114. Springer (1998)
14. Merlet, J.-P.: Wire-driven parallel robot: open issues. In: Padois, V., Bidaud, P., Khatib, O. (eds.) Romansy 19–Robot Design, Dynamics and Control, pp. 3–10. Springer (2013)
15. Pott, A.: Cable-driven parallel robots: theory and application. In: Siciliano, B., Khatib, O. (eds.) Springer (2018)
16. Gouttefarde, M., Krut, S., Company, O., Pierrot, F., Ramdani, N.: On the design of fully constrained parallel cable-driven robots. In: Lenarčič, J., Wenger, P. (eds.) Advances in Robot Kinematics: Analysis and Design, pp. 71–78. Springer (2008)

Design

Configuration Optimization of an Auto-reconfigurable Cable-Driven Upper-Limb Rehabilitation Robot

Yu Zhou, Bin Zhang(✉), Weiwei Shang, and Shuang Cong

Department of Automation, University of Science and Technology of China, Hefei 230027, China
bzhanghero@ustc.edu.cn

Abstract. Compared with traditional rigid structures, cable-driven parallel robots (CDPRs) have been widely used in rehabilitation training due to the lower motion inertia and cable flexibility. However, for the traditional CDPRs, configurations cannot be changed individually according to different patients and different scenes, thus the adaptability is limited and it is difficult to provide satisfactory training effects. In this paper, an auto-reconfigurable cable-driven upper-limb rehabilitation robot (ACURR) is put forward, which can change the configurations according to different rehabilitation training requirements automatically. Based on the traditional wrench feasible workspace, the performance space of the ACURR is defined according to the further analysis of cable tensions. When the ACURR moves in the performance space, patients can bear more comfortable tensions and the training system is more secure. An optimization problem is then formulated to obtain the optimal configurations of the ACURR by maximizing the performance space, and an optimization algorithm is proposed to solve it by combining the interval analysis method and the simulated annealing algorithm. The correctness and effectiveness of the algorithm are verified with the simulations, which also shows that the concept of the ACURR is reasonable and the optimal configurations can always be obtained in different scenarios.

Keywords: Configuration optimization · Rehabilitation treatment · Cable-driven parallel robot · Auto-reconfigurable

1 Introduction

Generally, rehabilitation treatment refers to making the patients with limb dyskinesia finish specific repetitive rehabilitation actions continuously with the help of the physiotherapists, whose skills will have a great influence on the rehabilitation effects [1]. However, after a period of treatment, the rehabilitation effects will gradually weaken which results from the energy decline of the physiotherapists, and the decline is mainly manifested as the weakening of the strength

M. Gouttefarde et al. (Eds.): CableCon 2021, MMS 104, pp. 145–157, 2021.
https://doi.org/10.1007/978-3-030-75789-2_12

and the deviations of the rehabilitation actions. Therefore, for such repetitive rehabilitation treatments, the robots without fatigue can be used to replace the physiotherapists [2].

Some rigid exoskeleton robots have been widely used in the field of rehabilitation treatments, and they can help patients to finish the rehabilitation training more effectively with real-time status monitoring and accurate motion assistance during the rehabilitation process [3]. However, the traditional rigid exoskeleton robots often have many rigid links and need to be customized for the patient, so its structure is relatively complex and has poor adaptability to different patients. Meanwhile, due to their relatively large weight and higher motion inertia, once collision happen between patients and robots during the rehabilitation process, a secondary injury may be caused. For these reasons, cable-driven parallel robots (CDPRs) can be used as the alternatives to meet the rehabilitation training requirements more conveniently of different patients in different scenarios, which benefits from the characteristics of lower motion inertia, larger workspace, simpler mechanical structure, and easier reconfiguration [4]. In addition, the soft cable can also ensure the patient's safety when robot is out of control. Therefore, the applications of CDPRs in rehabilitation treatments have a very broad research prospect. Wang *et al.* [5] designed a rigid-flexible hybrid cable-driven lower-limb rehabilitation robot, and the trajectory planning for different movements of the lower limb in simulations was presented to verify the effectiveness of the robot design. Daniele *et al.* [6] designed a cable robot for rehabilitation training of the upper limb, which has a light weight and a simple structure, and can be used for the clinical rehabilitation as well as the family training. Li *et al.* [7] proposed an admittance control strategy of the cable-driven upper-limb rehabilitation robot by using the trajectory tracking error and verified the improvement of the rehabilitation effects in the experiments. Kang *et al.* [8] designed a cable-driven rehabilitation robot to treat hemiplegia patients by applying a specific force to the pelvis to train the sick legs to work. However, the configurations and dimensions of these cable-driven rehabilitation robots are fixed, which means that these robots cannot make self-adjustments of their configurations or dimensions to achieve the suitable rehabilitation effects by considering the corresponding rehabilitation index from different patients and different rehabilitation tasks.

In order to improve the adaptability of CDPRs in different tasks, the concept of reconfiguration is proposed [9]. Gagliardini *et al.* [10] designed a discrete reconfigurable planning method for an 8–6 CDPR that uses Dijkstra's algorithm to select the optimal reconfiguration strategy. On such a basis, many researchers have taken advantage of CDPRs and made configuration optimization in the field of rehabilitation by considering different rehabilitation indexes. Mao *et al.* [11] optimized the positions of the cable connection points in the cable-driven exoskeleton robot by using the deviation area between the actual and desired trajectory of the robot arm as the optimization objective. Lamine *et al.* [12] made the structure optimization of a planar cable-driven gait rehabilitation robot using the discrete method by considering the cable tension index. Zou *et al.* [13]

optimized the configurations and the shape of the moving platform of the cable-driven lower-limb rehabilitation robot based on the stable workspace and cable tensions, and Hamida *et al.* [14] optimized the configurations of the cable-driven upper-limb rehabilitation robot by using the non-dominated sequencing genetic algorithm. Larib *et al.* [15] proposed an optimization design method for a CDPR with a specified workspace and parameters of the cable-driven upper-limb rehabilitation robot were optimized based on this workspace with the cable tensions as the optimization objective. However, the optimization index selected in these works come from CDPR's ordinary use in industrial which don't meet the requirements of the rehabilitation treatments to some extent. The rehabilitation treatments care more about the patient's feeling and the ability to adapt to different patients. What's more, the robots they studied can't change their configuration automatically which bring some trouble in rehabilitation treatments.

In order to make up for these shortages, this paper studies a novel 3-DOF auto-reconfigurable cable-driven upper-limb rehabilitation robot (ACURR) driven by 4 cables. We choose the two-norm of cable tension as the performance index which represents the patient's tolerance to tension. For different patients, we can change the performance index's volume so that all patients can be comfortable during the rehabilitation treatments. Then the corresponding performance space is defined based on the performance index and its volume is our optimization target. In different rehabilitation scenarios, we calculate ACURR's optimal configuration using the simulated annealing algorithm and the robot automatically converts the configuration to the optimal configuration so that the rehabilitation treatment can be carried out in the optimal condition. Finally, we verify the effectiveness of the optimization design in a variety of simulated rehabilitation scenarios, and analyze the influence of parameters to performance space and optimal configuration.

This paper is organized as follows. In Sect. 2, the model of the ACURR is established and the corresponding optimization problem is formulated. In Sect. 3, the optimization algorithm is clarified and designed, then in Sect. 4, simulations are carried out in a variety of scenarios to obtain the optimal configurations, and the effectiveness of the optimization algorithm is verified. Finally, Sect. 5 summarizes the whole paper.

2 Problem Formulation

2.1 Modeling

The structural schematic diagram of our proposed ACURR is shown in Fig. 1, which mainly includes a fixed frame, a moving platform, and four connecting cables between them. Four servo motors are connected to the vertically installed ball screws and each of them drives a slide block up and down along a slide rail, then the positions of the pulleys linked to blocks can be changed automatically. Four other servo motors are connected to the winches which can adjust the cable lengths by winding the cables. The moving platform is connected with the patient's wrist, which can be regarded as a mass point and has a 3-DOF motion in the workspace. The cable mass and cable sag are assumed to be negligible, and each cable can be regarded as a straight line.

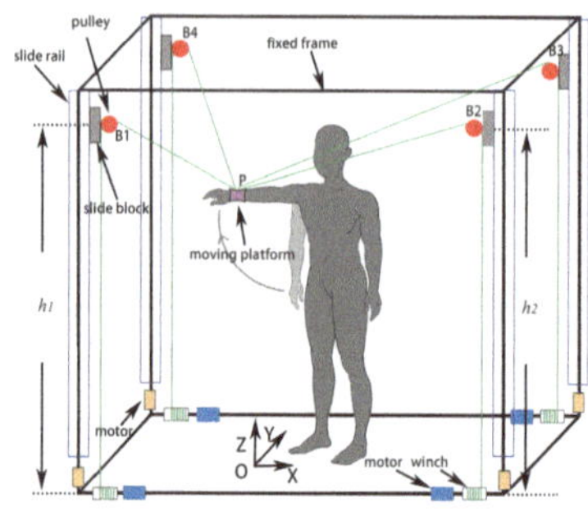

Fig. 1. The structural diagram of the ACURR.

A coordinate frame $O - XYZ$ is established and fixed on the base. $\boldsymbol{p} = (x, y, z)$ is the position vector of the moving platform P, and $h_i(i = 1, 2, 3, 4)$ represents the height of pulley B_i which determines the configuration of the ACURR: $\boldsymbol{c} = (h_1, h_2, h_3, h_4)$. Then, the static equilibrium equation of the ACURR can be formulated as:

$$\mathbf{A}(\boldsymbol{p}) \cdot \mathbf{T} = \mathbf{W}, \tag{1}$$

where $\mathbf{A} = [\mathbf{U}_1, \mathbf{U}_2, \mathbf{U}_3, \mathbf{U}_4]$ represents the structure matrix with $\mathbf{U}_i = \frac{\overrightarrow{\mathrm{PB}_i}}{|\overrightarrow{\mathrm{PB}_i}|} = \frac{\overrightarrow{\mathrm{OB}_i} - \overrightarrow{\mathrm{OP}}}{|\overrightarrow{\mathrm{OB}_i} - \overrightarrow{\mathrm{OP}}|}$ as the unit vector of the cable i, $\mathbf{T} = [T_1, T_2, T_3, T_4]$ is the cable tension vector with T_i as the tension of cable i, and $\mathbf{W}$ represents the applied force acting on the moving platform in rehabilitation training.

2.2 Optimization Problem

In the ACURR, for any applied force $\mathbf{W}$, if the position $\boldsymbol{p}$ satisfies:

$$\begin{cases} \mathbf{A}(\boldsymbol{p}) \cdot \mathbf{T} = \mathbf{W} \\ T_{\min} \leq T_i \leq T_{\max} \end{cases}, \tag{2}$$

where $T_{\min}$ and $T_{\max}$ represent the fixed upper and lower bounds of the cable tensions, then we can say that the position $\boldsymbol{p}$ satisfies the force feasibility condition, and all the positions that meet the conditions constitute the so-called wrench feasible workspace (WFW) [16]. Actually, the force feasible workspace only represents the area where the moving platform can maintain static balance with any applied force, which is an important characteristic for traditional CDPRs, but for the ACURR in rehabilitation training, it is not sufficient. In fact, the two-norm of the cable tension vector represents the comprehensive value of the cable tensions, which also represents the patients' intuitive feeling of the cable tensions to a certain extent in rehabilitation training. Different patients have different tolerance to the cable tensions, and on the basis of the fixed $T_{\min}$ and $T_{\max}$, a personalized adjustable parameter also needs to be introduced to adapt to the patients' tolerance. Hence, in our work, a parameter ζ is defined as the upper bound of the two-norm of the cable tension, which can be expressed as:

$$f = \sqrt{T_1^2 + T_2^2 + T_3^2 + T_4^2} \leq \zeta. \tag{3}$$

On the basis of the wrench feasible workspace, all the positions in the wrench feasible workspace that satisfy Eq. (3) constitute the performance space S_ζ, which has the mathematical expression as:

$$\boldsymbol{p} \in \mathrm{S}_\zeta \iff \begin{cases} \mathbf{A}(\boldsymbol{p}) \cdot \mathbf{T} = \mathbf{W} \\ T_{\min} \leq T_i \leq T_{\max} \\ f \leq \zeta \end{cases}. \tag{4}$$

With the configuration $\boldsymbol{c}$, the volume of the performance space is defined as $\mathrm{V}(\boldsymbol{c})$. Before the rehabilitation training of each patient, an individual ζ needs to be given to optimize the performance space, so that the patient can get a good rehabilitation effect. The optimization problem can be defined as follows:

$$\max_{\boldsymbol{c} \in \boldsymbol{c}_r} \mathrm{V}(\boldsymbol{c}), \quad s.t. \begin{cases} \mathbf{A} \cdot \mathbf{T} = \mathbf{W} \\ T_{\min} \leq T_i \leq T_{\max} \\ f \leq \zeta \\ \mathrm{I}_{\text{collision}} = 0 \end{cases}, \tag{5}$$

where $\boldsymbol{c}_r = \{(h_1, h_2, h_3, h_4) | \forall i, h_{\min} \leq h_i \leq h_{\max}\}$ represents the set of all possible configurations and $\mathrm{I}_{\text{collison}} = 0$ means that there are no collisions between the cables and the patients [14].

3 Configuration Optimization

3.1 Performance Space Solution Based on Interval Analysis

In order to optimize the configuration of the ACURR according to the above optimization problem in Eq. (5) and to obtain the performance space, the interval analysis method is used here. The interval $[m]$ represents the possible value of variable m and is a subset of real number set $\mathbb{R}$ defined as follows [17]:

$$[m] = [\underline{m}, \overline{m}] = \{m \in \mathbb{R} | \underline{m} \leq m \leq \overline{m}\}, \tag{6}$$

where $\underline{m}$ is the largest number on the left of the interval, $\overline{m}$ is the smallest number on the right of the interval, even if the interval is not closed, we still keep the notation $[m]$. The interval estimation of a variable can be extended to the interval estimation of a function by using interval calculation criteria as follows:

$$[g([m])] = \{g(m) | m \in [m]\}. \tag{7}$$

The interval analysis also can be extended to matrices, the interval matrix $[\mathbf{A}]$ will be written indifferently in any of the following forms:

$$[\mathbf{A}] = \begin{pmatrix} [\mathrm{A}_{11}] & \dots & [\mathrm{A}_{1n}] \\ \vdots & \ddots & \vdots \\ [\mathrm{A}_{m1}] & \dots & [\mathrm{A}_{mn}] \end{pmatrix}, \tag{8}$$

where $[\mathrm{A}_{ij}]$ is the projection of $[\mathbf{A}]$ onto the (i, j)th axis. For a pose $\boldsymbol{p}$, if it is in the performance space, it should satisfy Eq. (4), and for a box $[\boldsymbol{p}]$, if it is in the performance space, it should satisfy:

$$[\boldsymbol{p}] \in \mathrm{S}_\zeta \iff \begin{cases} [\mathbf{A}] \cdot \mathbf{T} = \mathbf{W} \\ T_{\min} \le T_i \le T_{\max} \\ [f] \le \zeta \end{cases}. \tag{9}$$

For a $n \times m$ interval matrix $[\mathbf{A}]$, let Y_n be the set of n-dimensional vectors $\boldsymbol{y}$, where $\boldsymbol{y}_i$ has a value of 1 or -1. Obviously, there are 2^n $\boldsymbol{y}$ in Y_n. For each $\boldsymbol{y}$, we can define the vertex matrix $\mathbf{A}_{\boldsymbol{y}}$, which satisfies:

$$\mathbf{A}_{\boldsymbol{y}ij} = \underline{\mathbf{A}_{ij}} + (\overline{\mathbf{A}_{ij}} - \underline{\mathbf{A}_{ij}})(1 - \boldsymbol{y}_i)/2, \tag{10}$$

that is, $\boldsymbol{y}_i$ fully defines the $\mathbf{A}_{\boldsymbol{y}ij}$, for example, when $\boldsymbol{y}_i = -1$, $\mathbf{A}_{\boldsymbol{y}ij} = \overline{\mathbf{A}_{ij}}$. Similarly, for an interval vector [b], we can also define its vertex vector $\boldsymbol{b}_{\boldsymbol{y}}$, which satisfies:

$$\boldsymbol{b}_{\boldsymbol{y}i} = \underline{\boldsymbol{b}_i} + (\overline{\boldsymbol{b}_i} - \underline{\boldsymbol{b}_i})(1 - \boldsymbol{y}_i)/2. \tag{11}$$

Theorem 1 *(Rohn's theorem). The system of interval linear equations* $\mathbf{A} \cdot \mathbf{T} = \mathbf{W}$ *is strongly feasible if and only if the* 2^n *systems of linear equations* $\mathbf{A}_{\boldsymbol{y}} \cdot \mathbf{T} = \mathbf{W}, \boldsymbol{y} \in Y_n$ *are all feasible.*

It is known by Rohn's theorem that the linear interval constraint system $\mathbf{\Gamma}$ is fully true if and only if its 2^n subconstraint system $\mathbf{\Gamma}_{\boldsymbol{y}}$ is fully true. Since the ACURR has three DOFs, Eq. (9) holds if and only if all its 2^3 subconstraint systems are true:

$$\begin{cases} \mathbf{A}_{\boldsymbol{y}} \cdot \mathbf{T} = \mathbf{W} \\ T_{\min} \le T_i \le T_{\max} \\ f_y \le \zeta \end{cases}. \tag{12}$$

Now, we need to solve 2^3 linear programming problems using the linear programming methods. If feasible solutions can be found for every linear programming problems in Eq. (13), then Eq. (9) is true and the box is completely located inside the performance space.

$$\min \sum T_i, \quad s.t. \begin{cases} \mathbf{A}_{\boldsymbol{y}} \cdot \mathbf{T} = \mathbf{W} \\ T_{\min} \le T_i \le T_{\max} \\ f_y \le \zeta \end{cases}. \tag{13}$$

If a box $[\boldsymbol{p}]$ is completely outside the performance space, it should satisfy:

$$\forall \mathbf{A} \in [\mathbf{A}], \forall \mathbf{T} \in [\mathbf{T}], \quad s.t.\ \mathbf{A} \cdot \mathbf{T} \neq \mathbf{W} \text{ or } f > \zeta. \tag{14}$$

It is not practical to verify all $\mathbf{A}$ in the box $[\mathbf{A}]$. An efficient solution method requires the consistency technique [18–20] applied to the continuous domain, which does not attempt to solve the constraint, but continuously compels the size of the box $[\mathbf{T}]$ without losing the solution. When the solution of Eq. (13) is $[\mathbf{T}] = \{\phi\}$, no pose $\boldsymbol{p}$ in the box $[\boldsymbol{p}]$ is inside the performance space, which means the box $[\boldsymbol{p}]$ is completely outside the performance space.

Once $[\boldsymbol{p}]$ can be judged whether inside or outside the performance space, the performance space can be calculated. By estimating $[\mathbf{T}]$, we can classify $[\boldsymbol{p}]$ as: completely located inside the performance space, completely located outside the performance space, and can not be determined. For $[\boldsymbol{p}]$ not be determined, it can be further subdivided into $[\boldsymbol{p}_1]$ and $[\boldsymbol{p}_2]$. $[\boldsymbol{p}_1]$ and $[\boldsymbol{p}_2]$ are then repeatedly processed until either the classification is complete or the termination condition is met. The pseudo code is as follows:

Algorithm1. The pseudo to calculate performance space

$\mathrm{V}(\boldsymbol{c})$ = Compute_PerformanceSpace $(\mathrm{s}, [\mathbf{W}], [\mathbf{T}], \varepsilon, \boldsymbol{c})$
Input : $\mathrm{s}, [\mathbf{W}], [\mathbf{T}], \varepsilon, \boldsymbol{c}$
Output : $\mathrm{V}(\boldsymbol{c})$
$\mathcal{L} \leftarrow \mathrm{s}$
while$(\mathcal{L} \neq \Phi)$**do**
  $[\boldsymbol{p}] \leftarrow$ Extrat$(\mathcal{L})$
  $[\mathbf{A}]$ = Compute_WM$([\boldsymbol{p}], \boldsymbol{c})$
  if My_Judge$([\mathbf{A}], [\mathbf{T}], \mathbf{W})$ = **inside** **then**
    $\mathrm{V_{in}} = \mathrm{V_{in}} +$ Compute_BoxVolume$([\boldsymbol{p}])$
  elseif My_Judge$([\mathbf{A}], [\mathbf{T}], \mathbf{W})$ = **outside** **then**
    $\mathrm{V_{out}} = \mathrm{V_{outn}} +$ Compute_BoxVolume$([\boldsymbol{p}])$
  else
    if Compute_BoxVolume$([\boldsymbol{p}]) > \varepsilon$ **then**
      $\mathcal{L} \leftarrow$ Cut$([\boldsymbol{p}])$
    endif
  endif
endwhile
$\mathrm{V}(\boldsymbol{c}) = \mathrm{V_{in}}$

In Algorithm 1, s is the space that the moving platform may involve in the rehabilitation process, ε is an artificial threshold. Function Compute_WM is used to calculate the interval matrix $[\mathbf{A}]$ under configuration $\boldsymbol{c}$ and in box $[\boldsymbol{p}]$. Function My_Judge is used to judge whether box $[\boldsymbol{p}]$ is inside the performance space. Function Compute_BoxVolume is used to calculate the volume of box $[\boldsymbol{p}]$. Function Cut is used to divide the box $[\boldsymbol{p}]$ into eight equal smaller box.

3.2 Configuration Optimization Based on Performance Space

Based on the optimization problem in Eq. (5) and the Algorithm 1, we use simulated annealing algorithm [21] to optimize the configuration of ACURR. After initialization, we keep changing $\boldsymbol{c}$ randomly, and accept the new $\boldsymbol{c}^*$ as the optimal configuration with the probability $\mathcal{P}$, where the mathematical definition of $\mathcal{P}$ is as follows:

$$\mathcal{P} = \begin{cases} 1, \mathrm{V}(\boldsymbol{c} + \Delta\boldsymbol{c}) < \mathrm{V}(\boldsymbol{c}) \\ \mathrm{e}^{-\frac{\mathrm{V}(\boldsymbol{c}+\Delta\boldsymbol{c})-\mathrm{V}(\boldsymbol{c})}{t}}, \mathrm{V}(\boldsymbol{c} + \Delta\boldsymbol{c}) \geq \mathrm{V}(\boldsymbol{c}) \end{cases}. \tag{15}$$

The current temperature t decreases with the progress of the iteration, so that in early iteration $\mathcal{P}$ is relatively large and the ability to search for the global optimal is strong. In later iteration, when it reaches the vicinity of the global optimal, $\mathcal{P}$ is relatively small to prevent jerk. The pseudo code is as follows:

Algorithm2. The pseudo to optimize configuration

```
Input : c0, cmin, cmax
Output : c*, V(c*)
initialization
while(k ≠ L)  do
  V(c) = Compute_PerformanceSpace(s, [W], [T], ε, c)
  r=Random(0,1)
  if  r < P  then
     c* = c
  endif
  for  j = 1 : 4  do
     Δh_j =Random(-1,1)
     h_j = |h_j + Δh_j| mod 2
   endfor
endwhile
V(c*) = Compute_PerformanceSpace(s, [W], [T], ε, c*)
```

In Algorithm 2, $\boldsymbol{c}_0$ is the initial configuration, $\boldsymbol{c}_{\min}$ and $\boldsymbol{c}_{\max}$ are the configurations where all the pulleys are at the lowest and highest position, $\boldsymbol{c}^*$ is the best configuration. L is the number of iterations. By using fuction Random(a, b), we can get a random number in the interval $[a, b]$.

4 Simulation

We take the upper-limb rehabilitation training of hemiplegia patients as the prototype for the design of the simulations. The dimension of the fixed frame in Fig. 1 is set as 3 m × 2 m × 2 m, and $h_{\min} = 0$ m and $h_{\max} = 2$ m. According to the physiological characteristics of the human upper limb, the motion of the upper limb on the $Y-Z$ plane is shown in Fig. 2, and two groups of s for different patients are designed as follows:

$$\mathrm{s}_1 = \{(x, y, z) | x \in [0.8, 2, 2], y \in [0.3, 1.7], z \in [0.7, 1.6]\} \tag{16}$$

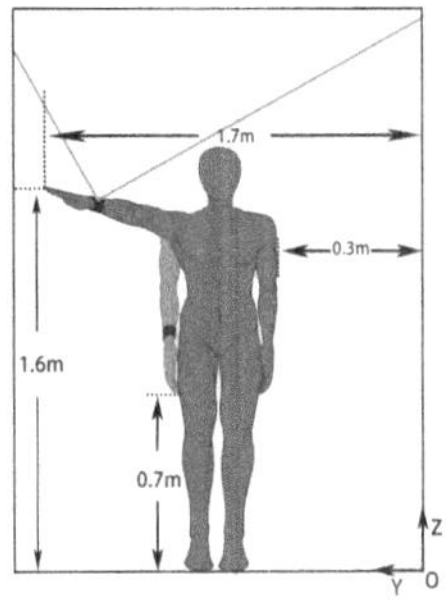

Fig. 2. The schematic diagram of rehabilitation treatment.

$$s_2 = \{(x, y, z)|x \in [0.8, 2, 2], y \in [0.3, 1.7], z \in [0.7, 1.3]\} \quad (17)$$

Moreover, to simulate the strength of the arm at different periods in the rehabilitation training, two groups of $\mathbf{W}$ are given: $\mathbf{W}_1 = [0, 5, 10]$, and $\mathbf{W_2} = [0, -5, -10]$. With $\mathbf{W}_1$, the ACURR generates a tractive force on the upper limb to make it easier to complete the rehabilitation training, while with $\mathbf{W}_2$, the ACURR generates a resistance force on the upper limb, thus more effort is needed to complete the training. In Eq. (5), $T_{\min} = 0\,\text{N}$, $T_{\max} = 100\,\text{N}$, and different optimizations are done with different ζ for comparison in the simulations. In Algorithm 1, the threshold $\varepsilon = 64 \times 10^{-6}\text{m}^3$. In Algorithm 2, the initial temperature $t = 100$ which drops exponentially as the iteration goes on, and the number of iterations at each temperature L= 100. $\boldsymbol{c}_0 = (1, 1, 1, 1)$, $\boldsymbol{c}_{\min} = (0, 0, 0, 0)$ and $\boldsymbol{c}_{\max} = (2, 2, 2, 2)$, the total number of iterations is set to 150. The simulations are carried out on Lenovo Xiaoxin 13 Pro (AMD Ryzen 7 4800U @1.8 GHz) with Matlab 2018b.

Table 1. Simulation results of different ζ, when $s = s_1$, $\mathbf{W} = \mathbf{W}_1$.

ζ	$h_1(m)$	$h_2(m)$	$h_3(m)$	$h_4(m)$	$V(m^3)$	Time(h)
10	1.9936	1.9831	1.9943	1.9985	0.1439	0.7305
15	1.9936	1.9926	1.9911	1.9982	1.0155	1.3442
20	1.9973	1.9967	1.9917	1.9942	1.3923	1.3837

Table 2. Simulation results of different ζ, when $s = s_1$, $\mathbf{W} = \mathbf{W}_2$.

ζ	$h_1(m)$	$h_2(m)$	$h_3(m)$	$h_4(m)$	$V(m^3)$	Time(h)
10	0.0019	0.0722	0.0329	0.1088	0.3592	0.9739
15	0.0433	0.0342	0.0321	0.0690	1.3032	1.2387
20	0.2581	0.3025	0.1030	0.0066	1.4638	1.2540

Table 3. Simulation results of different ζ, when s = s_2, $\mathbf{W} = \mathbf{W}_1$.

ζ	$h_1(m)$	$h_2(m)$	$h_3(m)$	$h_4(m)$	$V(m^3)$	Time(h)
10	1.9816	1.9995	1.9905	1.9971	0.1382	0.7718
15	1.9972	1.9962	1.9502	1.9787	0.9626	0.9657
20	1.9832	1.9691	1.7173	1.8682	1.0618	0.9783

Table 4. Simulation results of different ζ, when s = s_2, $\mathbf{W} = \mathbf{W}_2$.

ζ	$h_1(m)$	$h_2(m)$	$h_3(m)$	$h_4(m)$	$V(m^3)$	Time(h)
10	0.0026	0.0045	0.0057	0.0075	0.1395	0.8985
15	0.0164	0.0103	0.0159	0.0372	0.9619	1.1257
20	0.2431	0.1967	0.0064	0.0579	1.0683	1.1683

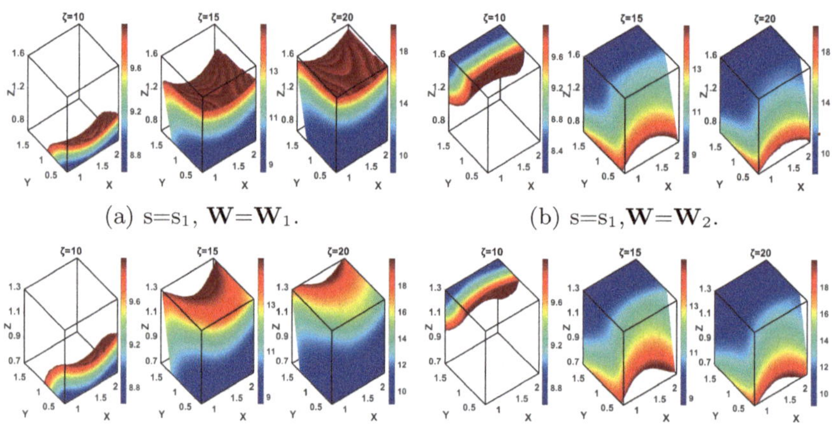

(a) s=s_1, $\mathbf{W}=\mathbf{W}_1$. (b) s=s_1, $\mathbf{W}=\mathbf{W}_2$.

(c) s=s_2, $\mathbf{W}=\mathbf{W}_1$. (d) s=s_2, $\mathbf{W}=\mathbf{W}_2$.

Fig. 3. The distribution of the performance space.

The Optimization results are given in Tables 1, 2, 3 and 4, and the distributions of the obtained corresponding performance space are also shown in Fig. 3. The black box means the reachable area of human's hand and it's defined in Eq. (16)–(17). The color changes in these figures represent the changes of f in Eq. (3). In Table 1 and Fig. 3(a), we first notice that the larger the performance index ζ is, the larger the performance space is and the longer the calculation time is. That is when ζ increases, the constraint of Eq. (4) is relaxed, and more poses can satisfy the constraint, and the performance space increases. In the optimization solution, when ζ increases, some boxes that can be directly determined outside the space become unable to be directly determined which need to be further subdivided, resulting in a corresponding increase in the calculation time.

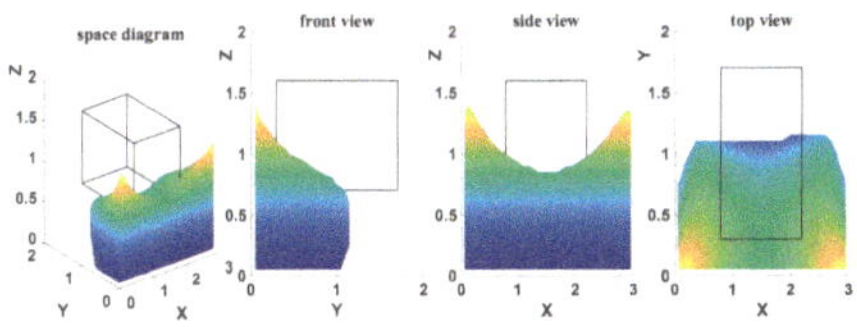

(a) $\zeta = 10, \mathbf{W} = \mathbf{W}_1$.

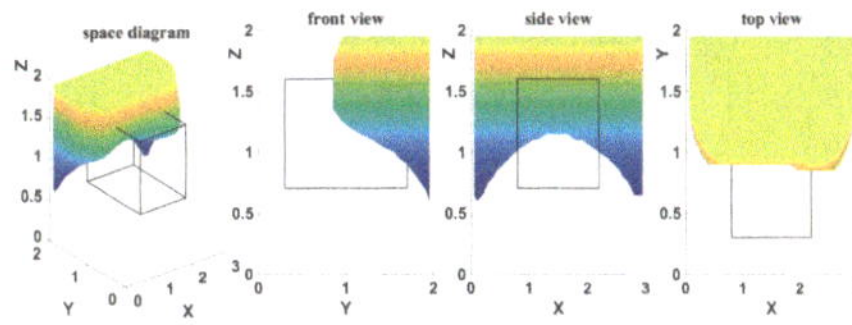

(b) $\zeta = 10, \mathbf{W} = \mathbf{W}_2$.

Fig. 4. Three views of the performance space under optimal configuration.

We also notice that, in Fig. 3(a) and Fig. 3(c), when $\mathbf{W}$ has an upward component of Z axis, f gradually increases from the bottom to the top in the space. The reason for this is that the lower the position of the moving platform is in the space, the greater the angle formed by the cable and the ground, and the smaller the tension is on the cable. When $\mathbf{W}$ has a downward component of Z axis, as shown in Fig. 3(b) and Fig. 3(d), similarly, f gradually increases from top to bottom. Just as we see in Table 1 and Table 2, in order for $\mathbf{W}$ to have an upward component, the cable must be above the arm, so h is large. Conversely, in order for $\mathbf{W}$ to have a downward component, the cable should be under the arm, and h is small. Besides, as we can see in Table 3 and Table 4, when $\mathbf{W}$ has the forward component of Y axis, h_3 and h_4 are lower than h_1 and h_2, so that the Angle between cable 3, 4 and the ground is smaller, resulting the cable tension in cable 3, 4 is smaller than the cable tension in cable 1, 2, which is more obvious when ζ is larger. Conversely, when $\mathbf{W}$ has the backward component of Y axis, h_1 and h_2 are higher than h_3 and h_4. However, we can see that there is no obvious height difference between h_3, h_4 and h_1, h_2 in Table 1. It is because that the s_1 is not symmetric in the whole space.

We further plot three views of the whole physical space ($x \in [0,3], y \in [0,2], z \in [0,2]$), $\zeta = 10$, with applied forces $\mathbf{W}_1$ and $\mathbf{W}_2$ under the corresponding optimal configuration, which is shown in Fig. 4. The black box means the space we give in Eq. (17). Compared with Fig. 4(a) and Fig. 4(b), When $\mathbf{W}_1 = -\mathbf{W}_2$, in the whole physical space, the performance space is symmetric about $(1.5, 1, 1)$, which also confirms that in Table 3 and Table 4. When the space s is symmetric about $(1.5, 1, 1)$, that is s_2, $\mathbf{W}_1 = -\mathbf{W}_2$, under the same ζ, the volume of the performance space with the optimal configuration is approximately equal. When the space s is not symmetric about $(1.5, 1, 1)$ and the whole space is close to $z = 2$, that is s_1, the performance space volume of $\mathbf{W}_2$ optimal configuration is always larger than that of $\mathbf{W}_1$ optimal configuration.

5 Conclusion

In this paper, we study an auto-reconfigurable cable-driven upper-limb rehabilitation robot, which can automatic change its configuration to meet the rehabilitation requirements. Based on the traditional wrench feasible workspace, the performance space of the ACURR is defined according to the further analysis of cable tensions. For different patients, we can adjust the performance index to make the rehabilitation treatment more comfortable. Then, we take the volume of the performance space as the optimization target to optimize the configuration of the robot by using simulated annealing algorithm. Based on this configuration optimization strategy, the robot can work in an optimal configuration in a variety of rehabilitation therapy scenarios. Finally, in the simulation experiment, the optimal configuration of the ACURR under different scenarios was solved, the performance space was plotted, and the influence of relevant parameters on the optimal configuration was analyzed. Future work will focus more on improving the efficiency of the algorithm and formulating appropriate control strategies to carry out actual rehabilitation training.

References

1. Major, K.A., et al.: Ranges of motion as basis for robot assisted poststroke rehabilitation. Hum. Vet. Med. **8**(4), 192–196 (2016)
2. Reinkensmeyer, D.J., Boninger, M.L.: Technologies and combination therapies for enhancing movement training for people with a disability. J. Neuroeng. Rehabil. **9**(17), 1–10 (2012)
3. Maciejasz, P., Eschweiler, J., Gerlach-Hahn, K., et al.: A survey on robotic devices for upper limb rehabilitation. J. Neuroeng. Rehabil. **11**(3), 1–29 (2014)
4. Zhang, B., Shang, W. W., Cong, S., and Li, Z.: Coordinated dynamic control in the task space for redundantly actuated cable-driven parallel robots. IEEE/ASME Trans. Mech. https://doi.org/10.1109/TMECH.2020.3038852
5. Wang, K., Yin, P., Yang, H., Wang, L.: Motion planning of rigid chain for rigid-flexible coupled robot. Int. J. Adv. Robot. Syst. **15**(3), 1–10 (2018)
6. Cafolla, D., Russo, M., Carbone, G.: CUBE, a cable-driven device for limb rehabilitation. J. Bionic Eng. **16**(3), 493–503 (2019)
7. Li, X., Yang, Q., Song, R.: Performance-based hybrid control of a cable-driven upper-limb rehabilitation robot. IEEE Trans. Biomed. Eng. **68**(4), 1351–1359 (2021)
8. Kang, J., Ghonasgi, K., Walsh, C.J., Agrawal, S.K.: Simulating hemiparetic gait in healthy subjects using TPAD with a closed-loop controller. IEEE Trans. Neural Syst. Rehabil. Eng. **27**(5), 974–983 (2019)
9. Gogu, G.: Structural Synthesis of Parallel Robots. Springer, Netherlands (2008). https://doi.org/10.1007/978-1-4020-5710-6
10. Gagliardini, L., Caro, S., Gouttefarde, M., Girin, A.: Discrete reconfiguration planning for cable-driven parallel robots. Mech. Mach. Theory **100**, 313–337 (2016)
11. Mao, Y., Agrawal, S.K.: Design of a cable-driven arm exoskeleton (CAREX) for neural rehabilitation. IEEE Trans. Robot. **28**(4), 922–931 (2012)
12. Lamine, H., Laribi, M.A., Bennour, S.: Structure optimization of the cable driven legs trainer. Mech. Mach. Sci. **85**(1), 691–698 (2018)
13. Zou, Y., Wang, N., Liu, K.: Design and analysis of movable cable-driven lower-limb rehabilitation robot. IEEE Access **7**, 2315–2326 (2019)
14. Hamida, I.B., Laribi, M.A., Mlika, A.: Multi-objective optimal design of a cable driven parallel robot for rehabilitation tasks. Mech. Mach. Theory **156**(104141), 1–24 (2021)

15. Laribi, M.A., Carbone, G., Zeghloul, S.: On the optimal design of cable driven parallel robot with a prescribed workspace for upper limb rehabilitation tasks. J. Bionic Eng. **16**, 503–513 (2019)
16. Gouttefarde, M., Daney, D., Merlet, J.: Interval-analysis-based determination of the wrench-feasible workspace of parallel cable-driven robots. IEEE Trans. Robot. **27**(1), 1–13 (2011)
17. Jaulin, L., Kieffer, M., Didrit, O., Walter, E.: Applied Interval Analysis. Springer, London (2001). https://doi.org/10.1007/978-1-4471-0249-6
18. Hansen, E.R.: Global optimization using interval analysis: the one-dimensional case. J. Optim. Theory Appl. **29**(3), 331–344 (1979)
19. Lhomme, O.: Consistency techniques for numeric CSPs. In: Proceeding of the 13th International Joint Conference On Artificial Intelligence, pp. 232–238. Morgan Kaufmann Publishers Inc., Chambery (1993)
20. Collavizza, H., Delobel, F., Rueher, M.: Comparing partial consistencies. Reliab. Comput. **5**, 213–228 (1999)
21. Boyd, S., Vandenberghe, L.: Convex Optimization. Cambridge University Press, Cambridge (2004)

Design of a Docking System for Cable-Driven Parallel Robot to Allow Workspace Reconfiguration in Cluttered Environments

Matthew Skopin[1(✉)], Philip Long[2], and Taskin Padir[1]

[1] Northeastern University, Boston, MA 02115, USA
{skopin.m,t.padir}@northeastern.edu
[2] Irish Manufacturing Research, Mullingar, Co. Westmeath, Ireland
philip.long@imr.ie

Abstract. Package couriers are exposed to a strenuous work environment, where up to one third of their time is spent loading and unloading vehicles manually. There is a clear need for robotic systems to augment the capacity of their vehicles while reducing costs and labor. The lightweight components and scalable workspaces of cable-driven Parallel Robots (CDPRs) make them a strong option for this application, however cable interference means deployment in cluttered environments is currently infeasible. In this paper, we propose a suspended CDPR consisting of a two stage end effector with a unique docking system. The design permits the end effector to latch itself into continuous docking struts fixed throughout the workspace, re-routing the cables and deploying a sub-CDPR which can operate freely between obstacles. The design of the mechanism and the robot are presented as well as its functional modes. Workspace analysis is conducted for both the docked and undocked configurations demonstrating the feasibility and utility of such a design.

Keywords: Cable-driven parallel robots · Collision avoidance · Mechanism design · Workspace analysis

1 Introduction

In spite of recent technological advances in autonomous vehicles, delivery of packages remains a largely manual process consisting of several sub-steps such as, loading, transport and unloading. Loading the truck is an ergonomically strenuous process, involving scanning, stacking, reorganizing, and optimizing to ensure that each parcel can be accessed at the appropriate destination. Typically, a driver is required to develop their own process and the mistakes committed optimizing this process costs time and effort. Meanwhile, package volumes continue to mount in firms like FedEx, seeing a 20% increase since 2016 [2]. There is a clear desire in the logistics industry to augment capacities while reducing costs, resulting in increased investment in diverse delivery systems [1].

M. Gouttefarde et al. (Eds.): CableCon 2021, MMS 104, pp. 158–169, 2021.
https://doi.org/10.1007/978-3-030-75789-2_13

An often overlooked aspect of the delivery operations is loading and unloading of the cargo area. Indeed, as much as 33.33% of a delivery person's times is spent on these operations [22]. In this paper, we propose deploying a cable-driven parallel robot (CDPR) in the cargo cabin to reduce the driver's ergonomic strain and optimize parcel placement. By implementing a robot and ceding the cargo volume as its work space, shelves and access pathways can be fully packed with cubical stacks of boxes, retrieved and presented to the delivery professional at each destination. However, this cluttered environment, in which the CDPR's workspace is slowly filled with objects, requires a unique, mechanical approach to avoid cable collision while still exploiting the cabin's volume.

1.1 Related Work

While the parallel structure of CDPR's enables the robot to cover a large workspace and achieve high accelerations, the possibility of cable collisions with the platform, other cables and o r dbstacles in the environment can greatly reduce the workspace. By using advanced path planning techniques such as grid based methods [12] or sampling algorithms e.g., *RRT* [4,24], collision free motions can be obtained. The complexity of the path planning increases with environment complexity and even the detection of these interference events requires significant computational effort [13]. Moreover, the feasible workspace is still defined by cable interference [3], indeed in [20] the robot's structure is designed explicitly for collision free workspace. An alternative method is to optimize the design of CPDRs, for instance the location of the base anchor points [8], for specific tasks. However, this method is limited to static environments. To overcome these limitations, the base anchor points can be relocated in order to change the robot's characteristics [11]. Discrete reconfiguration [9], displaces the base anchor points to optimize performance for a specific task at hand. In [7], a methodology for optimizing sequences of discrete reconfigurations of CDPR anchor points is proposed. The configuration is found through a graph search algorithm to enable the platform to follow a prescribed path. However discrete reconfiguration is limited to static environments with known parameters.

In [21], a CDPR consisting of three mobile cranes holding cable anchor points overhead is proposed for use in search and rescue environments. The anchor points for this application are moved via a telescoping boom on top of each mobile platform in order to enlarge the workspace and maintain equilibrium. While the booms allow the end effector to avoid obstructions of certain height and navigate around others, the additional planning complexity and size of the mobile bases limit the applications in which the system can be fielded. A CDPR using two mobile bases with four cables actuating a point mass end effector is proposed in [14,17], modifying workspace geometry by moving anchor points with mobile bases. This work has been extended to four mobile bases in [18] and performance extensively analysed in [16]. However, similar to the telescoping crane platforms, and indeed [5], the use of mobile bases requires additional footprint while increasing planning complexity. The floor space requirement and the need for dedicated pathways for mobile platforms render such systems inefficient for uses in cargo areas where space utilization is a critical criterion.

In this paper, we address the issue of cable interference with elements of the environment in which the CDPR is fielded rather than concerns with cable-on-cable interference. To achieve this, we propose a suspended cable-driven parallel robot consisting of a two stage end effector with a unique docking system suspended by eight overhead cables. Due to unidirectional force constraints associated with suspended CDPR's, we propose the use of latching mechanisms commonly referred to as *Push-Push Latches*. These systems enable connection and disconnection by applying a force in a single direction. It is desirable to avoid active methods which would increase the complexity of the end effector design while augmenting the weight. Passive mechanisms typically exploit a carotid (heart-shaped) path with a follower pin, or, in more complex cases, a series of 3 dimensional hinged cams that slip past each other in certain conditions to achieve the effect [15,23].

The proposed end effector latching method is inspired by the design of the retractable pen, due to effective transformation of a vertical translation into repeatable, incremented rotations of an inner assembly [19]. We propose a two stage end effector, where four of the eight suspending cables are threaded through to attach directly to the bottom stage. The top stage features four docking columns designed for passive, unpowered latching using upward force applied by the platform. This design permits the end effector to latch itself into continuous docking struts located on the top of the workspace. Once latched, the top stage is constrained allowing the smaller bottom stage to disengage. This creates what can be described as a smaller CDPR operating in a sub-workspace, in which the cable anchor positions are fixed to the docked top stage. This configuration allows the end effector to reach down into cluttered spaces by docking directly above the area of interest, constraining its cables to a narrow, workspace to avoid cable obstacle collisions. The docking columns exploit a unique repeating mechanism that turns an internal cam into one of two states. One stage permits the docking column to move into and out of the docking bracket, while the other keeps it captive and suspends the end-effector. The force needed to change docking column states can be adjusted, requiring as little as ten newtons of force depending on springs used.

The paper is structured as follows. Section 2 reviews the research methodology employed and presents the design details of the latching mechanism, the docking struts, and two stage end effector. Section 3 discusses feasible workspace and payload capacity for both configurations of the proposed CDPR. Section 4 draws conclusions and outlines forward future work.

2 Decoupling Mechanism Design

The goal is to maximize the workspace in cluttered environment. As such, two critical criteria have been identified:

- End effector - two stage end effector, minimizing friction in cable routing, held together by cable tension and gravity which can be released when top half is externally constrained.

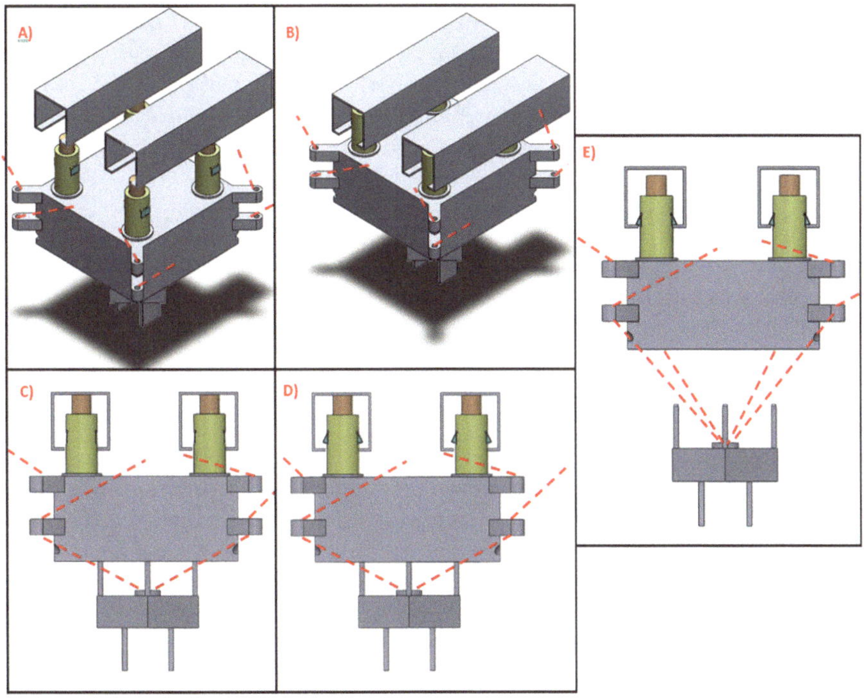

Fig. 1. Step by Step Docking. (A) End effector approaches ceiling (B, C) Buttons on each column are depressed (D) Depression triggers follower fins (E) lower stage of end effector is disengaged

– Decoupling mechanism - passive, low cost, repeatable and actuated with a one way input. Mechanism scales such that a larger assembly can be made to support heavier loads and end-effectors.

2.1 End Effector Docking

The proposed two stage end effector, docking columns, and ceiling-mounted docking struts work in together to achieve a narrow, column-shaped workspace when docked. An example step by step process is shown in Fig. 1.

The end effector is driven up into the cargo space ceiling, aligning the docking columns with the continuous struts (A), and depressing the buttons on top of each column using the internal face of the struts (B, C). These docking struts are placed at regular intervals along one axis of the work space such that the docking columns, as installed into the end effector, line up with any two struts. This action triggers a rotation of a cam within the column, pushing out the follower fins on the sides of column and effectively holding each column captive behind the opening of the docking strut (D). It should be noted that slackening the control cables at this point allows the lower stage of the robot end effector to disengage with the top stage, which is held up by the now captive docking columns (E). The force needed to depress the button, f_{latch} for its full travel and deploy/retract the fins is determined by the spring constant of the retaining spring used in the button component, $f_{latch} = 4k \cdot d$, where k is the spring constant and d the displacement. The placement of this spring is shown in Fig. 3, and its effect on system workspace illustrated in Sect. 3.

To aid in understanding, the steps are listed in order below:

1. The end effector drives up into the ceiling, the mounting columns entering the brackets and the bottom on top engaging with the internal surface of the bracket
2. The end effector is further driven up, depressing the button and turning the inner assembly, deploying fins and constraining the column in the bracket when driven lower.
3. The top stage of the end effector is now suspended by its now captive columns, permitting the lower stages to move independently in its own sub-workspace.
4. To disengage docking, the lower stage is driven up to meet the top stage, causing the top stage to be driven upwards and re-engage the column buttons.
5. By further driving the end effector up, the column button is depressed and the inner assembly turned once again, retracing the fins and permitting the end effector to travel the full workspace, so long as there is no cable interference.

The deployed mechanism operates in a narrow column sub-workspace as the cable routing holes in the top stage become the base anchor points for the new CDPR. This narrow workspace can be used to avoid cable collisions with objects in the larger cargo space, as demonstrated in Fig. 2.

2.2 End Effector Release

The release or undocking process is a reversal of the docking process such that it requires another push of similar force to trigger the docking column buttons. First, the bottom stage end effector is retracted back up to re-engage with the top stage, lifting the docking column's fins off the docking struts. This re-engagement is facilitated by sloped locating features on each end effector stage. The upward force from the cables again depresses the button on the top of the column, triggering another internal cam rotation and allowing the follower fins to retract. At this point, the control cables are slack and the columns are lowered from of the struts and resuming motion in the full size cargo workspace.

2.3 Docking Column Design

In the following, the design of the docking column's repeatable and passive mechanical response is detailed. The mechanism is inspired by the unique assembly of the retractable pen. As such, input is introduced to the system through the depression of the button on the top of the column when the end effector drives it into the docking bracket. During normal operation, the button is maintained

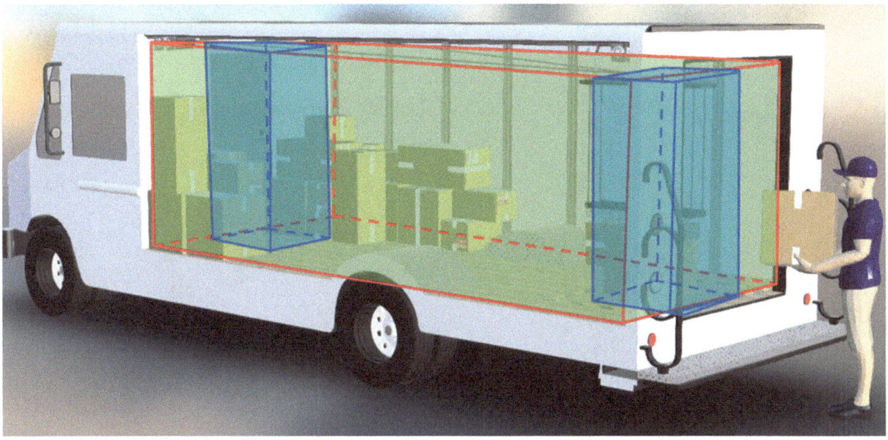

Fig. 2. Workspace Representation. The green box represents the undocked workspace, whereas the blue boxes represent workspaces for two different docked positions, subsspaces which can be accessed provided the end effector can latch.

at maximum height by a spring positioned at the top of the axle component. While in this position, the button component's base flange bears up against the column shell.

When the button is depressed, the four pawls positioned beneath the base flange meet the top ratchet teeth of the cam collar component, causing it to spin counter-clockwise on its seating geometry located on the axle component. The cam collar's geometry, designed to be a cam path, turns and pushes out into the fins. It should be noted that this collar component has two radial states, alternating between the two with each button depression. The fins are normally seated inside the shell, with a torsional spring at it's hinge bearing the fin's outer flange against the shell component. When the cam collar engages with the fins, they swing out of the shell, locking the column in its docking bracket. As the end effector descends, the button bounces back up to bear up against the shell, and a second group of four retaining pawls, as a result of four springs, push against the bottom ratchet teeth of the cam collar, turning the collar a couple degrees in order to align the button pawls with the next top ratchet tooth. At this point, the collar cam remains its second state, and the end effector has docked. A diagram of this mechanism and the positions of components throughout the docking process can be seen in Fig. 3 and Fig. 4.

The undocking process is completed with the same input. As the end effector is driven upwards into the docking brackets, the button depresses, turning the cam collar into its original state and permitting the fins to retract. When the button depresses, the retaining pawls reset the cam collar for it's next rotation, and end effector can slide out of the brackets and move freely.

The mechanism's use of springs and a simple docking geometry gives the column the advantage of complete passive actuation, requiring no electronic

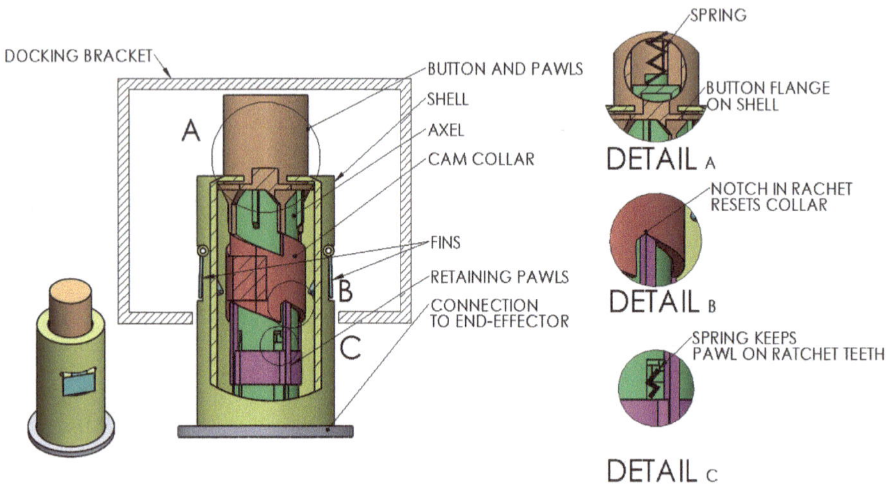

Fig. 3. Docking column mechanism diagram and component motion during docking

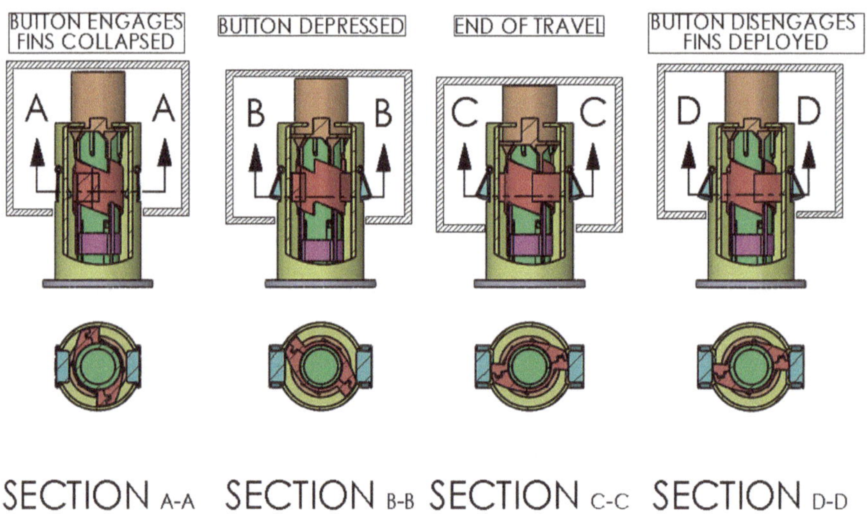

Fig. 4. Mechanism component motion during docking

communication or power to operate. In this way, the columns are sturdy, easy to repair, and do not require calibration.

The combination of the unique geometry of the bottom ratchet teeth of cam collar component and the retaining pawls allow the system to be repeatable and robust against unusual input. For example, a half button press will not cause the columns fins to change state: rather, the cam will only fully turn on a full extension. In such a case, the retaining pawls would force the cam collar to rotate back to its previous state.

3 Robot Analysis

To analyze the workspace, wrench feasibility is investigated. The wrench feasible workspace (WFW) is the set of poses for which the cables can exert a desired wrench on the platform [10]. A polytope, $\mathcal{P}$, can be defined as the volume bounded by a finite number of half-spaces ($\mathcal{H}$-representation)

$$\mathcal{P} = \mathbf{A}\mathbf{x} \leq \mathbf{b}, \tag{1}$$

where $\mathbf{A}$ contains the half-spaces' normals and $\mathbf{b}$ is the shifted distance from the origin along the normal. Thus the polytope encapsulating the feasible state of cable tensions, denoted $\mathbf{t}$, is given by

$$\begin{bmatrix} \mathbb{I}_n \\ -\mathbb{I}_n \end{bmatrix} \mathbf{t} \leq \begin{bmatrix} \mathbf{t}_{max} \\ -\mathbf{t}_{min} \end{bmatrix}, \tag{2}$$

where n is the number of cables, $\mathbb{I}_n$ is the $n \times n$ identity matrix and $\mathbf{t}_{max}$ and $\mathbf{t}_{min}$ denote the system's maximum and minimum cable tensions velocities, respectively. The equivalent polytope defined by its vertices is obtained using Fukuda's double description [6]. A vertex, $\mathbf{t}_i$ of the resulting ($\mathcal{V}$-representation) is transformed to the wrench space by

$$\mathbf{W}\mathbf{t}_i = \mathbf{w}_i, \tag{3}$$

where $\mathbf{W}$ is the $m \times n$ matrix mapping the vector of cable tensions $\mathbf{t}$ onto the m dimensional platform wrench $\mathbf{w}$. The polytope in wrench space representing all feasible wrenches, $\mathcal{W}$, is simply the convex hull of the transformed vertex set ($\mathcal{V}$-representation), i.e.,

$$\mathcal{W} = \{\mathbf{x} : \mathbf{x} = \sum_{i=1}^{2^n} \alpha_i \mathbf{w}_i \bigg| \alpha_i \geq 0, \sum_{i=1}^{n} \alpha_i = 1\}, \tag{4}$$

where $\mathbf{x}$ denotes any point inside $\mathcal{W}$.

To obtain the WFW, the space is discretized and at every point, we check if a set of desired wrenches lie inside $\mathcal{W}$. The effective workspace of the CDPR with docking brackets, docking columns, and two stage end effector is represented in two configurations: the large workspace available while undocked, and the smaller, tighter workspace available while docked and the stages are separated. The CDPR will shift between these configurations in order to complete pick and place tasks in a cluttered environment. In they undocked setting, $\mathbf{W}$ is 6×8 matrix and let a desired wrench, denote $\mathbf{w}^d$, comprise the total platform weight and the latching force required to overcome the spring resistance. in contrast in the docked case, $\mathbf{W}$ is 3×4 matrix $\mathbf{w}^d$ comprises the smaller platform weight and latching force. The analysis is shown for an example docking position in Fig. 5.

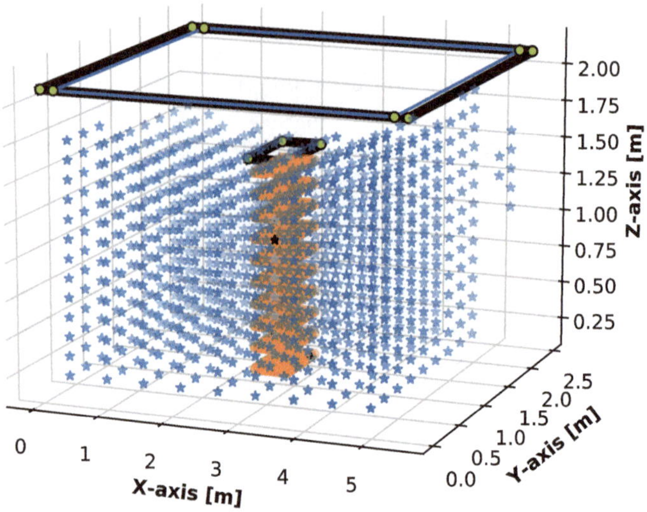

Fig. 5. Wrench feasible workspace analysis, where the blue points indicated wrench feasible for the larger system while the orange points denote the wrench feasibility for the sub-system. The yellow circles indicate the cable routing points. The size of the truck container is [5.5, 2.5, 2.1], with four attachment points in the corner and four offset ± 0.2 along the x axis and platform size [0.4, 0.3, 0.2] with attachment points in corners. Finally, the desired wrench for platform is $98N$ and who sub platform of $49N$, and maximum and minimum cable tension of 120 N and 2 N respectively.

It should be noted the docked workspace that Fig. 5 does not permit the bottom stage of the end effector to be lifted high enough to be flush with the top stage, in blue. For this reason, the dowel pin features on top of the bottom stage permit it to engage the top stage without having to leave its wrench feasible workspace. Aside for this consideration, the workspace is demonstrably sufficient for use in cluttered environments, as it can move freely within a column in excess of 0.5 m by 0.5 m, extending past a meter in height.

To further illustrate the usefulness of these narrow column shaped sub-workspaces, an example featuring the two stage end effector starting at idle position in (A), approaching (B), docking(C), and separating to approach a target location in between two tall obstructions (D) is outlined in Fig. 6.

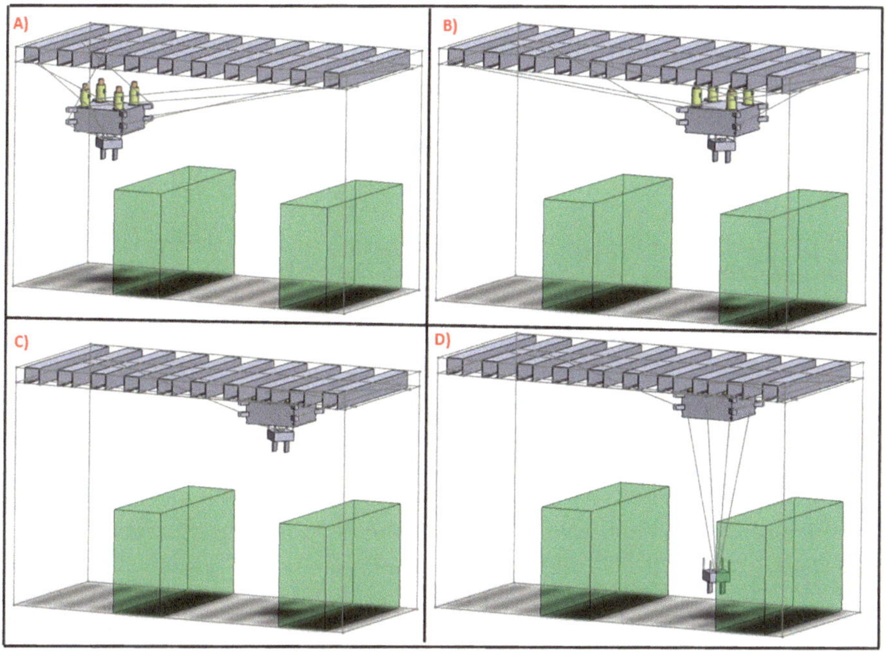

Fig. 6. Cable constraint example

4 Discussion and Conclusion

In order to meet the need for automating the strenuous loading and unloading operations for last-mile delivery couriers, we have developed a concept for a novel docking CDPR. The proposed system is capable of overcoming the constraints of a cluttered workspace while maintaining the breadth of workspace and lightweight components which make CDPRs effective. The docking CDPR consists of a two stage end effector with docking columns designed for use with simple docking brackets installed into the ceiling of the intended workspace. When the top stage is docked, the bottom stage can descend into a compact, column shaped sub-workspace eliminating the possibility of cable collision with the environment. To demonstrate the concept, a workspace analysis was conducted to ascertain wrench feasibility and payload capacity for both the larger workspace and the docked sub-workspace. To the best of our knowledge, this is the first time a system which enables passive reconfiguration during task execution has been proposed.

The undocked configuration can be thought of as a transition step between docked stations that can separated by high walls, stacks of materials, or other obstructions. The continuous rows of docking brackets permits these stations to be set up at any point along their length's direction, and in discrete rows based on the pair of brackets. For some applications, such as a delivery truck cargo space, vision or sensing can be introduced in order map a space, and avoid obstructions and stack and organize a cluttered space while minimizing wasted space.

The system allows continuous docking positioning in one direction and can be engaged and disengaged without actuation. However, there are limitations which will be investigated in future work. The docking brackets need to be placed high into the workspace such that they do not interfere with the cables. To ensure docking is feasible throughout the workspace, the height of the docking columns on the platform must be increased. This may alter the centre o f mass enough to lead to instability during the motions. Secondly, the system necessitates an overall frame and thus is suitable for fixed rather than deployable systems. In addition to this, we will develop an improved routing allowing additional cables through the top stage of the end effector. This will enable the bottom stage to be actuated with all 8 cables, rather than the existing 4 cables terminating in a point mass. The possibility of accelerating the end effector into it's docked position in order to reduce the torque needed to engage the docking columns is also being investigated, this will reduce the height of the columns currently required to avoid cable collisions during docking. This effect might be achieved by accelerating the end effector towards the mounting brackets, and making use of these dynamics to reduced the tension needed in the cables to dock. In addition, a 3D printed prototype for the docking column mechanism is currently in progress, which will be followed by end effector prototypes.

Acknowledgement. This material is based upon work supported by the National Science Foundation under Award No. 1928654.

References

1. Autonomous Delivery Robots Market – – – Growth, Trends, Forecasts (2020–2025). https://www.mordorintelligence.com/industry-reports/autonomous-delivery-robots-market
2. Topic: Courier, express and parcels (cep) market in the united states. https://www.statista.com/topics/4063/courier-express-and-parcels-cep-market-in-the-united-states/. Accessed 21 Apr 2021
3. Aref, M.M., Taghirad, H.D.: Geometrical workspace analysis of a cable-driven redundant parallel manipulator: KNTU CDRPM. In: 2008 IEEE/RSJ International Conference on Intelligent Robots and Systems, pp. 1958–1963. IEEE (2008)
4. Bak, J.H., Hwang, S.W., Yoon, J., Park, J.H., Park, J.O.: Collision-free path planning of cable-driven parallel robots in cluttered environments. Intell. Serv. Robot. **12**(3), 243–253 (2019)
5. Bosscher, P., Williams, R.L., Tummino, M.: A concept for rapidly-deployable cable robot search and rescue systems. In: International Design Engineering Technical Conferences and Computers and Information in Engineering Conference, vol. 47446, pp. 589–598 (2005)
6. Fukuda, K., Prodon, A.: Double description method revisited. In: Combinatorics and computer science, pp. 91–111. Springer (1996)

7. Gagliardini, L., Caro, S., Gouttefarde, M., Girin, A.: Discrete reconfiguration planning for cable-driven parallel robots. Mech. Mach. Theory **100**, 313–337 (2016)
8. Gagliardini, L., Caro, S., Gouttefarde, M., Wenger, P., Girin, A.: Optimal design of cable-driven parallel robots for large industrial structures. In: 2014 IEEE International Conference on Robotics and Automation (ICRA), pp. 5744–5749. IEEE (2014)
9. Gagliardini, L., Caro, S., Gouttefarde, M., Wenger, P., Girin, A.: A reconfigurable cable-driven parallel robot for sandblasting and painting of large structures. In: Cable-Driven Parallel Robots, pp. 275–291. Springer (2015)
10. Gouttefarde, M., Merlet, J.P., Daney, D.: Wrench-feasible workspace of parallel cable-driven mechanisms. In: Proceedings 2007 IEEE International Conference on Robotics and Automation, pp. 1492–1497. IEEE (2007)
11. Izard, J.B., Gouttefarde, M., Michelin, M., Tempier, O., Baradat, C.: A reconfigurable robot for cable-driven parallel robotic research and industrial scenario proofing. In: Cable-driven parallel robots, pp. 135–148. Springer (2013)
12. Lahouar, S., Ottaviano, E., Zeghoul, S., Romdhane, L., Ceccarelli, M.: Collision free path-planning for cable-driven parallel robots. Robot. Auton. Syst. **57**(11), 1083–1093 (2009)
13. Nguyen, D.Q., Gouttefarde, M.: On the improvement of cable collision detection algorithms. In: Cable-Driven Parallel Robots, pp. 29–40. Springer (2015)
14. Pedemonte, N., et al.: Fastkit: a mobile cable-driven parallel robot for logistics. In: Advances in Robotics Research: From Lab to Market, pp. 141–163. Springer (2020)
15. Quijano, J.J.: Heart shaped lock with sliding breakaway feature (Oct 28 2014), uS Patent 8,870,242
16. Rasheed, T., Long, P., Caro, S.: Wrench-feasible workspace of mobile cable-driven parallel robots. J. Mech. Robot. **12**(3) (2020)
17. Rasheed, T., Long, P., Marquez-Gamez, D., Caro, S.: Tension distribution algorithm for planar mobile cable-driven parallel robots. In: Cable-Driven Parallel Robots, pp. 268–279. Springer (2018)
18. Rasheed, T., Long, P., Roos, A., Caro, S.: Optimization based trajectory planning of mobile cable-driven parallel robots. In: The 2019 IEEE/RSJ International Conference on Intelligent Robots and Systems (IROS 2019), pp. 6788–6793 (2019)
19. Schultz, H.: Retractable pen (Jun 25 1974), uS Patent 3,819,282
20. Tadokoro, S., Matsushima, T., Murao, Y., Kohkawa, H.: A parallel cable-driven motion base for virtual acceleration. In: Proceedings 2001 IEEE/RSJ International Conference on Intelligent Robots and Systems. Expanding the Societal Role of Robotics in the the Next Millennium (Cat. No. 01CH37180), vol. 3, pp. 1700–1705. IEEE (2001)
21. Tan, H., Nurahmi, L., Pramujati, B., Caro, S.: On the reconfiguration of cable-driven parallel robots with multiple mobile cranes. In: 2020 5th International Conference on Robotics and Automation Engineering (ICRAE), pp. 126–130. IEEE (2020)
22. Wei, Y.: Feasibility analysis of parcel delivery pick-and-place system with cable driven parallel robots. Master's thesis, Gordon Institute, Northeastern University, Northeastern University (2019)
23. Weiland, N.R.: Push-push latch with clicker (Jun 19 2001), US Patent 6,247,733
24. Zhang, B., Shang, W., Cong, S.: Optimal RRT* planning and synchronous control of cable-driven parallel robots. In: 2018 3rd International Conference on Advanced Robotics and Mechatronics (ICARM), pp. 95–100. IEEE (2018)

Multi Objective Optimization of a Cable-Driven Robot with Parallelogram Links

Nasrollah Khodadadi[1], Mohammad Isaac Hosseini[1], S. Ahmad Khalilpour[1], Hamid D. Taghirad[1(✉)], and Philippe Cardou[2]

[1] Advanced Robotics and Automated Systems (ARAS), Faculty of Electrical Engineering, K. N. Toosi University of Technology, Tehran, Iran
taghirad@kntu.ac.ir

[2] Department of Mechanical Engineering, Robotics Laboratory, Laval University, Quebec City G1V 0A6, Canada

Abstract. In this paper, a new type of planar cable-driven parallel robot with parallelogram links is investigated. In such a design, the cables pass through multiple pulleys in order to make parallelogram links, instead of directly connecting to the moving platform. By using such configuration, one can greatly improve the robot kinematic performance without adding the cost of extra actuators. In this paper a multi objective optimization method is presented for this type of cable-driven robot using various kinematic indices. To this end, two planar robot configurations with parallelogram links are first introduced and the kinematic formulations are presented. Then, based on well-known kinematic indices like the size of the wrench closure workspace, the dexterity and the natural frequency, the optimal Pareto front of the design parameters is obtained. Finally, the above procedure is applied to the design of a planar cable-driven parallel robot and the optimal design is presented.

Keywords: Kinematic indices · Planar cable-driven robot · Parallelogram links · Multi objective optimization

1 Introduction

Since the late eighties, several desirable properties of cable-driven parallel robots (CDPRs) have stimulated the interest of researchers, namely, large workspace, fast motion, ease of assembly/disassembly, and low cost [1,2]. These features point to this class of manipulators as an exceptional choice in many real applications, such as large radio telescopes [3], rescue missions in dangerous environments [4], medical rehabilitation equipment [5] and video capture [6].

CDPR design poses important challenges, however, which restricts their use in many other applications. As an example, the need for actuation redundancy to keep the cables in tension is a challenge in many designs, [7,8], and increases the cost. To overcome this problem, in this paper, we present a new design that

M. Gouttefarde et al. (Eds.): CableCon 2021, MMS 104, pp. 170–181, 2021.
https://doi.org/10.1007/978-3-030-75789-2_14

eliminates the need for extra actuators. In this design, a parallelogram configuration is used rather than conventional cable links in the kinematic structure of the robot. Using this concept, the motion of the moving platform is constrained by redundant cables, while the number of actuators remains unchanged. Moreover, the robot's kinematic performances are vastly improved by employing transmission drums in the robot structure instead of conventional guiding pulleys.

The concept of parallelogram links in CDPRs has been previously developed for 3D spatial, suspended robots [9,10]. The authors in [11] have also proposed two new planar robots with parallelogram links and presented a general kinematic formulation for these robots. Comparing both proposed designs with corresponding traditional ones, authors in [11] demonstrate that by employing a parallelogram configuration, one can significantly improve the robot kinematic performance.

By assuming that the guiding pulleys have no friction, one can consider constant tension along with the cables [11]. In order to intentionally change the force along a cable, a novel idea is introduced in this paper. It consists in replacing the conventional pulleys with transmission drums, thus allowing to adjust cable forces by treating the transmission ratio of this drum as a design parameter in the optimization procedure.

Despite being accurately analyzed in terms of kinematic structure in [11], CDPRs with parallelogram designs have not yet been optimized due to their complicated structures. The main contribution of this paper is to modify the CDPR structure eliminate redundant actuators and present an optimal modified design based on the multi objective optimization techniques. The objectives of the optimization problem are selected such that they can adjust workspace volume as well as the accuracy and dexterity of the manipulator. In addition, the feasibility of manufacturing and mechanism performance are taken into account as other important factors in the manipulator design. Finally, the Pareto-optimal solution set is obtained and provided to the designer. Depending on the weights given to the design indices by the designer, one of the optimal Pareto points may be selected for real implementation.

The paper is organized as follows. Section 2 describes the cable configurations that are used to design two new planar CDPR structures. Section 3 is dedicated to the kinematic modeling of the proposed structures and to analyze the effect of employing transmission drums in the robot structure. Section 4 is devoted to the introduction of the kinematic performance indices including the workspace volume, robot dexterity, and the first natural frequency. Furthermore, optimization parameters are introduced in this section, the influence of these parameters on the objective functions is investigated, and the Pareto results are given. Finally, concluding remarks are given in Sect. 5. Note that in all the analysis, the cables are assumed as elements with negligible mass while their sagging is neglected. This assumption holds well for small -to medium- size CDPRs [12].

2 Parallelogram Configuration

In cable-driven parallel robots, the actuators are generally mounted on the base frame and the actuated cables from proximal anchor points are directly connected to the moving platform. In this conventional configuration, the end of each cable is simply fixed on a distal anchor point of the moving platform. Despite being a simple principle, this configuration requires redundancy in actuators to keep the cables in tension during the robot maneuvers. To overcome this problem, in this paper, we propose employing pulley systems that are guiding cables toward the moving platforms, instead of using a direct cable connection between distal and proximal anchor points. These pulley systems, shown in Fig. 1, can be classified into four categories as follows:

1. Π shape configuration: In this configuration, cables form a parallelogram structure connecting the actuator to the moving platform. This parallelogram link may be actuated with two drums. As shown in Fig. 1a, the two drums that are driven by one actuator are used to separately wind up cables, which are fixed on the moving platform. In this configuration, if the internal force in each cable is T, then the total cable force applied to the moving platform is $2T$, and the same force is applied to the double drum system.
2. U shape configuration: As shown in Fig. 1b in this configuration, in order to make a parallelogram link with only one actuated drum, one may arrange two pulleys mounted on the moving platform through which the cable is redirected to a fixed point on the base. In this design, if the internal force in this cable is equal to T, then the force applied to the moving platform is $2T$, twice the force applied to the drum. Thus, in the first type, an actuator with a larger torque output is needed to provide the same force for the moving platform, while higher velocity for the moving platform may be reached.
3. V shape configuration: This configuration is illustrated in Fig. 1c. As shown in Fig. 1c Unlike the U configuration, the cable coming out from the moving platform pulleys is not parallel to the incoming cable, while the end of the cable is also fixed on a point in the fixed frame.
4. $\mathbb{V}$ shape configuration: As shown in Fig. 1d, this configuration is like the V design except that the location of the fixing point, is placed on the moving platform rather than on the fixed frame. Therefore, a system of guiding pulleys is used in the base frame to change the cable direction and make parallelogram links.

Considering these different types of cable connections, in this paper, two planar CDPR structures with parallelogram links are investigated using configurations shown in Fig. 2. The first structure, which is shown in Fig. 2a, has a so-called $\mathbb{V}\mathbb{V}$ U structure, where two $\mathbb{V}$ configurations and one U configuration is used in this design. The first and second cables, which are respectively shown in blue and green, form two $\mathbb{V}$ configurations. The third cable, which is shown in red in Fig. 2a, forms a U configuration.

On the contrary, the second structure as shown in Fig. 2b consists of one V configuration and two $\mathbb{V}$. It is called V$\mathbb{V}\mathbb{V}$. In Fig. 2b, the first and third

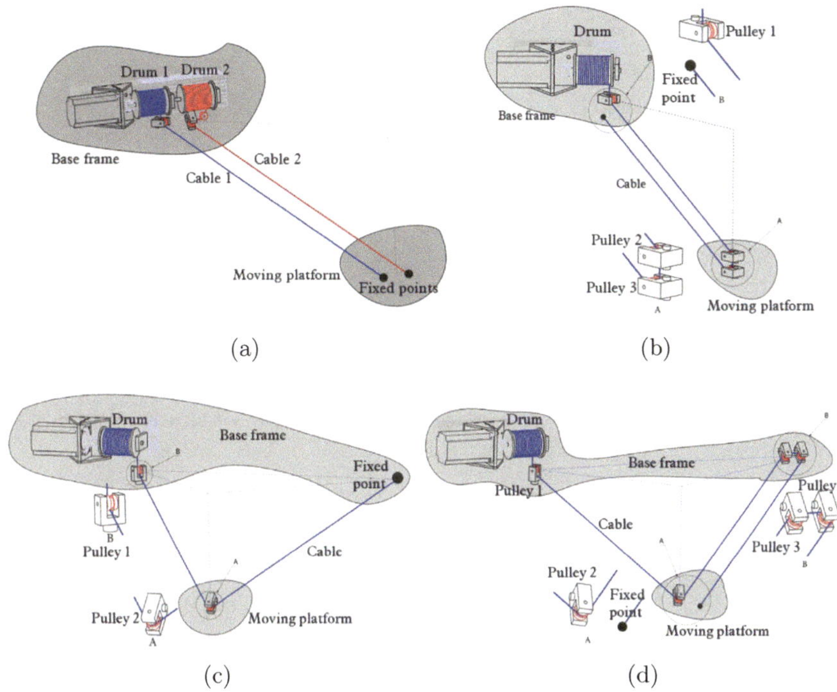

Fig. 1. Cable configurations which are used to design CDPRs with parallelogram link: (a) Π design (b) U configuration, (c) V structure, (d) $\mathbb{V}$ design.

cables are shown with blue and green colors, respectively, and both of them form $\mathbb{V}$ configurations. The kinetostatics and optimal design of these two types of parallelogram structures will be described in the following sections.

3 Kinetostatic Analysis

3.1 Structure Matrix Definition

This section is devoted to the derivation of the robot structure matrix, which maps the internal cable forces onto the task wrenches. As discussed in [11], the structure matrix for CDPRs with parallelogram links may be derived as follows:

$$\mathbf{A} = \mathbf{U}\mathbf{C}^T, \tag{1}$$

where $\mathbf{U} = [\boldsymbol{u}_1 \cdots \boldsymbol{u}_j \cdots \boldsymbol{u}_k]$ and $\boldsymbol{u}_j = \boldsymbol{l}_j / \|\boldsymbol{l}_j\|$. In this notation, $\boldsymbol{u}_j$ is a unit vector describing the direction of the jth cable, and the norm $\|\boldsymbol{l}_j\|$ is its the length. Furthermore, $\mathbf{C}$ is a transmission matrix mapping the actuator forces to the cable tensions [11]. The elements of $\mathbf{C}$ may be defined as $c_{i,j}$, $i = 1, \ldots, m$ and $j = 1, \ldots, k$. The value of $c_{i,j}$ is equal to one when the jth cable is connected to the ith actuator either directly or via some guiding pulleys. Otherwise, $c_{i,j}$ is

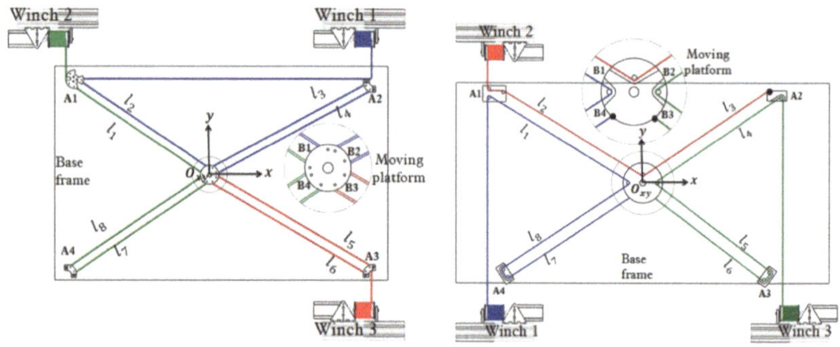

(a) (b)

Fig. 2. Planar CDPR with translational degrees of freedom structures with parallelogram links: (a) $\mathbb{V}\mathbb{V}$ U structure(b) $V\mathbb{V}\mathbb{V}$ structure.

equal to zero. For example, the structure matrices for the robots represented in Fig. 2a and 2b may be derived as

$$\mathbf{C}_{\mathbb{V}\mathbb{V}U} = \begin{bmatrix} 0\,1\,1\,1\,0\,0\,0\,0 \\ 1\,0\,0\,0\,0\,0\,1\,1 \\ 0\,0\,0\,0\,1\,1\,0\,0 \end{bmatrix}, \quad \mathbf{C}_{V\mathbb{V}\mathbb{V}} = \begin{bmatrix} 1\,0\,0\,0\,0\,0\,1\,1 \\ 0\,1\,1\,0\,0\,0\,0\,0 \\ 0\,0\,0\,1\,1\,1\,0\,0 \end{bmatrix}. \tag{2}$$

As stated in [11], the structure matrix is the transpose of the well-known Jacobain matrix, i.e. $\mathbf{J} = -\mathbf{A}^T$. The authors in [11] provide a method to calculate the structure matrices for CDPRs with parallelogram link structures.

3.2 Kinetostatic Analysis in Robots with Transmission Drums

In this part, we present the derivation of the transmission matrix when the robot is equipped with transmission drums. To this end, we redesign the CDPR structures displayed in Figs. 2a and 2b and put two transmission drums in the proximal anchor points. The modified designs are respectively shown in Figs. 3a and 3b.

As shown in these figures, transmission drums change the cable force with respect to the ratio of drum radii. This factor is called the transmission ratio which can be obtained as $c_e = \frac{r_1}{r_2}$, in which r_1 is the radius of the drum collecting the inserted cable and r_2 is the radius of the other drum. By this definition, the transmission matrices for the robots shown in Figs. 3a and 3b may be obtained respectively as

$$\mathbf{C}_{\mathbb{V}\mathbb{V}U} = \begin{bmatrix} 0\,1\,1\,c_e\,0\,0\,0\,0 \\ 1\,0\,0\,0\,0\,0\,1\,c_e \\ 0\,0\,0\,0\,1\,1\,0\,0 \end{bmatrix}, \quad \mathbf{C}_{V\mathbb{V}\mathbb{V}} = \begin{bmatrix} 0\,1\,1\,c_e\,0\,0\,0\,0 \\ 1\,0\,0\,0\,0\,0\,1\,c_e \\ 0\,0\,0\,0\,1\,1\,0\,0 \end{bmatrix}, \tag{3}$$

It is worth noting that the transmission drums may be mounted on the moving platform as well, but in such a design, the moving platform becomes larger, which might not be suitable in some applications.

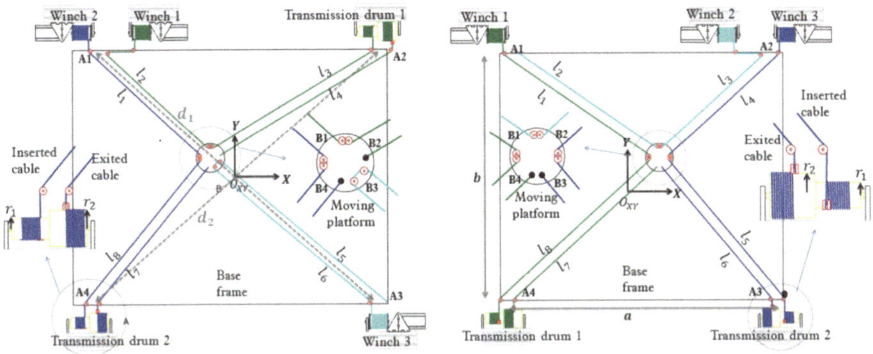

Fig. 3. CDPRs with parallelogram links and transmission drums: (a) modification of VV U structure with two transmission drums (b) promotion of VVV design with two transmission drums

4 Multi Objective Optimization

4.1 Design Indices

In this section, the design indices of CDPRs including the dexterity, the natural frequency, and the wrench-closure workspace size are briefly introduced. The analysis of these indices describes the kinematic performance of the robot in the controllable workspace. In the next section, the structural parameters of the robot are optimized by considering these indices as cost functions.

Dexterity Index. The dexterity index of a robot, which describes the uniformity of the robot's dexterity in its degrees of freedom, is defined as the condition number of the Jacobian matrix, $\kappa_{\mathbf{J}} = \sigma_{max}/\sigma_{min}$, where σ_{max} and σ_{min} denote the maximum and minimum singular values of the Jacobian matrix, respectively. Accordingly, the dexterity index ($\kappa_{\mathbf{J}}$) ranges from one to infinity, while its inverse ($\kappa_{\mathbf{J}}^{-1}$) varies between zero (singularity) and one (isotropy). Furthermore, $\kappa_{\mathbf{J}}^{-1}$ expresses only the local performance of the robot at a specific pose in the workspace. For this purpose, the global dexterity index (GDI) has been proposed by Gosselin et al. as: $GDI = \frac{\int_W \kappa_{\mathbf{J}}^{-1} dW}{\int_W dW}$, in which W denotes the volume of the robot workspace [13].

Natural Frequency. In cable robots, the cables may drastically vibrate when the system is excited near its natural frequency. Therefore, a higher value of this index leads to reduce the cable vibrations. In order to obtain the natural frequencies of the robot (ω_n) using the modal analysis, the following model may be considered [14]:

$$det[\mathbf{K}_p - \mathbf{M}\omega_n^2] = 0, \tag{4}$$

where $\mathbf{M}$ denotes the inertial matrix, and $\mathbf{K}_p$ is the stiffness matrix that is defined as [15]:

$$\mathbf{K}_p = \mathbf{J}^T \boldsymbol{\Omega} \mathbf{J}, \tag{5}$$

in which

$$\boldsymbol{\Omega} = diag(k_1, \cdots, k_n), \quad k_i = \frac{E_i S_i}{l_{T_i}}, \tag{6}$$

where E_i and S_i denote Young's modulus of elasticity, and the cross-section area of the ith cable, respectively. Furthermore, $\boldsymbol{l}_T = \mathbf{C}\boldsymbol{l}$ is the total length of the ith cable configuration [11]. The smallest value of ω_n obtained from Eq. (4) is considered to be ω_{n1}, which is configuration dependent in the robot workspace. Therefore, the global natural frequency (GNF) may be defined as GNF $= \min(\omega_{n1})$ over the whole workspace.

Wrench-Closure Workspace. The wrench-closure workspace (WCW) is an area of achievable workspace where the robot can apply any arbitrary force to its moving platform while all cables are in tension (without considering the external wrenches). Accordingly, this index is independent from the actuator characteristics and is determined only on the basis of the structure matrix and its null space. Given that all CDPR structures defined in this paper have one more actuator than degree of freedom, the null space of their structure matrices are spanned only by single vectors. Therefore, a pose of moving platform is in the WCW if and only if all elements of the null space vector have the same sign [16].

In this paper, in order to use the WCW as a global index which may be useful for optimal design of a CDPR, the achievable positions of the moving platform are determined through a grid of the CDPR workspace. In order to have a global index related to the size of the area of the WCW, the ratio of the number of controllable points to the total points of the workspace may be used.

4.2 Design Parameters and Objective Functions

This section is devoted to the optimization of the two robot designs illustrated in Fig. 3. The global kinematic indices defined in previous section are used as the objective functions of the optimization problem. All these kinematic indices are related to the structure matrix introduced in Eq. (1). As demonstrated before, this matrix depends on the position of the anchor points, and especially on the structure of the transmission matrix. In this paper, in order to simplify the optimization problem, the location of proximal anchor points is assumed to be on the corners of a symmetrical shape covering the robot's workspace. This symmetrical shape may be a rectangle for the VVV robot which is shown in Fig. 3b. Because of the symmetry of this CDPR structure with respect to Y, such a rectangle covers its whole workspace. The side length of this rectangle in the X and Y axes are labelled as a and b, respectively. Moreover, the area of this rectangle is considered to be one, while the ratio of a/b is chosen as an optimization parameter.

For the design of the VVU structure, which is illustrated in Fig. 3a, the line $Y = -X$ is the symmetry axis of the robot geometry. Therefore, the symmetric shape is a diamond whose corners are the proximal anchors. In this diamond, the length of diameters in the direction of $Y = -X$ and $Y = X$ are d_1 and d_2, respectively. Furthermore, the area of this diamond is constrained to be equal to one, and the ratio of d_1/d_2 is considered as an optimization parameter. The transmission ratio c_e, which is defined in Sect. 3.2, is used as another optimization parameter. For the sake of simplicity in the optimization problem and in order to preserve symmetry in the robot architecture, the drums are designed with identical radii. Depending on the construction constraints and implementation issues, the value of this radius is to be determined.

Figures 4 and 5 show the diagram of the variations of the kinematic indices with respect to the design parameters for both modified structures with transmission drums. In these figures for showing more detail in $\frac{a}{b} < 1$, and $\frac{d_1}{d_2} < 1$ the horizontal axis is divided into $log(a/b)$ for VVV structure, and $log(d_1/d_2)$ for VVU structure. The horizontal axes in these figures are of exponentially values in range $[-1.6, 1.6]$ which is correspondingly related to the variation interval of $[0.2, 5]$ for $\frac{a}{b}$ and $\frac{d_1}{d_2}$. This range is also used for the optimization procedure. The vertical axis is also allocated to changes of the c_e parameter in the range $[0, 2]$. On the other hand, bright colors in these figures represent the desired values for global design indices, while dark colors indicate the opposite of this issue. Furthermore, the level lines and their number are also displayed with black lines.

From Figs. 4 and 5, it can be concluded that:

1 Adding transmission drums to the CDPR structures with parallelograms links would improve kinematic indices. As can be seen in Fig. 4a and Fig. 5a, the GDI index for the first and second structures can be improved by selecting the ratio of transmission drums as $c_e = 0.5$ and $c_e = 0.6$. Furthermore, as illustrated in Fig. 4c and Fig. 5c, the area of the WCW can be increased by reducing the transmission ratios of both structures.
2 As shown in Fig. 4c and Fig. 5c, extending the robot along its symmetry expands the area of its WCW. It means that the VV U and VVV designs have a large workspace when d_1 and b are sufficiently large.
3 According to Fig. 4b and Fig. 5b, increasing the c_e may reduce the GNF index. In this situation, the robot is likely to become unstable due to a low natural frequency.

It should be noted that in most applications, a rectangular workspace for robots is more appropriate than other shapes. Therefore, the VVV design which has a symmetrical rectangular workspace is more suitable than the VVU structure offering a diamond workspace. Furthermore, comparing Figs. 4 and 5, it can be seen that the global kinematic indices in the second structure not only include better values but also provide a wider area than the design parameters for the robot. According to these facts, we aim to optimize VVV design in a multi-objective manner in the next section.

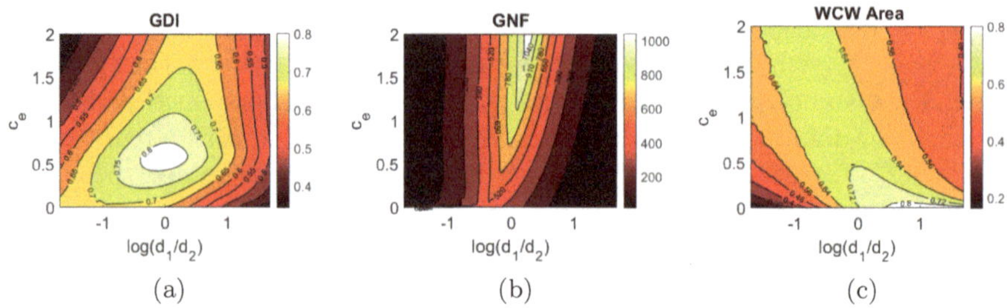

Fig. 4. Influence of the optimization parameters in: (a) GDI (b) GNF (c) WCW area for the VV U structure

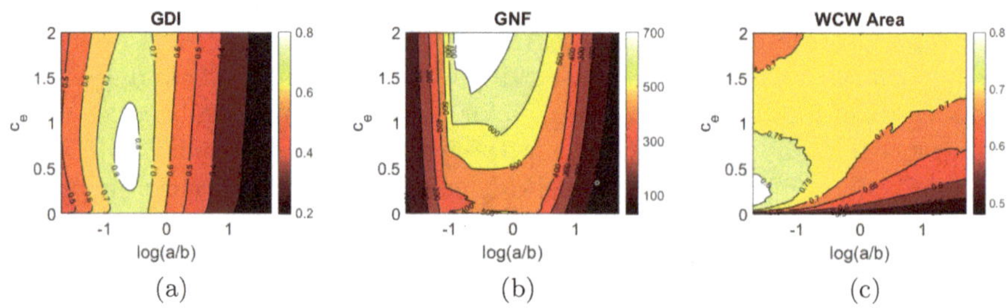

Fig. 5. Influence of the optimization parameters in: (a) GDI (b) GNF (c) WCW area for the VVV structure

4.3 Optimization Results

Some parameters need to be quantified before the optimization. These parameters include the moving platform mass, Young's modulus of elasticity, and the cross-section area of the cables, which are chosen as $m = 1\ kg$, $E = 1\ GPa$, and $S = 0.78\ \text{mm}^2$, respectively. The radius of the moving platform is also a constant parameter and equals to 70 mm. It should be noted that the robot structure used for the optimization is the modified VVV design shown in Fig. 3b. According to the optimization algorithm, the *paretosearch* function in MATLAB is used and the pareto size and the maximum iterations are set to 500 and 300, respectively. The Pareto front resulted from this algorithm is shown in Fig. 6. In this figure, the axes are kinematic indices GDI, WCW area and GNF and the points indicate the non-dominated optimal designs. Within the obtained pareto front, one may finally select an optimal design based on the practical considerations. This chosen point, which is shown with an arrow in Fig. 6, is selected by us such that the robot has a good GDI, that the wcw area approximately covers 75% of the robot workspace, and the natural frequency is as large as possible.

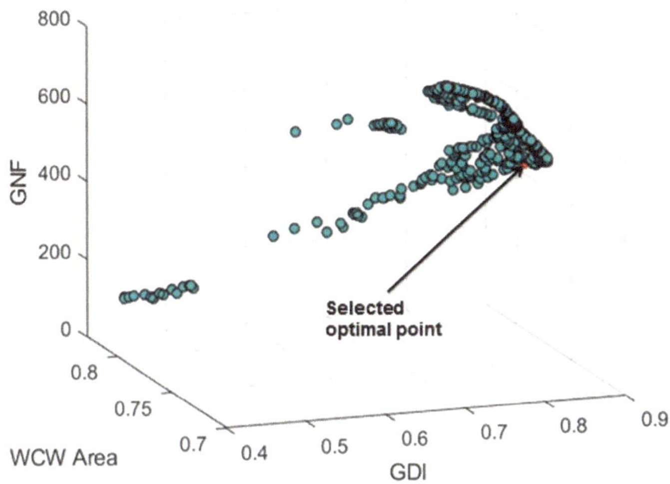

Fig. 6. The optimal pareto front obtained for the modified VVV.

The optimization parameters and the objective functions which are related to this optimal point are shown in Table 1. As demonstrated in this table, the ratio of a/b for the selected optimal point is about 0.5 which is close to the golden ratio [17]. This means that the length of the robot geometry in the Y direction should be twice as long as the length of that in the X direction. Furthermore, c_e is not equal to one in this design which shows that the transmission drums have a significant influence on the structure performance.

In Fig. 7, the dexterity index and ω_{n1} for the obtained optimal design are shown. As can be seen in Fig. 7a, the dexterity near the sides and center of the WCW workspace has suitable values, but deteriorates in the corners or near the proximal anchors. Moreover, it is obvious from Fig. 7b that in most parts of the workspace, the smallest natural frequency value is 1000 Hz but the GNF in the bottom proximal anchors is 518 Hz as mentioned in Table. 1.

Table 1. Parameters and objective functions of the selected optimal point.

Parameters		Objective functions		
a/b	c_e	GDI	GNF	WCW
0.5	0.63	0.84	518	0.75

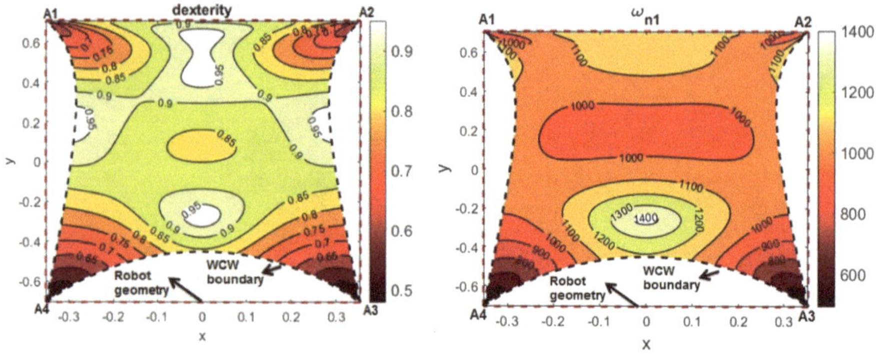

(a) (b)

Fig. 7. (a) Dexterity index (b) ω_{n1} in WCW for the optimal design described in Table 1.

5 Conclusions

In this paper, four cable configurations which are used to design CDPRs with parallelogram links were introduced. Based on these fundamental components, two planar structures were presented. These robots were equipped with transmission drums rather than proximal anchor pulleys in order to change cable force in different sections of a cable configuration. The ratio of these transmission drums and the dimensions of the robot are the parameters obtained through optimization. The objective functions considered in this problem are general kinematic indices such as GDI, GNF and the WCW area. Parameter selection of the geometrical shape of the manipulator and variation analysis of each index with respect to these design parameters is one of significant analysis given in this paper. This analysis is not only helpful to enlighten the design criteria and to analyze the dependency of each index to the design parameters, it can also be used for the validation of the optimization process. In order to balance between all the objectives, the optimal designs are derived from a pareto front computed through a multi-objective optimization algorithm. Finally, the solutions obtained are analyzed and, according to the relative importance of the design indices, a suitable solution is selected. Currently, this design solution is being used in the implementation of such a CDPR in the ARAS Laboratory.

References

1. Gosselin, C.: Cable-driven parallel mechanisms: state of the art and perspectives. Mech. Eng. Rev. **1**(1), DSM0004–DSM0004 (2014)
2. Hosseini, M.I., Harandi, M.J., Khalilpour, S.A., Taghirad, H.: Experimental performance of adaptive fast terminal sliding mode control on a suspended cable robot. J. Electr. Comput. Eng. Innov. (JECEI) (2018)

3. Zi, B., Duan, B., Du, J., Bao, H.: Dynamic modeling and active control of a cable-suspended parallel robot. Mechatronics **18**(1), 1–12 (2008)
4. Merlet, J.-P.: Marionet, a family of modular wire-driven parallel robots. In: Advances in Robot Kinematics: Motion in Man and Machine. Springer, pp. 53–61 (2010)
5. Zanotto, D., Rosati, G., Minto, S., Rossi, A.: Sophia-3: a semi adaptive cable-driven rehabilitation device with a tilting working plane. IEEE Trans. Robot. **30**(4), 974–979 (2014)
6. Khalilpour, S., Khorrambakht, R., Damirchi, H., Taghirad, H., Cardou, P.: Tip-trajectory tracking control of a deployable cable-driven robot via output redefinition. Multibody Syst. Dyn. 1–28 (2020)
7. Khalilpour, S., Loloei, A.Z., Masouleh, M.T., Taghirad, H.: Kinematic performance indices analyzed on four planar cable robots via interval analysis. In: 2013 First RSI/ISM International Conference on Robotics and Mechatronics (ICRoM), pp. 313–318. IEEE (2013)
8. Khalilpour, S.A., Taghirad, H.R., Masouleh, M.T., Shoorehdeli, M.A.: Multi-objective optimization of 6-degree-of-freedom cable-driven parallel robot using kinematic indices. J. Control, **7**(2). 43–56 (2013)
9. Filipovic, M., Kevac, L., Reljin, B.: Comparative analysis of two configurations of aerial robot. In: IEEE 10th Jubilee International Symposium on Intelligent Systems and Informatics, pp. 211–216. IEEE (2012)
10. Filipovic, M., Djuric, A., Kevac, L.: Influence of the construction type of a cable-suspended parallel robot on its kinematic and dynamic model. Sci. Tech. Rev. Mil. Techn. Inst. Belgrade, Serbia, ISSN **206**(1820) (2013)
11. Khodadadi, N., Hosseini, M.I., Khalilpour, S.A., Taghird, H.D., Cardou, P.: Kinematic analysis of planar cable-driven robots with parallelogram links. Submitted to CCToMM (2021)
12. Riehl, N., Gouttefarde, M., Krut, S., Baradat, C., Pierrot, F.: Effects of non-negligible cable mass on the static behavior of large workspace cable-driven parallel mechanisms. In: IEEE International Conference on Robotics and Automation, pp. 2193–2198. IEEE (2009)
13. Gosselin, C., Angeles, J.: A global performance index for the kinematic optimization of robotic manipulators (1991)
14. Diao, X., Ma, O.: Vibration analysis of cable-driven parallel manipulators. Multibody Syst. Dyn. **21**(4), 347–360 (2009)
15. Behzadipour, S., Khajepour, A.: Stiffness of cable-based parallel manipulators with application to stability analysis (2006)
16. Verhoeven, R.: Analysis of the workspace of tendon-based stewart platforms, Ph.D. dissertation, Universität Duisburg-Essen (2006)
17. Markowsky, G.: Misconceptions about the golden ratio. Coll. Math. J. **23**(1), 2–19 (1992)

Workspace Planning for In-Operation-Reconfiguration of Cable-Driven Parallel Robots

Felix Trautwein(✉), Thomas Reichenbach, Andreas Pott, and Alexander Verl

Institute for Control Engineering for Machine Tools and Manufacturing Units (ISW), University of Stuttgart, Stuttgart, Germany
felix.trautwein@isw.uni-stuttgart.de
http://www.isw.uni-stuttgart.de/

Abstract. This paper presents a planning method to systematically determine a geometry for a cable-driven parallel robot (short: cable robot) that satisfies a given workspace requirement. Furthermore, the method should be suitable for a fast workspace planning during In-Operation-Reconfiguration which implies that the algorithm should be executable fast, i.e. the aim is a computation time below one second. To do so, firstly the theoretical backgrounds for kinematics and workspace analysis of cable robots are introduced. Based on that, the parameter restrictions are modelled in terms of continuous closed sets, which are later used as bounds during the planning process. Afterwards, the theoretical problem of reconfiguration is presented and the solution strategy is proposed. In order to evaluate the given method, in the last section an exemplary reconfiguration situation is investigated, which shows that the method is suitable for the purpose to determine a cable robot for a given workspace requirement.

Keywords: Cable-driven parallel robot · Reconfiguration of cable robots · Design of cable robots · *COPacabana*

1 Introduction

A Cable-Driven Parallel Robot is a structural parallel robot, where cables are used to control the position and orientation of a moving platform. Due to cables as very lightweight components, cable robots exhibit several superior properties, such as high dynamics or a large and scalable workspace. The mostly known cable robot for industrial applications is the *Skycam* [1]. Beside that, there are further examples like *CoGiRo* [2] at *LIRMM*, the *Cablerobot Simulator* [3] at *Max-Planck* or the *IPAnema* [4] at *Fraunhofer IPA*. In addition to the above-mentioned advantages, cable robots offer a unique possibility of reconfiguration in contrast to conventional serial and parallel manipulators. Hereby, reconfiguration refers to the adaption of the existing robot to altered requirements. More specifically, the focus of this paper relies on geometrical In-Operation-Reconfiguration which aims at changing size and shape of the workspace while

M. Gouttefarde et al. (Eds.): CableCon 2021, MMS 104, pp. 182–193, 2021.
https://doi.org/10.1007/978-3-030-75789-2_15

largely maintaining payload and dynamical capabilities. Advantageous is, that a restart of the system can be avoided and the realizable properties of the robot can be varied during the runtime, which makes a cable robot even more flexible and modular. Consequently, this raises requirements, that all used hardware, software and control tools have to be flexible and adaptable. Practical measures of reconfiguration are for example adding or removal of actuated cables as well as altering positions of cable exit points on platform and frame. Two cable robots which are explicitly designed for reconfiguration experiments are *ReelAx* [5] and *COPacabana* [6], whereas the later one is applied in this paper as academic demonstrator system. The authors Zhou et al. [7,8] are investigating a redundant planar cable robot which winches are mounted on movable linear drives. Hereby, the authors are using this reconfiguration option to modulate stiffness and cable force distribution of the cable robot. Nguyen et al. presents in [9,10] approaches for the reconfiguration of cable robots during maintenance and repair work on aircraft components. In [10] the authors are presenting an approach for online reconfiguration of the cable robot. At this, the authors are using optimality with respect to the energy consumption along a trajectory as objective function to plan the reconfiguration of the cable robot during runtime of the system. Further essential works were presented by Gagliardini et al. in [11–13]. Hereby, the authors are focusing on planning a reconfiguration strategy for a cable robot with discrete reconfiguration options. This paper aims at presenting an approach for fast workspace planning for reconfiguration of cable robots. Therefore, Sect. 2 gives an overview of theoretical foundations and the used notation, followed by modelling of continuous reconfiguration options in Sect. 3. Based on the made definitions Sect. 3 concludes with transfer of the reconfiguration problem to a theoretical formulation, which is solved by the proposed solution strategy based on numerical optimization in Sect. 4. In order to show the suitability of given approaches Sect. 4 concludes with a simulation experiment, that studies a reconfiguration scenario of *COPacabana* robot and presents the results of the investigation. In the following section, kinematic and static foundations for cable robots are introduced briefly to ease the understanding of presented methods.

2 Theoretical Foundations

In Fig. 1, a structural model of a general CDPR is depicted [14]. Vector $\boldsymbol{a}_i$ describes the coordinates of the frames cable drawing point A_i given in coordinates of the global frame $\mathcal{K}_0$. Position $\boldsymbol{r}$ and orientation $\boldsymbol{R}$ are formulated as the displacement of the local platform coordinate frame $\mathcal{K}_P$ with reference to the global coordinate system. Starting from the origin of the platform local frame, the platform anchor B_i is addressed by the vector $\boldsymbol{b}_i$. The cable itself is expressed by the cable vector $\boldsymbol{l}_i$. With the given geometrical assumptions of straight-line cables, one can express a single kinematic loop as

$$\boldsymbol{l}_i = \boldsymbol{a}_i - \boldsymbol{r} - \boldsymbol{R}\boldsymbol{b}_i \,. \tag{1}$$

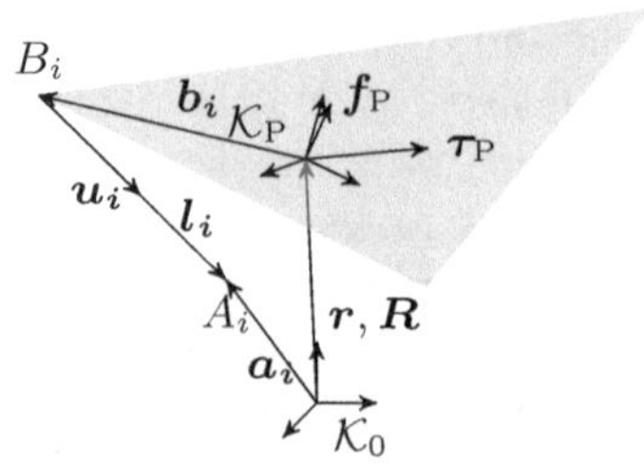

Fig. 1. Vector loop for kinematics of a general spatial cable robot

Hereby, the displacement between global and local frame is expressed by the rotation matrix $\boldsymbol{R}$. Using the Euclidean norm of the cable vectors, one can express the i-th unit vector pointing from the platform towards the frame as $\boldsymbol{u}_i = \frac{\boldsymbol{l}_i}{\|\boldsymbol{l}_i\|_2}$. At this point, the linear and angular static mechanical equilibrium can be stated as

$$\underbrace{\begin{bmatrix} \boldsymbol{u}_1 & \dots & \boldsymbol{u}_m \\ \boldsymbol{R}\boldsymbol{b}_1 \times \boldsymbol{u}_1 & \dots & \boldsymbol{R}\boldsymbol{b}_m \times \boldsymbol{u}_m \end{bmatrix}}_{\boldsymbol{A}^\mathrm{T}(\boldsymbol{r}, \boldsymbol{R})} \underbrace{\begin{bmatrix} f_1 \\ \vdots \\ f_m \end{bmatrix}}_{\boldsymbol{f}} + \underbrace{\begin{bmatrix} \boldsymbol{f}_\mathrm{P} \\ \boldsymbol{\tau}_\mathrm{P} \end{bmatrix}}_{\boldsymbol{w}_\mathrm{P}} = \boldsymbol{0}. \tag{2}$$

The collected unit directions of forces and moments are referred to as the structure matrix $\boldsymbol{A}^\mathrm{T}(\boldsymbol{r}, \boldsymbol{R}) \in \mathbb{R}^{n\times m}$. Its columns are scaled with the scalar cable forces f_i, which are collected to the vector of cable forces $\boldsymbol{f} \in \mathbb{R}^m$. Balanced with the acting external platform forces $\boldsymbol{f}_\mathrm{P}$ and moments $\boldsymbol{\tau}_\mathrm{P}$, the matrix sum is referred to as structure equation and is the starting point for numerous analyses and simulations of the properties of cable robots. The characterization of an usable workspace is a central aspect of the theoretical analysis of robots, and has the aim to describe a spatial domain, in what the robot can fulfil the requirements given by an application. In this work, the wrench feasibility criterion was chosen as a measure of whether or not a pose belongs to the workspace. The key property of the Wrench-Feasible-Workspace (WFW) relies on the proposition that all poses are part of the workspace in which a prescribed external wrench can be balanced by limited positive cable forces [15].

3 Modelling of Reconfiguration

The starting point for a reconfiguration of a cable robot is that the current system configuration can no longer cover a changed requirement. Thus, one of the essential question at reconfiguration of cable robots is to determine a geometry, that satisfies a requirement given by the application. Hence, the core problem of reconfiguration is closely related to the initial design of a cable robot. The main distinction between reconfiguration and design of cable robots is, that at reconfiguration, the robot already exists and the space of free design variables is strongly restricted due to the already existing hardware. In the following sections, it is described how these constraints are theoretically described and later transferred into a general problem formulation.

3.1 Reconfigurable Parameters

This section presents the applied modelling of the cable robot and the corresponding depiction of the geometrical reconfiguration options. Hereby, the standard model from Sect. 2 is applied, which leads us to the coordinate vectors for cable exit points as

$$\boldsymbol{a}_i = \left[a_i^x, a_i^y, a_i^z\right]^{\mathrm{T}}, \boldsymbol{b}_i = \left[b_i^x, b_i^y, b_i^z\right]^{\mathrm{T}} \in \mathbb{R}^3 \qquad \text{with} \qquad i = 1, \ldots, m. \tag{3}$$

Thereupon, the cable exit point vectors are put together as

$$\boldsymbol{g} = \left[\boldsymbol{a}_1^{\mathrm{T}}, \ldots, \boldsymbol{a}_m^{\mathrm{T}}, \boldsymbol{b}_1^{\mathrm{T}}, \ldots, \boldsymbol{b}_m^{\mathrm{T}}\right]^{\mathrm{T}}. \tag{4}$$

Thus, the geometry vector $\boldsymbol{g} \in \mathbb{R}^{6m}$ contains all coordinates which are defining the configuration of a cable robot according to the standard model. Subsequently, the vector $\boldsymbol{g}$ is split into two parts, stating as

$$\boldsymbol{g} = \bar{\boldsymbol{g}} + \Delta\boldsymbol{g}, \tag{5}$$

$$\bar{\boldsymbol{g}} = \left[\bar{\boldsymbol{a}}_1^{\mathrm{T}}, \ldots, \bar{\boldsymbol{a}}_m^{\mathrm{T}}, \bar{\boldsymbol{b}}_1^{\mathrm{T}}, \ldots, \bar{\boldsymbol{b}}_m^{\mathrm{T}}\right]^{\mathrm{T}}, \tag{6}$$

$$\Delta\boldsymbol{g} = \left[\Delta\boldsymbol{a}_1^{\mathrm{T}}, \ldots, \Delta\boldsymbol{a}_m^{\mathrm{T}}, \Delta\boldsymbol{b}_1^{\mathrm{T}}, \ldots, \Delta\boldsymbol{b}_m^{\mathrm{T}}\right]^{\mathrm{T}}. \tag{7}$$

Hereby, $\bar{\boldsymbol{g}}$ contains the reference geometry and $\Delta\boldsymbol{g}$ the corresponding relative displacement. For the next step it is assumed, that every component in $\boldsymbol{g}$ is part of a closed continuous set, which is exemplarily given for the x-component of the i-th kincmatic chain as,

$$a_i^x \in \left[\check{a}_i^x, \hat{a}_i^x\right], \qquad \text{with} \qquad \check{a}_i^x, \hat{a}_i^x \in \mathbb{R}. \tag{8}$$

Hereby, $\check{a}_i^x$ describes intervals lower bound and $\hat{a}_i^x$ upper bound, respectively. The introduced interval definition clearly applies for all components of the vector $\boldsymbol{g}$. Afterwards, the lower and upper bounds of every coordinate are collected in

$$\check{\boldsymbol{g}} = \left[\check{\boldsymbol{a}}_1^{\mathrm{T}}, \ldots, \check{\boldsymbol{a}}_m^{\mathrm{T}}, \check{\boldsymbol{b}}_1^{\mathrm{T}}, \ldots, \check{\boldsymbol{b}}_m^{\mathrm{T}}\right]^{\mathrm{T}} \qquad \text{and} \qquad \hat{\boldsymbol{g}} = \left[\hat{\boldsymbol{a}}_1^{\mathrm{T}}, \ldots, \hat{\boldsymbol{a}}_m^{\mathrm{T}}, \hat{\boldsymbol{b}}_1^{\mathrm{T}}, \ldots, \hat{\boldsymbol{b}}_m^{\mathrm{T}}\right]^{\mathrm{T}}. \tag{9}$$

As already mentioned, one substantial difference between initial design of cable robots and reconfiguration planning is that already existing hardware imposes restrictions on the set of geometrical parameters and not every parameter can be changed during reconfiguration. Taking this into account, the restricted set of reconfigurable parameters and respective bounds are defined as

$$\boldsymbol{g}_{\mathrm{r}} = \boldsymbol{S}\boldsymbol{g}, \qquad \check{\boldsymbol{g}}_{\mathrm{r}} = \boldsymbol{S}\check{\boldsymbol{g}} \qquad \text{and} \qquad \hat{\boldsymbol{g}}_{\mathrm{r}} = \boldsymbol{S}\hat{\boldsymbol{g}}. \tag{10}$$

where by the condition $\boldsymbol{g}_{\mathrm{r}} \subset \boldsymbol{g}$ holds true and the number of n_{r} reconfigurable parameters $\boldsymbol{g}_r$ are extracted from parameter vector $\boldsymbol{g}$ as subset with $\dim(\boldsymbol{g}_{\mathrm{r}}) < \dim(\boldsymbol{g})$. The selection matrix $\boldsymbol{S} \in \mathbb{R}^{n_{\mathrm{r}} \times 6m}$ is defined according to the hardware restrictions and has to be determined for each robot individually. A derivation of an exemplary selection matrix and thus the respective entries is given in Sect. 4.2. In order to express the recent and target state of the robot, the following section introduces workspace property and the workspace requirement, respectively.

3.2 Workspace Property and Requirement

In this work the decisive property is defined as the WFW of the robot, further properties like cable-cable collisions, cable-platform collisions or stiffness are not employed for reconfiguration planning. The workspace is expressed by the concept from [15]. Furthermore, the cable force calculation was carried out with Closed-Form-Method [16] and thereupon workspace is captured by the hull method [17]. This method relies on an ordered discretization of the search space and uses a line search algorithm for a fast and accurate determination of the workspace hull. In order to make both the workspace and requirement directly comparable, the same discretization approach was applied to the requirement, that is already used for the workspace. Since, we use the line search algorithm for border identification, actual shape of the body is irrelevant and can be chosen arbitrarily. As a result, each requirement body, regardless of the shape, is represented by a set of parametric lines and the respective centre point, see Sect. 3.3. But for sake of simplicity, in this work it is defined, that the required workspace is given by a cuboid, defined by position of its centre $\boldsymbol{r}_{\mathrm{req}}$, length l_{req}, width w_{req} and height h_{req}. Therefore, there are eight lines in total pointing from $\boldsymbol{r}_{\mathrm{req}}$ to every vertex of the cuboid. Furthermore, the described situation is illustrated in Fig. 2, where the current workspace of the cable robot does not satisfy the workspace requirement. Thus, the illustrated situation represents a motivation for reconfiguration, such that the adapted cable robot is able to cover the required cuboid target workspace. In this picture, the enveloping black frame is representing a steel frame on which the carriers of the pulleys can move freely. Hereby, the workspace of the robot is drawn by the green mesh. Consequently,

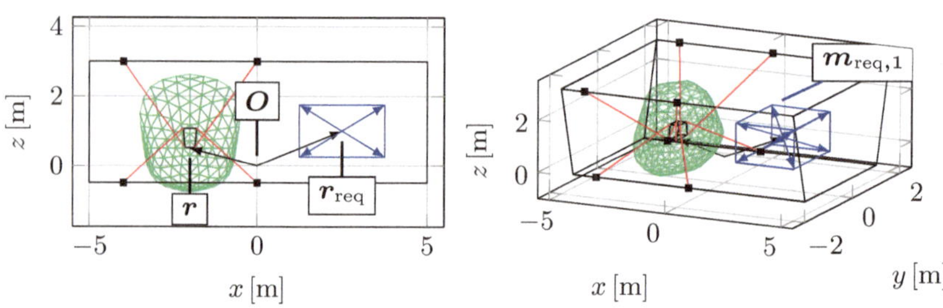

Fig. 2. An exemplary configuration of *COPacabana* robot, its current workspace and required target workspace.

the major challenge in this situation arises from finding a position of the pulleys, such that the blue cuboid is covered by the workspace entirely. Hence, to be able to overcome the challenge, the following section formulates the reconfiguration problem schematically.

3.3 Reconfiguration Problem

As we can see in Fig. 2 centre of actual workspace is displaced far away from given requirement and, additionally, the cuboid can not be covered by the cable robot in its current configuration. Based on these two essential observations, the process of reconfiguration is divided in two separate parts. Firstly, the robot structure as a whole has to be moved, such that the geometrical centres of workspace and requirement are as close as possible. After this step, workspace and requirement are matching roughly, however, it i s probable that the cuboid is not covered entirely by the workspace hull. In order to compensate the remaining difference in coverage of workspace and requirement the robot's internal geometrical relations have to be adapted next. For instance, a suitable measure is stretching or squeezing the robot by increasing or reducing the distance between opposite cable exit points. Based on these two steps, the following section defines the theoretical expression of translating and scaling the robot.

Translational Displacement. With reference to Fig. 2, it is defined that the robot's geometrical centre is expressed by the centre of the workspace hull $\boldsymbol{r}_{\mathrm{ws}}$. As the workspace is evaluated with the triangulation approach, the sought centre is determined by the average of all vertices as

$$\boldsymbol{r}_{\mathrm{ws}} = \frac{\sum_1^k \boldsymbol{v}_k}{n_{\mathrm{v}}} \qquad \text{with} \qquad k = 1, \ldots, n_{\mathrm{v}}. \tag{11}$$

where by the vector $\boldsymbol{v}_k$ represents a surface point and index variable k runs from first vertex $\boldsymbol{v}_1$ to last vertex $\boldsymbol{v}_{n_{\mathrm{v}}}$. Accordingly, the total number of vertices is given by $n_{\mathrm{v}} = 2^{2(n_{\mathrm{i}}+1)} + 2$ in terms of the number of iterations n_{i}. Consequently, the difference between centre's of current workspace Eq. (11) and requirement is expressed by the vector $\boldsymbol{d} = \boldsymbol{r}_{\mathrm{ws}} - \boldsymbol{r}_{\mathrm{req}}$ and the corresponding length $d = ||\boldsymbol{d}||_2$. Using the introduced definitions first part of reconfiguration problem is defined as functional mapping, that shifts the robot's structure, such that, distance measure is minimized. The function reads as,

$$\Delta \boldsymbol{g} = \boldsymbol{\Phi}^{\mathrm{t}}(\bar{\boldsymbol{g}}, \boldsymbol{g}_{\mathrm{r}}). \tag{12}$$

Hereby, the mapping $\boldsymbol{\Phi}^{\mathrm{t}} : \mathbb{R}^{2m} \times \mathbb{R}^{n_{\mathrm{r}}} \rightarrow \mathbb{R}^{2m}$ takes actual geometry $\bar{\boldsymbol{g}}$ and the reconfigurable parameters as inputs and gives back the sought geometry.

Inner Scaling. After the robot has been translatory shifted to the requirement centre, the next step is to adjust the geometry such that the entire requirement body is enclosed by the workspace hull. In order to check relation between requirement and workspace, the cuboid is discretized with parametric lines, that reads as,

$$\boldsymbol{m}_{\text{req},i} = \boldsymbol{u}_{\text{req},i}\lambda_{\text{req},i} + \boldsymbol{r}_{\text{req}} \qquad \text{and} \qquad \boldsymbol{m}_{\text{ws},i} = \boldsymbol{u}_{\text{ws},i}\lambda_{\text{ws},i} + \boldsymbol{p}. \tag{13}$$

At this, the i-th search direction is denoted with the unit vector $\boldsymbol{u}_{\text{req},i}$ and $\boldsymbol{u}_{\text{ws},i}$, respectively. The actual distance to the border of the requirement and workspace is defined by $\lambda_{\text{req},i}$ and $\lambda_{\text{ws},i}$. With definitions from Eq. (13) relation between both points are expressed as,

$$\boldsymbol{m}_{\text{req},i} - \boldsymbol{m}_{\text{ws},i} = \boldsymbol{u}_{\text{req},i}\lambda_{\text{req},i} + \boldsymbol{r}_{\text{req}} - \boldsymbol{u}_{\text{ws},i}\lambda_{\text{ws},i} - \boldsymbol{p}. \tag{14}$$

Since the centre of workspace and requirement were roughly matched during the elimination of translatory distance the following conditions $\boldsymbol{p} := \boldsymbol{r}_{\text{req}}$ and $\boldsymbol{u}_{\text{ws},i} := \boldsymbol{u}_{\text{req},i}$ are applying and Eq. (14) simplifies to

$$\boldsymbol{m}_{\text{req},i} - \boldsymbol{m}_{\text{ws},i} = \boldsymbol{d}_{\text{c},i} = \boldsymbol{u}_{\text{req},i}(\lambda_{\text{req},i} - \lambda_{\text{ws},i}). \tag{15}$$

Here, the difference vector $\boldsymbol{d}_c$ indicates for every search direction if the requirement is covered. The dimension of $\boldsymbol{d}_c \in \mathbb{R}^{n_{\text{req}} \times 3}$ depends on the number of search directions that were used to discretize the requirement body. Based on the introduced measure for requirement coverage, the functional mapping that adjusts the inner geometry is introduced as,

$$\Delta \boldsymbol{g} = \boldsymbol{\Phi}^{\text{s}}(\bar{\boldsymbol{g}}, \boldsymbol{g}_{\text{r}}). \tag{16}$$

So, the mapping $\boldsymbol{\Phi}^{\text{s}} : \mathbb{R}^{2m} \times \mathbb{R}^{n_r} \rightarrow \mathbb{R}^{2m}$ describes the function that stretches the robot until the requirement is covered. Following, introduced definitions and methods are summarized and finally the stepwise reconfiguration procedure i s presented.

Reconfiguration Procedure. During Sect. 2 and Sect. 3 foundations and basic methods about workspace description and the comparison with a given requirement were introduced. The following listing provides the essential steps to set up and solve a reconfiguration task for a cable robot.

1. Identify the reconfiguration options and define the selection matrix $\boldsymbol{S}$.
2. Determine requirement body according to application and then define discretization of requirement.
3. Set up translational function and determine the required shift.
4. Set up scaling function and determine the necessary scaling factors.
5. Validate if the determined geometry actually satisfies the workspace requirement.

In the following section, the introduced steps are applied to a real scenario, solved numerically and results are presented.

4 Solution of the Reconfiguration Problem

In the following, the made definitions and methods are summarized in the form of an optimization problem and the results obtained by a numerical solver are presented.

4.1 Solution Strategy

When we recall the definition from Eq. (7), the sought relative displacement $\Delta \boldsymbol{g}$ is replaced by the translational function Eq. (12) and the scaling function Eq. (16). Accordingly, the chained function calls are written as

$$\boldsymbol{g} = \bar{\boldsymbol{g}} + \boldsymbol{\Phi}^{\mathrm{s}} \circ \boldsymbol{\Phi}^{\mathrm{t}}(\bar{\boldsymbol{g}}, \boldsymbol{g}_{\mathrm{r}}). \tag{17}$$

The solution is determined by minimizing the functions $\boldsymbol{\Phi}^{\mathrm{t}}$ and subsequently $\boldsymbol{\Phi}^{\mathrm{s}}$. Hereby, the term chained refers to applying the translational displacement first and if the translational quality measure is minimized, the scaling function is applied on the set of reconfigurable parameters afterwards. In general, it is conceivable to choose all the reconfigurable parameters as independent and compute the solution with respect to all. However, we propose two further parametrization, which reads as,

$$\boldsymbol{g}_{\mathrm{r}} = \boldsymbol{\varphi}_{\mathrm{t}}(\boldsymbol{p}_{\mathrm{t}}) \quad \text{and} \quad \boldsymbol{g}_{\mathrm{r}} = \boldsymbol{\varphi}_{\mathrm{s}}(\boldsymbol{p}_{\mathrm{s}}). \tag{18}$$

The mappings $\boldsymbol{\varphi}_{\mathrm{t}}(\boldsymbol{p}_{\mathrm{t}}) \in \mathbb{R}^{n_{\mathrm{p,t}}}$ and $\boldsymbol{\varphi}_{\mathrm{s}}(\boldsymbol{p}_{\mathrm{s}}) \in \mathbb{R}^{n_{\mathrm{p,s}}}$ are taking a vector as input and distributes the entries component-wise on the reconfigurable parameters. Thus, by introducing these mappings it is possible to further decrease the number of design variables with respect to the hardware restrictions and additionally impose dependencies between the reconfigurable parameters to avoid obscure optimization results as well as robot geometries which are not applicable at the examined demonstrator system. Beside that, beneficial properties from initial design remaining untouched and are largely maintained through the reconfiguration planning. With the given information the final optimization problem is defined to determine the required translational displacement states as,

$$\begin{aligned} \min_{\boldsymbol{p}_{\mathrm{t}}} \quad & |\boldsymbol{d}(\boldsymbol{p}_{\mathrm{t}})|_2 \\ \text{s.t.} \quad & \check{\boldsymbol{g}}_{\mathrm{r},j} < \boldsymbol{\varphi}_{\mathrm{t}}(\boldsymbol{p}_{\mathrm{t}}) < \hat{\boldsymbol{g}}_{\mathrm{r},j} \qquad \text{with} \qquad j = 1, \ldots, n_{\mathrm{r}}. \end{aligned} \tag{19}$$

According to introduced definitions from Sect. 3.1 bounds on the parameters are implemented as constraints to the optimization variables. Following, the optimization problem for scaling of the robot reads as,

$$\begin{aligned} \min_{\boldsymbol{p}_{\mathrm{s}}} \quad & \sum_{k=1}^{n_{\mathrm{v}}} |\boldsymbol{d}_{c,k}(\boldsymbol{p}_{\mathrm{s}})|_2 \\ \text{s.t.} \quad & \check{\boldsymbol{g}}_{\mathrm{r},j} < \boldsymbol{\varphi}_{\mathrm{s}}(\boldsymbol{p}_{\mathrm{s}}) < \hat{\boldsymbol{g}}_{\mathrm{r},j} \qquad \text{with} \qquad j = 1, \ldots, n_{\mathrm{r}}. \end{aligned} \tag{20}$$

Here by, the objective functions consists of the single differences between requirement and workspace hull, which are summarized over all unit directions. Until this point, all definitions and declarations for a reconfiguration procedure were made and hereafter the formalism is applied to a reconfiguration scenario.

4.2 Simulation Experiment

The chosen theoretical setup was inspired and derived from existing *COPacabana* at *ISW* to assess suitability of the reconfiguration planning method to be implemented in the real system. For the simulation study a similar situation than illustrated in Fig. 2 is applied. Hereby, the requirement box is defined by $\boldsymbol{r}_{\text{req}} = [4\,\text{m}, 0\,\text{m}, 1\,\text{m}]^{\text{T}}$ and the dimensions $h_{\text{req}} = 1.5\,\text{m}$, $w_{\text{req}} = 1.5\,\text{m}$ and $l_{\text{req}} = 2.5\,\text{m}$. Additionally, it is assumed, that the anchor points on the frame are mounted on linear drives, such that, every cable exit point A_i can move freely between the frame's bounds. It is assumed, that the upper an lower bounds for the x directions of the frame exit points are $[\check{a}_i^x, \hat{a}_i^x] = [-7.5\,\text{m}, 7.5\,\text{m}]$. The assumed limits $[\check{a}_i^x, \hat{a}_i^x]$ are exceeding the given parameters from [6]. However, *COPacabana* was just used as an inspiration not to exactly represent the robot's dimensions. According to the proposed procedure from Sect. 3.3 next step is to determine the reconfigurable parameters, followed by defining the functions to translate and scale the robot. For that reason, the following listing provides essential steps for the exemplary system from Fig. 2.

- The selection matrix $\boldsymbol{S}$ is defined, such that, just the x-directions of every A_i is part of the vector $\boldsymbol{g}_{\text{r}}$.
- The translational displacement is implemented by moving all eight drives simultaneously. Accordingly, the relations are given as,

$$\varphi_t(p_{\text{t}}) = \boldsymbol{S}_{\text{t}} p_{\text{t}}, \quad \text{with} \quad \boldsymbol{S}_{\text{t}} = \begin{cases} 1 & \text{if} \quad g_i \in \boldsymbol{g}_{\text{r}} \quad \text{for} \quad i = 1, \ldots, 6m \\ 0 & \text{else.} \end{cases} \tag{21}$$

  Hereby, the optimization variable $p_t \in \mathbb{R}$ is just a scalar factor that shifts the structure as a whole. The vector $\boldsymbol{S}_t \in \mathbb{R}^{6m}$ ensures that the shift is added on all components from $\boldsymbol{g}_r = [a_1^x, a_2^x, a_3^x, a_4^x, a_5^x, a_6^x, a_7^x, a_8^x]^{\text{T}}$ simultaneously.
- Afterwards, the robot structure is now modified, such that, the exit points on the left and on the right are moving towards or away from each other. Therefore, the functions are reading as,

$$\varphi_{\text{s}}(\boldsymbol{p}_{\text{s}}) = \boldsymbol{S}_{\text{s}} \boldsymbol{p}_{\text{s}}, \quad \text{with,} \tag{22}$$

$$[\boldsymbol{S}_{\text{s}}]_1 = \begin{cases} 1 & \text{if} \quad g_i \in \{a_1^x, a_4^x, a_5^x, a_8^x\} \quad \text{for} \quad i = 1, \ldots, 6m \\ 0 & \text{else,} \end{cases} \tag{23}$$

$$[\boldsymbol{S}_{\text{s}}]_2 = \begin{cases} 1 & \text{if} \quad g_i \in \{a_2^x, a_3^x, a_6^x, a_7^x\} \quad \text{for} \quad i = 1, \ldots, 6m \\ 0 & \text{else.} \end{cases} \tag{24}$$

  So, the matrix $\boldsymbol{S}_{\text{s}} \in \mathbb{R}^{6m \times 2}$ is defined column-wise, whereby each column contains four entries with 1 marking the x direction of the respective cable exit point. Accordingly, vector $\boldsymbol{p}_{\text{s}} \in \mathbb{R}^2$ consists of two variables, which are distributed on both sides of the cable robot. Using the given definition, the signs of the entries of $\boldsymbol{S}_{\text{s}}$ are not important, because the sign of the final coordinates are determined by the results of the optimization procedure.

Beforehand, it is to mention that all presented results were achieved by an implementation in *Python2.7* using cable robots algorithms from the free library *WireX* [18] and optimization tools from *SciPy* [19]. For the numerical solution of the reconfiguration task, two different solvers for constraint optimization were tested. Firstly, a Trust Region Method (TRM) solver and secondly a Sequential Least Square Quadratic Programming solver (SLSQP) were applied to the problem. Both algorithms determined consistent and accurate results with stable convergency. Since, the computation with SLSQP solver was much faster, which leads to the decision to apply the SLSQP solver for the studied example. According to the definition of optimization problems in Eq. (19) and Eq. (20) limits of the steel frame are implemented as constraints on the Cartesian coordinates of the cable exit points. The computational results from solving the exemplary reconfiguration problem are illustrated in Fig. 3, whereas the resulting geometrical data is given in Table 1. The left image shows the cable robot that was moved in positive x-direction such that the distance between centre of workspace and requirement is minimal. Coming from there, the right picture illustrates the robot structure after the inner scaling of the robot. Here it is visible that the robot was stretched and finally the requirement box is cov-

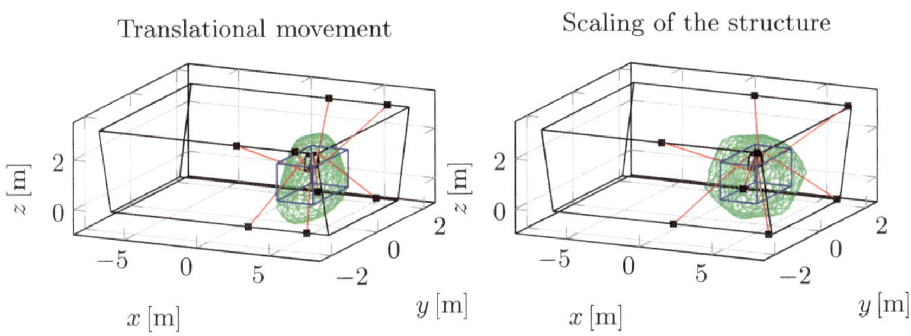

Fig. 3. Results of the reconfiguration planning method. Applied parameters for workspace computation are $f_{\min} = 50\,\text{N}$, $f_{\max} = 500\,\text{N}$ and $n_i = 3$.

Table 1. Geometrical parameters

Cable i [−]	b_i [m]	$a_{i,\text{trans}}$ [m]	$a_{i,\text{scaled}}$ [m]
1	$[-0.15, 0.24, 0.0225]$	$[2.00, 2.00, 3.00]$	$[0.73, 2.00, 3.00]$
2	$[0.15, 0.24, 0.0225]$	$[6.00, 2.00, 3.00]$	$[7.27, 2.00, 3.00]$
3	$[0.15, -0.24, 0.0225]$	$[6.00, -2.00, 3.00]$	$[7.27, -2.00, 3.00]$
4	$[-0.15, -0.24, 0.0225]$	$[2.00, -2.00, 3.00]$	$[0.73, -2.00, 3.00]$
5	$[-0.2, 0.15, 0.5735]$	$[2.00, 1.50, -0.50]$	$[0.73, 1.50, -0.50]$
6	$[0.2, 0.15, 0.5735]$	$[6.00, 1.50, -0.50]$	$[7.27, 1.50, -0.50]$
7	$[0.2, -0.15, 0.5735]$	$[6.00, -1.50, -0.50]$	$[7.27, -1.50, -0.50]$
8	$[-0.2, -0.15, 0.5735]$	$[2.00, -1.50, -0.50]$	$[0.73, -1.50, -0.50]$

ered by the workspace hull. For the given scenario, the procedure of finding the translatory distance between requirement took approximately $t_t \approx 0.1$ s and the determination of the required stretch was calculated in $t_s \approx 0.3$ s. Thus, the total computation time of the approach was mostly less than 0.5 s for a given configuration in different combinations of starting position and requirement box. The time durations were determined on an *Intel Core i7-7500U* CPU with 2.7 GHz and *Python2.7* 32 bit. Thus, even with non-optimized Python code, the presented method enables potential users to interactively redesign a cable robot.

5 Conclusion and Outlook

In this paper, a rapid strategy for determination of a cable robot's geometry for In-Operation-Reconfiguration is presented. At the beginning, continuous reconfiguration options and their theoretical depiction are introduced, followed by a description of the regarded workspace and target requirement workspace. Afterwards, Sect. 3 describes a reconfiguration as a two-step procedure and proposes a solution strategy, which is applied to a simulative reconfiguration situation that involves a workspace planning for *COPacabana* robot. This paper concludes with the simulation results, which are showing that it is possible to plan a workspace reconfiguration for an existing cable robot with parameter restrictions in a processing time below 0.5 s. Conclusively, we emphasize that the presented method can be applied in a framework for In-Operation-Reconfiguration as a tool to rapidly plan a new geometry of the robot according to a workspace requirement. This motivates us to further work in this direction. Thus, the next steps are to investigate a scenario where the robot is reconfigurable in two dimensions and the translation and scaling procedures have to be performed two times each and to improve computation times by analyses of the solver as well as optimizing the performance of implementation.

Acknowledgements. This work was supported by the German Research Foundation (DFG-project number: 317440765) at the University of Stuttgart.

References

1. Cone, L.L.: Skycam, an aerial robotic camera system. Byte Mag. **10**, 122–132 (1985)
2. Tempel, P., Hervé, P.-E., Tempier, O., Gouttefarde, M., Pott, A.: Estimating inertial parameters of suspended cable-driven parallel robots: use case on cogiro. In: 2017 IEEE/RAS International Conference on Robotics and Automation (ICRA 2017), pp. 6093–6098. IEEE (2017)
3. Miermeister, P., et al.: The cablerobot simulator large scale motion platform based on cable robot technology. In: 2016 IEEE/RSJ International Conference on Intelligent Robots and Systems (IROS 2016), pp. 3024–3029. IEEE (2016)

4. Pott, A., Mütherich, H., Kraus, W., Schmidt, V., Miermeister, P., Verl, A.: Ipanema: a family of cable-driven parallel robots for industrial applications. In: Bruckmann, T., Pott, A. (eds.) Proceedings of the First International Conference on Cable-Driven Parallel Robots, vol. 12 of Mechanisms and Machine Science, pp. 119–134. Springer, Berlin and Heidelberg (2012)
5. Izard, J.-B., Gouttefarde, M., Michelin, M., Tempier, O., Baradat, C.: A reconfigurable robot for cable-driven parallel robotic research and industrial scenario proofing. In: Bruckmann, T., Pott, A. (eds.) Proceedings of the First International Conference on Cable-Driven Parallel Robots Mechanisms and Machine Science, pp. 135–148. Springer, Berlin and Heidelberg (2012)
6. Trautwein, F., Reichenbach, T., Tempel, P., Pott, A., Verl, A.: Copacabana: ein modularer paralleler seilroboter. In: Pfurner, M., Dohnal, F. (eds.) Sechste IFToMM D-A-CH Konferenz 2020 (2020)
7. Zhou, X., Jun, S.-K., Krovi, V.: Tension distribution shaping via reconfigurable attachment in planar mobile cable robots. Robotica **32**(02), 245–256 (2014)
8. Zhou, X., Jun, S.-K., Krovi, V.: Stiffness modulation exploiting configuration redundancy in mobile cable robots. In: 2014 IEEE International Conference on Robotics and Automation (ICRA), pp. 5934–5939. IEEE, 31.05.2014–07.06.2014
9. Nguyen, D.Q., Gouttefarde, M.: Study of reconfigurable suspended cable-driven parallel robots for airplane maintenance. In: 2014 IEEE/RSJ International Conference on Intelligent Robots and Systems (IROS 2014), pp. 1682–1689. IEEE (2014)
10. Nguyen, D.Q., Gouttefarde, M., Company, O., Pierrot, F.: On the analysis of large-dimension reconfigurable suspended cable-driven parallel robots. In: 2014 IEEE/RAS International Conference on Robotics and Automation (ICRA 2014), pp. 5728–5735. IEEE (2014)
11. Gagliardini, L., Caro, S., Gouttefarde, M., Girin, A.: A reconfiguration strategy for reconfigurable cable-driven parallel robots. In: Robotics and Automation (ICRA 2014): 2014 IEEE/RAS International Conference on, pp. 1613–1620. IEEE (2014)
12. Gagliardini, L.: Discrete Reconfigurations of Cable-Driven Parallel Robots. Université Nantes-Angers-Le Mans, Nantes, France, Ph.D thesis (2016)
13. Gagliardini, L., Caro, S., Gouttefarde, M., Girin, A.: Discrete reconfiguration planning for cable-driven parallel robots. Mech. Mach. Theor. **100**, 313–337 (2016)
14. Pott, A.: Cable-Driven Parallel Robots: Theory and Application, vol. 120 of Springer Tracts in Advanced Robotics. Springer, Cham (2018)
15. Gouttefarde, M., Merlet, J.-P., Daney, D.: Wrench-feasible workspace of parallel cable-driven mechanisms. In: 2007 IEEE/RAS International Conference on Robotics and Automation (ICRA 2007), pp. 1492–1497 (2007)
16. Pott, A., Bruckmann, T., Mikelsons, L.: Closed-form force distribution for parallel wire robots. In: Kecskeméthy, A., Müller, A. (eds.) Computational Kinematics, pp. 25–34. Springer, Berlin and Heidelberg (2009)
17. Pott, A.: Efficient computation of the workspace boundary, its properties and derivatives for cable-driven parallel robots. In: Zeghloul, S., Romdhane, L., Laribi, M.A. (eds.) Computational Kinematics, vol. 50 of Mechanisms and Machine Science, pp. 190–197. Springer International Publishing, Cham (2018)
18. Pott, A.: Wirex – An open source initiative scientific software for analysis and design of cable-driven parallel robots
19. Virtanen, P., et al.: SciPy 1.0: fundamental algorithms for scientific computing in python. Nat. Methods **17**, 261–272 (2020)

Heuristic-Based Design Framework for the Cable Arrangement of Cable-Driven Parallel Robots

Yaoxin Guo(✉) and Darwin Lau

The Chinese University of Hong Kong, Sha Tin, Hong Kong
yxguo@mae.cuhk.edu.hk, darwinlau@cuhk.edu.hk

Abstract. It is important that the design of cable arrangements for the cable-driven parallel robot (CDPR) is a systematic process. Not only because of the unique inherent characteristic of cables, such as unilateral forces, but also the multiple objectives and requirements defined according to specific projects. Although some geometry-based and optimization methods have been studied, they are mostly suitable for specific cases with well-defined demands, where the parameters of the structure need to be designed repeatedly. In this work, a new heuristic-based design framework is proposed, comprising of a set of simple design heuristics to generate a set of designs to be selected through an assessment and evaluation framework. By using such a framework, the structure of CDPR can be effectively and efficiently designed for different needs.

1 Introduction

Cable-driven parallel robot (CDPR) is a type of parallel robot actuated by cables from a base frame connected to the end-effector on the other. This type of robot has attracted increasingly more attention in recent years, owing to their many advantages, such as low inertia, large potential workspace, high payload to weight ratio, simple robot structure and high ease in reconfiguration. These advantages have allowed a range of interesting applications, such as high-speed manipulation [1] and large-scale manipulation [2].

For CDPRs, the attachment points of the actuating cables, both at the base frame and the end-effector, are vital to the performance of the robot. Most importantly, the attachments significantly influence the robot's Jacobian matrix [3], and other robot properties such as the workspace, required maximum cable forces and stiffness. However, the determination of desirable attachment points is challenging for two reasons. First, it is not easy to define desired CDPR properties (*objective function*) as it varies depending on the project needs. Second, the determination of optimal cable attachments is particularly challenging due to the non-linear and complex relationship between the objective and attachments.

The determination of desirable cable attachment points can be broadly categorized as *geometric-based* and *optimization-based* methods. Geometric-based methods use the geometric understanding of the CDPR Jacobian, manipulability, workspace and stiffness [4,5] to determine the attachment points. Although

M. Gouttefarde et al. (Eds.): CableCon 2021, MMS 104, pp. 194–205, 2021.
https://doi.org/10.1007/978-3-030-75789-2_16

existing geometry-based methods can be used to design the fully constrained [6], under-constrained [7] and suspended [8] CDPRs, they are intuitive design strategies for specific needs rather than being a generalized and systematic framework.

Optimization-based design methods [9–11] formulate the cable attachment design as an optimization problem, where the attachments are searched for to optimize a mathematically defined objective function while satisfying various system constraint requirements. Widely used optimization methods in the design of CDPRs include Genetic Algorithms (GA) [12] and Particle Swarm Optimization (PSO) [13]. Although such methods are generically applicable to all CDPRs in a systematic manner, the non-linear and non-convex optimization problem is generally difficult to solve and may result in non-intuitive attachment locations.

In this paper, a new *heuristic-based* design framework for the cable attachments of CDPRs is proposed. The framework consists of two primary components: 1) a set of intuition-based heuristic design strategies that generate different candidate cable attachment designs; and 2) an evaluation framework to analyze and assess the performance of the heuristic generated cable attachment designs for selection. The ultimate selection of cable attachments can be done either manually by designer or automated through the performance assessments, such as the *Volume Percentage* of the static workspace, maximum cable force, tension factor and so on. To illustrate the concept of the heuristic-based design, several example heuristics are presented along with the evaluation framework for the candidate designs. Simulations were conducted for the design of a planar CDPR to illustrate the usage of the proposed framework. It should be noted that this work provides the framework ideas and the heuristics included are simply examples, which only consider the planar two degree-of-freedom CDPR with the *Volume Percentage* of static Workspace and maximum cable force, but can be applied to any type of CDPR and consider an expanded library that is comprehensive for different project needs.

2 Background of CDPR

Figure 1 shows the general model of a spatial cable-driven parallel robot actuated by m cables, where $\mathbf{r}_{\mathrm{OE}}$ is the translation from the origin of the base frame $\{F_0\}$ to the origin of the end-effector frame $\{F_{\mathrm{E}}\}$. For cable i, $\mathbf{r}_{\mathrm{Ai}}$ and $\mathbf{r}_{\mathrm{Bi}}$ represent the cable attachment at the base and the end-effector in $\{F_{\mathrm{E}}\}$, respectively. For a spatial CDPR, the Jacobian matrix, which is a function with respect to the CDPR pose $\mathbf{q} \in \mathbb{R}^6$ and the attachment points $\mathbf{r}_{\mathrm{Ai}}$ and $\mathbf{r}_{\mathrm{Bi}}$, can be expressed as

$$J(\mathbf{q}) = \begin{bmatrix} \hat{\mathbf{l}}_1(\mathbf{q}) & \mathbf{r}_{\mathrm{B1}}(\mathbf{q}) \times \hat{\mathbf{l}}_1(\mathbf{q}) \\ \vdots & \vdots \\ \hat{\mathbf{l}}_{\mathrm{m}}(\mathbf{q}) & \mathbf{r}_{\mathrm{Bm}}(\mathbf{q}) \times \hat{\mathbf{l}}_{\mathrm{m}}(\mathbf{q}) \end{bmatrix}, \tag{1}$$

where $\hat{\mathbf{l}}_{\mathrm{i}}(\mathbf{q})$ is the unit vector of the cable vector $\mathbf{l}_{\mathrm{i}}(\mathbf{q}) = \mathbf{r}_{\mathrm{OE}}(\mathbf{q}) + \mathbf{r}_{\mathrm{Bi}}(\mathbf{q}) - \mathbf{r}_{\mathrm{Ai}}$ for cable i. The equation of motion for such a general cable driven manipulator can be expressed as

$$M(\mathbf{q})\ddot{\mathbf{q}} + C(\mathbf{q}, \dot{\mathbf{q}}) + G(\mathbf{q}) = -J^{\mathrm{T}}(\mathbf{q})\mathbf{f} \tag{2}$$

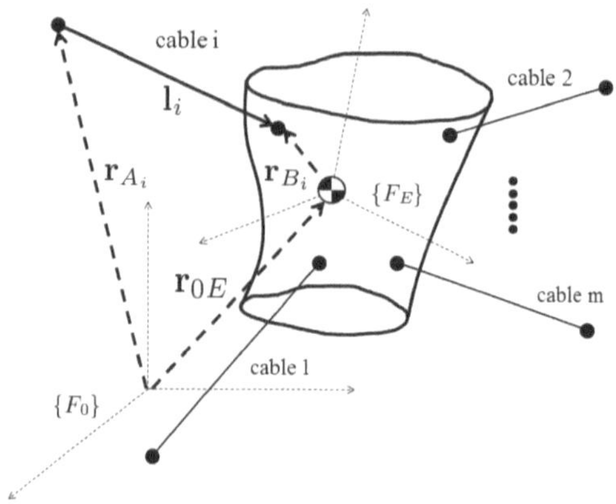

Fig. 1. General Model for Cable Driven Manipulator

where M is the mass inertia matrix, C is the centrifugal and Coriolis force vector, G is the gravitational vector, and the cable force vector is $\mathbf{f} \in \mathscr{R}^m$. From the Eq. (1), it can be observed that the cable attachment points $\mathbf{r}_{\mathrm{Ai}}$ and $\mathbf{r}_{\mathrm{Bi}}$ will greatly influence the CDPR's Jacobian matrix and hence the equations of motion (2).

3 Heuristic Design Framework for CDPR

The proposed heuristic design framework for CDPR is a complete design process to determine the cable attachment points on the base. As shown in Fig. 2, the design process is comprised of the following steps:

1. The requirements and specifications of the project are defined based on the particular project or application. These include information such as the site boundaries where cable attachments must lie within and the expected locations the CDPR is required to operate at.
2. The requirements become the input into the library of k heuristic algorithms, which then generates a set of k candidate CDPR designs (cable attachments and maximum cable forces to allow the selection of actuating motors).
3. The candidate CDPR designs are then evaluated and assessed using different criteria, such as maximum cable force, workspace volume and so on, in which a desired design is selected.
4. The selected design can then be re-evaluated to ensure that it meets all requirements defined by the designer or user. If the requirements are met, then the design is a suitable and favourable final CDPR design.

As will be described in Sect. 4, the library of heuristics is important and is responsible in generating a set of candidate CDPR designs. Under this framework, each heuristic can be a simple but intuitive design strategy. The simplicity

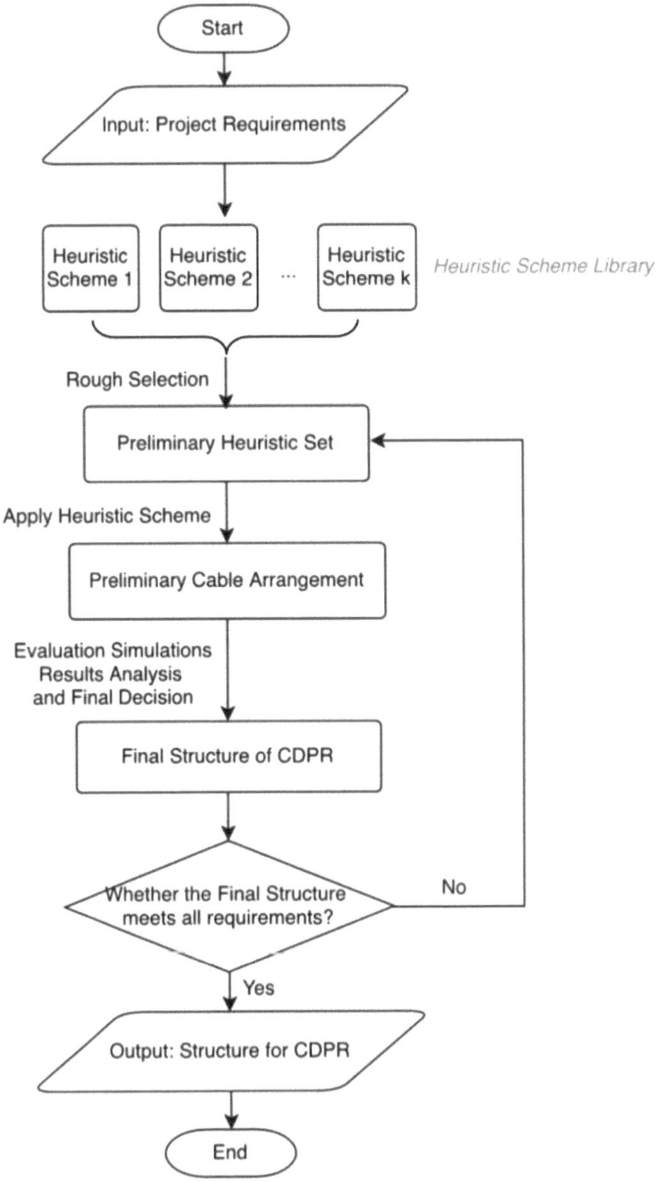

Fig. 2. Heuristic design framework for CDPR

is acceptable as each heuristic may only need to be effective under particular circumstances, rather than an "one size fits all" complex design strategy that needs to be effective on all scenarios and applications.

Moreover, it is important that the formation of the heuristic library can cover sufficiently different desirable goals within the design of CDPR. Furthermore, it is expected that this library can be expanded by designers based on their experiences over time, creating a convenient to use design library. As such, the set of heuristics proposed in this work simply serve as an example on the types of heuristics that can be included and should be expanded upon.

4 Heuristics Library

The creation of a library of heuristic scheme is essential for the heuristic design framework. Each heuristic is responsible for generating the locations of cable attachments based on the requirements. In order to maximize the advantages of this design method, a new heuristic should have good performance for certain characteristics of CDPRs and application needs. It needs to be emphasized that the heuristic does not refer to a structure for an independent project, each heuristic is a design idea. For each heuristic, different number of cables can be considered using the same design idea. The result will contain the number of cables and the corresponding attachment points.

For the purpose of illustration, a 2-DoF (degree-of-freedom) planar CDPR with four heuristic schemes will be considered to explain how to define a heuristic. In this work, only the *Volume Percentage* of the static workspace and maximum cable force are taken into consideration as example methods to evaluate the scheme. However, in general, other characteristics, such as stiffness, and interference free regions, should be included for more comprehensive evaluation of the heuristics. Furthermore, an ideal heuristics set should have heuristics that have similarly equal chance to be selected when applied on a large set of different application settings. A s such, the evolving heuristics library should monitor and remove or upgrade those that are consistently producing undesired designs.

In the following, several samples of heuristic schemes will be presented to simply show the concept and potential of heuristic designs. As shown in Fig. 3, the black dots represent the desired poses, and the outer convex polygon means the constraints of installation space of the CDPR, or *worksite constraints*, that the final cable attachments should locate on it.

4.1 Center-Based Scheme

As shown in Fig. 3 (a), this scheme mainly considers all the desired configurations of the CDPR.

1. Find the center point (C) of all the desired configurations.
2. With C as the center, find the smallest regular n-sided polygon, which can just contain all the desired configurations, where n is the number of cables, here $n = 5$.

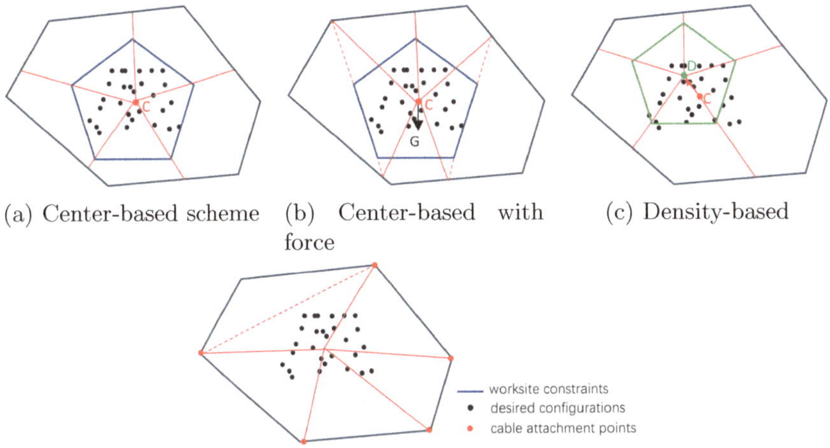

(a) Center-based scheme (b) Center-based with force (c) Density-based

(d) Installation space-based scheme

Fig. 3. Samples of Heuristic Scheme

3. Taking C as the starting point, the rays passing through each vertex of the regular n-sided polygon intersect the edges of installation space respectively.
4. The intersection points are the final cable attachment points.

By using this heuristic scheme, the cables are evenly distributed in the installation space, and the pre-given advantage of this heuristic scheme is that the *Volume Percentage* of the workspace should be larger than the heuristic scheme that considering with force.

4.2 Center-Based Scheme with Consideration of Force

As shown in Fig. 3 (b), in addition to considering the poses of desired configurations, the influence of force is also added. Here takes gravity as an example.

1. Find the center point (C) of all the desired configurations.
2. With C as the center, find the smallest regular n-sided polygon, which can just contain all the desired configurations, where n is the number of cables.
3. Considering with the gravity (G), find two sides of the n-sided polygon which have the minimum included angle with the gravity direction.
4. Extend the two sides separately to intersect the edges o f installation space; for the remaining vertices of the regular n-sided polygon that are not involved, taking C as the starting point, the rays passing through each of them intersect the edges of installation space respectively.
5. The intersection points are the final cable attachment points.

Because the gravity is taken in account, it has reason to assume that if the structure of the 2-DoF CDPR is decided by this way, the maximum cable force would be smaller.

4.3 Density-Based Scheme

As shown in Fig. 3 (c), not only the poses of configurations are considered, but also the density distribution of the configurations in space.

1. Find the density center point (D) of all the desired configurations.
2. With D as the center, find the smallest regular n-sided polygon, which can just contain all the desired configurations.
3. Taking D as the starting point, the rays passing through each vertex of the regular n-sided polygon intersect the edges of installation space respectively.
4. The intersection points are the final cable attachment points.

Obviously, this heuristic scheme is similar to *Center-based scheme without Gravity*, but here considering the spatial distribution of desired configurations. Based on the main working area (black dots in Fig. 3(c)), the *Volume Percentage* of the workspace could be larger. Therefore, this is more suitable for the projects that where the main working area is determined, but the necessity of the rest space is uncertain.

4.4 Installation Space-Based Scheme

As shown in Fig. 3 (d), which is only based on the constraints of installation space to maximize the area of workspace.

1. Choose n vertices of the installation space as the cable attachment points on the base.
2. Make sure all the desired configurations are included in the new polygon composed of selected cable attachment points.
3. Then, the final structure can be decided, which means the cable attachment points are at the selected n vertices.

This heuristic scheme can be regarded as an open solution, it may or may not have the larger *Volume Percentage* of the workspace or the smaller maximum cable force. But, other requirements can be easily added as the standard to determine the final structure of CDPR.

5 Selection Indicators for Evaluation Process

As mentioned above, the preliminary heuristic set can be selected based on a sufficient heuristic scheme library and the experience of designers. In order to allow the design results to be selected, both manually or automatically, in a reasonable and effective manner, selection indicators should be used to evaluate the performance of different heuristic schemes. There are three requirements for choosing or defining a selection indicator: 1) the selection indicators should have corresponding physical meaning to the CDPR; 2) the selection indicators can be represented numerically; and 3) that it can be applied to different designs resulting from the heuristic schemes.

Here are two simple selection indicators as example, and designers can define a broader range of selection indicators for different requirements when using this proposed framework.

1. *Volume Percentage* of the static workspace can be defined as

$$V = \frac{A_{\mathrm{sw}}}{A_{\mathrm{is}}} \tag{3}$$

where A_{sw} and A_{is} are the area of static workspace and installation space, respectively. If the final structure of CDPR is defined by using this selection indicator, the workspace volume of CDPR would be the largest.
2. *Maximum Cable Force* to maintain static equilibrium can be computed by

$$\begin{aligned} \mathbf{f}_{\mathrm{i}} = \underset{\mathbf{f}}{\operatorname{argmin}} \quad & \mathbf{f}^{\mathrm{T}}\mathbf{f} \\ \mathrm{s.t} \quad & -J^{\mathrm{T}}(\mathbf{q})\mathbf{f} = G(\mathbf{q}) \\ & \mathbf{f} \geqslant \mathbf{0} \end{aligned} \tag{4}$$

$$f_{\mathrm{max}}^{\mathrm{i}} = \max(\mathbf{f}_{\mathrm{i}}) \quad i = 1...k \tag{5}$$

$$f_{\mathrm{max}} = \max(f_{\mathrm{max}}^{1}, ..., f_{\mathrm{max}}^{\mathrm{k}}) \tag{6}$$

where $\mathbf{f}_{\mathrm{i}} = [f_1 f_2 ... f_{\mathrm{m}}]^{\mathrm{T}} \in \mathbb{R}^{\mathrm{m}}$ is the cable force vector of i-th (total k) desired configuration. The maximum cable force can be calculated for different heuristic schemes, and then is simple in comparing these schemes to choose the structure of CDPR with minimum maximum cable force.

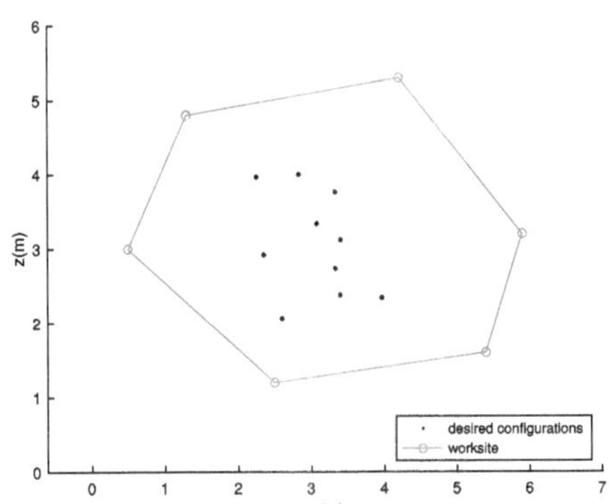

Fig. 4. A 2-DoF planar example

6 Example of Heuristic Design

According to the proposed heuristic design framework for CDPR, an example to design a 2-DoF planar CDPR (in XZ-Plane) with the gravity of the point mass is $[0; -20]N$ is presented, as shown in Fig. 4. The requirements for the CDPR are as below, where the first requirement is the constraints of installation space and the remaining three are the desired performances of CDPR.

1. The installation space is given as the outer convex polygon in Fig. 4.
2. There are ten desired poses, which are the black dots in Fig. 4.
3. It is desired to achieve a larger workspace and smaller maximum cable force.
4. When other features are similar, less number of cables (motors) is desired.

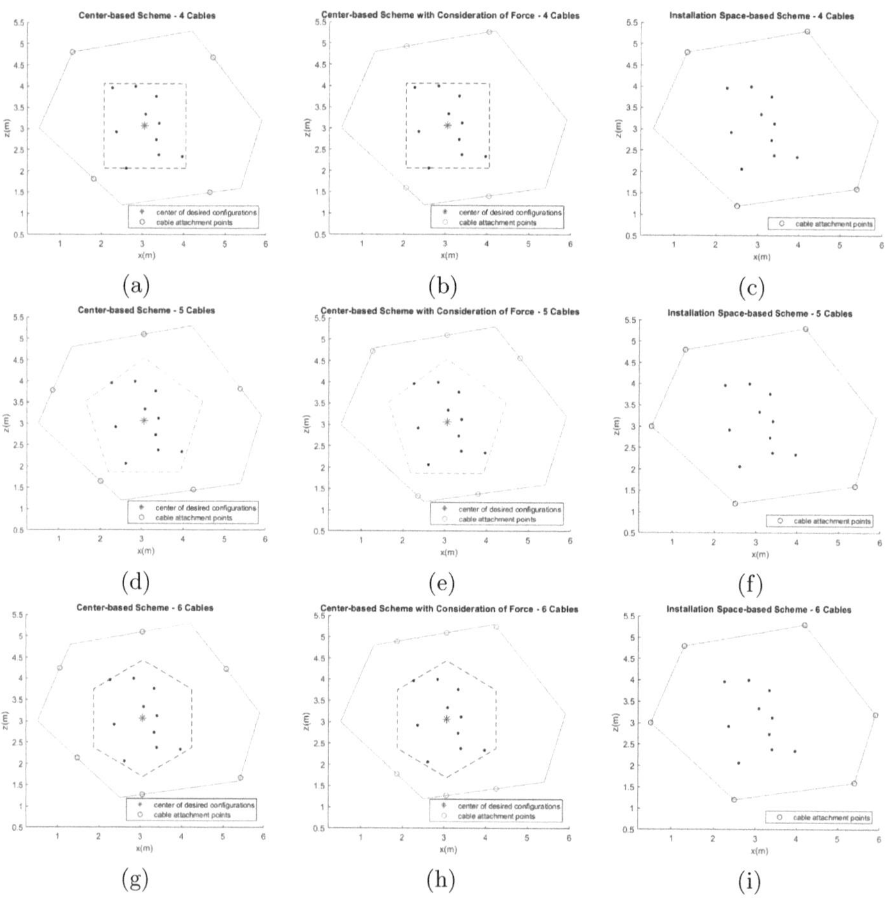

Fig. 5. Examples with three preliminary heuristics

6.1 Step 1: Rough Selection

The four heuristic schemes shown in Fig. 3 are regarded as a heuristic scheme library, and by considering the advantages of different heuristic schemes and the needs of the project at the same time. *Density-based scheme* is not suitable for this example, because all the ten desired configurations need to be considered. Then, *Center-based scheme without Gravity*, *Center-based scheme with Gravity* and *Installation Space-based scheme* are chosen as the preliminary heuristic set. For this example, 4, 5 and 6 cables are considered, following the design method of each preliminary heuristic scheme, different cable arrangements (cable attachment points) for the 2-DoF planar CDPR can be determined, see in Fig. 5.

6.2 Step 2: Simulations and Results Analysis

Taking *Volume Percentage* and Maximum Cable Force as selection indicators, which are consistent with the project requirements. Simulations were performed on the different structures resulting from the heuristic library. In Fig. 6, the *q-sequence* (x-axis) refers to the set of desired poses to allow all poses to be conveniently presented in a single graph for clarity. Comparing with the other two heuristic schemes, the maximum cable forces of *center-based scheme with gravity* are always smaller for the most desired poses. Thus, we firstly decide to use *center-based scheme with gravity* to generate the final structure.

Besides, by setting the upper limit of cable force to 30 N to make all desired poses satisfy the static equilibrium condition in all the three heuristic schemes, see in Fig. 6. Then the static workspace can be shown in Fig. 7b, separately. From Fig. 7b, Fig. 7e and Fig. 7h, it is easy to see that the volume of workspace of 5 cables is the largest, the *Volume Percentage* is 78.37%. Moreover, comparing *Center-based scheme with Gravity* with 5 cables and *Installation Space-based scheme* with 6 cables, see in Fig. 7e and Fig. 7i, the former one has a smaller maximum cable force for the central area of the workspace, and the *Volume Percentages* are 78.37% and 87.44%, respectively. As such, the workspace of the former one has met the requirements with a smaller number of cables.

Finally, the *Center-based scheme with Gravity* is the most suitable for this example with the number of cable is 5. The final structure of 2-DoF planar

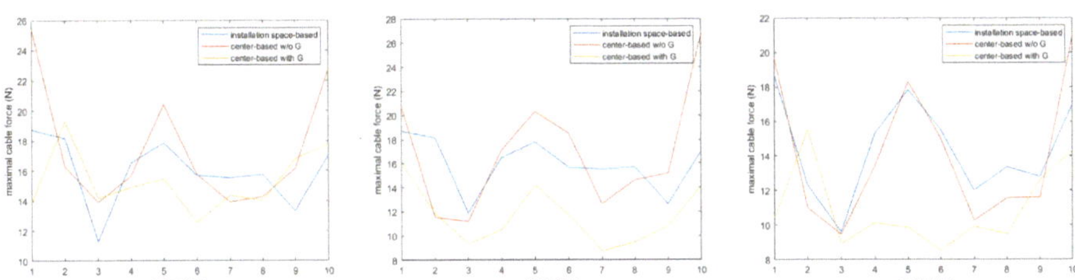

(a) 4 cables (b) 5 cables (c) 6 cables

Fig. 6. Comparison of maximum cable forces

CDPR is shown as Fig. 5e, with the maximum cable force is 16.0644 N and the matrix of cable attachment points at the base frame is

$$A = \begin{bmatrix} 1.26 & 2.355 & 3.8 & 4.8 & 3.05 \\ 4.72 & 1.33 & 1.38 & 4.57 & 5.1 \end{bmatrix}.$$

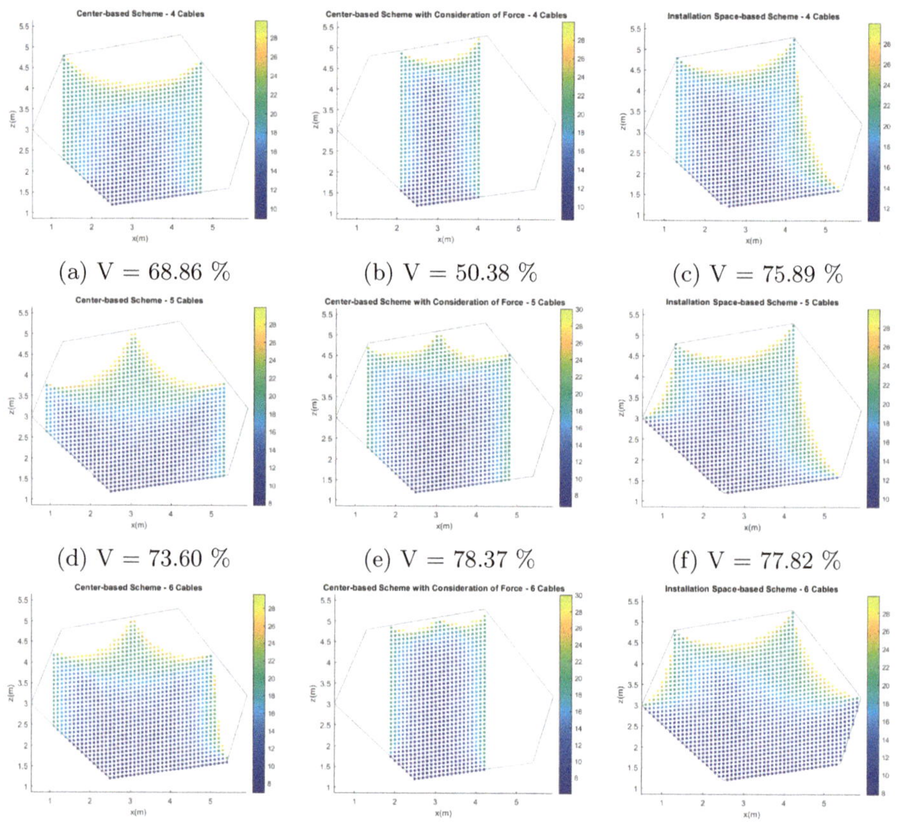

(a) V = 68.86 % (b) V = 50.38 % (c) V = 75.89 %

(d) V = 73.60 % (e) V = 78.37 % (f) V = 77.82 %

(g) V = 78.05 % (h) V = 62.79 % (i) V = 87.44 %

Fig. 7. Static workspace for each structure with the upper limit of cable force is 30 n. the color bar shows the maximum cable force.

7 Conclusion and Future Work

This paper proposed a new heuristic design framework for CDPR. There are two key elements of this framework, the first one is the library of heuristic scheme, and the other is an appropriate set of indicators to evaluate the corresponding performances of different heuristic schemes. Future works will focus on the development of a comprehensive heuristics library and an evolutionary framework where the library of heuristics will be autonomously updated and evolved.

References

1. Bostelman, R., Albus, J., Dagalakis, N., Jacoff, A., Gross, J.: Applications of the NIST RoboCrane. In: Proceedings of the 5th International Symposium on Robotics and Manufacturing, vol. 5 (1994)
2. Pott, A., Mutherich, H., Kraus, W., Schmidt, V., Verl, A.: Cable-driven parallel robots for industrial applications: the IPAnema system family. In: 2013 44th International Symposium on Robotics (ISR) (2013)
3. Lau, D., Bhalerao, K., Oetomo, D., Halgamuge, S.K.: On the task specific evaluation and optimisation of cable-driven manipulators. In: Advances in Reconfigurable Mechanisms and Robots I, pp. 707–716. Springer (2012)
4. Gallina, P., Rosati, G.: Manipulability of a planar wire driven haptic device. In: Mechanism Machine Theory, vol. 37, no. 2, pp. 215–228 (2002)
5. Hao, X., Xiumin, D.: Geometric isotropy indices for workspace analysis of parallel manipulators. In: Mechanism Machine Theory, vol. 128, pp. 648–662 (2018)
6. Williams, R.L., Xin, M., Bosscher, P.: Contour-crafting-cartesian-cable robot system concepts: workspace and stiffness comparisons. In: International Design Engineering Technical Conferences and Computers and Information in Engineering Conference, vol. 43260, pp. 31–38 (2008)
7. Carricato, M., Merlet, J.P.: Direct geometrico-static problem of underconstrained cable-driven parallel robots with three cables. In: IEEE International Conference on Robotics and Automation, ICRA 2011, Shanghai, China, 9–13 May 2011 (2011)
8. Gouttefarde, M., Collard, J.-F., Riehl, N., Baradat, C.: Geometry selection of a redundantly actuated cable-suspended parallel robot. IEEE Trans. Robot. (2015)
9. Gagliardini, L., Caro, S., Gouttefarde, M., Wenger, P., Girin, A.: Optimal design of cable-driven parallel robots for large industrial structures. In: IEEE International Conference on Robotics Automation (2014)
10. Hanafie, J., Nurahmi, L., Caro, S., Pramujati, B.: Design optimization of spatial four cables suspended cable driven parallel robot for rapid life-scan (2018)
11. Zhang, F., Shang, W., Zhang, B., Cong, S.: Design optimization of redundantly actuated cable-driven parallel robots for automated warehouse system. IEEE Access **8**, 56867–56879 (2020)
12. Amine, L.M.: On the optimal design of cable driven parallel robot with a prescribed workspace for upper limb rehabilitation tasks. 16.3 (2019)
13. Jin, X., Agrawal, S.K., Bryson, J.T.: Optimal design of cable- driven manipulators using particle swarm optimization. J. Mech. Robot. Trans. Asme (2016)

Variable-Structure Cable-Driven Parallel Robots

Mitchell Rushton(✉) and Amir Khajepour

University of Waterloo, Waterloo On., Canada
{mmrushto,a.khajepour}@uwaterloo.ca

Abstract. Cable-driven parallel robots (CDPR) have seen great advances over the past two decades; however, they remain plagued by their inability to cover non-convex installation-spaces and low workspace to installation-space ratios relative to their rigid counterparts. These fundamental limitations arise because of the need to avoid collisions between cables and the environment, or the mobile platform. Allowing cables to collide with a set of permissible objects fixed in the environment addresses these issues. Still, the idea of allowing collisions along the length of cables is a concept that has not been well studied or discussed in the literature. In this paper, we hope to promote the idea of variable-structure CDPRs (VSCR) and lead a discussion on the possibilities for future research and applications. We identify three classifications of CDPRs based on anchor-point behavior and define what is meant by variable-structure within the context of CDPRs. We also provide a summary of our recent experimental observations and ongoing work on the study of VSCRs. We end the paper with a discussion of some of the major challenges that remain to be solved for VSCRs.

Keywords: Variable-structure · Cable-driven parallel robots · Collision avoidance

1 Introduction

Over the past two decades, our knowledge of cable-driven parallel robots (CDPR) has grown significantly. With CDPRs, large scale robotic systems have become possible and have been developed and deployed in several applications; the FAST radio telescope and stadium camera systems such as Skycam and Spidercam are prime examples of CDPRs and are some of the largest robotic manipulators in the world [11]. At smaller scales, the low inertia and high tensile strength of cables have led to the development of high-speed, high-payload industrial robotic manipulators such as the DeltaBot from AEMK systems.

CDPRs have many advantages over rigid mechanisms; however, the list of potential applications which can reap those benefits is at least partially reduced by the fundamental design limitation of the potential for interference between cables and the environment. Because platform rigidity comes from antagonistic cable forces, due to the fact that cables are only able to transmit force while under

M. Gouttefarde et al. (Eds.): CableCon 2021, MMS 104, pp. 206–214, 2021.
https://doi.org/10.1007/978-3-030-75789-2_17

tension, the collision-free workspace of CDPRs is much smaller than their rigid counterparts for a given installation volume. Since the span of a cable is always radially convex [15], it is impossible for a fully constrained CDPR to interact with two opposing sides of an object fixed in the environment, or to cover non-convex installation spaces. Not surprisingly, cable interference has been a major point of interest in the CDPR research community with numerous studies being published on topics such as collision detection [2,3,9,12], determining the collision free workspace [4,10,21], and collision avoidance strategies [1,5,8]; however, there has been very little discussion in the literature about the possibility of allowing collisions to occur. In [14,20], it was found that permitting collisions between cables can extend the useable workspace of CDPRs; however, they also found that allowing cable-cable contacts introduced significant control challenges and concerns for cable wear at the points of contact. In [18], an alternative CDPR architecture was studied that allowed for cables to collide with two inactive pulleys as a way of extending the dynamic workspace. In [19] a topological approach was used to investigate CDPR workspace design in the presence of polygonal obstacles.

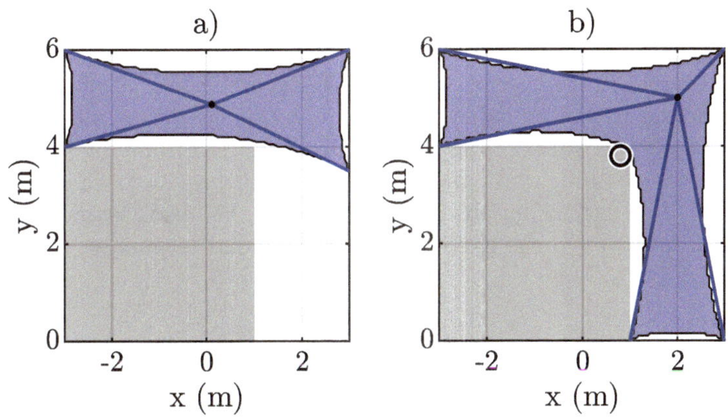

Fig. 1. A simple example to demonstrate the enhanced design freedom that VSCRs provide over traditional CDPRs.

One of the unique advantages that cables have over rigid links is their ability to bend around points of contact. Regardless, there has been little in terms of systematic studies on collidable CDPRs. In our previous work, [17], we introduced the concept of variable-structure CDPRs (VSCR) as a way of addressing some of the current limitations regarding CDPR workspaces and in order to extend the set of applications in which CDPRs can be successfully deployed. In that paper we presented a general extended cable model and inverse kinematics method for dealing with collidable cables. It was shown that by allowing for collisions between cables and rigid objects fixed in the environment, VSCRs can be used to cover non-convex installation spaces that would otherwise be impossible using standard CDPRs. Consider, for example, the design scenario shown in Fig. 1, where it is desired to cover the maximum amount of a ⌝ shaped installation space, shown in white, while avoiding cables from entering the disallowed area, shown in gray. With a 4-cable constant structure CDPR, the maximum workspace that can be covered is shown in Fig. 1 a). By adding an additional cable and placing a collidable idler pulley at the interior corner, as shown in Fig. 1 b), the entire installation space can be covered.

In this paper, we identify three separate CDPR classifications and define what is meant by variable-structure within the context of CDPRs. Additionally, we present some of our current and ongoing work on the study of VSCRs in hopes to inspire an interest in, and provoke discussion about, VSCRs among the larger CDPR research community. In Sect. 2, we identify three separate CDPR classifications based on anchor point behavior. Section 3 provides a summary of our recent observations using an experimental testbed built specifically for studying VSCRs. In Sect. 4, we discuss our current ongoing research and highlight some of the main challenges that remain to be solved for VSCRs.

2 Generalized Modeling and Classification of CDPRs

We begin by revisiting the standard CDPR kinematic model, but with a wider lens, considering the entire cable paths. In its most basic form, a CDPR consists of a rigid mobile platform connected to a fixed ground frame by a set of cables (see Fig. 2a). We assume that cables are massless, always under tension, and able to be collected or released at one of their two ends. We also assume that one end of a cable terminates in the fixed frame, and one end terminates on the mobile platform.

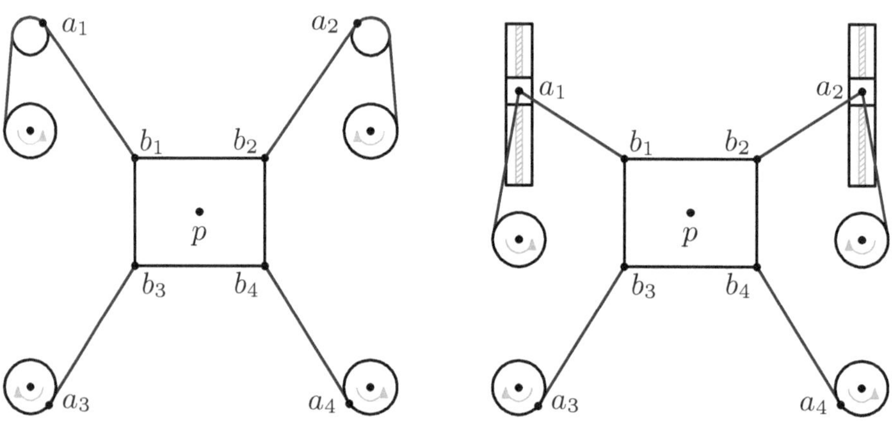

(a) Constant-structure CDPR (b) Reconfigurable CDPR

Fig. 2. Example instances of different CDPR classifications.

As a cable may be in contact with various objects (such as pulleys) in either the fixed frame or on the mobile platform, the complete path of a cable is an alternating sequence of straight-line and wrapping segments. Traveling along the length of a cable, we find that there is always a single line-segment that has one end in the fixed frame, and one end on the mobile platform. We call this segment the active segment. The end-points of the active segment are commonly referred to in the literature as the 'anchor-points' of a cable, although this term is a bit of a misnomer: a cable must be free to slide through at least one of the two active segment end-points. Nonetheless, it is a convenient term which we will continue to use throughout this paper. For the ith cable, given the fixed-frame anchor, $\boldsymbol{a}_i$, defined in the ground frame, and the platform fixed anchor, $\boldsymbol{b}_i$, defined in a platform body-fixed frame, we obtain a vector $\boldsymbol{c}_i$, pointing along the active segment from the platform to the fixed-frame anchor:

$$\boldsymbol{c}_i = \boldsymbol{a}_i - (\boldsymbol{p} + \boldsymbol{R}\boldsymbol{b}_i) \tag{1}$$

where $\boldsymbol{R}$ is a rotation matrix from the body frame to the ground frame, and $\boldsymbol{p}$ is the position of the platform. The length of the active segment is then simply

$$l_i = \|\boldsymbol{c}_i\|_2 \tag{2}$$

and the wrench applied to the platform by the ith cable is

$$\boldsymbol{w}_i = \begin{bmatrix} \hat{\boldsymbol{c}}_i \\ \boldsymbol{R}\boldsymbol{b}_i \times \hat{\boldsymbol{c}}_i \end{bmatrix} \tau_i \tag{3}$$

where τ_i is the cable tension. $\hat{\boldsymbol{c}}_i$ is the unit vector pointing from the platform to the fixed-frame along the active segment, found as

$$\hat{\boldsymbol{c}}_i = \frac{\boldsymbol{c}_i}{l_i} \tag{4}$$

From Eq. (2), we see that the length of the active segment for the ith cable is a function of $\boldsymbol{a}_i$, $\boldsymbol{b}_i$, and the platform pose. If we assume for now that the cable anchor points are constant, then the length of all other segments along the cable path must also be constant, since they are fully constrained by fixed rigid objects at both ends. Therefore, variations in cable length come entirely from variations in l_i and the platform pose is a sufficient set of coordinates to uniquely describe the configuration of the platform and all cables. It is for this reason that parts of the cables beyond the active segment are usually not considered during modeling: if the anchor points are constant, which is commonly the case, the other segments have no bearing on the platform kinematics or dynamics; however, the cable anchors may not necessarily be constant, and in such cases, knowledge of the platform pose alone is no longer sufficient to completely describe the system configuration. Depending on how and if the anchor points change, we identify three classifications of CDPRs, each defined and briefly described below:

Definition 1 (Constant Structure CDPRs). *Also simply referred to as CDPRs. Anchor point locations are constant or known given the platform pose and a set of geometric constraints (for example, a cable exiting a pulley).*

Definition 2 (Reconfigurable CDPRs). *The precise naming comes from its previous use in the literature, [6,13,16]. Achievable with the use of additional actuators in either the fixed-frame or on the mobile platform that provide direct control over cable anchor points.*

Definition 3 (Variable-Structure CDPRs). *Result from permitting collisions between cables and fixed, rigid objects in the environment. As the platform moves, any collisions that occur along the active segment result in an instantaneous change in the location of one of the cable anchors, depending on whether the collided object is fixed in the ground frame or on the mobile platform. Unlike reconfigurable CDPRs, there is no direct control over the anchor point locations.*

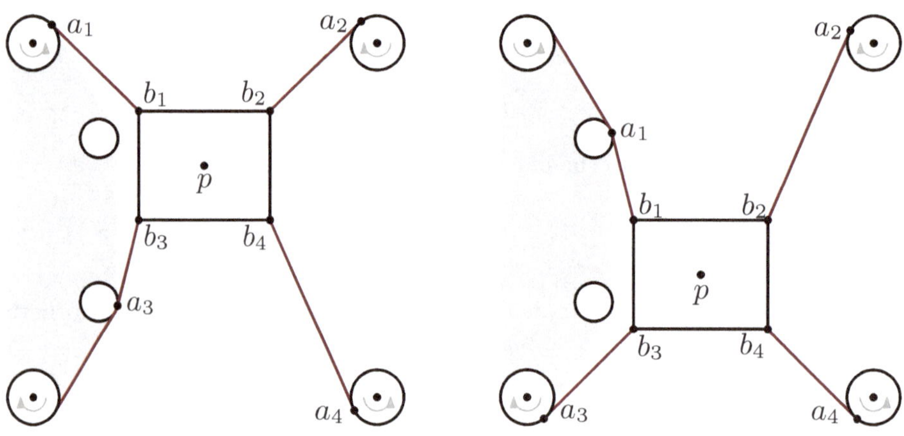

Fig. 3. Example of a variable-structure CDPR. Cables are prevented from entering the gray region by the addition of two collidable idler pulleys.

Visual examples of the three separate classifications are presented in Fig. 2a, Fig. 2b, and Fig. 3 respectively. In Fig. 2b, The addition of two linear actuators along the path of cables 1 and 2 allow for direct control over $\boldsymbol{a}_1$ and $\boldsymbol{a}_2$. For the VSCR shown in Fig. 3, the addition of two collidable idler pulleys causes $\boldsymbol{a}_1$ and $\boldsymbol{a}_3$ to passively vary across the workspace as a function of the platform position; in so doing, cables 1 and 3 are prevented from entering the gray region on the left-side of the workspace.

Of the three classifications, VSCRs are the only ones capable of modifying the sequences of cable segments, with segments being added or removed after each collision or separation. It is this unique property that allows VSCRs to cover workspaces impossible for constant-structure or reconfigurable CDPRs.

3 Planar VSCR Testbed

In our lab we have recently developed an experimental testbed, shown in Fig. 4, for studying VSCRs and validating our work on real hardware. The setup was built to be low-cost and highly flexible. In total, there are eight identical winches and a number of collidable idler pulleys. Each component has a common base plate designed to fit into the regularly spaced holes on a sheet of pegboard. This was done to allow for easy and rapid reconfiguration. The winches use a Turnigy 4250–350KV brushless DC motor, CUI AMT102-V incremental encoder, and are directly coupled to a drive pulley. The drive pulleys, motor mount plates, and control cabinet were constructed out of laser-cut MDF. Spectra fishing line was used for the cables due to its high strength-to-weight ratio and low elasticity. All winches are controlled using ODrive motor controllers. The control software was developed using a mix of Python and Matlab. The complete build cost was under US$2000 and has so far demonstrated itself to be quite versatile [7,17].

(a) Control Cabinet (b) Winches and Collidable Idlers

Fig. 4. Planar VSCR experimental testbed

In [17], our setup was used to demonstrate two design concepts (Fig. 5) showing some of the potential applications where VSCRs may excel. Notably, the results demonstrate how VSCRs can achieve workspaces that are impossible using traditional constant-structure CDPRs. Two videos of the setup in motion are available in the footnotes.[1, 2]

Our plan was to use motor torque sensing for estimating cable tensions directly without the added cost and complexity of using loadcells; however, in using low-cost hobby-grade brushless DC motors (intended for use in drones), we have found the high cogging torque makes it difficult to implement smooth and precise tension control using direct motor torque feedback. We plan to add anti-cogging compensation in the future in hopes that it will help to improve

[1] https://youtu.be/7N8DHFy3cDM.
[2] https://youtu.be/a8WY_QiS3_0.

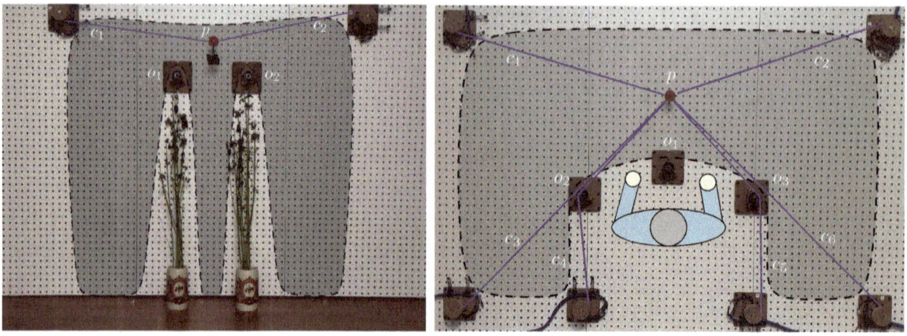

(a) Horticulture (b) Rehabilitation

Fig. 5. Two VSCR design concepts [17]. The gray area shows the approximate workspace.

the tension tracking performance. We have otherwise not observed any major practical issues regarding the implementation or control of VSCRs.

4 Ongoing Work and Remaining Challenges

With very little having been published on the subject, the topic of variable-structure cable-driven parallel robots is currently a wide open field. Many fundamental problems remain completely untouched, such as forward kinematics, calibration, pose estimation, workspace analysis, and trajectory planning.

The focus of our current ongoing research is on producing a general method for identifying the set of structures reachable for a given VSCR and the exact regions where each structure is active. From there, we identify the adjacency relationships among the various structures and construct a graph-based atlas to cover and identify the entire reachable workspace. We have found the structure atlas to be useful tool for analyzing VSCRs that greatly simplifies problems such as workspace analysis and feasible trajectory planning. Another part of our ongoing work is devoted to developing effective VSCR control strategies and studying the impact of collisions on cable tension distribution.

With regards to the design of VSCRs, there is currently no systematic way of determining where collidable objects should be placed and how many are required. Additionally, cable tolerance to non-ideal contact conditions and long-term cable wear that may occur from repeated collision, separation, and sliding against an object surface should be investigated to better understand the practical limitations. To date, past demonstrations of VSCRs have been limited to planar examples. Extending these results to general spatial systems will likely be a significant modelling challenge. In the planar case, we are safe to assume that slippage occurs only in the tangential direction; however, this cannot be guaranteed for a spatial contact and surface stability must be considered (for example, the unstable contact between a cable and a sphere). Using cylindrical columns may be a useful starting point. The existing results have also made the assumption that the exact location and geometry of the collidable objects are known beforehand. This is reasonable for mechanisms where the variable-structure nature is intentionally designed to serve its desired application, but it is not hard to imagine that this may not always be the case: for deployable CDPRs or those working in dynamic environments, this information is usually not available. There is still much research that needs to be done in terms of detecting collisions along the length of a cable, estimating the object geometry, and updating the mechanism's internal kinematic model in order to maintain a desired platform motion; however, the ability to operate in the presence of unknown objects has an exciting potential and is a problem that if solved could greatly enhance the usability and adaptability of CDPRs in unstructured or unknown environments.

5 Conclusion

In this paper, we have shown that variable-structure cable-driven parallel robots (VSCR) can be seen as a generalization of regular cable-driven parallel robots (CDPR). VSCRs that result from allowing collisions between cables and the environment create an exciting potential for the future versatility of CDPRs because of their unique ability to cover non-convex installation spaces. To date, VSCRs have not been well studied and many open problems remain, creating an exciting potential for future research.

Acknowledgment. The authors would like to acknowledge the financial support of the Natural Sciences and Engineering Research Council of Canada (NSERC) in this research.

References

1. Bak, J.H., Hwang, S.W., Yoon, J., Park, J.H., Park, J.O.: Collision-free path planning of cable-driven parallel robots in cluttered environments. Intell. Serv. Robot. 1–11 (2019)
2. Blanchet, L., Merlet, J.P.: Interference detection for cable-driven parallel robots (CDPRs). In: 2014 IEEE/ASME International Conference on Advanced Intelligent Mechatronics, pp. 1413–1418. IEEE (2014)
3. Bury, D., Izard, J.B., Gouttefarde, M., Lamiraux, F.: Continuous collision detection for a robotic arm mounted on a cable-driven parallel robot. arXiv preprint arXiv:1909.10857 (2019)
4. Fabritius, M., Martin, C., Pott, A.: Calculation of the cable-platform collision-free total orientation workspace of cable-driven parallel robots. In: International Conference on Cable-Driven Parallel Robots, pp. 137–148. Springer (2019)
5. Gagliardini, L., Caro, S., Gouttefarde, M., Girin, A.: Discrete reconfiguration planning for cable-driven parallel robots. Mech. Mach. Theory **100**, 313–337 (2016)
6. Izard, J.B., Gouttefarde, M., Michelin, M., Tempier, O., Baradat, C.: A reconfigurable robot for cable-driven parallel robotic research and industrial scenario proofing. In: Cable-driven parallel robots, pp. 135–148. Springer (2013)

7. Jamshidifar, H., Rushton, M., Khajepour, A.: A reaction-based stabilizer for nonmodel-based vibration control of cable-driven parallel robots. IEEE Trans. Robot. (2020)
8. Lahouar, S., Ottaviano, E., Zeghoul, S., Romdhane, L., Ceccarelli, M.: Collision free path-planning for cable-driven parallel robots. Robot. Auton. Syst. **57**(11), 1083–1093 (2009)
9. Lesellier, M., Gouttefarde, M.: A bounding volume of the cable span for fast collision avoidance verification. In: International Conference on Cable-Driven Parallel Robots, pp. 173–183. Springer (2019)
10. Martin, A., Caro, S., Cardou, P.: Geometric determination of the cable-cylinder interference regions in the workspace of a cable-driven parallel robot. In: Cable-Driven Parallel Robots, pp. 117–127. Springer (2018)
11. Nan, R., Li, D., Jin, C., Wang, Q., Zhu, L., Zhu, W., Zhang, H., Yue, Y., Qian, L.: The five-hundred-meter aperture spherical radio telescope (fast) project. Int. J. Mod. Phys. D **20**(06), 989–1024 (2011)
12. Nguyen, D.Q., Gouttefarde, M.: On the improvement of cable collision detection algorithms. In: Cable-Driven Parallel Robots, pp. 29–40. Springer (2015)
13. Nguyen, D.Q., Gouttefarde, M., Company, O., Pierrot, F.: On the analysis of large-dimension reconfigurable suspended cable-driven parallel robots. In: 2014 IEEE International Conference on Robotics and Automation (ICRA), pp. 5728–5735. IEEE (2014)
14. Otis, M.J.D., et al.: Determination and management of cable interferences between two 6-dof foot platforms in a cable-driven locomotion interface. IEEE Trans. Syst. Man Cybern. Syst. Hum. **39**(3), 528–544 (2009)
15. Pott, A.: Determination of the cable span and cable deflection of cable-driven parallel robots. In: Cable-Driven Parallel Robots, pp. 106–116. Springer (2018)
16. Rasheed, T., Long, P., Caro, S.: Wrench-feasible workspace of mobile cable-driven parallel robots. J. Mech. Robot. **12**(3) (2020)
17. Rushton, M., Khajepour, A.: Planar variable structure cable-driven parallel robots for circumventing obstacles. J. Mech. Robot. **13**(2), (2021)
18. Tang, L., Tang, X., Jiang, X., Gosselin, C.: Dynamic trajectory planning study of planar two-dof redundantly actuated cable-suspended parallel robots. Mechatronics **30**, 187–197 (2015)
19. Wang, X., Bhattacharya, S.: A topological approach to workspace and motion planning for a cable-controlled robot in cluttered environments. IEEE Robot. Autom. Lett. **3**(3), 2600–2607 (2018)
20. Wischnitzer, Y., Shvalb, N., Shoham, M.: Wire-driven parallel robot: permitting collisions between wires. Int. J. Robot. Res. **27**(9), 1007–1026 (2008)
21. Wu, Y., Cheng, H.H., Fingrut, A., Crolla, K., Yam, Y., Lau, D.: Cu-brick cable-driven robot for automated construction of complex brick structures: from simulation to hardware realisation. In: 2018 IEEE International Conference on Simulation, Modeling, and Programming for Autonomous Robots (SIMPAR), pp. 166–173. IEEE (2018)

Control

An Emergency Strategy for Cable Failure in Reconfigurable Cable Robots

Roland Boumann[(✉)] and Tobias Bruckmann

Universität Duisburg-Essen, Forsthausweg 2, 47057 Duisburg, Germany
{roland.boumann,tobias.bruckmann}@uni-due.de

Abstract. This paper investigates cable failure and an emergency strategy for reconfigurable cable-driven parallel robots (CDPRs). Reconfiguration of CDPRs is a key feature to avoid collisions, which is currently applied e.g. in prototypes for automated construction tasks. This paper proposes to utilize this ability within an emergency strategy based on minimization of the kinetic energy, which is extended to include movable pulleys on actuated linear slides. The paper analyzes the static equilibrium workspace with constant orientation of a prototype both pre- and post-failure. Based on global numerical optimization, a geometric reconfiguration is proposed to restore most of the workspace after failure. Furthermore, within a dynamic simulation, the extended strategy is evaluated for dynamically moving pulleys. Finally, the results are compared to a simulation with fastened pulleys.

Keywords: Cable-driven parallel robot · Reconfiguration · Cable break · Cable failure · Force distribution · Emergency strategies · Nonlinear model prediction

1 Introduction

A cable-driven parallel robot (CDPR) consists of a set of cables, wound on digitally controlled winches and connected to a common payload. The range of applications for CDPRs is increasing since years – featuring e.g. logistics or automation in construction – implying a strongly advancing industrial usage. Consequently, industrial safety requirements must be considered to avoid damage to the system and surroundings. Standards and guidelines are known for components like e.g. pulleys or cables, but still a risk of a cable failure remains [5]. Several works regarding failure modes and error detection within CDPRs exist and some strategies to handle a cable failure have been proposed already. First considerations of possible errors in CDPRs and error tolerance were formulated in [15], investigating the effects of a missing cable on the nullspace of the robot's Jacobian. [11] examines the effects of slack or defect cables resulting in incorrect cable forces. A restoration of the force and momentum capacity at the end effector is aspired and criteria for the feasibility of the latter are formulated. An emergency strategy involving dynamic trajectory planning based on

M. Gouttefarde et al. (Eds.): CableCon 2021, MMS 104, pp. 217–229, 2021.
https://doi.org/10.1007/978-3-030-75789-2_18

an elliptical path was introduced by [1]. Herein, the platform might be outside its workspace after cable break. In [2], the post-failure workspace of a given system was investigated and an algorithm for planning a feasible trajectory along a straight line path back into the post-failure workspace was introduced. The work was extended in [3] considering drivetrain models and tested successfully on a prototype in [4], using four cables and three degrees-of-freedom. [12] proposes algorithms for dynamical trajectory planning of a cable driven camera system after a cable failure, aiming at stopping the system within the post-failure workspace, avoiding collisions. Also favorable goal positions have been determined. The effects of repositioning of the cable anchor points on the post-failure behaviour of the end effector was investigated in [8]. The cables of the proposed deep sea manipulator were mounted on ships. Their positions were optimized to enhance the fault tolerance of the system. Yet, a dynamic strategy to control the system in case of cable failure was not proposed.

In contrast to the described emergency strategies, which focus mainly on planning dynamic trajectories back into the post-failure workspace or redistribution of the forces as long as the end effector is still within the workspace, the emergency strategies proposed in former work of the authors of this paper [5] seek to safeguard the end effector into a safe position without pre defining a trajectory. The two strategies base either on potential fields or on minimization of the kinetic energy of the system [5]. Former work of the authors [6,14] feature also CDPRs with dynamically reconfigurable pulleys that can be moved on linear axis, actuated by additional drives. A strategy for dynamic reconfiguration of the SEGESTA prototype throughout a trajectory to enhance the mechanical stiffness of the system was proposed in [14]. More details on CDPR reconfiguration can be found in [7]. Summarizing, several emergency strategies to handle cable failure exist, but none of them involve reconfigurable pulley mechanisms that are dynamically moved. Therefrom the question arises if an active movement of the pulleys using extra drives can be utilized in case of cable failure. To investigate this, the emergency strategy of kinetic energy minimization is extended in this work to simulatively study the effect and applicability of movable pivoting pulleys in case of cable failure.

The paper is structured as follows: At first, an introduction about CDPRs and existing emergency strategies in case of cable failure is given. In Sect. 2, the kinematic and dynamic robot models are introduced. In Sect. 3, the emergency strategy of kinetic energy minimization, based on nonlinear model predictive control, is presented and extended assuming reconfigurable pulleys. An examination of the post-failure workspace for a prototype at hand is carried out in Sect. 4, in particular considering reconfiguration. Using numerical optimization, an optimal set of pulley positions is found to restore most of the static equilibrium workspace after cable failure. Furthermore and within a dynamic simulation based on the introduced models, the proposed cable break strategy is applied to the given robot in a cable failure scenario. The extended strategy utilizing dynamically movable pulleys is compared to the approach limited to fixed pulleys. In the end a summary and an outlook are given.

2 Kinematic and Dynamic Cable Robot Modeling

Kinematics. The robot platform as the end-effector possesses m cables and n degrees-of-freedom, resulting in a degree of redundancy $r = m - n$. The platform is referenced within the inertial coordinate system $\underline{B}$ and has the fixed coordinate system $\underline{P}$. The pose ${}^{B}x_P = [{}^{B}\boldsymbol{r}_P\ \boldsymbol{\Phi}]^{\mathrm{T}}$ contains position ${}^{B}\boldsymbol{r}_P$ and orientation $\boldsymbol{\Phi}$, both defined as row vectors. Employing yaw-pitch-roll angles, the orientation of the end effector with respect to the inertial system is described by the rotation matrix ${}^{B}\boldsymbol{R}_P$. Let ${}^{P}\boldsymbol{p}_i$ be the vectors where the cables are connected to the platform, denoted in $\boldsymbol{P} = [{}^{P}\boldsymbol{p}_1 \ldots {}^{P}\boldsymbol{p}_i]$, and ${}^{B}\boldsymbol{b}_{d,i}$ the points, where the cables exit the pulleys and are guided into the robot's workspace. The latter can be derived from inverse pulley kinematics [13], which is not further detailed here. The pulley radii are r_d and the base vectors of the robot are ${}^{B}\boldsymbol{b}_i$, see Fig. 1, denoted in $\boldsymbol{B} = \left[{}^{B}\boldsymbol{b}_1 \ldots {}^{B}\boldsymbol{b}_i\right]$. From inverse kinematics, the cable vectors ${}^{B}\boldsymbol{l}_i$ can be obtained:

$$ {}^{B}\boldsymbol{l}_i = {}^{B}\boldsymbol{b}_{d,i} - \underbrace{({}^{B}\boldsymbol{r}_P + {}^{B}\boldsymbol{R}_P\,{}^{P}\boldsymbol{p}_i)}_{{}^{B}\boldsymbol{p}_{B_i}}, \quad 1 \le i \le m. \tag{1} $$

Assuming a reconfigurable system, the position of the pulley along the linear axis is s_j and its velocity is $\dot{s}_j$. Accordingly, the base vectors of the latter change in z-direction, adding an increment of s_j to the z-component of ${}^{B}b_{j,z}$.

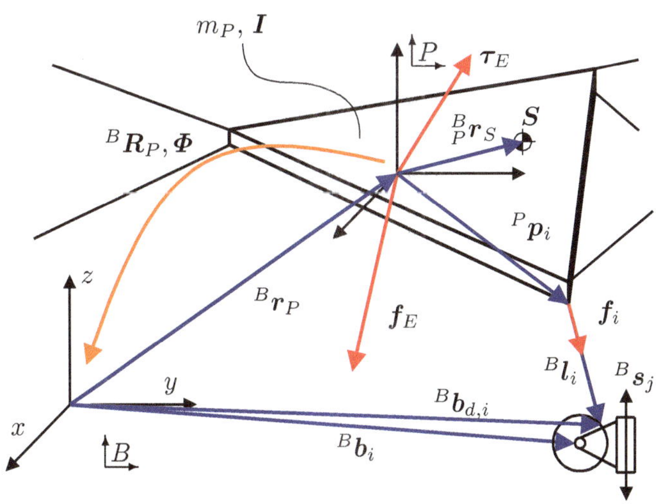

Fig. 1. Cable-robot model parameters

Forces and Dynamics. All vectors are decomposed and referenced in coordinate system $\underline{B}$ in the following, whereas for simplicity, the top-left index is left out. In the cable directions – each represented by unit vector $\boldsymbol{\nu}_i$ – every cable is tensioned by a force f_i. The cable force vector $\boldsymbol{f} \in \mathbb{R}^{m\times 1}$ contains all cable tensions. Each force $\boldsymbol{f}_i$ onto the end effector, exerted by the ith cable, is

$$\boldsymbol{f}_i = f_i \cdot \frac{\boldsymbol{l}_i}{\|\boldsymbol{l}_i\|_2} = f_i \cdot \boldsymbol{\nu}_i, \quad 1 \leq i \leq m. \tag{2}$$

$\boldsymbol{A}^{\mathrm{T}}$ is the structure matrix of the robot, corresponding to the transpose of its Jacobian. $\boldsymbol{w}_E$ contains all forces $\boldsymbol{f}_E$ and torques $\boldsymbol{\tau}_E$ acting on the platform except from the cable forces $\boldsymbol{f}_i$. Therefrom, the static force equilibrium at the platform is modeled as

$$-\boldsymbol{w}_E = \begin{bmatrix} -\boldsymbol{f}_E \\ -\boldsymbol{\tau}_E \end{bmatrix} = \begin{bmatrix} \boldsymbol{\nu}_1 & \dots & \boldsymbol{\nu}_m \\ \boldsymbol{p}_1 \times \boldsymbol{\nu}_1 & \dots & \boldsymbol{p}_m \times \boldsymbol{\nu}_m \end{bmatrix} \begin{bmatrix} f_1 \\ \vdots \\ f_m \end{bmatrix} = \boldsymbol{A}^{\mathrm{T}} \boldsymbol{f}. \tag{3}$$

The mass of the platform is m_P and its inertia tensor is ${}^{B}\boldsymbol{\theta}_P$, derived from the inertia tensor in the center of gravity ${}^{P}\boldsymbol{\theta}_S$ and the Steiner-Huygens relation. $\boldsymbol{E}_3$ is a 3×3 identity matrix. From kinematic Kardan equations, $\boldsymbol{\mathcal{H}}$ and $\dot{\boldsymbol{\mathcal{H}}}$ can be derived [16]. ${}^{B}_{P}\boldsymbol{r}_S$ is the vector from the platform coordinate system to its center of gravity. Thereby, the skew-symmetric matrix ${}^{B}_{P}\boldsymbol{R}_S$ is defined as

$${}^{B}_{P}\boldsymbol{R}_S = \begin{pmatrix} 0 & -{}^{B}_{P}r_{S,z} & {}^{B}_{P}r_{S,y} \\ {}^{B}_{P}r_{S,z} & 0 & -{}^{B}_{P}r_{S,x} \\ -{}^{B}_{P}r_{S,y} & {}^{B}_{P}r_{S,x} & 0 \end{pmatrix}, \tag{4}$$

Now setting up Newton-Euler equations [10] expressed in $\underline{B}$ leads to

$$\begin{aligned} &\underbrace{\begin{bmatrix} m_P\,\boldsymbol{E}_3 & -m_P\,{}^{B}_{P}\boldsymbol{R}_S \boldsymbol{\mathcal{H}} \\ m_P\,{}^{B}_{P}\boldsymbol{R}_S & {}^{B}\boldsymbol{\theta}_P \boldsymbol{\mathcal{H}} \end{bmatrix}}_{\boldsymbol{M}(\boldsymbol{x}_P)} \underbrace{\begin{bmatrix} {}^{B}\ddot{\boldsymbol{r}}_P \\ \ddot{\boldsymbol{\phi}} \end{bmatrix}}_{\ddot{\boldsymbol{x}}_P} + \dots \\ &+ \underbrace{\begin{bmatrix} m_P \left[(\dot{\boldsymbol{\mathcal{H}}}\dot{\boldsymbol{\phi}}) \times {}^{B}_{P}\boldsymbol{r}_S + (\boldsymbol{\mathcal{H}}\dot{\boldsymbol{\phi}}) \times ((\boldsymbol{\mathcal{H}}\dot{\boldsymbol{\phi}}) \times {}^{B}_{P}\boldsymbol{r}_S) \right] \\ {}^{B}\boldsymbol{\theta}_P \dot{\boldsymbol{\mathcal{H}}}\dot{\boldsymbol{\phi}} + (\boldsymbol{\mathcal{H}}\dot{\boldsymbol{\phi}}) \times ({}^{B}\boldsymbol{\theta}_P \boldsymbol{\mathcal{H}}\dot{\boldsymbol{\phi}}) \end{bmatrix}}_{\boldsymbol{K}(\boldsymbol{x}_P, \dot{\boldsymbol{x}}_P)} \dots \\ &\dots + \underbrace{-{}^{B}\boldsymbol{w}_E}_{\boldsymbol{Q}(\boldsymbol{x}_P, \dot{\boldsymbol{x}}_P)} = {}^{B}\boldsymbol{A}^{\mathrm{T}} \boldsymbol{f}. \end{aligned} \tag{5}$$

Herein, the first and second time derivatives of the end effector pose $\boldsymbol{x}_P$ are denoted $\dot{\boldsymbol{x}}_P$ and $\ddot{\boldsymbol{x}}_P$. $\boldsymbol{M}(\boldsymbol{x}_P)$ is the mass matrix of the platform. Coriolis and centrifugal forces and torques are defined in $\boldsymbol{K}(\boldsymbol{x}_P, \dot{\boldsymbol{x}}_P)$. All remaining forces and torques including gravitational forces, disturbances acting on the platform and friction are summarized in $\boldsymbol{Q}(\boldsymbol{x}_P, \dot{\boldsymbol{x}}_P)$. Cables are assumed to be massless and non-sagging. As they can only exert pulling forces, a minimum cable tension $f_{\min}$ must be maintained. To consider motor limits and mechanical load capabilities of the components, a maximum tension $f_{\max}$ must not be exceeded. For given wrenches $\boldsymbol{w}$ resulting from a given end effector trajectory, Eq. (5) must be solved for $\boldsymbol{f}$. To find solutions, several commonly known methods exist, differing e.g. in force level or computation time [13]. Note, that modeling of friction and inertia within the drivetrains as well as disturbances is neglected within this paper.

3 Minimization of Kinetic Energy in Case of Cable Failure

Cable failure in cable driven parallel robots might have fatal consequences, such as e.g. loss of carried payload, machine damage, harm to workers or the environment due to uncontrolled movement of the robot. Therefore, strategies of action in case of cable failure are essential. Assuming the case of a cable break, commonly known methods to calculate cable forces might fail if the robot is outside its workspace [5]. One strategy to handle cable failure in CDPRs, which was introduced in former work of the authors [5], will be briefly described in the following and extended for application in reconfigurable CDPRs.

3.1 Nonlinear Model Predictive Control Approach

After a cable break, a robot structure with $m-1$ cables still exists, assuming the failure of one single cable for simplicity. In the following, the remaining cable force distribution is denoted $\boldsymbol{f}^* \in \mathbb{R}^{(m-1)\times 1}$ and the structure matrix of the remaining system, reduced by the column of the broken cable, is $\boldsymbol{A}^{\mathrm{T}*} \in \mathbb{R}^{n\times m-1}$. As long as some cables remain after a failure, the platform motion can be influenced and the motion might be stopped in a controlled manner. However, the workspace of the reduced system might be remarkably smaller, promoting the situation that the end effector is outside the workspace of the remaining system.

One strategy to stop the system in a controlled manner is a model predictive approach [9], aiming at minimization of the systems kinetic energy after a cable failure [5]. The method predicts the system state utilizing the robot's nonlinear dynamic model to find cable forces that will minimize the kinetic energy of the platform (and thus its velocity). This necessitates a stop of the system. Inherently, the robot will be automatically within the post-failure workspace if it is able to stop. One major advantage of this method is that the final position does not need to be predefined. The kinetic energy assuming a rigid body is given by

$$E_{Kin} = \frac{1}{2} m_p {}^B\dot{\boldsymbol{r}}_P^{\mathrm{T}}\, {}^B\dot{\boldsymbol{r}}_P + \frac{1}{2}\boldsymbol{\Omega}^{\mathrm{T}}\, {}^B\boldsymbol{\theta}_P \boldsymbol{\Omega}. \tag{6}$$

To minimize the kinetic energy, the magnitude of $\dot{\boldsymbol{x}}_P = [\dot{\boldsymbol{r}}_P\, \dot{\boldsymbol{\phi}}]^{\mathrm{T}}$ must be minimized, since the angular velocity $\boldsymbol{\Omega} = \boldsymbol{\mathcal{H}}\dot{\boldsymbol{\phi}}$ is directly depending on $\dot{\boldsymbol{\phi}}$ [16].

The nonlinear discrete state equations using the nonlinear system function $f(\boldsymbol{x}(k), \boldsymbol{u}(k))$ based on Eq. (5), describe the system by

$$\begin{aligned} \boldsymbol{x}(k+1) &= f(\boldsymbol{x}(k), \boldsymbol{u}(k)) \\ \boldsymbol{y}(k+1) &= \boldsymbol{C}\,\boldsymbol{x}(k+1), \end{aligned} \tag{7}$$

with the state vector of velocity and position $\boldsymbol{x}(k)$ and the output vector $\boldsymbol{y}(k)$:

$$\begin{aligned} \boldsymbol{x}(k) &= [\,\dot{\boldsymbol{r}}_p(k), \dot{\boldsymbol{\phi}}(k), \boldsymbol{r}_p(k), \boldsymbol{\phi}(k)\,]^{\mathrm{T}} \\ \boldsymbol{y}(k) &= [\,\dot{\boldsymbol{r}}_p(k), \dot{\boldsymbol{\phi}}(k)\,]^{\mathrm{T}}. \end{aligned} \tag{8}$$

With the 3×3 identity matrix $\boldsymbol{K}$, the output matrix is given by

$$\boldsymbol{C} = \begin{bmatrix} \boldsymbol{K} & 0 & 0\,0 \\ 0 & \boldsymbol{K} & 0\,0 \end{bmatrix}. \tag{9}$$

The vector of remaining cable forces is defined as input vector

$$\boldsymbol{u}(k) = \boldsymbol{f}^{*}(k). \tag{10}$$

$f(\boldsymbol{x}(k), \boldsymbol{u}(k))$ returns the state vector in the next time step $\boldsymbol{x}(k+1)$ using numerical integration based on the Euler-Cromer method with a prediction step size of Δt_c:

$$\begin{aligned} \begin{bmatrix} \dot{\boldsymbol{r}}_p(k+1) \\ \dot{\boldsymbol{\phi}}(k+1) \end{bmatrix} &= \begin{bmatrix} \dot{\boldsymbol{r}}_p(k) \\ \dot{\boldsymbol{\phi}}(k) \end{bmatrix} + \left[\boldsymbol{M}(\boldsymbol{x}_P)^{-1}\left(\boldsymbol{A}^{\mathrm{T}*}\boldsymbol{f}^{*} - \boldsymbol{K}(\boldsymbol{x}_P, \dot{\boldsymbol{x}}_P) - \boldsymbol{Q}(\boldsymbol{x}_P, \dot{\boldsymbol{x}}_P)\right)\right](k)\Delta t_c \\ \begin{bmatrix} \boldsymbol{r}_p(k+1) \\ \boldsymbol{\phi}(k+1) \end{bmatrix} &= \begin{bmatrix} \boldsymbol{r}_p(k) \\ \boldsymbol{\phi}(k) \end{bmatrix} + \begin{bmatrix} \dot{\boldsymbol{r}}_p(k+1) \\ \dot{\boldsymbol{\phi}}(k+1) \end{bmatrix} \Delta t_c. \end{aligned} \tag{11}$$

With the prediction horizon n_p and the control horizon n_c, a cost function J can be set up, which is minimized at every time step to generate set-point cable forces $\boldsymbol{f}^{*}$ considering the cable force limits. To minimize the kinetic energy, the reference output $\boldsymbol{y}_r$ is set to zero. Therefore, the function minimum will be obtained when the robot comes to a full stop and the cable forces remain constant. The parameters $\boldsymbol{Q}$ and r_1 can be used to weight the criteria included in the optimization problem.

$$\begin{aligned} J = &\sum_{i=1}^{n_p} \left(\boldsymbol{y}(k+i) - \boldsymbol{y}_r(k+i)\right)^{\mathrm{T}} \boldsymbol{Q} \left(\boldsymbol{y}(k+i) - \boldsymbol{y}_r(k+i)\right) + \dots \\ &\sum_{i=i}^{n_c} \left(\boldsymbol{f}^{*}(k+i-1) - \boldsymbol{f}^{*}(k+i-2)\right)^{\mathrm{T}} r_1 \left(\boldsymbol{f}^{*}(k+i-1) - \boldsymbol{f}^{*}(k+i-2)\right) \end{aligned} \tag{12}$$

3.2 Extension of the Strategy Using Movable Pulleys

Considering movable pulleys in vertical direction, denoted within $\boldsymbol{s}(k)$, the input vector is extended to

$$\boldsymbol{u}(k) = [\boldsymbol{f}^{*}(k),\ \boldsymbol{s}(k)]^{\mathrm{T}}. \tag{13}$$

This only holds for failure of one of the cables without a movable pulley. Otherwise, the corresponding entry needs to be removed from $\boldsymbol{s}$, leading to $\boldsymbol{s}^*$. Note, as $\boldsymbol{B}$ changes over time accordingly to the pulley movement , the structure matrix $\boldsymbol{A}^{\mathrm{T}*}$ is now a function of values contained in the extended input within the optimization problem and Eq. (11) is adjusted correspondingly. Applying a weighting parameter r_2 for the pulley movement, the cost function is extended to

$$\begin{aligned} J = &\sum_{i=1}^{n_p} \big(\boldsymbol{y}(k+i) - \boldsymbol{y}_r(k+i)\big)^{\mathrm{T}} \boldsymbol{Q} \big(\boldsymbol{y}(k+i) - \boldsymbol{y}_r(k+i)\big) + \dots \\ &\sum_{i=i}^{n_c} \big(\boldsymbol{f}^*(k+i-1) - \boldsymbol{f}^*(k+i-2)\big)^{\mathrm{T}} r_1 \big(\boldsymbol{f}^*(k+i-1) - \boldsymbol{f}^*(k+i-2)\big) + \dots \\ &\sum_{i=i}^{n_c} \big(\boldsymbol{s}(k+i-1) - \boldsymbol{s}(k+i-2)\big)^{\mathrm{T}} r_2 \big(\boldsymbol{s}(k+i-1) - \boldsymbol{s}(k+i-2)\big). \end{aligned} \tag{14}$$

4 Examination of the Post-failure Workspace

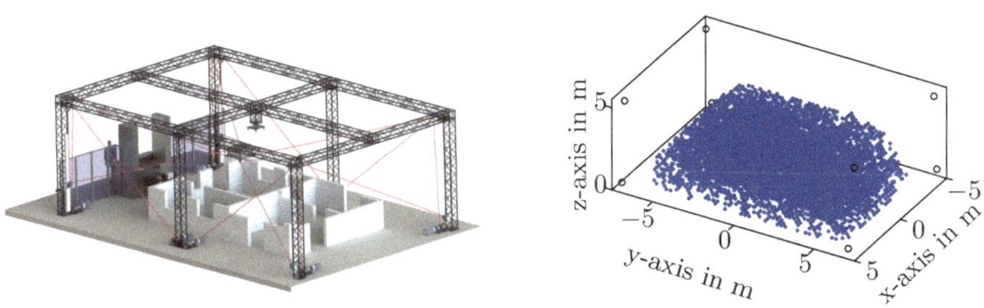

(a) CAD drawing of the robot prototype (b) Workspace of simulation model

Fig. 2. Static equilibrium workspace of the system with movable pulleys at lowest position before a cable failure considering only gravitational forces. The circles in the corners indicate the pulley positions.

As shown in [5], the workspace of a CDPR after a cable break might be remarkably smaller. In the following, the effects of a cable failure on the static equilibrium workspace with constant orientation ($\mathcal{SEW}$) will be investigated, considering a CDPR prototype for automated masonry work that is currently being developed. The robot features a gripper for masonry units and eight cables. The four lower pulleys $j \in \{1, 2, 5, 6\}$ are pivoting and movable in vertical direction on actuated linear slides, mainly to cope for collision avoidance in the construction process of a rising building. Figure 2 shows the CAD drawing of the prototype and its corresponding workspace with the lower four pulleys at their lowest position. The platform mass $m_P = 132.8125$ kg, inertia and center of gravity ${}^{B}_{P}\boldsymbol{r}_S = [-0.0078,\ -0.0074,\ -0.1152]^{\mathrm{T}}$ m include a masonry unit carried by the platform. The remaining essential parameters of the prototype are displayed within Eq. (15). Further details on the prototype can be found in [6]. In the following, the failure of one of the upper cables is considered. Figure 3(a) shows that the post-failure workspace is remarkably smaller after a cable break. This promotes the chance of the end effector being outside of the remaining workspace, see also former work [5]. To investigate if pulley reconfiguration might counteract this problem, an optimization was carried out to find pulley positions that restore as much ratio of the static equilibrium workspace as possible. Global optimization using MATLAB's `particleswarm()` algorithm was employed for this purpose. First, the pre-failure $\mathcal{SEW}$ was discretized with a grid of points to obtain an optimal number of workspace points within the frame. The total number of evaluated grid points was 27489. After cable-break, the optimizer was allowed to change all values of $\boldsymbol{s}$ within the borders $[s_{\min}, s_{\max}] = [0, 3]$ m. The cost function was defined as the difference between the number of discrete $\mathcal{SEW}$ points pre-failure and the number of discrete $\mathcal{SEW}$ points after reconfiguration that match the pre-failure points. The swarm size was set to 100, and the maximum iterations were set to 200. The optimization converged after 80 iterations. After cable break, 27.7% of the pre-failure $\mathcal{SEW}$ remained. After optimization, 62.1% of the pre-failure $\mathcal{SEW}$ can be restored, which is displayed in Fig. 3(b). This result provides an indication that reconfigurable pulleys might be utilized to support an emergency strategy in case of cable failure and to counteract the immense loss of workspace until the robot can be serviced and repaired.

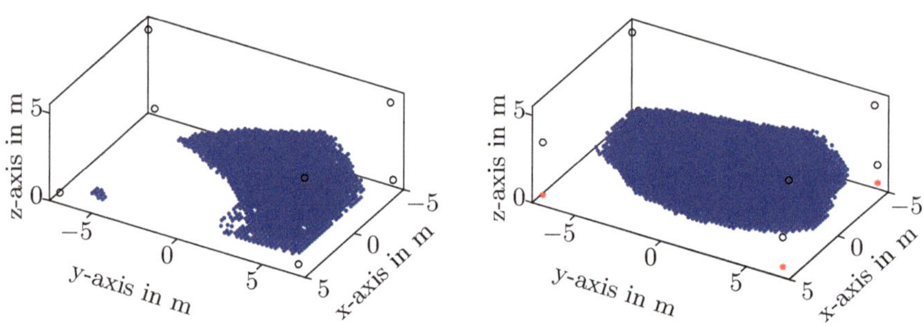

(a) Workspace with failure of upper cable (b) Workspace after reconfiguration

Fig. 3. Static equilibrium workspace of the system after cable failure of cable 3 considering only gravitational forces. The circles in the corners indicate the pulley positions, red asterisks indicate the initial pulley positions, final pulley shift $\boldsymbol{s} = [2.9998, 0.0102, 1.6549, 1.0028]^{\mathrm{T}}$ m (right side).

5 Dynamic Simulation and Results

As demonstrated in Sect. 4, the reduction of $\mathcal{SEW}$ is remarkable which causes the platform likely to be outside of the post-failure workspace after a cable breakage. This calls out for emergency strategies in order to guide the end effector into a stable static position. In the following, the proposed strategy from Sect. 3 will be tested within simulation, including the ability of the robot for rapid reconfiguration of the lower four pulley positions. A dynamic simulation based on Sect. 2 is set up for testing purposes, employing the robot parameters according to Eq. (15). The model has six degrees of freedom, eight cables and is subject to gravity.

$$
\begin{aligned}
\boldsymbol{P} &= \begin{bmatrix} 0.51 & -0.51 & 0.51 & -0.51 & 0.51 & -0.51 & 0.51 & -0.51 \\ -0.0315 & -0.0315 & -0.445 & -0.445 & 0.0315 & 0.0315 & 0.445 & 0.445 \\ 0.215 & 0.215 & -0.175 & -0.175 & 0.215 & 0.215 & -0.175 & -0.175 \end{bmatrix} \mathrm{m}, \\
\boldsymbol{B} &= \begin{bmatrix} 4.8 & -4.8 & 4.33 & -4.33 & 4.8 & -4.8 & 4.33 & -4.33 \\ -6.996 & -6.996 & -7.096 & -7.096 & 6.996 & 6.996 & 7.096 & 7.096 \\ 0.5 & 0.5 & 5.1875 & 5.1875 & 0.5 & 0.5 & 5.1875 & 5.1875 \end{bmatrix} \mathrm{m}, \\
{}^{P}\boldsymbol{\theta}_S &= \begin{bmatrix} 11.0424 & 0.3203 & 0.4188 \\ 0.3203 & 10.8053 & 0.3696 \\ 0.4188 & 0.3696 & 9.8330 \end{bmatrix} \mathrm{kg\,m^2}, \\
& f_{\min} = 50\ \mathrm{N},\ f_{\max} = 4500\ \mathrm{N},\ r_d = 0.126\,\mathrm{m},\ r_1 = 0.0000005,\ r_2 = 75, \\
& \boldsymbol{Q} = diag(10, 1, 5, 10, 50, 10),\ \Delta t_c = 1\ \mathrm{ms},\ n_p = n_c = 1,\ y_r = \boldsymbol{0}, \\
& \boldsymbol{x}_{P,\mathrm{Start}} = [2.5\ \mathrm{m}, -4\ \mathrm{m}, 3.25\ \mathrm{m}, 0\ \mathrm{rad}, 0\ \mathrm{rad}, 0\ \mathrm{rad}]^{\mathrm{T}}, \\
& \dot{\boldsymbol{x}}_{P,\mathrm{Start}} = [0\ \mathrm{m/s}, 0\ \mathrm{m/s}, 0\ \mathrm{m/s}, 0\ \mathrm{rad/s}, 0\ \mathrm{rad/s}, 0\ \mathrm{rad/s}]^{\mathrm{T}}.
\end{aligned} \tag{15}
$$

The cables are assumed to be massless and non-sagging, thus a dynamic effect from a snapping cable is not modeled. Modeling of friction, disturbances and drivetrains is neglected. The time-discrete model is integrated numerically using the Euler-Cromer Method as well with a fixed stepsize of $\Delta t = 1$ ms to determine velocity and position in the next time step $k+1$. Note, only the results for a failure of one upper cable (cable three) are presented here, due to paper length limitations. Moreover, collisions are not checked within the simulation. The platform is placed near the border of the pre-failure $\mathcal{SEW}$ at $\boldsymbol{x}_{P,\mathrm{Start}}$ with velocity $\dot{\boldsymbol{x}}_{P,\mathrm{Start}}$ and thus far away from the post-failure $\mathcal{SEW}$. The cable forces are set to $f_{\min}$ at the beginning of the simulation. Two simulations (pulleys fastened and pulleys movable, according to Eq. (12) and (14)) are carried out using the minimization of kinetic energy approach with an interior-point-algorithm. To increase numerical efficiency, the solution from the previous time step is used for a warm start of the optimization problem. The results of both simulations are shown in Fig. 4. The results show that in the first scenario without moving lower pulleys at 1.8 s, the platform might be subject to collision with the frame in y-direction, assuming that it is limited by the maximum values of $\boldsymbol{B}$. The platform is stopped after approx. 2.6 s. In the scenario with moving pulleys, the latter can be avoided. The platform is stopped in a stable position earlier after approx. 1.5 s, while its travel distance through the workspace is shorter. The maximum occurring speed in y-direction is 6 m/s, which is half of the maximum speed in the first scenario of 12 m/s. The cable force limits are reached but not exceeded in both scenarios, as well as the boundaries of the vertical moving pulleys. The maximum occurring pulley velocity is about 3 m/s, which needs to be achieved by the linear slide drives. Noteworthy, the corresponding accelerations may lead to vibrations in the system, which has to be addressed in future work. Moreover, it can be observed that in the scenario at hand, the lower pulley 1 in the corner of the broken cable moves near to its upper boundary, partially compensating the missing corresponding upper cable, which leads to a better capability of stopping the platform. Pulleys 5 and 6 move upwards by small distances, while pulley 2 tends to stay at its initial position. This is consistent with the observations from Fig. 3(b), since this configuration is beneficial in restoring most of the $\mathcal{SEW}$. Finally, tilting of the platform can be reduced from 0.5 rad to 0.12 rad around the x-axis in the second scenario, which might counteract the risk of losing a payload. The simulation studies indicate that the use of actively movable lower pulleys in case of a cable failure can be reasonable and basis of extended emergency strategies for cable breaks.

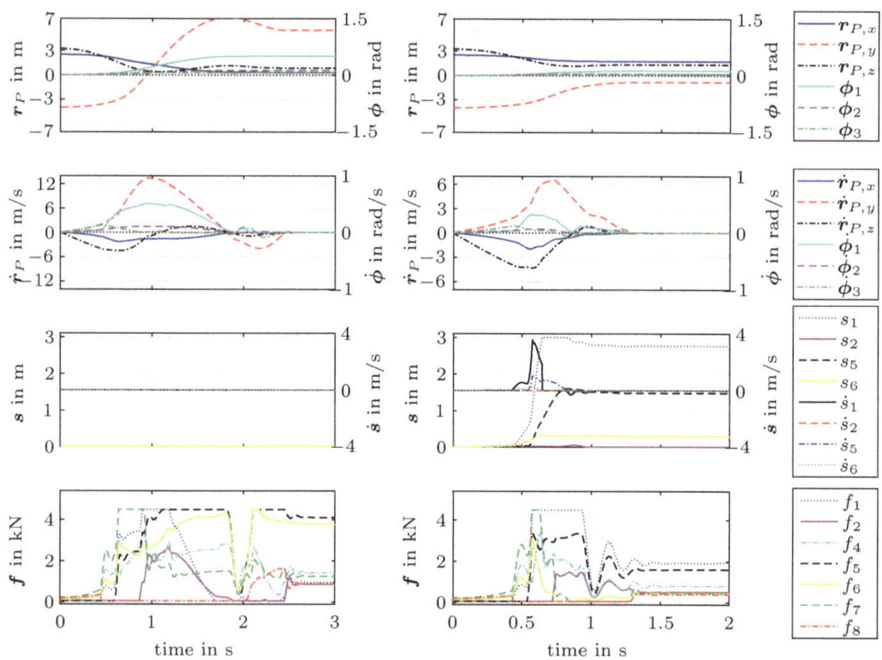

Fig. 4. Pose, velocity, movable pulley positions, movable pulley velocities and cable forces for an emergency scenario after failure of cable 3. Left side: Pulleys fastened. Right side: Pulleys movable.

6 Summary and Outlook

In this work, a recovery strategy minimizing kinetic energy was represented and extended towards use of movable pulleys on actuated linear slides. First, a given prototype for automated masonry was analyzed regarding its static equilibrium workspace with constant orientation. It was demonstrated that the $\mathcal{SEW}$ is strongly reduced after a cable failure. Using global optimization and movable lower pulleys, a geometrical reconfiguration could be identified which restores as much $\mathcal{SEW}$ as possible. Second, dynamic simulations were carried out to investigate the potential of movable pulleys for application in the proposed emergency strategy. The simulation results were compared to a scenario with fastened pulleys. The extended strategy featuring movable pulleys was proven to be beneficial in terms of breaking capability, prevention of a crash and tilting of the system.

In future work, the proposed strategy should be analyzed in terms of systematical tilting avoidance and thus loss of payload prevention. The same accounts for systematical collision avoidance. In more detailed simulation, collisions between cables or platform with obstacles or self-collisions need to be considered. Moreover, simulations should be extended with friction and disturbance modeling as well as modeling of drivetrains. Finally, experiments on a prototype need to be carried out, investigating real-time capabilities of the algorithms. Also acceleration and jerk of the slides need to be limited. Additionally, the proposed model-predictive approach allows also for tracking set point positions over time, taking into account collision avoidance. To apply the proposed strategy within a system with lower redundancy or a suspended system is part of future work. For future prototype design, it has to be decided whether the feature of movable pulleys powered by external drives is reasonable. Studies in this work indicate a benefit in case of cable failure but further investigation is required. As the simulation data show, certain speed and torque capabilities of the linear slide actuators would be needed. All in all this work serves to improve the safety of CDPRs. This might help to pave the way for the introduction of cable robots in industrial practice.

Acknowledgments. This work was supported by the Federal Ministry of Economic Affairs and Energy (BMWi) via the Arbeitsgemeinschaft industrieller Forschungsvereinigungen "Otto von Guericke" e.V. (AiF) within the program Industrial Collective Research (IGF), funding no. 20061 BG - "Entwicklung von Seilrobotern für die Erstellung von Kalksandstein-Mauerwerk auf der Baustelle", based on a resolution of the German Bundestag. It was also funded by the Ministry of Regional Identity, Communities and Local Government, Building and Gender Equality of the Land of North Rhine-Westphalia as part of the funding "Langfristige experimentelle Untersuchung und Demonstration von automatisiertem Mauern und 3D-Druck mit Seilrobotern".

References

1. Berti, A., Gouttefarde, M., Carricato, M.: Dynamic recovery of cable-suspended parallel robots after a cable failure. In: Lenarčič, J., Merlet, J.P. (eds.) Advances in Robot Kinematics 2016, pp. 331–339. Springer (2018). https://doi.org/10.1007/978-3-319-56802-7_35
2. Boschetti, G., Passarini, C., Trevisani, A.: A strategy for moving cable driven robots safely in case of cable failure. In: Boschetti, G., Gasparetto, A. (eds.) Advances in Italian Mechanism Science, pp. 203–211. Springer (2017)
3. Boschetti, G., Minto, R., Trevisani, A.: Improving a cable robot recovery strategy by actuator dynamics. Appl. Sci. **10**(20) (2020). https://doi.org/10.3390/app10207362
4. Boschetti, G., Minto, R., Trevisani, A.: Experimental investigation of a cable robot recovery strategy. Robotics **10**(1) (2021). https://doi.org/10.3390/robotics10010035
5. Boumann, R., Bruckmann, T.: Development of emergency strategies for cable-driven parallel robots after a cable break, pp. 269–280 (2019). https://doi.org/10.1007/978-3-030-20751-9_23
6. Boumann, R., Lemmen, P., Heidel, R., Bruckmann, T.: Optimization of trajectories for cable robots on automated construction sites. In: Proceedings of the 37th ISARC. International Association for Automation and Robotics in Construction (IAARC), Kitakyushu, Japan (2020)
7. Gagliardini, L., Caro, S., Gouttefarde, M., Wenger, P., Girin, A.: A reconfigurable cable-driven parallel robot for sandblasting and painting of large structures. In: Pott, A., Bruckmann, T. (eds.) Cable-Driven Parallel Robots, pp. 275–291. Springer (2015)
8. Ghaffar, A., Hassan, M.: Failure analysis of cable based parallel manipulators. In: Recent Trends in Materials, Mechanical Engineering, Automation and Information Engineering. Applied Mechanics and Materials, vol. 736, pp. 203–210. Trans Tech Publications Ltd (2015). https://doi.org/10.4028/www.scientific.net/AMM.736.203
9. Grüne, L., Pannek, J.: Nonlinear Model Predictive Control, pp. 45–69. Springer (2017). https://doi.org/10.1007/978-3-319-46024-6_3
10. Hahn, H.: Rigid Body Dynamics of Mechanisms: 1 Theoretical Basis. Springer, 1 edn. (2002). https://doi.org/10.1007/978-3-662-04831-3
11. Notash, L.: Wrench recovery for wire-actuated parallel manipulators. In: Padois, V., Bidaud, P., Khatib, O. (eds.) RoManSy 19, pp. 201–208. Springer (2013). https://doi.org/10.1007/978-3-7091-1379-0_25
12. Passarini, C., Zanotto, D., Boschetti, G.: Dynamic trajectory planning for failure recovery in cable-suspended camera systems. J. Mechan. Robot. **11**(2) (2019). https://doi.org/10.1115/1.4041942
13. Pott, A.: Cable-Driven Parallel Robots: Theory and Application. Springer Tracts in Advanced Robotics, Springer (05 2018). https://www.doi.org/10.1007/978-3-319-76138-1
14. Reichert, C., Glogowski, P., Bruckmann, T.: Dynamische Rekonfiguration eines seilbasierten Manipulators zur Verbesserung der mechanischen Steifigkeit. In: Bertram, T., Corves, B., Janschek, K. (eds.) Fachtagung Mechatronik, pp. 91–96. Aachen, Germany (2015). https://doi.org/10.17877/DE290R-7388

15. Roberts, R.G., Graham, T., Lippitt, T.: On the inverse kinematics, statics, and fault tolerance of cable-suspended robots. J. Robot. Syst. **15**(10), 581–597 (1998). https://doi.org/10.1002/(SICI)1097-4563(199810)15:10⟨581::AID-ROB4⟩3.0.CO;2-P
16. Schramm, D., Hiller, M., Bardini, R.: Modellbildung und Simulation der Dynamik von Kraftfahrzeugen. Springer Vieweg, 3 edn. (2018). https://doi.org/10.1007/978-3-662-54481-5

Velocity Based Hybrid Position-Force Control of Cable Robots and Experimental Workspace Analysis

Thomas Reichenbach(✉), Kathrin Rausch, Felix Trautwein, Andreas Pott, and Alexander Verl

Institute for Control Engineering for Machine Tools and Manufacturing Units (ISW), University of Stuttgart, Stuttgart, Germany
thomas.reichenbach@isw.uni-stuttgart.de
http://www.isw.uni-stuttgart.de/

Abstract. One important design criteria of a cable-driven parallel robot is its workspace. Disturbance effects on drive systems, pulley mechanisms and cables have a major impact on the workspace due to their influence on cable forces. Thus, the computed wrench-feasible workspace with ideal kinematic constraints may not be valid while operating the platform of a cable robot. To ensure that the platform moves within the workspace, cable forces can be controlled, whereby kinematic inaccuracies such as calibration errors affecting the platform's positional accuracy. Hence, in this paper a concept of the velocity based hybrid position-force controller is presented in order to pretension cable forces on one hand, and ensure positioning accuracy on the other hand. The controller stability is investigated on extreme poses outside the wrench-feasible workspace hull of a dynamic model of a planar cable robot in simulation environment. Thereupon, the hybrid controller was implemented in a real-time control system and analyzed on the cable robot setup *COPacabana* with experiments inside and outside the computed workspace border. The analysis of measured signals is based on statistical investigation by performing multiple measurements of the same platform trajectory. Both, simulation and experiments show that the proposed velocity based hybrid position-force controller is feasible and can be applied to move the platform outside the workspace.

Keywords: Cable-driven parallel robot · Hybrid position-force control · Experimental workspace analysis · COPacabana

1 Introduction

Cable-driven parallel robots (short: cable robots) are a special type of parallel mechanism which uses cables instead of rigid prismatic actuators to control a platform. This structure of a cable robot consists of a single platform which can be manipulated by varying the lengths of attached cables. Due to their flexibility, cables can be spooled on drums and therefore cover large distances. Conventionally, a drum or winch attached to a fixed frame is used for winding. In addition, cables can be redirected via pulley mechanisms. This leads to advantages

M. Gouttefarde et al. (Eds.): CableCon 2021, MMS 104, pp. 230–242, 2021.
https://doi.org/10.1007/978-3-030-75789-2_19

of cable robots to operate the platform within a large geometrical workspace which is affected by minimal and maximal cable length limitations. However, manipulating the platform is also affected by applied cable forces. Based on the static equilibrium of a cable robot, the cable force distribution can be calculated with position and orientation of the platform. Depending on the number of cables, infinite possible cable force distributions exist for the same pose. The determined force per cable must fulfill certain constraints. For example, depending on the material of cables, a maximum cable force can be applied without violating the cable's breaking load. Damage on the robot can also occur if the permissible force on pulleys or drive systems is exceeded. Another constraint is the lower cable force limit which keeps each cable under pretension and ensures a well-defined stiffness of the robot. A cable which is not preloaded cannot apply forces, and potentially resulting in the loss of degrees of freedom (DOF) of the robot platform. If at least one cable force distribution exists for a considered pose that satisfies the constraint conditions, this pose belongs to the wrench-feasible workspace (WFW) of a cable robot. Conversely, without considering disturbance effects and errors, controlling cable robots within valid cable force distributions ensures that the platform is located or moves in the robot's workspace.

However, considering disturbance effects and errors while positioning cables, leads to a conflict between position controlled drives and controlling cable forces. In the following, a cable robot manipulated with position-controlled drive systems is considered. Applied cable forces on the platform are not reproducible for identical poses, which is also shown later in this work. It is assumed that nonlinear cable slip (for example creep, elongation, and ovalization effects) on the drum and pulleys cause varying forces, Schmidt et al. [13]. Also, hysteresis effects of actuated fibre cables are investigated by Miermeister et al. [5]. Changing cable forces becomes important when moving the platform at extreme poses, for example at the workspace border. In case of manipulating the platform by only controlling cable forces will result in low path accuracy because nonlinear time-invariant dynamics of platform and actuated plastic fiber cables could not be considered in real-time set-point generation. Thus, hybrid position-force control was early investigated by Fang et al. [1] which uses torque feed-forward based on cable force limits computed with nonlinear time-invariant platform dynamics. With a small size and low platform weight of 151 g, platform position errors of up to 2.5 mm may result from the time-variant impact of disturbances while manipulating cable forces. Another hybrid controller design was investigated by Kraus et al. [3] where cable force set-points are computed with Closed-Form method [9] and Quadratic Programming [4]. The cable force controller uses the applied Cartesian wrench on the platform which is computed with cable force sensor signals and delivers motor position offset as output. Kraus et al. describes that the lower frequency of the position controller is mainly applicable for winches with a small bandwidth. With this hybrid controller, the cable robot accuracy at the WFW hull of cable robot *IPAnema3* was experimentally investigated. However, cable force target values must be generated with high frequency. Therefore complex nonlinear dynamics cannot be used in real-time environment. Furthermore, Santos et al. [11,12] describes a method for model predictive control of cable robot dynamics with steel cables which is only possible by assuming a linear time-invariant system at low velocities. Additionally, a hybrid position-force controller was also investigated by Piao et al. [6] using artificial neural network to determine the applied wrench on the platform.

In contrast to the described control strategies, this paper presents a alternative concept of a hybrid position-force controller which uses an angular velocity offset as input for both position and force signals. The controller design is investigated with dynamic simulation of a cable robot and experiments at workspace hull. In order to prove feasibility of the controller concept, multiple experiments were performed for statistical evaluation. With the position-force controlled operation, the WFW of cable robot *COPacabana* [16] is analyzed. Using the velocity control layer for cable force control may also be beneficial for high frequent stimulation of the platform, for example grinding or polishing process.

2 Concept of Hybrid Position-Force Controller

This section focuses on the design of the hybrid position-force controller. Later, the model is used for simulation to find suitable control parameters, proof stability of the force controller of each cable actuator, and developing a procedure for the following experiments to analyze the controller at extreme positions outside the WFW. Considering slip effects while actuating cables leads to a conflict between position controlled drives and controlling cable forces as described Sect. 1. These slip effects appear decentrally on each single cable actuator. Thus, the proposed hybrid controller will deliver output signals into existing cascaded controller of the drive system. To do so, the velocity layer interface is used to achieve a higher bandwidth for controlling cable forces. The hybrid controller was designed acc. to [10], however, position and force feedback were directly used on the axis layer. An overview of the proposed hybrid controller concept is shown in Fig. 1. The cable robot system can be segmented in electronic and mechanic subsystems: drive controller, winch mechanics, cable dynamics and platform dynamics.

Starting with set-point generation, the hybrid controller uses the Cartesian target pose $\boldsymbol{p}_\mathrm{d}$ of the platform as input for inverse kinematics (IK) and to calculate the force distribution of the cable robot. The inverse kinematics delivers the target cable length $l_{\mathrm{d},i}$ for each winch which is calculated to a target angular position $\varphi_{\mathrm{d},i}$ for the attached motor, with $i = 1...m$ number of actuated cables. With the geometrical properties of the cable robot, the inverse kinematics is defined acc. to [8] as

$$l_{\mathrm{d},i} = \|\boldsymbol{l}_{\mathrm{d},i}\|_2 = \|\boldsymbol{a}_i - \boldsymbol{r}_\mathrm{d} - \boldsymbol{R}_\mathrm{d}\boldsymbol{b}_i\|_2, \tag{1}$$

where $\boldsymbol{a}_i$ is the vector from reference coordinate system to frame anchor, $\boldsymbol{b}_i$ is the vector from platform coordinate system to platform anchors, $\boldsymbol{r}_\mathrm{d}$ and $\boldsymbol{R}_\mathrm{d}$ are target position and orientation of the platform, and $\|\cdot\|_2$ as Euclidean Norm. Hereby, $\boldsymbol{p}_\mathrm{d}$ contains Cartesian position $\boldsymbol{r}_\mathrm{d}$ and orientation $\boldsymbol{R}_\mathrm{d}$ as rotation matrix

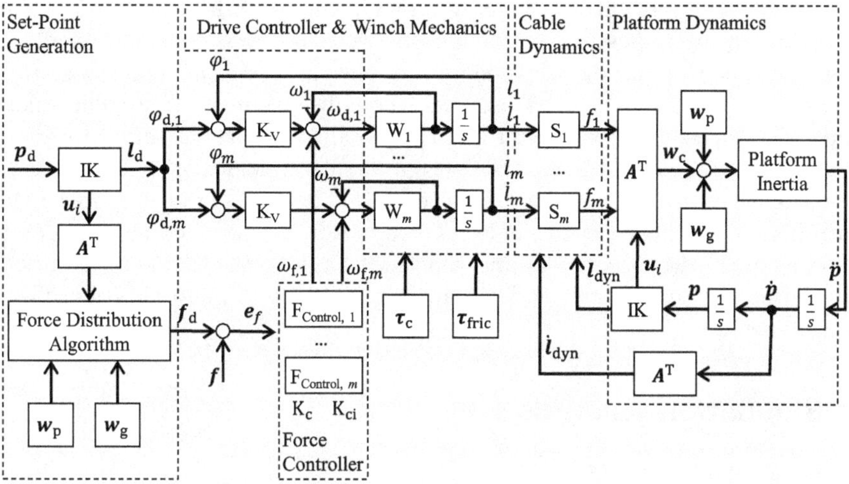

Fig. 1. Schematic overview of a cable robot system with hybrid position-force controller concept

of the cable robot. The position controlled part delivers an angular velocity output

$$\omega_{\mathrm{q},i} = \mathrm{K}_{\mathrm{v}} \mathrm{t}_{\mathrm{W}} \mathrm{t}_{\mathrm{G}} (l_{\mathrm{d},i} - l_i) = \mathrm{K}_{\mathrm{v}} (\varphi_{\mathrm{d},i} - \varphi_i), \tag{2}$$

with φ_i for the actual motor position, K_{v} as position controller gain, t_{G} as transmission ratio for the planetary gear, and t_{W} as transmission ratio from the winch. For set-point generation of target cable forces, the structure matrix $\boldsymbol{A}^{\top}$ (transposed Jacobian) with cable unit vectors $\boldsymbol{u}_i = \frac{\boldsymbol{l}_{\mathrm{d},i}}{\|\boldsymbol{l}_{\mathrm{d},i}\|}$ is used. With reference to [8], the target force $\boldsymbol{f}_{\mathrm{d}} = [f_{\mathrm{d},1}, ..., f_{\mathrm{d},m}]^{\top}$ for each cable are computed by solving static equilibrium of the platform with,

$$\boldsymbol{A}^{\top} \boldsymbol{f}_{\mathrm{d}} + (\boldsymbol{w}_{\mathrm{p}} + \boldsymbol{w}_{\mathrm{g}}) = \boldsymbol{0}, \tag{3}$$

with $\boldsymbol{w}_{\mathrm{p}}$ as applied external wrench vector, and $\boldsymbol{w}_{\mathrm{g}}$ as gravity wrench vector on the platform. To solve the static equilibrium, the Closed-Form method is used because of high computation speed due t o low cycle time $\mathrm{T}_{\mathrm{cyc}} = 0.5$ m s of the control system of *COPacabana*. Thus, dynamic effects are not considered for cable force set-point generation. In addition, using Closed-Form method is beneficial because the solutions are continuous along a trajectory [9]. With cable force error $\boldsymbol{e}_f$ as difference between measured cable forces $\boldsymbol{f} = [f_1, ..., f_m]^{\top}$ on the winch output and target cable forces $\boldsymbol{f}_{\mathrm{d}}$, a second inner closed loop control system can be built. The cable force errors will be minimized by using a PI-controller ($\mathrm{F}_{\mathrm{Control},i}$) with K_{c} and K_{ci} as gain factors. Later, a P-controller is used for hybrid controlled operation ($\mathrm{K}_{\mathrm{ci}} = 0$) and PI-controller is used for a fully force controlled operation. Hence, the cable force controllers deliver angular velocity offset $\omega_{f,i}$ as output. Thus, with adapting of position gain K_v or cable force gain K_c, it is possible to balance between position controlled and cable force controlled operation while manipulating the platform. Following, target velocity $\omega_{d,i} = (\omega_{q,i} + \omega_{f,i})$ is used for the subsequent drive angular velocity controller. This controller is also PI-controlled and outputs the target torque $\tau_{d,i}$ for the drive system, with ω_i as the actual angular velocity of the motor. Later, the current controller uses the target torque as input. As the proposed hybrid controller concept affect each single cable actuator, the next section focus on the modeling of the drive controller and winch mechanics (W_i). Cable Dynamics (S_i) and platform dynamics are modeled according to [14,15], and also described in the next section.

3 Simulation Analysis

Following, platform, cable, winch and drive dynamic models are described to estimate control parameters and stability of the proposed hybrid position-force controller in simulation environment. In addition, the controller behavior was investigated at extreme poses outside the WFW. In Sect. 3.2 the determined controller parameters and simulation results will be described and discussed.

3.1 Subsystem Modeling

Current Controller and Drive Electronics. Figure 2 shows the block diagram of drive electronics and its current controller. In case of a synchronous motor, which is used for cable robot *COPacabana*, the torque generating current is estimated as an PT1 transmission function. The gain of the PT1 element is reciprocal to winding resistance R of the motor. The time constant T_{el} results from winding inductance L and winding resistance R according to [18]. The PI current controller (K_{pi}, T_{ni}) sets target motor voltage $U_{d,i}$ with the error between target motor current $I_{d,i}$ and actual current I_i as input. The target motor voltage is reduced by the value of induced voltage $U_{ind,i}$ which is generated during the movement. $U_{ind,i}$ is calculated with the actual angular motor velocity ω_i and voltage constant K_s. With the motor constant K_m, motor current can be converted into motor torque and vice versa.

Winch Mechanics. Winch mechanics are schematically shown in Fig. 3. From the motor's current controller on the input side, motor torque $\tau_{res,i}$ can be calculated with friction torque $\tau_{fric,i}$ and converted disturbance torque of the cables due to preloading $\tau_{c,i}$ with drum radius R_D and transmission ratio of the planetary gear t_G. With inertia of the winch system J_W, resulting acceleration α_i and angular speed ω_i of drive system and drum, cable velocity $\dot{l}_i$ and position l_i can be calculated by integration ($\frac{1}{s}$) and multiplication with transmission ratio of winch t_W and planetary gear t_G, respectively.

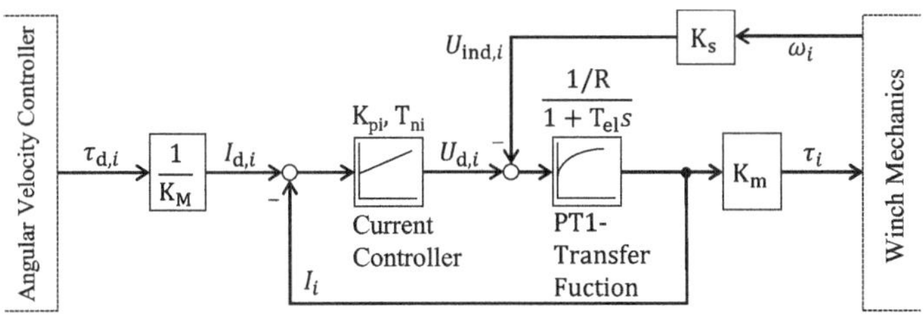

Fig. 2. Block diagram of current controller and drive electronics

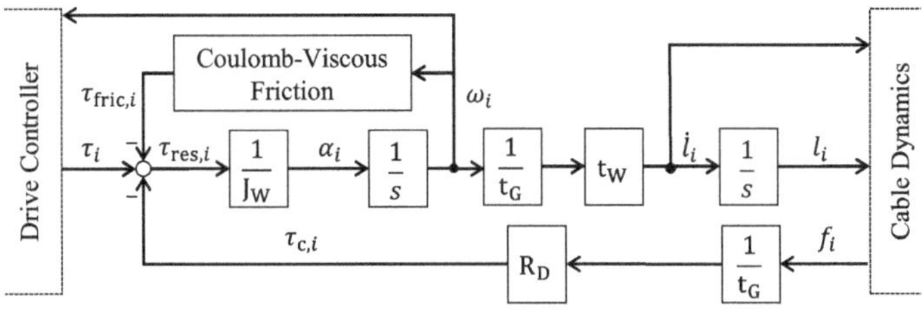

Fig. 3. Block diagram of winch mechanics with $\mathrm{t_W} = 2\pi\mathrm{R_D}$

Cable Dynamics. Dynamic behavior of fibre cables can be simplified to a parallel spring-damper system (Kelvin-Voigt material), according to Tempel [14]. Therefore, output of this subsystem are applied cable forces on the platform, which are computed with variable cable lengths $\Delta l_i = l_i - l_{\mathrm{dyn},i}$ and its time derivative $\Delta \dot{l}_i = \dot{l}_i - \dot{l}_{\mathrm{dyn},i}$, with stiffness constant $\mathrm{E_c}$ and damping constant $\mathrm{D_c}$, neglecting time-variant effects like ovalization and creep.

$$f_{\mathrm{c},i} = \mathrm{E_c}\Delta l_i + \mathrm{D_c}\Delta \dot{l}_i, \tag{4}$$

Platform Dynamics. According to Pott [8] forward platform dynamics can be formulated with Newton-Euler-Equations to

$$\boldsymbol{M}\ddot{\boldsymbol{p}} - \boldsymbol{D}\dot{\boldsymbol{p}} + \boldsymbol{g}^{\mathrm{C}} = \boldsymbol{w}_{\mathrm{c}} + \boldsymbol{w}_{\mathrm{p}} + \boldsymbol{w}_{\mathrm{g}} \tag{5}$$

where $\boldsymbol{M}$ is the mass matrix, $\ddot{\boldsymbol{p}}$ is the second order time derivative of the platforms generalized coordinates, $\boldsymbol{w}_{\mathrm{c}} = \boldsymbol{A}^{\top}\boldsymbol{f}$ is the applied wrench on the platform due to cable forces, $\boldsymbol{w}_{\mathrm{p}}$ is the applied external wrench, and $\boldsymbol{w}_{\mathrm{g}}$ is the resulting wrench of gravity forces. In Fig. 1 the actual platform pose $\boldsymbol{p}$ can be calculated by integration $(\frac{1}{s})$ of its derivatives. Since the platform moves slowly and in a thin medium like air, for example, Coriolis effects $\boldsymbol{g}^{\mathrm{C}}$ and damping $\boldsymbol{D}$ will be neglected.

3.2 Simulation Results

In order to investigate control parameters and stability of the proposed hybrid position-force controller, the presented models are used for simulation. Therefore, a planar cable robot model with four cables is implemented in MATLAB

Simulink 2020b. The cable robot frame and platform trajectory is shown in Fig. 5, further properties of the cable robot model can be checked out on Github[1]. The dynamic properties of the cable actuators have been selected according to information given in datasheets of *COPacabana*, see Trauwein et al. [16]. Figure 4 shows the bode diagram of an open-loop and closed-loop force controlled system. The force controlled system shows stable behavior with a bandwidth 150 Hz and a gain margin of 42 dB. With the control of drive position and cable force, extreme points outside the computed WFW of cable robot are also experimentally investigated, which is shown in Fig. 5. The platform moves from reference point in the center to upper left cable outlet point. The red cross marks the position where at least one cable force exceeds or falls below the cable force limit. While moving further, the minimum or maximum cable force will be hold at the limit value. Moving the platform beyond the red crossed coordinate will result in lower path accuracy while manipulating, as expected. The red crossed coordinate is outside the workspace because a low stiffness of the cables was chosen to investigate the stability of the controller at extreme values.

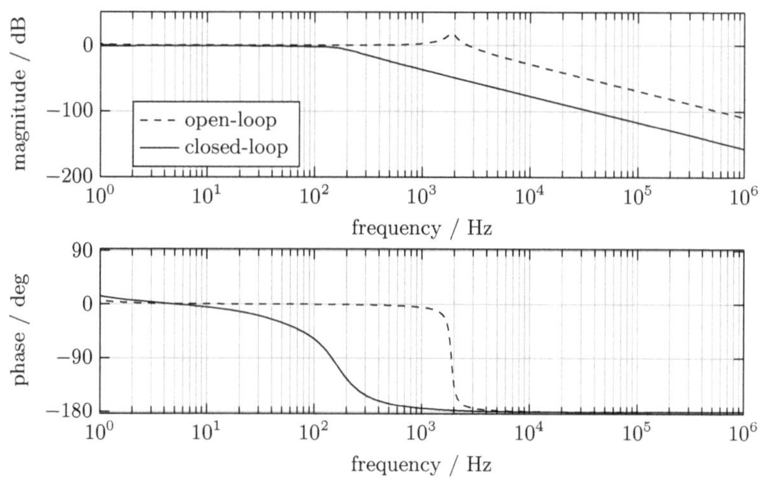

Fig. 4. Bode diagram of the force controlled system in open-loop and closed-loop

4 Experimental Analysis at the Workspace Border

4.1 Experimental Setup

Cable robot *COPacabana* at the *ISW* is a redundantly restrained positioning mechanism (RRPM) with a 3R3T motion pattern, according to [17]. Depending on the requirements, the cable robot can be manipulated with up to 16 built-in

[1] https://github.com/iswunistuttgart/cable-robot-hybrid-position-force-control-simulation.git.

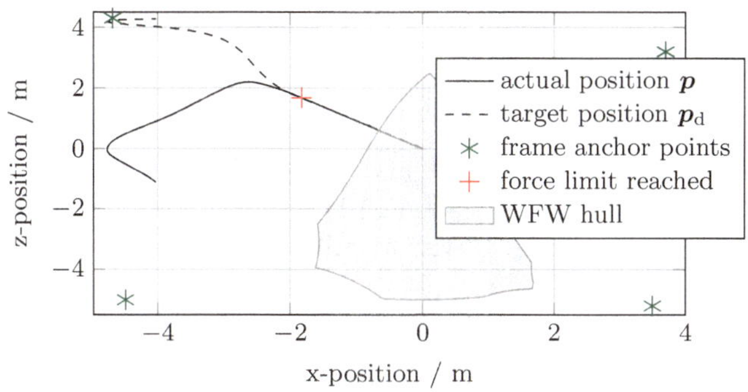

Fig. 5. Dynamic behavior of the platform with hybrid position-force controller

winch systems. For this work, eight winches are used for the movement of the cuboid platform. The control program *TwinCAT 3* from *Beckhoff* is installed on an industrial PC with Windows operating system and communicates via *EtherCAT* bus with *Rexroth* drive components. Winch systems manipulate the robot platform with a mass of 11.6 kg and dimension of $(0.3 \times 0.4 \times 1)$ m^3 within the workspace. Inverse and forward kinematics as well as the proposed hybrid position force controller are integrated via *TwinCAT-TcCOM* interface programmed in C++. The platform of cable robot can be manipulated using standardized G-Code and cable force controller can be simply activated using M-function calls, see [2]. The characteristics of *COPacabana* were presented in [16], which can be used as reference to describe the overall experimental setup.

4.2 Experiment Procedure

To move the platform to extreme points, the WFW is computed with hull method, according to [7]. The Closed-Form method is used to compute cable force distribution for the WFW which can be assumed as non-singular since no cable force becomes infinite or negative. Hull points (grey) of the ideal computed WFW is shown in Fig. 6. The blue crossed coordinates were chosen to extreme poses to analyze the proposed hybrid position-force controller. Green points symbolize frame anchors of *COPacabana* and Fig. 6 shows the platform pose in the origin of the reference coordinate system. To investigate the hybrid position-force controller, the following procedure was developed:

1. The platform is pretensioned and moves from the origin to an inner or outer pose of the wrench feasible workspace hull. Inner and outer poses have a predefined distance with a tolerance of $\Delta|p| = 0.1$ m to each investigation point. Inner and outer poses are reached with hybrid position-force controlled operation.
2. Position part of hybrid controller is deactivated by using PI-controller. Hence, cables are fully force controlled.

3. The Cable robots pose is observed in operational space, and the drive system position as well as the cable force are observed in joint space.
4. The Platform will be moved back to the origin and will be pretensioned.

For five investigation points, inner and outer pose from worksapce with ten iterations are considered which leads to 100 performed measurements. With ten iterations, measurement data is statistically analyzed to show repeatability of results.

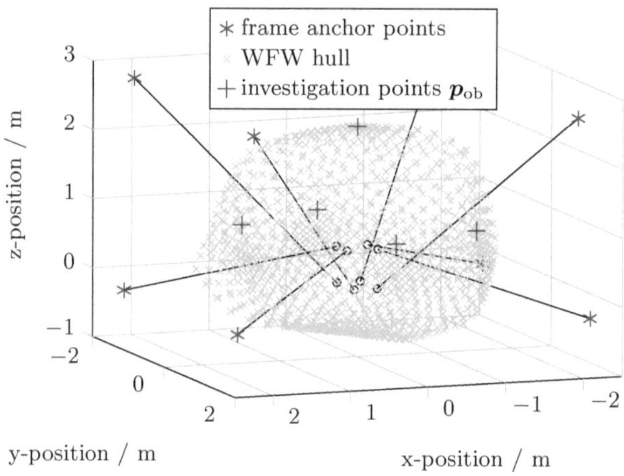

Fig. 6. WFW hull of COPacabana with investigation points

4.3 Results and Discussion

First, the measured cable forces and motor positions in joint space are analyzed. As for each cable similar effects were observed, only data of cable #5 is shown in Fig. 7 for investigation point $\boldsymbol{p}_{\text{ob}} = [0, 0, 2.33]^{\top}$m. The graphs show the mean values of all experiments for measured cable forces and motor position at inner and outer pose of workspace hull. Additionally, winch position error from target is shown. Hereby, it has to be noticed that the initial error of approx. 17 mm results from cable pretension as mentioned in Sect. 4.2.

The target cable force of the outer pose reaches and holds the lower force limit at about 55.5 s and thus violates the workspace condition. While hybrid controlled, the actual cable force f_5 falls below force limit at about 40 s but is still tensioned due to generated offset by gain of force controller. With fully force controlled cables at about 65 s, motor position error $\Delta\varphi_5$ increases by up to 27 mm and hence cable force is increasing. At inner poses, the motor position error increases only by 19.5 mm to 20 mm which is comparatively small to the motor position error at outer poses. Thus, the measured values show a loss of accuracy at motor position when reaching the force limit while robot is fully force controlled. As in the simulation model, this loss of accuracy can be

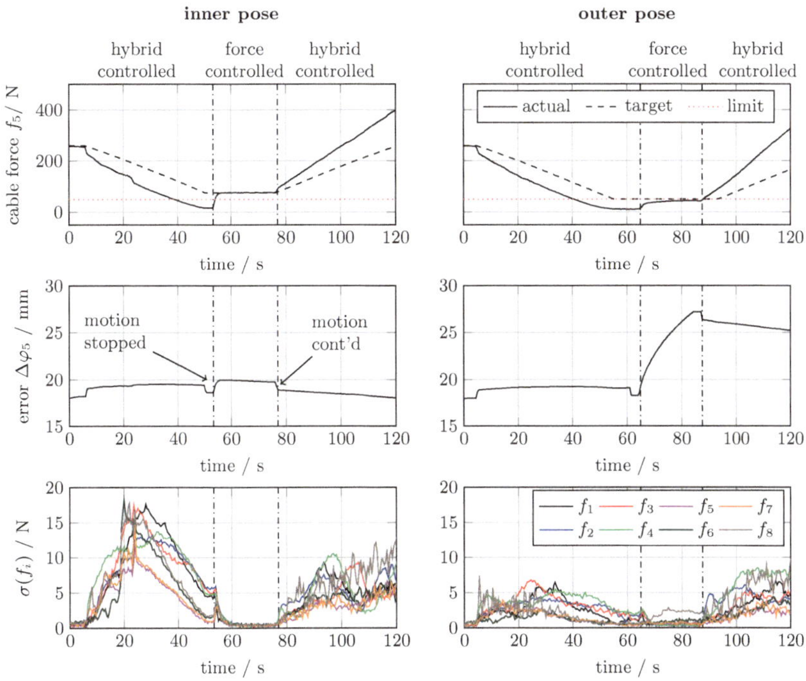

Fig. 7. Results of measured cable forces f_5, position error $\Delta\varphi_5 = \mathrm{t_W t_G}(\varphi_{\mathrm{d},5} - \varphi_5)$, and standard deviation of all measurements per cable force $\sigma(f_i)$

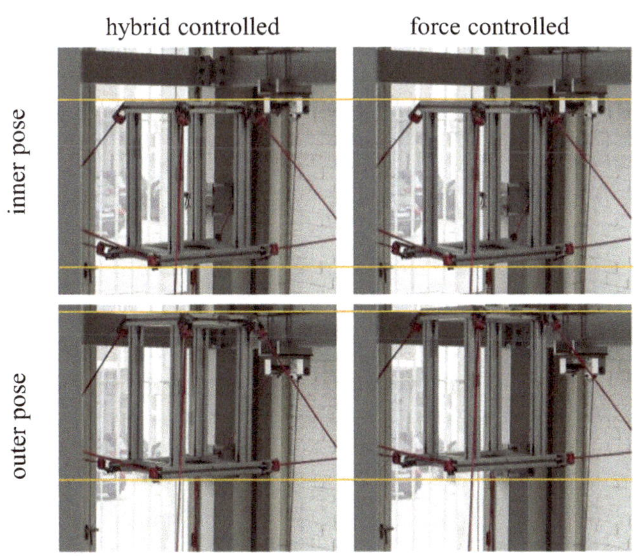

Fig. 8. Camera measurement for inner and outer pose with hybrid controlled and fully force controlled system at $\boldsymbol{p}_{\mathrm{ob}} = [0, 0, 2.33]^\top \mathrm{m}$

explained by the limitation of determined cable force distribution limits $\boldsymbol{f}_{\text{lim}} = [f_{\min}, f_{\max}]^\top = [50, 700]^\top\text{N}$. As soon as a cable force exceeds or falls below the range of the permissible cable forces, the corresponding cable force is limited and thus no longer fits the specified pose. The angular velocity offset value which is determined by the force controller, also generates a change in the drive position of the robot that does not match the target pose. In all performed experiments, only the lower and never the upper force limit is reached. Thus, in the next step, the gain parameters of the hybrid controller can be increased and investigated. Because with ease of balancing between position and force, this hybrid controller concept may also conceivable for design of adaptive control via gain parameters with considering platform pose and velocity.

Analyzing standard deviations of the experiments, it is noticeable that the scatter of the measured cable forces has its minimum for all experiments as soon as the cables are fully force controlled. In hybrid controlled operation, the scatter varies differently for each cable and experiment. The maximum standard deviation of the cable forces for all experiments is about 32 N which is 21% of the measured mean cable force with 150 N and occurs when investigating the workspace at inner pose. Standard deviations of the cable forces show that with the hybrid position-force controller, different cable forces are applied for each measurement for the same given trajectory of the robot platform. Thus, inside the WFW the position part of the hybrid controller predominates, while outside the WFW the force part predominates, since the force error increases compared to the position error. Furthermore, platform pose is qualitatively observed with a camera due to lack of cost-intensive absolute coordinate measuring device. Figure 8 shows the platform for inner and outer pose for both, hybrid position-force controlled and fully force controlled. The yellow lines mark the static position of the platform in hybrid controlled operation mode for the inner and outer pose. At outer poses, while fully force controlled, the platform pose is slightly higher than in hybrid controlled operation, whereas platform poses not change at inner poses for both operation modes. Hence, the platform can be operated outside the workspace hull with the hybrid controller which allows to slightly pretension the cables with higher accuracy compared to fully force control the platform.

5 Conclusion

In this work, a hybrid position-force controller concept is presented which uses angular velocity offset as output. The controller concept is parameterized and firstly investigated with a dynamic model of a planar cable robot. First results show, that the proposed controller concept may achieve high bandwidth which has to be investigated in detail in the next steps of research. Additionally, the controller was integrated in real-time control system of *COPacabana*. Cable forces and motor positions were observed at geometrical extreme points outside the hull of the WFW. In addition, the platform pose was observed qualitatively. Finally, the experimental results shows that the hybrid controller allows a properly manipulation of the platform inside the workspace, and also pretensioned cables outside the workspace, which may be beneficial in case of emergency operation of a cable robot. Additionally, this concept allows the possibility to adaptively change gain factors of position and force controllers while in operation to trust either motor position or cable forces.

Acknowledgments. This work was supported by the Graduate School of Excellence advanced Manufacturing Engineering (GSaME) at the University of Stuttgart.

References

1. Fang, S., Franitza, D., Torlo, M., Bekes, F., Hiller, M.: Motion control of a tendon-based parallel manipulator using optimal tension distribution. IEEE/ASME Trans. Mechatron. **9**(3), 561–568 (2004)
2. International Organization for Standardization: ISO 6983. Automation systems and integration - Numerical control of machines - Program format and definitions of address words, December 2009
3. Kraus, W., Miermeister, P., Schmidt, V., Pott, A.: Hybrid position-force control of a cable-driven parallel robot with experimental evaluation. Mech. Sci. **6**(2), 119–125 (2015)
4. Mattingley, J., Boyd, S.: CVXGEN: a code generator for embedded convex optimization. Optim. Eng. **13**(1), 1–27 (2012). https://doi.org/10.1007/s11081-011-9176-9
5. Miermeister, P., Kraus, W., Lan, T., Pott, A.: An elastic cable model for cable-driven parallel robots including hysteresis effects. In: Pott, A., Bruckmann, T. (eds.) Proceedings of the Second International Conference on Cable-Driven Parallel Robots (CableCon 2014). Mechanisms and Machine Science, vol. 32, pp. 17–28. Springer, Cham (2015)
6. Piao, J., et al.: Indirect force control of a cable-driven parallel robot: tension estimation using artificial neural network trained by force sensor measurements. Sensors (Basel Switz.) **19**(11), 2520 (2019). https://www.mdpi.com/1424-8220/19/11/2520
7. Pott, A.: Forward kinematics and workspace determination of a wire robot for industrial applications. In: Lenarčič, J., Wenger, P. (eds.) Proceedings of the 11th International Conference on Advances in Robot Kinematics (ARK 2008), pp. 451–458. Springer (2008)
8. Pott, A.: Cable-Driven Parallel Robots: Theory and Application. Springer Tracts in Advanced Robotics, vol. 120. Springer, Cham (2018)
9. Pott, A., Bruckmann, T., Mikelsons, L.: Closed-form force distribution for parallel wire robots. In: Kecskeméthy, A., Müller, A. (eds.) Computational Kinematics, pp. 25–34. Springer, Berlin (2009)
10. Raibert, M.H., Craig, J.J.: Hybrid position/force control of manipulators. J. Dyn. Syst. Meas. Contr. **102**(2), 126–133 (1981)
11. Santos, J.C., Chemori, A., Gouttefarde, M.: Model predictive control of large-dimension cable-driven parallel robots. In: Pott, A., Bruckmann, T. (eds.) Proceedings of the Fourth International Conference on Cable-Driven Parallel Robots (CableCon 2019). Mechanisms and Machine Science, vol. 74, pp. 221–232. Springer, Cham (2019)

12. Santos, J.C., Chemori, A., Gouttefarde, M.: Redundancy resolution integrated model predictive control of CDPRS: concept, implementation and experiments. In: IEEE/RAS International Conference on Robotics and Automation (ICRA 2020). IEEE (2020)
13. Schmidt, V., Pott, A.: Investigating the effect of cable force on winch winding accuracy for cable-driven parallel robots. Proc. Inst. Mech. Eng. Part K J. Multibody Dyn. **24**, 315–323 (2015)
14. Tempel, P.: Dynamics of cable-driven parallel robots with elastic and flexible, time-varying length cables. Ph.D. thesis, Stuttgart (2019)
15. Tempel, P., Miermeister, P., Lechler, A., Pott, A.: Modelling of kinematics and dynamics of the IPAnema 3 cable robot for simulative analysis. Appl. Mech. Mater. **794**, 419–426 (2015)
16. Trautwein, F., Reichenbach, T., Tempel, P., Pott, A., Verl, A.: COPacabana: a modular cable-driven parallel robot. In: Pfurner, M., Dohnal, F. (eds.) Sechste IFToMM D-A-CH Konferenz (2020)
17. Verhoeven, R.: Analysis of the workspace of tendon-based stewart platforms. Ph.D. thesis, University of Duisburg-Essen, Duisburg, Germany (2004)
18. Zirn, O., Weikert, S.: Modellbildung und Simulation hochdynamischer Fertigungssysteme: Eine praxisnahe Einführung. Springer, Heidelberg (2006)

Constrained Optimal Control of Cable-Driven Parallel Robots Based on SDRE

Hanie Marufkhani(✉) and Mohammad A. Khosravi(✉)

Amirkabir University of Technology (Tehran Polytechnic), Tehran, Iran
{marufkhanie,m.a.khosravi}@aut.ac.ir

Abstract. This paper addresses constrained optimal control of *Cable-Driven Parallel Robots* (CDPRs) in details. As it is known, in a CDPR mechanism, cables should remain in tension at all movement maneuvers in contrast to conventional robots. In this paper, considering this constraint, an optimal control scheme is presented for this class of robots. To this end, first, the general dynamic model of the CDPR is converted to state-dependent coefficient (SDC) linear structure. In this model, uncertainties of the cable robot are included as a vector in the SDC linear structure. Next, a constrained optimal control algorithm based on the Hamiltonian and *Karush-Kuhn-Tucker* (KKT) conditions is proposed using the state-dependent coefficient parameterization form. The proposed control scheme is formed based on the *state-dependent Riccati equation* (SDRE), and the stability of the closed-loop system is proved using Lyapunov second method. Simulation results show the effectiveness of the proposed control algorithm in terms of tracking and coping with the uncertainties and external disturbances.

Keywords: CDPR · SDRE · Optimality · Lyapunov

1 Introduction

Cable-driven parallel robots are a special type of parallel robot that the pose of the end-effector is affected by changing the length of the cables. In this structure, rigid links are replaced by the cables, and as a result, the weight of the robot is greatly reduced [2]. Therefore, this structure enables the robot to gain high velocity and acceleration compared to the conventional robots [7,9,17]. The most significant challenge in control of the CDPRs is that the cables can apply only tensile forces [7,12,17]. Due to this constraint, cable-driven parallel robots are classified into two types: fully-constrained and under-constrained. In under-constrained CDPRs, passive force ensures that all cables are under tension. While in the fully-constrained type, the usage of the internal force concept remains the cables on tension [1,9].

Supported by Electrical Engineering Department of Tehran Polytechnic.

M. Gouttefarde et al. (Eds.): CableCon 2021, MMS 104, pp. 243–259, 2021.
https://doi.org/10.1007/978-3-030-75789-2_20

Since the cables can only apply tensile forces, the control of cable robots is a challenging problem. Several methods have been proposed to control cable-driven parallel robots until now, such as PID [18], a combination of PI and fuzzy, improved feedback linearizing and Lyapunov based controllers [16], PD controller with gravity compensator [6], robust adaptive [1,5], and robust PD controllers [9].

In the control schemes as mentioned above, the internal force concept is utilized to ensure that the tensionability condition is satisfied for all the cables. However, it should be noted that according to the kinematic uncertainties, employing the internal force concept, which is obtained by calculating the null space of transpose of the Jacobian matrix, will cause problems in precisely controlling the movement of the end-effector. Therefore, we face the optimization problem in the design of the CDPR controller. Based on this fact and to avoid using internal force in the control algorithm as well as considering the optimization of force distribution, constrained optimal control can be a suitable alternative for control of the CDPRs. This control scheme can be formed based on *Hamilton-Jacobi-Bellman* (HJB) equations along with KKT conditions.

In the optimal algorithms, a trade-off must take place between minimizing control inputs and errors. To implement the optimal controller, it is necessary to express the nonlinear dynamics of the system as state space. If nonlinear constraints, including complex and intricate dynamics, are added to the Hamiltonian function, nonlinear terms and complexity become more by differentiating from the Hamiltonian function. Therefore, obtaining a linear model or structure of a nonlinear system is one of the most important problem theory aspects. State-Dependent Riccati Equation strategy is well known, advantageous, developed, and one of the most popular algorithms in controlling nonlinear feedback due to nonlinearities in the system. Besides, it offers a superb flexible design with regard to the use of state-dependent weighted matrices [3,14]. In recent years, the SDRE method has been efficient in designing nonlinear controllers, observers, and filters because it has a flexible design, simplicity, and real-time execution [10]. Therefore, the SDRE approach can be a great method to implement a controller for the CDPR to satisfy all optimization goals.

Since we do not want to use a quasi-positive internal force to control CDPRs, we turn to optimal control by defining the Hamiltonian function. As expressed before, the main input of the CDPR dynamic system is constrained. Thus, only the first Hamiltonian condition can not be used to design the optimal controller. Also, the KKT conditions should be used as generalized conditions of the first condition of Hamiltonian. The second and third conditions of Hamiltonian in the optimal control design procedure are used. It is noteworthy that the SDRE necessary condition for optimality is negligible and close to zero. Thus, the SDRE optimality criterion can be ignored. As a result, the suboptimal SDRE is used in the present paper.

In this paper, the dynamics of cable-driven parallel robots and the control of such robots are discussed. Section 2 is dedicated to the dynamics of the fully-constrained cable-driven parallel robot. In Sect. 3, The mathematical formulation of a nonlinear CDPR system will be expressed in the form of a state-dependent coefficient, which is necessary to use the SDRE method. In Sect. 4, designing an optimal constrained controller is described using the Karush Kuhn Tucker theorem. After that, the stability of the proposed methods is proved using the Lyapunov function. In Sect. 5, the system is simulated with variable weighted matrices in the proposed control algorithm in the presence of uncertainty and disturbance. Finally, the internal force method is simulated based on just the first Hamiltonian condition. These methods are implemented on the 3 DOF cable robot. At the end of Sect. 5, the proposed control method and internal force results are compared. The final Sect. 6 will present the summarized contents and conclusions.

2 CDPR Dynamics

Schematic diagram of a CDPR is shown in Fig. 1. Based on this structure and using [9], dynamic model of an n DOF fully-constrained CDPR with nonelastic massless cables can be written as,

$$M(x)\ddot{x} + C(x,\dot{x})\dot{x} + G(x) + T_d = -J^T(x)\tau_{total} = f_{total} \tag{1}$$

where $x = [x_p, y_p, z_p, \alpha, \beta, \varphi]^T$ is generalized coordinates vector that $\theta = [\alpha, \beta, \varphi]^T$ represents Euler-angles. $M(x)$ is the mass matrix, $C(x,\dot{x})$ refers to the Coriolis/centripetal matrix, $G(x)$ denotes the gravity vector, and $J(x)$ introduce the Jacobin matrix of the robot. T_d denotes the effects of disturbance and unmodeled dynamics. $\tau_{total} \geq [\mathbf{0}]_{n\times 1}$ is the vector of cable tension. f_{total} is the vector of the applied force by the cables on the platform.

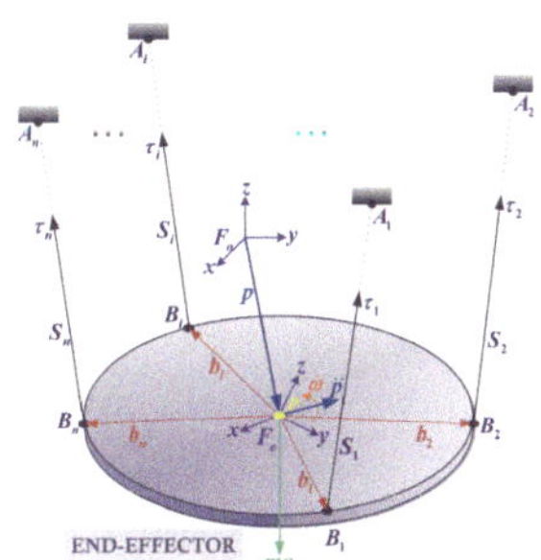

Fig. 1. General Dynamic Schematic of CDPR

2.1 Nominal Model of CDPR

In this paper, the estimated model of the CDPR is used because the precise model of the system is not available. Moreover, the uncertainties are taken into account in the nominal model. The CDPR dynamics (1) is rewritten as follows.

$$\hat{M}(x)\ddot{x} + \hat{C}(x,\dot{x})\dot{x} + \hat{G}(x) + T_d = -J^T(x)\tau_{total} = f_{total} \tag{2}$$

where $\hat{M}(x) = M(x) + \underline{\overline{\delta M}}$, $\hat{C}(x,\dot{x}) = C(x,\dot{x}) + \underline{\overline{\delta C}}$, and $\hat{G}(x) = G(x) + \underline{\overline{\delta G}}$ are the dynamic uncertainties in the model of inertia, Coriolis/centripetal matrices, and gravity vector, respectively. As a result, the CDPR system (2) becomes:

$$M(x)\ddot{x} + C(x,\dot{x})\dot{x} + G(x) + D = -J^T(x)\tau_{total} = f_{total} \tag{3}$$

in which,

$$D = T_d + \underline{\overline{\delta M}}\ddot{x} + \underline{\overline{\delta C}}\dot{x} + \underline{\overline{\delta G}} \tag{4}$$

The vector D is a model of the uncertainties and disturbances vector. The nominal motion equation (3) has the following essential properties (**Property 2.1** to **Property 2.5**) [8]:

$$0 < \underline{m} \le \|M(x)\| \le \overline{m} < \infty \tag{5}$$

$$\|C(x,\dot{x})\| \le \xi_c \|\dot{x}\| \tag{6}$$

$$\chi^T(\dot{M}(x) - 2C(x,\dot{x}))\chi = 0 \tag{7}$$

$$\|G(x)\| \le \xi_g \tag{8}$$

$$\|T_d\| \le \xi_{td} \tag{9}$$

See [8] for more information. Besides, the following property is satisfied.
Property 2.6: The vector D has an upper bound.

$$\|D\| \le \xi_d + \xi_{\underline{\overline{\delta M}}} \|\ddot{x}\| + \xi_{\underline{\overline{\delta C}}} \|\dot{x}\| \tag{10}$$

where $\left\|T_d + \underline{\overline{\delta G}}\right\| \le \xi_d$. ξ_d, $\xi_{\underline{\overline{\delta M}}}$, and $\xi_{\underline{\overline{\delta C}}}$ are known and positive scalar values.

3 SDC Parameterization

A nonlinear time-varying system with uncertainty and disturbance is considered in the form of the state space structure.

$$\dot{z}(t) = f(z(t)) + g(z(t), u(t)) + \Gamma(z(t), t) \tag{11}$$

in which $z(t) \in \Re^{n\times 1}$ is the state vector of the system, $u(t) \in \Re^{m\times 1}$ is the input vector, the vector $\Gamma(z(t),t) : \Re^{n\times 1}\times\Re \to \Re^{n\times 1}$ models all system uncertainties, unknowns, and disturbances, and $t \in \Re$ is time. The vectors $f(z(t)) : \Re^{n\times 1} \to \Re^{n\times 1}$ and $g(z(t), u(t)) : \Re^{n\times m} \times \Re^{m\times 1} \to \Re^{n\times 1}$ are continuously differentiable. Also, they satisfy Lipschitz conditions.

The state-dependent coefficient parameterization of (11) is accomplished as,

$$\dot{z}(t) = A(z(t))z(t) + B(z(t))u(t) + \Gamma(z(t), t) \tag{12}$$

where $A(z(t)) : \Re^{n\times n} \to \Re^{n\times n}$ is the system matrix, and $B(z(t)) : \Re^{n\times m} \to \Re^{n\times m}$ is the input matrix. The matrices $A(z(t))$ and $B(z(t))$ are continuous and at least one time differentiable. Also, pair $\langle A(z(t)), B(z(t))\rangle$ has to be controllable (stabilizable) for all $z(t) \in \Re^{n\times 1}$ and $u(t) \in \Re^{m\times 1}$ [11].

3.1 SDC Parameterization of CDPR

The applied wrench f_{total} to the moving platform is considered as the input of the SDC to obtain the linear structure state space of the estimated dynamic system of the cable robot. Moreover, in dynamic model (3), the applied wrench is considered as (13) to eliminate the gravitational effect $G(x)$. Since we aim to convert the nonlinear dynamic equation of the CDPR to the linearized SDC structure using the factorization method, the gravity vector effect is eliminated in the first design step. In general, the mentioned vector depends on the position of the cable robot. In the end, the effect of the gravity vector is added to the main control input (i.e., the vector τ_{total}) when the design is complete.

$$-J^T(x)\tau_{total} = f_{total} = f + G(x) \tag{13}$$

Substituting (13) in (3), results in,

$$M(x)\ddot{x} + C(x,\dot{x})\dot{x} + D = f \tag{14}$$

Or,

$$\ddot{x} = M^{-1}(x)(f - C(x,\dot{x})\dot{x} - D) \tag{15}$$

Where,

$$x = [x_p, y_p, z_p, \alpha, \beta, \varphi]^T = [x_1, x_2, x_3, x_4, x_5, x_6]^T \tag{16}$$

With this notation, the state vector can be defined as,

$$z = [x_1, x_2, x_3, x_4, x_5, x_6, \dot{x}_1, \dot{x}_2, \dot{x}_3, \dot{x}_4, \dot{x}_5, \dot{x}_6]^T \tag{17}$$

$$z = [z_1, z_2, z_3, z_4, z_5, z_6, z_7, z_8, z_9, z_{10}, z_{11}, z_{12}]^T \tag{18}$$

and as a result,

$$\dot{z} = \left[z_7, z_8, z_9, z_{10}, z_{11}, z_{12}, M^{-1}(z)(f - C(z)z - D)\right]^T \tag{19}$$

With respect to (19), the SDC parameterization form is concluded as,

$$\dot{z}(t) = A(z(t))z(t) + B(z(t))u(t) + \Gamma(z(t),t) \tag{20a}$$

$$A_{2j\times 2j}(z(t)) = \begin{bmatrix} 0_{j\times j} & I_{j\times j} \\ 0_{j\times j} & -\left(M^{-1}(z(t))C(z(t))\right)_{j\times j} \end{bmatrix} \tag{20b}$$

$$B_{2j\times j}(z(t)) = \begin{bmatrix} 0_{j\times j} \\ \left(M^{-1}(z(t))\right)_{j\times j} \end{bmatrix} \tag{20c}$$

$$\Gamma_{2j\times 1}(z(t),t) = \begin{bmatrix} 0_{j\times 1} \\ -\left(M^{-1}(z(t))D\right)_{j\times 1} \end{bmatrix} \tag{20d}$$

$$u_{j\times 1}(t) = f_{j\times 1} = -\left(J^T(z(t))\right)_{j\times n}\tau_{n\times 1} \tag{20e}$$

where I is the identity matrix. According to the dynamic characteristics in Appendix A, pair $\langle A(z(t)), B(z(t))\rangle$ is controllable. In (20b) to (20d), j represents the degrees of freedom of the fully-constrained CDPR. It should be noted that the constrained input is $\tau_{total} \geq 0$. The purpose of this statement is to point out that the constraint should be applied to the main cable tension, i.e., τ_{total}. "*The tension of the cables is always non-negative*" is the essential constraint. As can be seen in (20d), the vector $\Gamma(z(t), t)$ includes uncertainties and unwanted disturbances.

4 Proposed Control Algorithm

In this section, due to the constraint on the force exerted by the actuators, we will control the cable-driven parallel robot based on HJB and KKT conditions within the controllable workspace.

4.1 Constrained Optimal Control of CDPR

In order to evaluate the system performance quantitatively in the optimization problem, the objective function with the infinite horizon is defined as Eq. (21) to minimize input and tracking error in the optimization problem.

$$J(e, f) = \frac{1}{2}\int_0^\infty \left[e^T \cdot Q(z) \cdot e + f^T \cdot R(z) \cdot f\right] dt \tag{21}$$

Matrices $Q(z) = Q^T(z);\;\; Q(z) \geq 0$ and $R(z) = R^T(z);\;\; R(z) > 0$ are weighted matrices that determine the cost and worth of each element. Given system (20a), the tracking error vector is defined as (22).

$$e = w - z \tag{22}$$

where $w \in \Re^{n\times 1}$ is the reference trajectory.

The Hamiltonian function is defined to obtain the control law that minimizes the cost function (21) and is related to the state-dependent coefficient state space (20a).

$$\begin{aligned} H(e, f, P) = &\tfrac{1}{2}(e^T \cdot Q(z) \cdot e + \tau^T \cdot J(z) \cdot R(z) \cdot J^T(z) \cdot \tau) \\ &+ P^T \cdot \left(A(z) \cdot z - B(z) \cdot J^T(z) \cdot \tau + \Gamma(z, t)\right) \end{aligned} \tag{23}$$

The Lagrange multiplier P is defined as follows:

$$P = s(z) \cdot z - \nu(z) + \gamma \tag{24}$$

where the variable $s(z)$ is the solution of SDRE, which is derived from the third Hamiltonian condition. It is worth noting that in the design of the controller, two important goals are considered. These two goals are precisely tracking, or in other words, converging tracking error e to zero and counteracting disturbance and uncertainty $\Gamma(z,t)$. Based on the mentioned design goals, two variables $\nu(z)$ and γ are used in the definition of P. The differentiable variable $\nu(z)$ causes the tracking error to converge to zero. Simultaneously, the differentiable variable γ counteracts the uncertainties and disturbances in the system and thwarts their effect. The optimal conditions are obtained by the HJB [13] as follows:

$$f = -R^{-1}(z) \cdot B^T(z) \cdot (s(z) \cdot z - \nu(z) + \gamma) \tag{25}$$

$$\dot{z} = A(z) \cdot z + B(z) \cdot f + \Gamma(z,t) \tag{26}$$

$$\begin{aligned}
&\dot{s}(z) \cdot z - \dot{\nu}(z) + \dot{\gamma} \\
&= \left[-s(z) \cdot A(z) - A^T(z) \cdot s(z) + s(z) \cdot B(z) \cdot R^{-1}(z) \cdot B^T(z) \cdot s(z) - Q(z)\right] \cdot z \\
&\quad + \left[A(z) - B(z) \cdot R^{-1}(z) \cdot B^T(z) \cdot s(z)\right]^T \cdot \nu(z) + Q(z) \cdot w \\
&\quad - \left[A(z) - B(z) \cdot R^{-1}(z) \cdot B^T(z) \cdot s(z)\right]^T \cdot \gamma - s(z) \cdot \Gamma(z,t) \\
&\quad - \tfrac{1}{2}\left[\left[e^T \cdot \tfrac{\partial Q(z)}{\partial z} \cdot e\right]\right] - \tfrac{1}{2}\left[\left[f^T \cdot \tfrac{\partial R(z)}{\partial z} \cdot f\right]\right] \\
&\quad - \left(\left(\left\{\tfrac{\partial A(z)}{\partial z} \cdot z\right\}\right)^T + \left(\left\{\tfrac{\partial B(z)}{\partial z} \cdot f\right\}\right)^T\right) \cdot (s(z) \cdot z - \nu(z) + \gamma)
\end{aligned} \tag{27}$$

See Appendix B for clarity on derivative calculations.

Considering (27), four equations are constructed.

$$\dot{s}(z) = \left[-s(z) \cdot A(z) - A^T(z) \cdot s(z) + s(z) \cdot B(z) \cdot R^{-1}(z) \cdot B^T(z) \cdot s(z) - Q(z)\right] \tag{28}$$

$$-\dot{\nu}(z) = \left[A(z) - B(z) \cdot R^{-1}(z) \cdot B^T(z) \cdot s(z)\right]^T \cdot \nu(z) + Q(z) \cdot w \tag{29}$$

$$\dot{\gamma} = -\left[A(z) - B(z) \cdot R^{-1}(z) \cdot B^T(z) \cdot s(z)\right]^T \cdot \gamma - s(z) \cdot \Gamma(z,t) \tag{30}$$

$$\begin{aligned}
&-\tfrac{1}{2}\left[\left[e^T \cdot \tfrac{\partial Q(z)}{\partial z} \cdot e\right]\right] - \tfrac{1}{2}\left[\left[f^T \cdot \tfrac{\partial R(z)}{\partial z} \cdot f\right]\right] \\
&-\left(\left(\left\{\tfrac{\partial A(z)}{\partial z} \cdot z\right\}\right)^T + \left(\left\{\tfrac{\partial B(z)}{\partial z} \cdot f\right\}\right)^T\right) \cdot (s(z) \cdot z - \nu(z) + \gamma) = 0
\end{aligned} \tag{31}$$

The first Eq. (28) is the suboptimal SDRE. The second Eq. (29) and third Eq. (30) are called the suboptimal tracking equation and the suboptimal disturbance rejection one, respectively. The fourth Eq. (31) is called the SDRE necessary condition for optimality or the SDRE optimality criterion that converges to zero [4]. Given Eq. (28), when $t \to \infty$, $\dot{s}_{ss}(z) = 0$. The state variable vector has tended the reference trajectory vector, and the tracking error has converged to zero at the steady-state response. Hence, Eq. (28) becomes (32).

$$s(z) \cdot A(z) + A^T(z) \cdot s(z) - s(z) \cdot B(z) \cdot R^{-1}(z) \cdot B^T(z) \cdot s(z) + Q(z) = 0 \tag{32}$$

Also, as $t \to \infty$, $\dot{\nu}_{ss}(z) = 0$ and $\dot{\gamma}_{ss} = 0$. Therefore, Eqs. (33) and (34) are solutions of Eqs. (29) and (30), respectively.

$$\nu(z) = -\left(\left[A(z) - B(z) \cdot R^{-1}(z) \cdot B^T(z) \cdot s(z)\right]^T\right)^{-1} \cdot Q(z) \cdot w \tag{33}$$

$$\gamma = -\left(\left[A(z) - B(z) \cdot R^{-1}(z) \cdot B^T(z) \cdot s(z)\right]^T\right)^{-1} \cdot s(z) \cdot \Gamma(z,t) \tag{34}$$

The defined objective function (21) concluded that the weighted matrix $Q(z)$ could have a high impact on track. The matrix $Q(z)$ appears in (33), which clearly shows the effect of this matrix. Concerning variable γ, vector $\Gamma(z,t)$ in (34) enables the control input to reciprocate the effects of uncertainty and disturbance. With regards to solutions (32) to (34), the value of vector τ is obtained from (25).

Based on Eq. (3), it can be concluded that,

$$\tau_{total} = -\left(J^T(z)\right)^{\dagger} \cdot f_{total} \tag{35}$$

Considering Eq. (13), we can conclude:

$$\tau_{total} = -\left(J^T(z)\right)^{\dagger} \cdot \left(-J^T(z) \cdot \tau + G(z)\right) \tag{36}$$

Control input (36) is optimized, but there is no guarantee that it is constrained and greater than zero. Therefore, it is not suitable for application to the CDPR systems. In fact, the main control input ($\tau_{total} \geq [\mathbf{0}]_{n\times 1}$) can not be constrained using the Hamiltonian function and its optimal conditions. Therefore, Karush-Kuhn-Tucker theorem [15] is needed to complete the problem definition. Respecting the KKT theorem [15], the Lagrangian function is defined as follows:

$$\nabla_\tau \mathcal{L}(\tau, \underline{\lambda}, \overline{\lambda}) = H(z, f, P, e)\big|_{f=-J^T\tau} - \underline{\lambda}^T \cdot (\tau - \tau_{\min}) - \overline{\lambda}^T \cdot (\tau_{\max} - \tau) \tag{37a}$$

$$(\tau_{total})_i - (\tau_{\min})_i \geq 0 \tag{37b}$$

$$(\tau_{\max})_i - (\tau_{total})_i \geq 0 \tag{37c}$$

where $\underline{\lambda} \in \Re^{n\times 1}_{\geq 0}$ and $\overline{\lambda} \in \Re^{n\times 1}_{\geq 0}$ are Lagrange multipliers. i denotes the i_{th} elements of each vector such that $i \in \{1, 2, ..., n\}$. $\tau_{\min} \in \Re^{n\times 1}$ and $\tau_{\max} \in \Re^{n\times 1}$ are the minimum and maximum control input that the motor can supply and can be applied to the CDPR system, respectively. Equation (36) is converted to,

$$\tau_{total} = \tau - \left(J^T(z)\right)^{\dagger} \cdot G(z) \tag{38}$$

The optimization conditions are specified in the KKT method as,

$$\begin{aligned} &J(z) \cdot R(z) \cdot J^T(z) \cdot \tau - J(z) \cdot B^T \cdot s(z) \cdot z \\ &+ J(z) \cdot B^T \cdot \nu(z) - J(z) \cdot B^T \cdot \gamma - \underline{\lambda} + \overline{\lambda} = 0 \end{aligned} \tag{39a}$$

$$(\tau)_i - (\tau_{\min})_i \geq \left(\left(J^T(z)\right)^{\dagger} \cdot G(z)\right)_i \tag{39b}$$

$$(\tau_{\max})_i - (\tau)_i \geq -\left(\left(J^T(z)\right)^{\dagger} \cdot G(z)\right)_i \tag{39c}$$

$$\underline{\lambda}_i \geq 0 \tag{39d}$$

$$\overline{\lambda}_i \geq 0 \tag{39e}$$

$$\underline{\lambda}_i^T \cdot (\tau_i - (\tau_{\min})_i) = 0 \tag{39f}$$

$$\overline{\lambda}_i^T \cdot ((\tau_{\max})_i - \tau_i) = 0 \tag{39g}$$

Eqs. (39) is the constrained form of Eq. (25). By satisfying Eqs. (32) to (34) and calculating the vector τ through conditions (39), the constrained optimal control problem of the fully-constrained CDPR is completed. Equation (38) can be used to calculate the cable tension vector.

4.2 Stability Analysis

The general form of the SDC linear structure is written as (11). When the designer's goal is tracking, the designer should put the variable $\nu(z)$ in the control input. Therefore, considering state space (11), the control input is applied to the system as relation (40).

$$f = -R^{-1}(z) \cdot B^T(z) \cdot (s(z) \cdot z - \nu(z) + \gamma) - \underline{\lambda} + \overline{\lambda} \tag{40}$$

After placing (40) in (12), Eq. (41) appears.

$$\begin{aligned}\dot{z} = &\left[A(z) - B(z) \cdot R^{-1}(z) \cdot B^T(z) \cdot s(z)\right] \cdot z + B(z) \cdot R^{-1}(z) \cdot B^T(z) \cdot \nu(z) \\ &- B(z) \cdot R^{-1}(z) \cdot B^T(z) \cdot \gamma + \Gamma(z,t) + B(z) \cdot (\overline{\lambda} - \underline{\lambda})\end{aligned} \tag{41}$$

The non-zero reference trajectory is obtained from Eq. (42).

$$\dot{w} = A_w(w) \cdot w + B_w(w) \cdot r \tag{42}$$

where the matrices $A_w(w)$ and $B_w(w)$ denote system and input matrices of the state space of reference trajectory, respectively. The vector r is the input of the state space of the reference trajectory system. The designer has to select setpoint r such that the desired reference trajectory is created. The time derivative of the tracking error constructs as Eq. (43).

$$\dot{e} = \dot{w} - \dot{z} \tag{43}$$

Eq. (44) is created by substituting Eqs. (41) and (23) in (43). The terms $\pm\left[A(z) - B(z) \cdot R^{-1}(z) \cdot B^T(z) \cdot s(z)\right] w$ are added to Eq. (43) to complete the analysis. The sign $\pm$ indicates the plus and minus of the specific term.

$$\begin{aligned}\dot{e} = &\left[A(z) - B(z) \cdot R^{-1}(z) \cdot B^T(z) \cdot s(z)\right] \cdot e \\ &+ \left[A_w(w) - A(z) + B(z) \cdot R^{-1}(z) \cdot B^T(z) \cdot s(z)\right] \cdot w \\ &+ B_w(w) \cdot r - B(z) \cdot R^{-1}(z) \cdot B^T(z) \cdot \nu(z) \\ &+ B(z) \cdot R^{-1}(z) \cdot B^T(z) \cdot \gamma - \Gamma(z,t) + B(z) \cdot (\underline{\lambda} - \overline{\lambda})\end{aligned} \tag{44}$$

It is worth noting that matrices $A(z)$ and $B(z)$ are system and input matrices, which norm and the value of their elements are known for the designer. Matrices $Q(z)$ and $R(z)$ are error and input weighted matrices, respectively. The designer chooses them. Thus, their norm is known for the designer. Knowing the system, input, and weighted matrices, the SDRE equation is solved. Since the positive definite matrix $s(z)$ is the solution of the SDRE, its value and norm are known. Once the matrix $s(z)$ is computed, Eq. (30) can be solved. Therefore, the vector

γ and its derivative are calculated. Vectors $\underline{\lambda}$ and $\overline{\lambda}$ are Lagrange multipliers that have non-negative and limited elements. The elements of vectors $\overline{\lambda}$ and $\underline{\lambda}$ are obtained by solving the optimization problem with respect to the KKT conditions. Therefore, the norms of the mentioned vectors are available for the designer. Finally, the tracking error equation is attained as (45).

$$\dot{e} = A_e \cdot e + B_e u_e + \Gamma_e \tag{45a}$$

$$A_e = A(z) - B(z) \cdot R^{-1}(z) \cdot B^T(z) \cdot s(z) \tag{45b}$$

$$B_e = \begin{bmatrix} B_e^{11} \; B_e^{12} \; B_e^{13} \; B_e^{14} \end{bmatrix} \tag{45c}$$

$$u_e = \begin{bmatrix} w^T \; r^T \; \nu^T(z) \; \gamma^T \end{bmatrix}^T \tag{45d}$$

$$\Gamma_e = -\Gamma(z,t) + B(z) \cdot \left(\underline{\lambda} - \overline{\lambda}\right) \tag{45e}$$

in which,

$$\begin{aligned} B_e^{11} &= A_w(w) - A(z) + B(z) \cdot R^{-1}(z) \cdot B^T(z) \cdot s(z) \\ B_e^{12} &= B_w(w) \\ B_e^{13} &= -B(z) \cdot R^{-1}(z) \cdot B^T(z) \\ B_e^{14} &= B(z) \cdot R^{-1}(z) \cdot B^T(z) \end{aligned}$$

The structure of Eq. (45a) is very similar to Eq. (12). The difference is that instead of the state vector z, the error vector e is substituted. Hence, the existence of a variable such as $\tilde{\nu}(e)$ is not necessary for the control input u_e because we want the elements of vector e to track the zero reference trajectory. Thus, control inputs (46) can be designed for Eq. (45a) and the vector e to converge to zero.

$$u_e = -\tilde{R}^{-1}(e) \cdot B_e^T \cdot (\tilde{s}(e) \cdot e + \tilde{\gamma}) \tag{46}$$

The positive definite matrix $\tilde{s}(e)$ is the solution of the SDRE (47). Also, $\tilde{R}(e)$ and $\tilde{Q}(e)$ are weighted matrices that the designer chose them.

$$\dot{\tilde{s}}(e) + \tilde{s}(e) \cdot A_e + A_e^T \cdot \tilde{s}(e) - \tilde{s}(e) \cdot B_e \cdot \tilde{R}^{-1}(e) \cdot B_e^T \cdot \tilde{s}(e) + \tilde{Q}(e) = 0 \tag{47}$$

When the matrix $\tilde{s}(e)$ is obtained, the vector $\tilde{\gamma}$ will be achieved from Eq. (48).

$$\dot{\tilde{\gamma}} + A_e^T \cdot \tilde{\gamma} + \tilde{s}(e) \cdot \Gamma_e - \tilde{s}(e) \cdot B_e \cdot \tilde{R}^{-1}(z) \cdot B_e^T \cdot \tilde{\gamma} = 0 \tag{48}$$

From Eq. (45d), it is understood that the control input u_e has absolute values. The vectors w and r correspond to the reference trajectory system (42), and the vectors $\nu(z)$ and γ correspond to the solution of Eqs. (30) and (31), respectively. Therefore, the weighted matrices $\tilde{R}(e)$ and $\tilde{Q}(e)$ have to be adjusted such that after calculating the vectors $\tilde{s}(e)$ and $\tilde{\gamma}$ from Eqs. (47) and (48), Eq. (49) is satisfied.

$$u_e = \begin{bmatrix} w^T \; r^T \; \nu^T(z) \; \gamma^T \end{bmatrix}^T = -\tilde{R}^{-1}(e) \cdot B_e^T \cdot (\tilde{s}(e) \cdot e + \tilde{\gamma}) \tag{49}$$

The Lyapunov function is proposed as (50).

$$V_e(e) = \frac{1}{2} e^T \cdot \tilde{s}(e) \cdot e \tag{50}$$

The time derivative of Lyapunov function (50) leads to,

$$\dot{V}_e(e) = \frac{1}{2}\left[\dot{e}^T \cdot \tilde{s}(e) \cdot e + e^T \cdot \dot{\tilde{s}}(e) \cdot e + e^T \cdot \tilde{s}(e) \cdot \dot{e}\right] \tag{51}$$

Respecting (45a), the derivative of Lyapunov (51) is rewritten as,

$$\begin{aligned}\dot{V}_e(e) = \tfrac{1}{2}e^T \cdot &[\dot{\tilde{s}}(e) + \tilde{s}(e) \cdot A_e + A_e^T \cdot \tilde{s}(e) - \tilde{s}(e) \cdot B_e \cdot \tilde{R}^{-1}(e) \cdot B_e^T \cdot \tilde{s}(e) \\ &-\tilde{s}(e) \cdot B_e \cdot \tilde{R}^{-1}(e) \cdot B_e^T \cdot \tilde{s}(e)] \cdot e + e^T \cdot \tilde{s}(e) \cdot \left[\mathit{\Gamma}_e - B_e \cdot \tilde{R}^{-1}(e) \cdot B_e^T \cdot \tilde{\gamma}\right]\end{aligned} \tag{52}$$

Defining $\psi_e = B_e^T \cdot \tilde{s}(e) \cdot e$ and based on Eq. (48), it can be concluded:

$$\dot{V}_e(e) = -\tfrac{1}{2}e^T \cdot \tilde{Q}(e) \cdot e - \tfrac{1}{2}\psi_e^T \cdot \tilde{R}^{-1}(e) \cdot \psi_e - e^T \cdot \left[\dot{\tilde{\gamma}} + A_e^T \cdot \tilde{\gamma}\right] \tag{53}$$

We can write,

$$-e^T \cdot \left[\dot{\tilde{\gamma}} + A_e^T \cdot \tilde{\gamma}\right] \leq \left\|\dot{\tilde{\gamma}} + A_e^T \cdot \tilde{\gamma}\right\| \|e\| \tag{54}$$

Thus, Eq. (53) converts to the following inequality.

$$\dot{V}_e(e) \leq -\tfrac{1}{2}e^T \cdot \tilde{Q}(e) \cdot e - \tfrac{1}{2}\psi_e^T \cdot \tilde{R}^{-1}(e) \cdot \psi_e + \left\|\dot{\tilde{\gamma}} + A_e^T \cdot \tilde{\gamma}\right\| \|e\| \tag{55}$$

System stability is guaranteed by applying control input (49) when the following condition satisfied:

$$\dot{V}_e(e) \leq 0 \tag{56}$$

According to inequality (55), it can be written:

$$\left\|\dot{\tilde{\gamma}} + A_e^T \cdot \tilde{\gamma}\right\| \|e\| \leq \frac{1}{2}e^T \cdot \tilde{Q}(e) \cdot e + \frac{1}{2}\psi_e^T \cdot \tilde{R}^{-1}(e) \cdot \psi_e \tag{57}$$

On the other hand, it can be written:

$$\tfrac{1}{2}e^T \cdot \tilde{Q}(e) \cdot e + \tfrac{1}{2}\psi_e^T \cdot \tilde{R}^{-1}(e) \cdot \psi_e \leq \tfrac{1}{2}\left(\left\|\tilde{Q}(e)\right\| + \left\|\tilde{R}^{-1}(e)\right\| \|B_e\|^2 \|\tilde{s}(e)\|^2\right) \|e\|^2 \tag{58}$$

If the two inequalities (57) and (58) are considered together, inequality (59) can be written.

$$\frac{2\left\|\dot{\tilde{\gamma}} + A_e^T \cdot \tilde{\gamma}\right\|}{\left\|\tilde{Q}(e)\right\| + \left\|\tilde{R}^{-1}(e)\right\| \|B_e\|^2 \|\tilde{s}(e)\|^2} \leq \|e\| \tag{59}$$

Derivative of Lyapunov function (50) is not negative (semi) definite outside of set (59). Hence, the solution is uniformly ultimately bounded. In this problem, the negative definiteness of $\dot{V}_e(e)$ is guaranteed by the uniformly ultimately bounded concept. When $t \to \infty$, the tracking error is limited to a trivial and negligible value. In this case, the tracking is done with high accuracy. By changing the weighted matrices, the error can be reached zero. It should be noted that condition (59) is not heavy and unacceptable. If condition (59) is satisfied, the stability of the system is guaranteed. Therefore, controller (46) is robust against any uncertainty in Eq. (45a). Consequently, controller (40) is robust against any uncertainty in Eq. (12). Therefore, the stability of the system and the robustness of the controller to any uncertainty and disturbance were proved.

5 Simulation Results

In the presence of disturbance and considering the uncertainty of the dynamic model of the cable robot, the vector D is created. The modeled uncertainty and disturbance D creates the vector $\Gamma(z,t)$ that plays the role of disturbance in the linear structure of SDC (20a). The reference trajectory is assumed as Eqs. (60)

$$w = \begin{bmatrix} 0.1\cos(0.25t) & 0.1\sin(0.25t) & 0 & -0.025\sin(0.25t) & 0.025\cos(0.25t) & 0 \end{bmatrix}^T \tag{60}$$

The state variables are initialization as $z(0) = \begin{bmatrix} 0.1\,(m)\;0\;0\;0\;0\;0 \end{bmatrix}^T$. The dynamic matrices of the CDPR are given in Appendix A. The variable weighted matrices in terms of state variables are selected as Eqs. (61) and (62) to track the reference trajectory (60).

$$R(z(t)) = \left[r_{ik}(t)\right]_{\substack{3\times 3\\ i,k\in\{1,2,3\}}} \tag{61a}$$

$$r_{ik}(t) = \begin{cases} 0 & if \quad i \neq k \\ |z_1(t)| + |z_2(t)| & if \quad i = k \in \{1,2\} \\ |z_1(t)\cdot z_2(t)| + |z_4(t)\cdot z_5(t)| & if \quad i = k = 3 \end{cases} \tag{61b}$$

$$Q(z(t)) = \left[q_{ik}(t)\right]_{\substack{6\times 6\\ i,k\in\{1,2,\ldots,6\}}} \tag{62a}$$

$$q_{ik}(t) = \begin{cases} 0 & if \quad i \neq k \\ \frac{500}{1+|z_i(t)|} & if \quad i = k \in \{1,2,3\} \\ \frac{5}{1+|z_i(t)|} & if \quad i = k \in \{4,5,6\} \end{cases} \tag{62b}$$

According to the reference trajectory, $|z_1(t)| + |z_2(t)| \leq 0.1$. Therefore, it is a suitable choice for the weight matrix $R(z(t))$ because it makes the norm of matrix $R(z(t))$ less than 0.1. By applying coefficients of 500 and 5 to the diagonal elements of the matrix $Q(z(t))$, the minimizing the elements $(e_1(t), e_2(t), e_3(t))$ is more valuable than $(e_4(t), e_5(t), e_6(t))$. A fractional relation is used to reduce its norm. Since $\underset{i\in\{1,2,\ldots,6\}}{|z_i(t)|} \leq 0.1$, $\|Q(z(t))\| < 500$. The specified disturbance T_d in (63), enters the robotic system at 20 s and remains.

$$T_d = \begin{bmatrix} 0.06\cos(5.75t) & 0.06\sin(5.75t) & 0 \end{bmatrix}^T \tag{63}$$

Figure 2a shows the state variables $(z_1(t), z_2(t), z_4(t), z_5(t))$. In this figure, the uncertainty compensating variable $\gamma(z(t))$ is not applied to the input vector. As a result, the effect of disturbance is seen. After 20 seconds, the state variables $z_1(t), z_2(t)$ lose their natural form of trigonometric signal. Also, the presence of disturbance affects the behavior of the fourth and fifth state variables

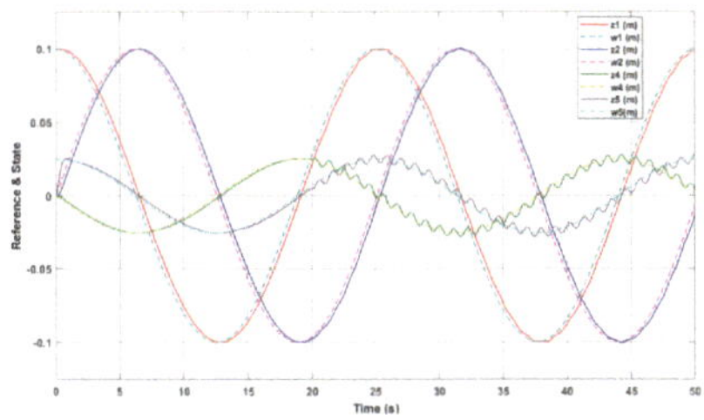

(a) No counteract with Uncertainty in Tracking Responses

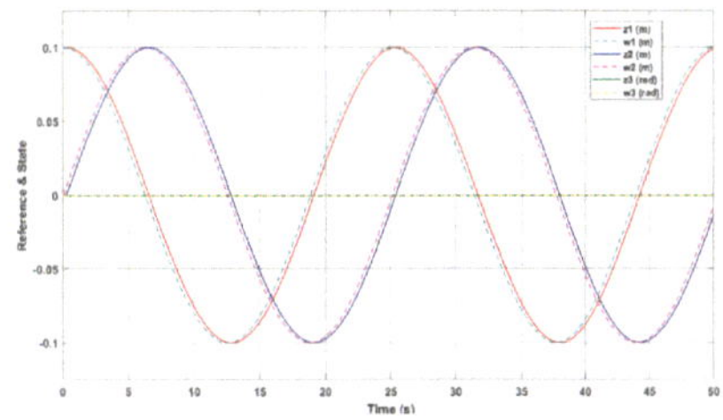

(b) Counteracting with Uncertainty in Tracking Responses

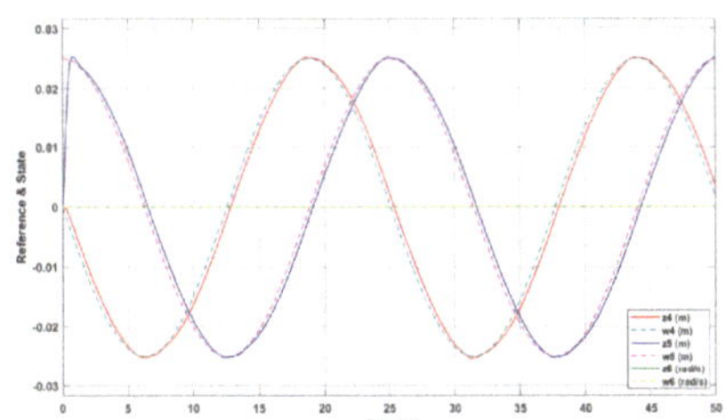

(c) Counteracting with Uncertainty in Tracking Responses

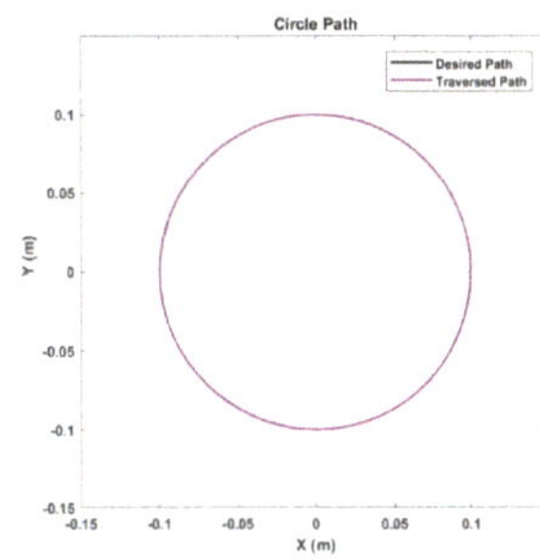

(d) Circle Path

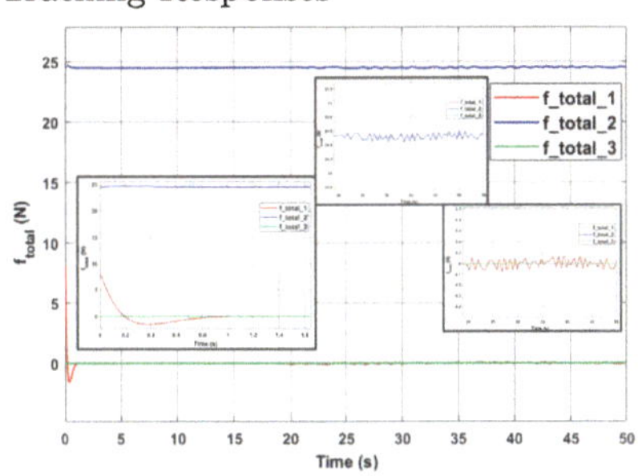

(e) Forces

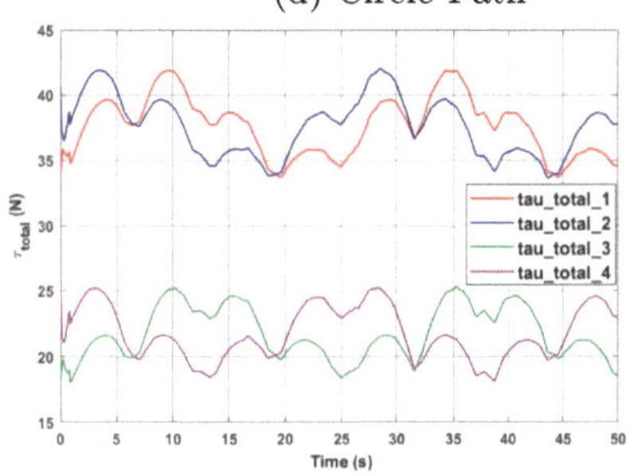

(f) Tensions

Fig. 2. Constrained optimal control method

$(z_4(t), z_5(t))$. This effect indicates the model of vector $\Gamma(z(t), t)$ (20d), which includes vector D.

By applying the variable $\gamma(z(t))$ to the control input, the disturbance effect is eliminated. Figures 2b and 2c show the response of $z(t)$. Figure 2b shows the response of the first to third state variables that the presence of disturbance does not cause to fail the tracking. Figure 2c, specify the first derivative of the state variables $(z_1(t), z_2(t), z_3(t))$, respectively. When the controller confronts the uncertainties and disturbances, the circular path traversed is precisely on the reference circular path. According to Fig. 2d, the disturbance could not deflect x_p and y_p from the circular path. The force vector f_{total}, is shown in Fig. 2e.

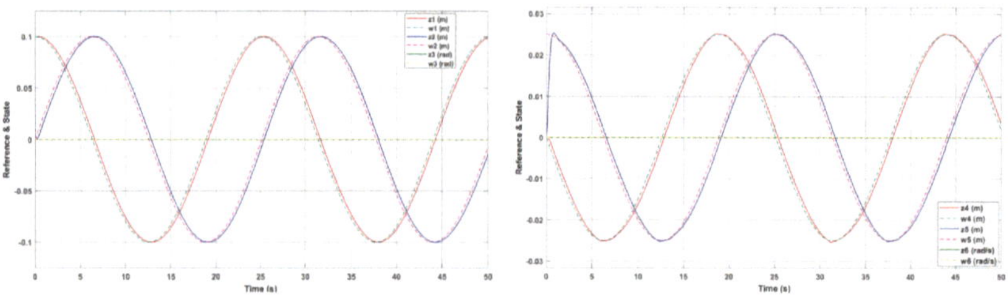

(a) Counteracting with Uncertainty in Tracking Responses

(b) Counteracting with Uncertainty in Tracking Responses

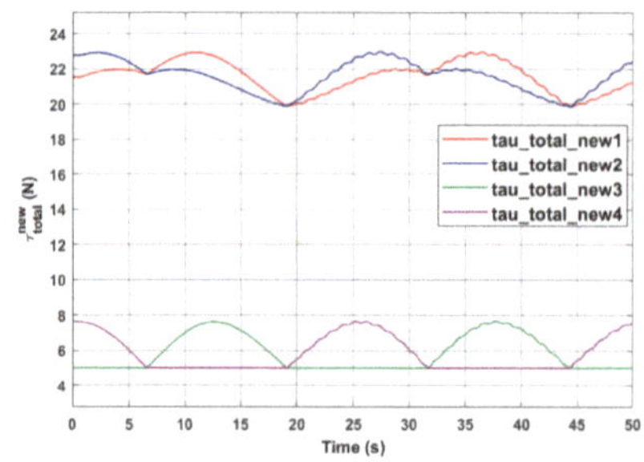

(c) Tensions

Fig. 3. Internal force control method

The elements of f in Fig. 2e have irregular oscillations as the disturbance enters. The behavior of the force vector f_{total} is affected by the presence of the gravity vector. The tensile vector τ_{total} is shown in Fig. 2f. The behavior of τ_{total} has changed from 20 seconds onwards, with the onset of disturbance. This change occurred due to the counteraction of the disturbance effect.

Another approach to control a cable-driven parallel robot is presented. Using the null space of the transpose of the Jacobian matrix, internal force is defined. Internal force is used to satisfy the tensionability condition of the cables [7]. In internal force method, Figs. 3a and 3b show the reference trajectory and response of state variables. The disturbance rejection variable has been added to the control input, and the elements of vector $z\,(t)$ have been able to accurately track the reference trajectory even after disturbance existence in the system. Figure 3c shows the force τ_{total}, which is calculated using the internal force method. The response of this force oscillates on the amplitude with the arrival of disturbance.

Controlling a planar fully-constrained CDPR based on the state-dependent Riccati equation is the aim of this section. Two control methods were simulated considering the constraint of the main control input of the cable robot system, namely τ_{total}. In the first method, the Karush Kuhn Tucker conditions were considered to constrain the cable tension in addition to using Hamiltonian. In the design of the second method, the force vector applied from the actuators to the cables was constrained based on the null space of the transpose of the

Jacobian matrix with applying the Hamiltonian conditions. The internal force method always forces the smallest negative element of the unconstrained vector τ_{total}^{new} (3c) to be placed on $\tau_{\min}$. This matter can limit the designer's authority. While, in the optimal control method based on KKT conditions, the cable tension is obtained from solving the optimization problem, and there is no requirement in quantifying the forces. Moreover, the internal force method uses the null space of transpose of the Jacobian matrix directly. Since the Jacobian matrix is not available, using its estimation creates an additional incorrect force. As a result, the tracking error increases. Nevertheless, in the constrained optimization control method, the forces and tensions are obtained with a suitable distribution.

6 Conclusion

In this paper, first the cable robot model was used. Then, the dynamic model of the CDPR became the state-dependent coefficient linear controllable structure. In this parameterized form, the applied force by the cables on the platform was considered as the input of the system. Therefore, we could comment on the controllability of the system. It is worth noting that the main constraint control input is the exerted force by the actuators to the cables. If the cable tension was considered the input of the SDC system, one could not be expressed anything about the system controllability. Next, the control input was calculated based on the Hamiltonian and KKT conditions. Then, the stability of the system and the robustness of the controller were proved. After that, the proposed control method was simulated with uncertainty existence, and the results were discussed. Finally, our proposed method was compared with the internal force method.

A Appendix

The dynamic matrices of the 3 DOF planar CDPR are presented as follows [9]:

$$M = \begin{bmatrix} m & 0 & 0 \\ 0 & m & 0 \\ 0 & 0 & I_z \end{bmatrix}; \qquad C(x, \dot{x}) = 0; \qquad G = \begin{bmatrix} 0 \\ mg \\ 0 \end{bmatrix};$$

where parameters $m = 2.5\,(\boldsymbol{kg})$, $I_z = 0.03\,(\boldsymbol{kg\,m^2})$, and $g = 9.806\,(\boldsymbol{m/s^2})$ are considered. The position of the cable attachment points is considered as Article [9].

B Appendix

Consider the following notations:

Notation B.1: The symbol $[\![.]\!]$ indicates that the inside parameter is a vector, e.g.,

$$[\![\theta]\!] = \left[\theta_1^T\ \theta_2^T\ \ldots\ \theta_N^T\right]^T \tag{64}$$

which N is the number of the vector elements.
Notation B.2: The symbol $\{.\}$ indicates the tensor space in which the inside parameter dimensions must be homogenized according to the problem.
Notation B.3: $\theta_{i:}$ and $\theta_{:i}$ represent the i_{th} row and i_{th} column of matrix θ, respectively.

Since algebraically, the derivatives of the matrices $R(z(t))$ and $Q(z(t))$ in terms of $z(t)$ create more extensive dimensions than $R(z(t))$ and $Q(z(t))$, the vectors $\left[\left[e^T(t) \cdot \frac{\partial Q(z(t))}{\partial z(t)} \cdot e(t)\right]\right]$ and $\left[\left[f^T(t) \cdot \frac{\partial R(z(t))}{\partial z(t)} \cdot f(t)\right]\right]$ are used. With regards to Notation B.1, after multiplying their internal parameters, an answer vector is formed for each of them. Hence, the dimensions of the problem are not affected. According to the discussed topics and taking advantage of Notation B.2, the tensor space $\left\{e^T(t) \cdot \frac{\partial Q(z(t))}{\partial z(t)} \cdot e(t)\right\}$ and $\left\{f^T(t) \cdot \frac{\partial R(z(t))}{\partial z(t)} \cdot f(t)\right\}$ can be calculated. The exact amount of vectors $\left[\left[e^T(t) \cdot \frac{\partial Q(z(t))}{\partial z(t)} \cdot e(t)\right]\right]$ and $\left[\left[f^T(t) \cdot \frac{\partial R(z(t))}{\partial z(t)} \cdot f(t)\right]\right]$ are obtained as Eqs. (65) and (66).

$$\left[\left[e^T(t) \cdot \frac{\partial Q(z(t))}{\partial z(t)} \cdot e(t)\right]\right] = \begin{bmatrix} e^T \frac{\partial Q_{:1}}{\partial z_1} & \cdots & e^T \frac{\partial Q_{:n}}{\partial z_1} \\ \vdots & \ddots & \vdots \\ e^T \frac{\partial Q_{:1}}{\partial z_n} & \cdots & e^T \frac{\partial Q_{:n}}{\partial z_n} \end{bmatrix} e(t) \tag{65}$$

$$\left[\left[f^T(t) \cdot \frac{\partial R(z(t))}{\partial z(t)} \cdot f(t)\right]\right] = \begin{bmatrix} f^T \frac{\partial R_{:1}}{\partial z_1} & \cdots & f^T \frac{\partial R_{:n}}{\partial z_1} \\ \vdots & \ddots & \vdots \\ f^T \frac{\partial R_{:1}}{\partial z_n} & \cdots & f^T \frac{\partial R_{:n}}{\partial z_n} \end{bmatrix} f(t) \tag{66}$$

According to Notation B.3, the mentioned tensor spaces are computed.

$$\left\{\frac{\partial A(z(t))}{\partial z(t)} \cdot z(t)\right\} = \begin{bmatrix} \frac{\partial A_{1:}}{\partial z_1} z & \cdots & \frac{\partial A_{1:}}{\partial z_n} z \\ \vdots & \ddots & \vdots \\ \frac{\partial A_{n:}}{\partial z_1} z & \cdots & \frac{\partial A_{n:}}{\partial z_n} z \end{bmatrix} \tag{67}$$

$$\left\{\frac{\partial B(z(t))}{\partial z(t)} \cdot f(t)\right\} = \begin{bmatrix} \frac{\partial B_{1:}}{\partial z_1} f & \cdots & \frac{\partial B_{1:}}{\partial z_n} f \\ \vdots & \ddots & \vdots \\ \frac{\partial B_{n:}}{\partial z_1} f & \cdots & \frac{\partial B_{n:}}{\partial z_n} f \end{bmatrix} \tag{68}$$

References

1. Babaghasabha, R., Khosravi, M.A., Taghirad, H.D.: Adaptive robust control of fully-constrained cable driven parallel robots. Mechatronics **25**, 27–36 (2015)
2. Bruckmann, T., Mikelsons, L., Brandt, T., Hiller, M., Schramm, D.: Wire robots part i: Kinematics, analysis & design (2008)
3. Çimen, T.: State-dependent Riccati equation (SDRE) control: a survey. IFAC Proc. Vol. **41**(2), 3761–3775 (2008)

4. Cloutier, J.R., D'Souza, C.N., Mracek, C.P.: Nonlinear regulation and nonlinear H control via the state-dependent Riccati equation technique: Part 1, theory. In: Proceedings of the International Conference on Nonlinear Problems in Aviation and Aerospace, pp. 117–131. Embry Riddle University (1996)
5. Hervé, J., Sparacino, F.: Structural synthesis of parallel robots generating spatial translation. In: Proceedings of the 5th IEEE International Conference on Advanced Robotics, pp. 808–813 (1991)
6. Kawamura, S., Kino, H., Won, C.: High-speed manipulation by using parallel wire-driven robots. Robotica **18**(1), 13–21 (2000)
7. Khosravi, M.A., Taghirad, H.D.: Dynamic analysis and control of cable driven robots with elastic cables. Trans. Can. Soc. Mech. Eng. **35**(4), 543–557 (2011)
8. Khosravi, M.A., Taghirad, H.D.: Experimental performance of robust PID controller on a planar cable robot. In: Cable-Driven Parallel Robots, pp. 337–352. Springer (2013)
9. Khosravi, M.A., Taghirad, H.D.: Robust PID control of fully-constrained cable driven parallel robots. Mechatronics **24**(2), 87–97 (2014)
10. Kilicaslan, S.: Tracking control of elastic joint parallel robots via state-dependent Riccati equation. Turk. J. Electr. Eng. Comput. Sci. **23**(2), 522–538 (2015)
11. Korayem, M.H., Nekoo, S.R.: Nonlinear optimal control via finite time horizon state-dependent Riccati equation. In: 2014 Second RSI/ISM International Conference on Robotics and Mechatronics (ICRoM), pp. 878–883. IEEE (2014)
12. Landsberger, S.E.: Design and construction of a cable-controlled, parallel link manipulator. Ph.D. thesis, Massachusetts Institute of Technology (1984)
13. Lewis, F.L., Vrabie, D., Syrmos, V.L.: Optimal Control. Wiley, Hoboken (2012)
14. Menon, P., Lam, T., Crawford, L., Cheng, V.: Real-time computational methods for SDRE nonlinear control of missiles. In: Proceedings of the 2002 American Control Conference (IEEE Cat. No. CH37301), vol. 1, pp. 232–237. IEEE (2002)
15. Nocedal, J., Wright, S.: Numerical Optimization. Springer, Heidelberg (2006)
16. Oh, S.R., Agrawal, S.K.: Generation of feasible set points and control of a cable robot. IEEE Trans. Rob. **22**(3), 551–558 (2006)
17. Schmidt, V.L.: Modeling techniques and reliable real-time implementation of kinematics for cable-driven parallel robots using polymer fiber cables. Fraunhofer Verlag, Stuttgart (2017)
18. Zi, B., Duan, B., Du, J., Bao, H.: Dynamic modeling and active control of a cable-suspended parallel robot. Mechatronics **18**(1), 1–12 (2008)

Model-Based Control of a Planar 3-DoF Cable Robot Using Exact Linearization

Marcus Hamann(✉) and Christoph Ament

Faculty of Applied Computer Science, University of Augsburg,
86159 Augsburg, Germany
{marcus.hamann,christoph.ament}@uni-a.de
https://www.uni-augsburg.de/de/fakultaet/fai/informatik/prof/rt/

Abstract. The following work shows the control of a planar cable-driven parallel robot with three degrees of freedom based on exact input-output linearization. For this purpose, first, a model of the cable robot is presented. The model of the robot is then modified to be able to apply the control to a broad class of other parallel robots. Then, the control concept is briefly introduced to subsequently apply it to the cable robot. Finally, the controller is tested in simulation. For the control, it is essential that the pose of the platform can be measured. Therefore, a laser-based measuring system is proposed for the following implementation.

Keywords: Cable-driven parallel robot · Control · Exact input-output linearization

1 Introduction

Cable robots form a class of robots, which is flexible and dynamic. They can cover a large workspace and can have a modular design. However, one disadvantage addressed in this work, which applies to many cable robots, is positioning accuracy. The primary reason for this is that mainly internal sensors such as position encoders or accelerometers are used. Figure 1 shows a version of the modular cable robot of the University of Augsburg.

The presented work deals with the model-based control of a planar cable robot with three degrees of freedom, as can be seen in the figure. These degrees of freedom must be measurable in real-time in the control process. Therefore, for the realization of the approach presented here, a laser tracker is proposed, as it is already successfully used for this kind of control. This modular laser tracker consists of four tracker modules. Each of these modules can track a retroreflector both dynamically and precisely. While tracking the reflector, changes in the length of the laser beam can be measured with high precision using an interferometer. For the determination of the absolute position of the reflector, it must

M. Gouttefarde et al. (Eds.): CableCon 2021, MMS 104, pp. 260–270, 2021.
https://doi.org/10.1007/978-3-030-75789-2_21

Fig. 1. System consisting of laser tracker and cable robot

be tracked simultaneously by four trackers. Since two reflectors must be used to determine the rotation of the platform, this is not possible without further effort. This problem can be solved by having the laser beams hop cyclically from one target to another.

The control is based on an exact input-output linearization which implies that the closed-loop system will be able to decouple the cartesian outputs from each other. Thus, each output follows a predefined differential equation. This approach avoids competing for cable forces and, with some extensions, provides a closed-loop control without permanent control error even for an incompletely known model of the robot. Also, the state-dependent transformation of forces from joint space to operational space is performed. Besides, a straightforward approach is presented through which it is possible to use the controller presented here for other parallel robots as well.

The concept of exact input-output linearization has been applied to a cable robot once so far (see [6]). In this work, a spatial cable robot with three degrees of freedom is controlled. The inputs of the system are the cable forces. The controlled variables are the cartesian components of the position vector. In [4], control of the prototype SEGESTA is introduced. Two approaches are presented. They differ in the choice of the control variables. First, motion control in joint space and then a motion control in operational space is presented. The local controllers are PD controllers. In addition, an optimal cable force distribution is strived for. The local PID controller approach is often used because it is very easy to implement (see [3,18] or [10]). However, a disadvantage of this approach is that the control can become dynamically slow and possible vibrations of the platform cannot be counteracted because the effect of the cable forces on the platform is not considered. Furthermore, flatness-based approaches similar to this approach have been presented in the past. Examples are [7,8,12,17] or [16]. If flat outputs can be found for a system, they can be used to design an open-loop control. The controller can be designed to calculate control variables based on the outputs and their derivatives. One drawback of this approach is finding these flat outputs. This is implicitly done in the approach presented here. Some further control approaches addressed Lyapunov-based control strategies, which also guarantee a stable tracking control, but cannot be easily determined based on a given model of the robot. For this, we refer to [2,13,14] or [11]. The main drawback of this approach is to find a suitable control-Lyapunov function for the given model.

For this purpose, the modeling of the cable robot is presented in Sect. 2. Subsequently, Sect. 3 presents the structure of the control system to apply it to the model of the cable robot. Section 4 shows simulative results of the control and Sect. 5 concludes the presented work.

2 Modeling

In the following the kinematics and dynamics of the cable robot are presented. Figure 2 shows a sketch of the platform.

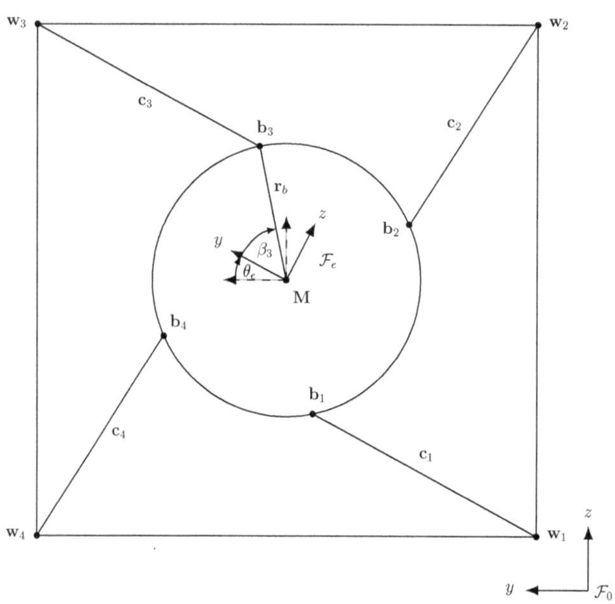

Fig. 2. Sketch of the platform with all geometric parameters

The center of the platform $\mathbf{M}$ ist given by

$$\mathbf{M} = \begin{bmatrix} y_e \\ z_e \end{bmatrix} . \tag{1}$$

The orientation of the platform in the reference frame $\mathcal{F}_0$ is given by θ_e. Variable β_i represents the orientation of connection point $\mathbf{b}_i$ in the endeffector frame $\mathcal{F}_e$. The connection point $\mathbf{b}_i$ in the endeffector frame $\mathcal{F}_e$ is given as

$$\mathbf{b}_i = \mathbf{R}(\beta_i) \cdot \begin{bmatrix} r_b \\ 0 \end{bmatrix} = r_b \cdot \begin{bmatrix} \cos\beta_i \\ \sin\beta_i \end{bmatrix} \tag{2}$$

with rotation matrix

$$\mathbf{R}(\beta_i) = \begin{bmatrix} \cos\beta_i & -\sin\beta_i \\ \sin\beta_i & \cos\beta_i \end{bmatrix} \tag{3}$$

and distance r_b. The connection point ${}^0\mathbf{b}_i$ in the reference frame $\mathcal{F}_0$ is given as

$$\begin{aligned} {}^0\mathbf{b}_i &= \begin{bmatrix} y_e \\ z_e \end{bmatrix} + \mathbf{R}(\theta_e) \cdot \mathbf{b}_i & (4) \\ &= \begin{bmatrix} y_e \\ z_e \end{bmatrix} + r_b \cdot \begin{bmatrix} \cos\beta_i \cos\theta_e - \sin\beta_i \sin\theta_e \\ \cos\beta_i \sin\theta_e + \sin\beta_i \cos\theta_e \end{bmatrix} & (5) \\ &= \begin{bmatrix} y_e \\ z_e \end{bmatrix} + r_b \cdot \begin{bmatrix} \cos(\beta_i + \theta_e) \\ \sin(\beta_i + \theta_e) \end{bmatrix} & (6) \end{aligned}$$

with rotation matrix

$$\mathbf{R}(\theta_e) = \begin{bmatrix} \cos\theta_e & -\sin\theta_e \\ \sin\theta_e & \cos\theta_e \end{bmatrix} . \tag{7}$$

The cable vector $\mathbf{c}_i$ is composed of the detachment point $\mathbf{w}_i = [w_{i,y}, w_{i,z}]^T$ and the connection point $\mathbf{b}_i$.

$$\begin{aligned} \mathbf{c}_i &= \mathbf{w}_i - {}^0\mathbf{b}_i & (8) \\ &= \mathbf{w}_i - \mathbf{R}(\theta_e) \cdot \mathbf{b}_i & (9) \\ &= \begin{bmatrix} w_{i,y} - y_e - r_b \cdot \cos(\beta_i + \theta_e) \\ w_{i,z} - z_e - r_b \cdot \sin(\beta_i + \theta_e) \end{bmatrix} & (10) \end{aligned}$$

The static equilibrium of cable forces is written as

$$\begin{aligned} \sum F_y = 0 &\Rightarrow \sum_{i=1}^{4} F_i \cdot \cos\theta_i = 0 \\ \sum F_z = 0 &\Rightarrow \sum_{i=1}^{4} F_i \cdot \sin\theta_i - mg = 0 \\ \sum M_x = 0 &\Rightarrow \sum_{i=1}^{4} F_i \cdot q_i = 0 \end{aligned} \tag{11}$$

with cable forces F_i, normalized cable vector

$$\mathbf{v}_i = \frac{\mathbf{c}_i}{||\mathbf{c}_i||} = \begin{bmatrix} \cos\theta_i \\ \sin\theta_i \end{bmatrix} \tag{12}$$

and the normal distance between the center of the platform and the cable axis

$$q_i = (\mathbf{R}(\theta_e) \cdot \mathbf{b}_i) \times \mathbf{v}_i \,. \tag{13}$$

From this follows the structure matrix

$$\mathbf{A}(\mathbf{x}) = \begin{bmatrix} \mathbf{v}_1 \cdots \mathbf{v}_4 \\ q_1 \cdots q_4 \end{bmatrix} \tag{14}$$

with state vector

$$\mathbf{x} = \begin{bmatrix} y_e \\ z_e \\ \theta_e \end{bmatrix} . \tag{15}$$

Eq. (11) can now be rewritten as

$$\mathbf{A}(\mathbf{x}) \cdot \underbrace{\begin{bmatrix} F_1 \\ F_2 \\ F_3 \\ F_4 \end{bmatrix}}_{\mathbf{u}} = \begin{bmatrix} F_y \\ F_z \\ M_x \end{bmatrix} = \mathbf{F} \,. \tag{16}$$

with input vector $\mathbf{u}$ and Cartesian forces $\mathbf{F}$. During motion

$$\mathbf{F} = \begin{bmatrix} m\ddot{y}_e \\ m(\ddot{z}_e + g) \\ I_x\ddot{\theta}_e \end{bmatrix} \tag{17}$$

where m is the mass and I_x is the moment of inertia of the plattform and g is the force of gravity. The dynamics equations can be written as

$$\mathbf{D}\ddot{\mathbf{x}} + \mathbf{G} = \mathbf{A}(\mathbf{x})\mathbf{u} \tag{18}$$

with

$$\mathbf{D} = \begin{bmatrix} m & 0 & 0 \\ 0 & m & 0 \\ 0 & 0 & I_x \end{bmatrix} \tag{19}$$

and

$$\mathbf{G} = \begin{bmatrix} 0 \\ mg \\ 0 \end{bmatrix} . \tag{20}$$

Rearranging Eq. (18) yields to

$$\ddot{\mathbf{x}} = -\mathbf{D}^{-1}\mathbf{G} + \mathbf{D}^{-1}\tilde{\mathbf{u}}\,. \tag{21}$$

with

$$\tilde{\mathbf{u}} = \mathbf{A}(\mathbf{x})\mathbf{u}\,. \tag{22}$$

In this form Eq. (21) fits into a state space representation

$$\dot{\mathbf{x}} = \mathbf{a}(\mathbf{x}) + \mathbf{b}(\mathbf{x})\tilde{\mathbf{u}} \tag{23}$$

or

$$\begin{bmatrix} \ddot{y} \\ \ddot{z} \\ \ddot{\theta} \end{bmatrix} = \begin{bmatrix} 0 \\ -g \\ 0 \end{bmatrix} + \begin{bmatrix} 1/m \\ 1/m \\ 1/(m \cdot r^2) \end{bmatrix} \tilde{\mathbf{u}} \tag{24}$$

respectively where r is the radius of the platform. The output equations are given by

$$\mathbf{y} = \mathbf{c}(\mathbf{x}) = \begin{bmatrix} y \\ z \\ \theta \end{bmatrix}\,. \tag{25}$$

This requires that we can measure the pose of the platform.

3 Control

The concept of exact linearization is described by [1,9] and [5]. For more detailed information, please refer to the cited literature. At this point only a brief introduction to the topic is given. Besides, in [6], a control based on exact linearization for a spatial 3-DoF cable robot was presented.

For the cable robot presented in Sect. 2 we assume the following *MIMO* (Multiple Input Multiple Output) state space representation with n states and m inputs.

$$\begin{aligned} \dot{\mathbf{x}} &= \mathbf{a}(\mathbf{x}) + \textstyle\sum_{i=1}^{m} \mathbf{b}_i(\mathbf{x})\tilde{u}_i \\ \mathbf{y} &= \mathbf{c}(\mathbf{x}) \end{aligned} \tag{26}$$

This notation is equivalent to the formulation of Eq. (23) with $u_i = F_i$. A requirement for the design is that the number of inputs and outputs are equal. As part of the controller design we need the output y_i with $i = 1 \ldots m$ and its derivatives with respect to time. These are defined as

$$
\begin{aligned}
y_i &= c_i(\mathbf{x}) \\
\dot{y}_i &= L_{\mathbf{a}} c_i(\mathbf{x}) + \sum_{k=1}^{m} L_{\mathbf{b}_k} c_i(\mathbf{x}) u_k \\
\ddot{y}_i &= L_{\mathbf{a}}^2 c_i(\mathbf{x}) + \sum_{k=1}^{m} L_{\mathbf{b}_k} L_{\mathbf{a}} c_i(\mathbf{x}) u_k \\
&\vdots \\
y_i^{(\delta_i-1)} &= L_{\mathbf{a}}^{\delta_i-1} c_i(\mathbf{x}) + \sum_{k=1}^{m} L_{\mathbf{b}_k} L_{\mathbf{a}}^{\delta_i-2} c_i(\mathbf{x}) u_k \\
y^{(\delta)} &= L_{\mathbf{a}}^{\delta_i} c_i(\mathbf{x}) + \sum_{k=1}^{m} L_{\mathbf{b}_k} L_{\mathbf{a}}^{\delta_i-1} c_i(\mathbf{x}) u_k
\end{aligned}
\tag{27}
$$

with

$$
L_{\mathbf{f}} h(\mathbf{x}) = \frac{\partial h(\mathbf{x})}{\partial \mathbf{x}} \mathbf{f}(\mathbf{x}) . \tag{28}
$$

In MIMO case, the term

$$
\tilde{y}^{(\delta)} = \sum_{k=1}^{m} L_{\mathbf{b}_k} L_{\mathbf{a}}^{\delta_i-1} c_i(\mathbf{x}) u_k \tag{29}
$$

of the last line of Eq. (27) becomes unequal to zero. The degree of derivative of y is called relative degree δ. The term for the delta-times derivative of y with respect to time can be noted compactly in vectorial notation as

$$
\mathring{\mathbf{y}} = \mathring{\mathbf{c}}(\mathbf{x}) + \mathbf{D}(\mathbf{x}) \cdot \mathbf{u} \tag{30}
$$

with

$$
\mathring{\mathbf{y}} = \begin{bmatrix} y^{(\delta_1)} \\ y^{(\delta_2)} \\ \vdots \\ y^{(\delta_m)} \end{bmatrix} , \tag{31}
$$

$$
\mathring{\mathbf{c}}(\mathbf{x}) = \begin{bmatrix} L_{\mathbf{a}}^{\delta_1} c_1(\mathbf{x}) \\ L_{\mathbf{a}}^{\delta_2} c_2(\mathbf{x}) \\ \vdots \\ y L_{\mathbf{a}}^{\delta_m} c_m(\mathbf{x}) \end{bmatrix} \tag{32}
$$

and

$$
\mathbf{D}(\mathbf{x}) = \begin{bmatrix} L_{\mathbf{b}_1} L_{\mathbf{a}}^{\delta_1-1} c_1(\mathbf{x}) & L_{\mathbf{b}_2} L_{\mathbf{a}}^{\delta_1-1} c_1(\mathbf{x}) & \dots & L_{\mathbf{b}_m} L_{\mathbf{a}}^{\delta_1-1} c_1(\mathbf{x}) \\ L_{\mathbf{b}_1} L_{\mathbf{a}}^{\delta_2-1} c_2(\mathbf{x}) & L_{\mathbf{b}_2} L_{\mathbf{a}}^{\delta_2-1} c_2(\mathbf{x}) & \dots & L_{\mathbf{b}_m} L_{\mathbf{a}}^{\delta_2-1} c_2(\mathbf{x}) \\ \vdots & \vdots & \ddots & \vdots \\ L_{\mathbf{b}_1} L_{\mathbf{a}}^{\delta_m-1} c_m(\mathbf{x}) & L_{\mathbf{b}_2} L_{\mathbf{a}}^{\delta_m-1} c_m(\mathbf{x}) & \dots & L_{\mathbf{b}_m} L_{\mathbf{a}}^{\delta_m-1} c_m(\mathbf{x}) \end{bmatrix} \tag{33}
$$

For the cable robot the relative degree for each output is $\delta_i = 2$. The sum of the relative degrees of each output corresponds to the system order of the cable robot. Thus, there are no uncontrolled system dynamics and the controlled system is stable.

The matrices $\overset{\circ}{\mathbf{c}}(\mathbf{x})$ and $\mathbf{D}(\mathbf{x})$ result in

$$\overset{\circ}{\mathbf{c}}(\mathbf{x}) = \begin{bmatrix} 0 \\ -g \\ 0 \end{bmatrix} \tag{34}$$

and

$$\mathbf{D}(\mathbf{x}) = \begin{bmatrix} 1/m & 0 & 0 \\ 0 & 1/m & 0 \\ 0 & 0 & 1/(m \cdot r^2) \end{bmatrix}. \tag{35}$$

The control law results by rearranging Eq. (30) in

$$\mathbf{u}_x = -\mathbf{D}^{-1}(\mathbf{x})\overset{\circ}{\mathbf{c}}(\mathbf{x}) + \mathbf{D}^{-1}(\mathbf{x})\mathbf{v}\,. \tag{36}$$

where $\mathbf{v}$ corresponds to the differential equation of the desired output trajectory. The resulting control signal $\mathbf{u}$ consists of a feedforward control

$$\mathbf{r}(\mathbf{x}) = \mathbf{D}^{-1}(\mathbf{x})\overset{\circ}{\mathbf{c}}(\mathbf{x}) \tag{37}$$

and a feedback control

$$\mathbf{M}(\mathbf{x}) = \mathbf{D}^{-1}(\mathbf{x})\mathbf{v}\,. \tag{38}$$

At this point it is essential that matrix $\mathbf{D}(\mathbf{x})$ is invertible. For the invertibility the following must apply

$$\det \mathbf{D}(\mathbf{x}) \neq 0\,. \tag{39}$$

Therefore, the column vectors of the matrix from Eq. (35) must be linearly independent. In this case $\mathbf{D}^{-1}$ is easily calculable. The reason for this is the suitable choice of $\tilde{\mathbf{u}}$ since we have isolated the coordinate transformation presented in Eq. (22) from the controller design.

This completes the design of an exact linearization via feedback with the consequence that each reference variable w_i affects the corresponding output variable y_i through a transfer function $G_i(s) = \frac{Y_i(s)}{W_i(s)}$. For a simulation study, this approach is quite sufficient. In practice, however, some adjustments will usually be necessary. These include

- a *trajectory planning* to avoid step changes of the reference signal,
- a *tracking control* reducing tracking errors,
- an *output-feedback control* to compensate for permanent control errors due to nonlinear effects such as friction or other disturbances.

Figure 3 shows a block diagram of a system controlled by an exact linearization including the three extensions mentioned above.

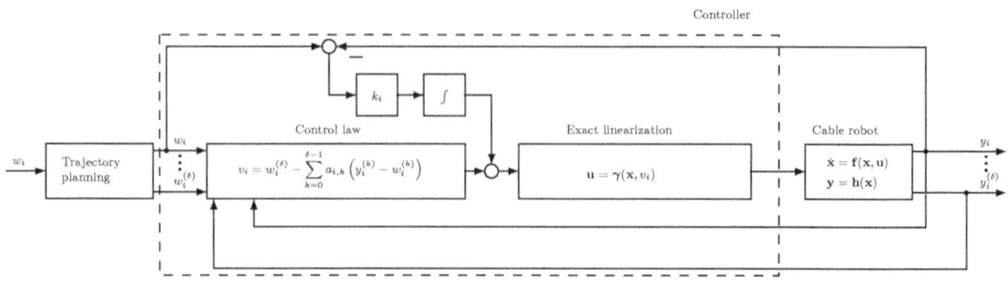

Fig. 3. Exact linearization of one output y_i with trajectory planning, tracking control and output-feedback control

4 Simulation

For the simulation depicted in Fig. 4, we have defined step changes for each state (dashed line). At this point, the trajectory planning, as well as the output-feedback control, is not active. The shape of the states (solid line) results solely from the choice of the dynamics of the different outputs. The dynamics of the outputs were determined by pole placement. Other conceivable approaches are a linear-quadratic regulator or a robust design. The figure shows that the states reach their desired value smoothly. The required forces are converted into positive cable forces based on the approach presented in [15]. In this way, it is guaranteed that the cables are in tension.

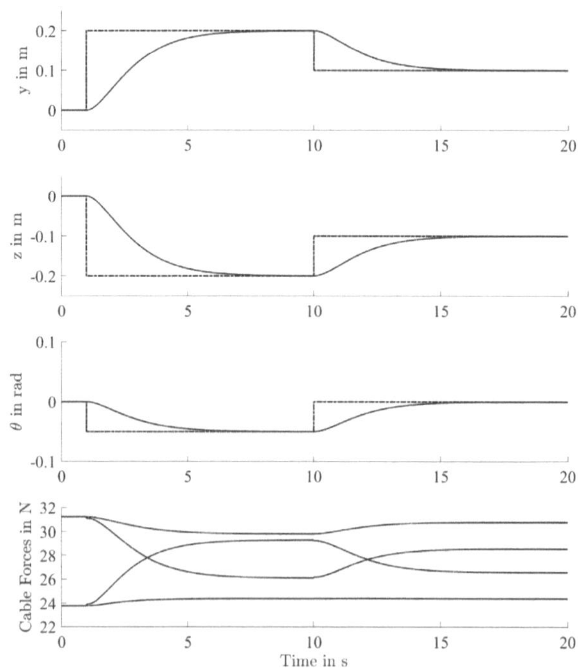

Fig. 4. Simulation results

In an experiment, adjustments will be necessary due to non-modeled effects such as friction. This includes, in particular, using the two extensions mentioned in Sect. 3 and deactivated for the simulation - trajectory planning and output-feedback linearization. Otherwise, a permanent control error would result in the event of a disturbance.

5 Conclusion

The presented paper described an approach to control a fully constrained, planar cable-driven parallel robot with three degrees of freedom. The control is based on an exact input-output linearization. This approach is particularly suitable for cable robots since the state-dependent transformation of forces from joint space to operational space is performed. As a result of the controller design, the cable forces applied to the platform are not competing with each other as they would, for example, in the case of motion control in joint space. For this purpose, a model of the cable robot was built. The input of this model was then combined with the structure matrix to separate the geometric coordinate transformation from the dynamics of the robot. In this way, a generic model is created. The resulting control law is simple and the approach can be applied to a broad variety of other parallel robots. Possible extensions of the control law have also been proposed which are necessary to compensate for disturbances. Finally, simulative results were shown.

For the future, we want to test the control approach in experiments to verify its performance. For this, we will use the laser tracker as described in the introduction. Also, we want to determine the dynamics of the closed-loop control system based on a robust approach to compensate for non-modeled effects such as friction.

References

1. Adamy, J.: Nichtlineare Systeme und Regelungen. Springer Vieweg, Berlin, 3., aktualisierte Auflage edn. (2018). https://doi.org/10.1007/978-3-662-55685-6
2. Alp, A.B., Agrawal, S.K.: Cable suspended robots: feedback controllers with positive inputs. In: Proceedings of the 2002 American, pp. 815–820 (2002). https://doi.org/10.1109/ACC.2002.1024915
3. Fang, S., Franitza, D., Torlo, M., Bekes, F., Hiller, M.: Motion control of a tendon-basedparallel manipulator using optimal tension distribution. IEEE/ASME Trans. Mechatron. **9**(3), 561–568 (2004). https://doi.org/10.1109/TMECH.2004.835336
4. Fang, S.: Design, modeling and motion control of tendon-based parallel manipulators: Zugl.: Duisburg-Essen, Univ., Diss., 2005, Fortschritt-Berichte VDI Reihe 8, Mess-, Steuerungs- und Regelungstechnik, vol. 1076. VDI-Verl., Düsseldorf, als ms. gedr edn. (2005)
5. Föllinger, O.: Nichtlineare Regelungen 2: Harmonische Balance, Popow- und Kreiskriterium, Hyperstabilität, Synthese im Zustandsraum: mit 18 Übungsaufgaben mit Lösungen (1993). https://doi.org/10.1524/9783110406146

6. Hamann, M., Winter, D., Ament, C. (eds.): Model-Based Control of a Pendulum by a 3-DoF Cable Robot Using Exact Linearization (2020). https://doi.org/10.1016/j.ifacol.2020.12.2130
7. Heyden, T., Maier, T., Woernle, C.: Trajectory tracking control for a cable suspension manipulator. In: Lenarčič, J., Thomas, F. (eds.) Advances in Robot Kinematics, vol. 19, pp. 125–134. Springer, Dordrecht (2002). https://doi.org/10.1007/978-94-017-0657-5_14
8. Heyden, T., Woernle, C.: Dynamics and flatness-based control of a kinematically undetermined cable suspension manipulator. Multibody Syst. Dyn. **16**(2), 155–177 (2006). https://doi.org/10.1007/s11044-006-9023-5
9. Isidori, A.: Nonlinear control systems. Communications and Control Engineering Series, Springer, Berlin, 3. ed., [nachdr.] edn. (2002)
10. Khosravi, M.A., Taghirad, H.D., Oftadeh, R.: A positive tensions pid controller for a planar cable robot: an experimental study. In: 2013 First RSI/ISM International Conference, pp. 325–330 (2013). https://doi.org/10.1109/ICRoM.2013.6510127
11. Kino, H., Yahiro, T., Takemura, F., Morizono, T.: Robust PD control using adaptive compensation for completely restrained parallel-wire driven robots: translational systems using the minimum number of wires under zero-gravity condition. IEEE Trans. Robot. **23**(4), 803–812 (2007). https://doi.org/10.1109/TRO.2007.900633
12. Maier, T., Woernle, C.: Flachheitsbasierte bahnsteuerung von seilgeführten handhabungssystemen (flatness-based control of cable suspension manipulators). at - Automatisierungstechnik **51**(6-2003), 265–273 (2003). https://doi.org/10.1524/auto.51.6.265.22449
13. Oh, S.R., Agrawal, S.K.: Nonlinear sliding mode control and feasible workspace analysis for a cable suspended robot with input constraints and disturbances. In: Proceedings of the 2004 American Control Conference ACC, pp. 4631–4636 vol.5. American Automatic Control Council, Evanston, Ill and Piscataway, N.J (2004). https://doi.org/10.23919/ACC.2004.1384041
14. Oh, S.R., Agrawal, S.K.: Generation of feasible set points and control of a cable robot. IEEE Trans. Robot. **22**(3), 551–558 (2006). https://doi.org/10.1109/TRO.2006.870646
15. Pott, A., Bruckmann, T., Mikelsons, L.: Closed-form force distribution for parallel wire robots. In: Müller, A., Kecskeméthy, A. (eds.) Computational Kinematics, vol. 40, pp. 25–34. Springer, s.l. (2009). https://doi.org/10.1007/978-3-642-01947-0_4
16. Stoltmann, M., Froitzheim, P., Fuchs, N., Woernle, C.: Flatness-based feedforward control of a crane manipulator with four load chains. In: Corves, B., Wenger, P., Hüsing, M. (eds.) EuCoMeS 2018, Mechanisms and Machine Science, vol. 59, pp. 61–68. Springer, Cham (2019). https://doi.org/10.1007/978-3-319-98020-1_8
17. Woernle, C.: Trajectory tracking for a three-cable suspension manipulator by nonlinear feedforward and linear feedback control. In: Bruckmann, T., Pott, A. (eds.) Cable-Driven Parallel Robots, Mechanisms and Machine Science, vol. 12, pp. 371–386. Springer, Berlin and Heidelberg (2013). https://doi.org/10.1007/978-3-642-31988-4_23
18. Yingjie, L., Wenbai, Z., Gexue, R.: Feedback control of a cable-driven Gough-Stewart platform. IEEE Trans. Robot. **22**(1), 198–202 (2006). https://doi.org/10.1109/TRO.2005.861459

Motor Current Based Force Control of Simple Cable-Driven Parallel Robots

Jonas Bieber(✉), David Bernstein, Micha Schuster, Konrad Wauer, and Michael Beitelschmidt

Chair of Dynamics and Mechanism Design, Technische Universität Dresden, Dresden, Germany
jonas.bieber@tu-dresden.de

Abstract. State of the art cable-driven parallel robots achieve high precision and reliability using a complex setup. The complexity results largely from the integration of force sensors which are omitted in the presented design. Instead, a motor current based approach is proposed to determine the cable forces. Therefore, methods are employed to compensate the cogging torque of the motor and frictional forces along the transmission path from the motor torque to the acting force at the platform. Despite the compensation, it is useful to reduce the overall frictional forces. Therefore, a gearless design scheme with a single pivoting pulley for cable guidance is realized. The proposed approach enables the application of advanced control methods for simple cable-driven robots.

Keywords: Cable-driven parallel robot · Force control · Motor current · Cogging compensation · Friction compensation · Gearless

1 Introduction

Advantages of cable-driven parallel robots are a large workspace, high achievable speeds and high payload. Intensive research in the last decades has lead to robots with high precision and reliability [4,11,17]. Many advanced approaches rely on a force or torque based control to stay within cable force limits and enable high pose control performance or haptic force feedback [7,9,10]. To determine and to control the acting forces, force sensors are used, by which however the system complexity is increased notably. This complexity is reflected in the number of mechanical and electrical components, as well as the effort and cost of integrating them into the system.

In order to make the technology of cable-driven robots accessible for a broader range of applications, reducing the complexity and costs is an important issue. The costs can be reduced by taking advantage of the increasing availability of inexpensive yet powerful components from the hobbyist sector such as single board computers and brushless motors. On the other hand, a high potential for a reduction of complexity can be found in omitting force sensors. To maintain a high control performance by force based control, the motor currents can be used to determine the forces acting on the robot platform. As a consequence, a

M. Gouttefarde et al. (Eds.): CableCon 2021, MMS 104, pp. 271–283, 2021.
https://doi.org/10.1007/978-3-030-75789-2_22

setup with minimized friction and a model of the remaining friction and other disturbing effects is necessary, and will be presented in this paper.

2 Relationship Between Motor Current and Cable Force

There is an approximately linear relationship between motor current and cable force acting on the platform. However, nonlinearities and discontinuities appear in the detailed consideration of the transmission path. These are described below.

2.1 From Current to Torque

For ideal electric motors, the proportional relation between the armature current i and the generated torque τ_0 is given with the motor dependent torque constant k_τ as

$$\tau_0 = k_\tau i. \tag{1}$$

For brushless motors with three phases, the torque generating motor current i corresponds to the quadrature current component i_q orthogonal to the rotor magnetic field. It can be obtained from the phase currents via direct-quadrature-zero transformation. Utilizing field oriented control the torque generating current i_q is set to the desired value. On real brushless motors, angle-dependent deviations from the linear relation in Eq. (1) occur. The main cause of these torque-ripples is the cogging phenomenon. It results from the non-constant interaction force between the permanent magnets on the rotor and the stator slots inside the motor. The cogging torque $\tau_\mathrm{c}(\varphi)$ leads to the mechanical output torque τ_m

$$\tau_\mathrm{m} = \tau_0 + \tau_\mathrm{c}(\varphi). \tag{2}$$

This effect is independent of the motor current i and depends only on the motor angle φ, which results in a deterministic offset with

$$\tau_\mathrm{c}(\varphi) = \tau_\mathrm{c}(\varphi + 2\pi n) \quad \text{with } n \in \mathbb{Z}. \tag{3}$$

Thus, at low load, cogging contributes a high proportion of the total torque τ_m, whereas its contribution decreases with higher loads. Furthermore, at high speeds, the system inertia reduces the impact of the effect. The magnitude of the cogging torque depends on the motor design, manufacturing variations and tolerances [5]. Therefore, the effect is in general more significant for inexpensive motors [14].

2.2 From Motor Torque to Cable Force

For the idealized transmission, the relationship between motor torque τ_m and cable force f holds

$$f = \nu \cdot \tau_\mathrm{m} \tag{4}$$

with the transmission ratio ν.

In reality, the load is transmitted via the motor, gearbox, winch and several other mechanical components as pulleys. Along the transmission path, additional forces apply. Considering the most important effects, the resulting cable force f_{act} can be described as

$$f_{\text{act}} = \nu \cdot \tau_{\text{m}} - \nu \cdot J\ddot{\varphi} - \operatorname{sgn}(\dot{\varphi}) \cdot f_{\text{f}}. \tag{5}$$

Accordingly, inertial forces apply depending on the motor acceleration $\ddot{\varphi}$. The inertias of all transmission path elements are represented by the total inertia J. The frictional force f_{f} depends on the direction of motion $\operatorname{sgn}(\dot{\varphi})$. Within the frictional force, three main components can be identified:

$$f_{\text{f}} = f_{\text{f,i}} + f_{\text{f,l}}(\tau_{\text{m}}) + f_{\text{f,v}}(\dot{\varphi}) \tag{6}$$

The frictional force mainly results from losses in the bearings of the rotating components. It is composed of:

a) $f_{\text{f,i}}$ = const.: a constant frictional force that is independent of cable force and movement. It can be considered as internal friction caused by permanent normal forces on the bearings such as gravity and preloads.
b) $f_{\text{f,l}} \approx \mu \cdot \nu\tau_{\text{m}}$: a load dependent frictional force. Approximately, a linear dependency on the mechanical torque τ_{m} is assumed. A more precise approach as presented in [6] also takes into account the force change due to friction along the transmission path and the pose-dependent wrapping angles of the pulleys.
c) $f_{\text{f,v}}(\dot{\varphi})$: a frictional force depending on the cable velocity, typically described as viscous friction.

While $f_{\text{f,v}}$ and the inertial loads $J\ddot{\varphi}$ only occur during movements of system, $f_{\text{f,i}}$ and $f_{\text{f,c}}$ are also relevant in the static case. The dependence of friction on the direction of motion leads to a hysteresis effect. Therefore, a specific motor torque results in higher cable forces when uncoiling compared to coiling up the cable. This leads to a discontinuity in the relationship between motor torque and cable force. In reality, however, the discontinuous signum function in Eq. (5) is smoothened by elasticity effects.

3 Considerations for Motor Current Based Force Control

The mechanical and electrical integration of accurate force sensors and A/D converters increases costs and complexity of advanced cable-driven robots. Furthermore, integrating the measured cable forces into cascaded force control loops leads to latencies due to signal processing [8]. In order to simplify force based control, under certain conditions, the cable force can be determined via the motor currents instead. The drawbacks of this current based force estimation are systematic deviations mainly caused by friction and cogging torque. These deviations can generally be described based on a simple propagation of uncertainty of the cable force from Eq. (5), which describes the deviation Δf of the acting cable force f_{act} from the desired cable force f_{des} at the platform as

$$\Delta f = f_{\text{act}} - f_{\text{des}} = \nu \cdot \Delta\tau_{\text{m}} + \Delta f_{\text{f}} + f_{\text{res}}. \tag{7}$$

Since every uncompensated frictional force as well as every modeling error is incorporated in the term $\Delta f_{\text{f}} = f_{\text{f}} - f_{\text{comp}}$, it is suitable to take measures to reduce the overall friction f_{f} in the setup and develop a model for the friction compensation with $f_{\text{comp}} \approx f_{\text{f}}$. On the other hand, the error torque $\Delta\tau_{\text{m}}$ comprises the cogging torque τ_c that has to be compensated, as well as measurement errors of the current and motor parameter uncertainties. All residual effects are summarized in f_{res}. In the following, a robot design scheme and compensation methods to reduce the impact of cogging torque and friction are proposed.

3.1 Design

As mentioned above, it is crucial to minimize the effects of cogging and friction to maintain a high precision in the prediction of the cable force even in the presence of modeling errors or further unmodeled phenomena. Thus, to reduce the frictional force, a minimalistic design is beneficial. The consequences of this approach are discussed below.

Absence of a Gearbox. Typically, gearboxes are an integral part of the drive system design of cable-driven robots. Since high-torque motors are more expensive in general, gearboxes allow high cable forces using low-torque motors. But especially the use of gearboxes increases the frictional forces. As described in [16], the efficiency factor of gearboxes of up to 95% at nominal load descreases at low cable forces. In particular, the frictional force f_{f} resulting from friction in the gearbox can be higher than low actual cable forces [7]. Those low cables forces typically occur when uncoiling the cable, often combined with high velocities increasing the viscous frictional force $f_{\text{f,v}}$.

Due to the design goal of reducing the frictional force f_{f} as much as possible, gearboxes are not used. Instead, there is a direct coupling between the motor shaft and the winch.

Simple Cable Guidance. For cable-driven robots, cable guidance is used to ensure a controlled coiling of the cable on the drum. A cable guidance with multiple redirections of the cable allows a flexible positioning of the drive units, in addition. Typically, for this purpose mechanisms using pulleys or eyelets are deployed. The mechanical force loss inside of pulleys due to friction is about 3% [16] and is likely even higher in eyelets. Therefore, especially a cable guidance with many redirections of the cable leads to increased mechanical losses.

Figure 1 shows a setup where a single pivoting pulley is sufficient to achieve a single-layer coiling. Since the winch consists solely of a drum on a shaft without

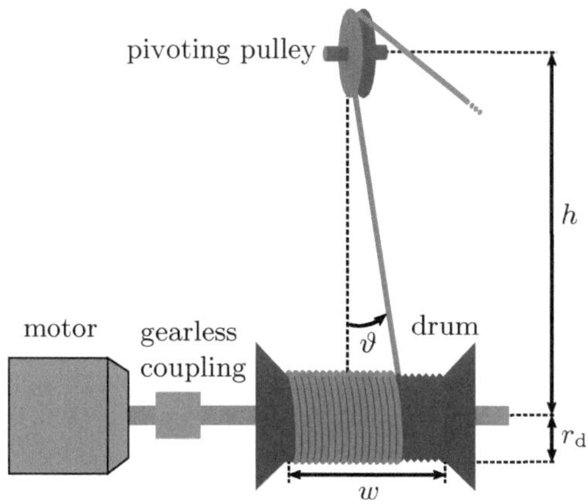

Fig. 1. Basic design of the drive unit and the simple cable guidance using a simple pivoting pulley

dedicated cable guidance, this imposes restrictions on the placement o f the winch, especially regarding the deflection angle ϑ. The angle ϑ is defined relative to the pivoting pulley and increases when coiling up the cable. To prevent cable twisting and allow single-layer coiling, the angle ϑ has to meet the limits:

$$\vartheta_{\text{min}} \leq \vartheta \leq \vartheta_{\text{max}}. \tag{8}$$

The absolute value of the maximum deflection angle is typically larger than the absolute value of the minimum ($|\vartheta_{\text{max}}| > |\vartheta_{\text{min}}|$) due to the support of the already coiled cable. The absolute value of the minimum deflection angle $|\vartheta_{\text{min}}|$ can be increased by a guiding helical groove on the drum. The angle restrictions result in a maximum width w_{max} of the drum of

$$w_{\text{max}} = h \cdot (\tan(\vartheta_{\text{max}}) + \tan(|\vartheta_{\text{min}}|)) \tag{9}$$

for large distances $h \gg r_{\text{d}}$ between pulley and drum.

Radius of the Drum. In the simplified setup, the radius of the drum r_{d} plays an important role. Since no gearbox is used, the transmission $\nu = \frac{1}{r_{\text{d}}}$ only depends on the radius r_{d} of the drum. Therefore, a large drum radius can be used to minimize the effect of the torque errors $\Delta\tau_{\text{m}}$ by reducing its scale factor ν in Eq. (7). On the other hand, an enlarged drum radius leads to higher motor torques for the same cable force and thus reduces the relative torque errors $\Delta\tau_{\text{m}}/\tau_{\text{m}}$ due to the load-independent cogging torque τ_{c}. As a side effect, a high drum radius increases the maximum usable cable length. With a given maximum drum width w_{max} (Eq. (9)) and diameter of the cable d_c, the maximum cable length that can be coiled is

$$\Delta\ell_{\text{max}} = 2\pi r_{\text{d}} \frac{w_{\text{max}}}{d_{\text{c}}}. \tag{10}$$

On the other hand, typical applications for cable-driven robots require relatively high cable forces and rather lower speeds. Thus, with a large radius r_d more expensive high-torque motors are needed. The proposed simple design scheme therefore requires careful adjustment of the distance between winch and pulley, the drum radius and the maximum motor torque. Due to their high torque-density, especially brushless outrunner motors are suitable. Here, appropriate hobbyist motors are available at moderate price.

3.2 Cogging Compensation

Besides frictional forces, the second source of disturbances that has to be minimized is cogging. Figure 2 shows the effect of cogging on the cable force for the prototype presented in Sect. 4.1. The quasi-static cable force at the winch is measured with a force sensor over one motor revolution for a constant motor current i_q at very low speed with about one rotation per minute.

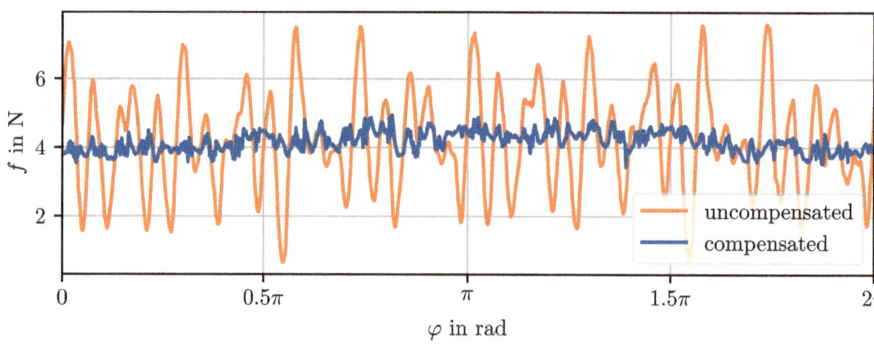

Fig. 2. Quasi-static cable force f directly behind the winch over one motor revolution with and without cogging compensation for $\tau_\mathrm{des} = \mathrm{const}$

Since the cogging torque occurs systematically and depends only on the rotation angle of the motor, it can be largely compensated in the motor controller. For this purpose, the desired torque τ_des is adjusted with an offset $\tau_\mathrm{comp} \approx \tau_\mathrm{c}$ to cancel out the cogging torque

$$\tau_\mathrm{set} = \tau_\mathrm{des} - \tau_\mathrm{comp}(\varphi). \tag{11}$$

A method to find this angle-dependent offset without requiring external measurement equipment is proposed in [14]. Here, the offset torque $\tau_\mathrm{comp}(\varphi)$ is determined as the motor torque required to maintain the position of the unloaded motor. The method is implemented as a feature in the firmware of the *ODrive* motor controllers [1] that are used in the experimental setup. Figure 2 also shows the improvements achieved by the anti-cogging feature. The fluctuations of the motor torque can be reduced noticeably.

3.3 Friction Compensation

In general, to compensate the relevant frictional forces, all contributions between platform and sensors for force determination have to be considered. For systems where the cable force is measured at the winch or pulleys, friction estimation and compensation methods are proposed in [6,10,12,13]. If the force determination is based on the motor current as proposed in Sect. 3.1, the friction in the motor and winch bearings has to be considered as well. In the following, a method is proposed to compensate main parts of the frictional force f_f. As mentioned in Sect. 2.2, the resulting friction along the whole transmission path is analyzed rather than examining the individual components bottom-up.

Since friction acts against the direction of motion, it can be calculated from the measured force difference $f_{\text{act,diff}} = f_{\text{act,uncoil}} - f_{\text{act,coil}}$ between the cable forces during coiling $f_{\text{act,coil}}$ and uncoiling $f_{\text{act,uncoil}}$ while the same desired force f_{des} is applied. Figure 3 shows the resulting cable forces in relation to the desired cable force f_{des} measured on the prototype (see Sect. 4.1).

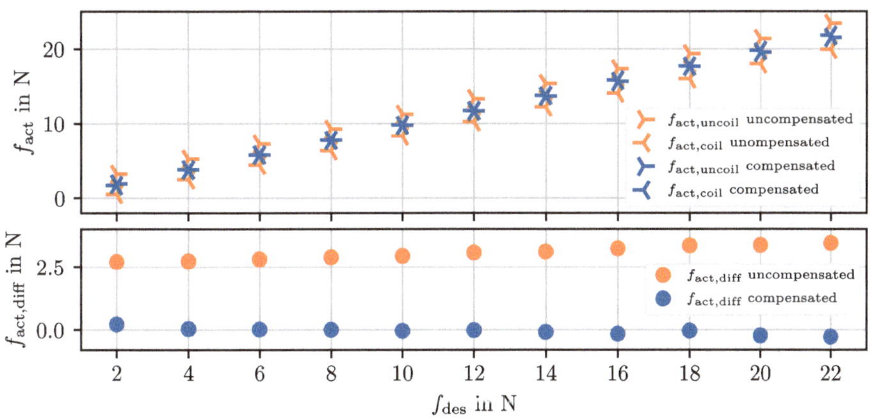

Fig. 3. Cable force f_{act} and cable force difference $f_{\text{act,diff}}$ between coiling and uncoiling, with and without friction compensation

For the measurement, the cable force f_{act} is determined using a force sensor close to the center of the workspace. It thereby includes all friction components from motor, winch and pulley. The force sensor is manually moved in direction of the cable neutral axis while f_{des} is hold constant. To be able to neglect the influence of cogging, measurements are averaged over one motor revolution. Since the force sensor is moved at very low and constant speed, the cable force components resulting from inertia J and velocity-dependent friction $f_{\text{f,v}}$ (see Eq. (5)) can be neglected. Under the corresponding assumption of a static case, it holds

$$f_{\text{f}} = f_{\text{f,i}} + \mu f_{\text{des}} = \frac{f_{\text{act,diff}}}{2}. \tag{12}$$

The frictional force is calculated from the cable force difference in the uncompensated case. From the measurement the parameters $f_{\mathrm{f,i}} \approx 1.3\,\mathrm{N}$ and $\mu \approx 0.018$ are approximated.

After compensating the friction based on Eq. (12), the lower part of Fig. 3 shows the cable force difference $f_{\mathrm{act,diff}}$ for the uncompensated case and a compensation based on Eq. (12). It is shown that velocity independent frictional forces are compensated well.

To compensate the frictional force, the set cable force f_{set} is calculated as

$$f_{\mathrm{set}} = f_{\mathrm{des}} + f_{\mathrm{comp}}. \tag{13}$$

Apart from direction changes of the cable motion $\mathrm{sgn}(\dot{\varphi})$, f_{comp} corresponds to the frictional force f_{f}. However, to avoid control instability resulting from the discontinuity of the frictional force when changing the direction of motion, the frictional force is integrated in the compensational force as

$$f_{\mathrm{comp}} = f_{\mathrm{f}} \cdot \mathrm{sgn}(\dot{\varphi}) + (f_{\mathrm{comp,0}} - f_{\mathrm{f}} \cdot \mathrm{sgn}(\dot{\varphi})) \cdot e^{-\frac{\sigma}{f_{\mathrm{f}}} \cdot |\varphi - \varphi_0|} \tag{14}$$

instead of using a signum function as in Eq. (5). When the direction of motion changes, the variables $\varphi_0 = \varphi$ and $f_{\mathrm{comp,0}} = f_{\mathrm{comp}}$ are set. The parameter σ determines how stiff a direction change reverses the frictional force. At the time of the paper submission it is set based on empirical investigation.

The approach implies the consideration of the system elasticity using the Dahl model [2] as already used in [6] and [12] for cable-driven robots. However, systematical errors occur when the cable force is calculated in a reactive manner from the encoder values. Trajectory based prediction of direction changes might minimize those errors.

As a further approach, the dynamic forces in the Eqs. (5) and (6) can be taken into account. The current procedure requires the use of a force sensor for calibration. Possibilities for sensorless identification of friction parameters are part of future research.

4 Prototype

A cable-driven parallel robot prototype has been developed at the TU Dresden in the project *RopeBot*. In the following, the hardware used and the implemented control loop are described and a measurement on a trajectory is shown.

4.1 Hardware

The prototype is a redundant, suspended setup with 8 cables and 6 degrees of freedom (see Fig. 4). The design corresponds to the simple scheme presented in Sect. 3.1. Two drive units, consisting of a motor and winch, are mounted in a common modular motor block. A single *ODrive* motor controller controls both brushless motors of a motor block. Apart from the 3D-printed drums and pulleys

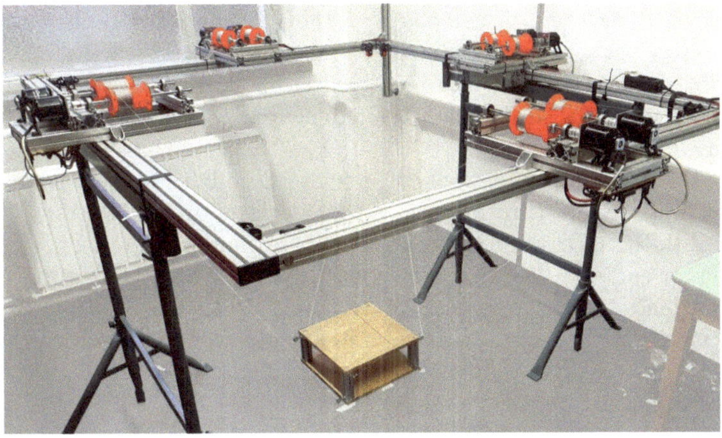

Fig. 4. Prototype of a spacial redundant cable-driven robot

made of PETG plastic, only mechanical standard parts are used to assemble the motor blocks and pivoting pulleys. Those are attached to an aluminum frame.

With a maximum motor torque of $\tau_{\text{max}} = 3.86\,\text{Nm}$ and a drum radius of $r_{\text{d}} = 30\,\text{mm}$, the maximum cable force is $f_{\text{act,max}} = 128\,\text{N}$. Further design details are summarized in Table 1. For a future application with an installation space of about $6\,\text{m} \times 6\,\text{m} \times 4\,\text{m}$, the drums are designed with the specified width w providing a distance $h \geq 2\,\text{m}$.

Table 1. Components and parameters of the prototype

Cable:	*Dyneema*® 1 mm	**Frame:**	$2\,\text{m} \times 1.7\,\text{m} \times 1.2\,\text{m}$
Control computer:	*Raspberry Pi 4 Model B*	**Platform:**	$0.4\,\text{m} \times 0.4\,\text{m} \times 0.12\,\text{m}$
Motor:	*ODrive D6374 150KV*		5.1 kg
Motor controller:	*ODrive V3.6*	**Drum:**	$r_{\text{d}} = 30\,\text{mm}$
Encoder:	*CUI Devices AMT-102*		$w = 80\,\text{mm}$

4.2 Control

Figure 5 shows the implemented control scheme. The estimated pose of the platform $\mathbf{p}_{\text{est}}$ is determined from the motor angles φ using iterative forward kinematics. Starting with a known pose, the platform pose is observed through an iterative scheme using Newton-Raphson algorithm [3]. The initial pose is estimated with the interval analysis inspired technique given in [16].

A desired wrench $\mathbf{w}_{\text{des}}$ is calculated in the PID controller from the deviation $\mathbf{e} = \mathbf{p}_{\text{des}} - \mathbf{p}_{\text{est}}$ between the desired pose and the actual pose. This is distributed to the cable forces $\mathbf{f}_{\text{des}}$ using the improved closed-form solution [15]. Although minimum cable forces are targeted, it is ensured that the cable force is kept within the limits $1\,\text{N} \leq f_{\text{des}} \leq 100\,\text{N}$.

Using the relationships described in Sect. 2.2 and the friction compensation from Sect. 3.3, the motor torques $\boldsymbol{\tau}_{\text{des}}$ are calculated from the cable forces and transferred to the *ODrive* motor controllers. The motor controllers calculate the desired motor current i_{q} according to Sect. 2.1 and the adjusted torque according to the cogging compensation in Sect. 3.2. In a decentralized cascaded control loop, the motor voltages of the three phases are adjusted by field-oriented control.

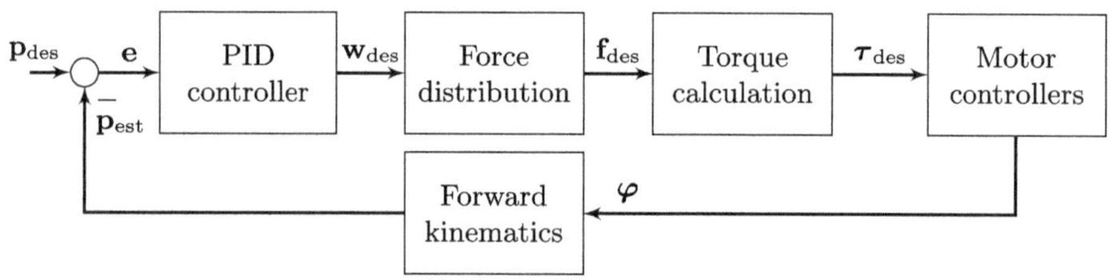

Fig. 5. Diagram of the control loop

4.3 Experimental Validation

The control of cable forces via motor currents is tested using a simple trajectory. Starting from a central point in the workspace, the platform performs a linear, translational back and forth movement. For a single cable, the motor angle φ and the desired cable force f_{des} are recorded. The actual cable force f_{act} is measured by a force sensor simultaneously. The results are shown in Fig. 6.

In measurements of the uncompensated case, a mean absolute deviation of $\Delta f_{\text{mean}} = |f_{\text{act}} - f_{\text{des}}|_{\text{mean}} = 2.37\,\text{N}$ occurs. With compensation of friction and cogging, the mean absolute deviation of $\Delta f_{\text{mean}} = 0.57\,\text{N}$ is reduced by more than a factor of 4. Deviations mainly occur when the direction of motion changes at about 2.2 s and 6.4 s. This results from not considering inertia forces at the transmission path and from the reactive behavior of the Dahl implementation in Eq. (14).

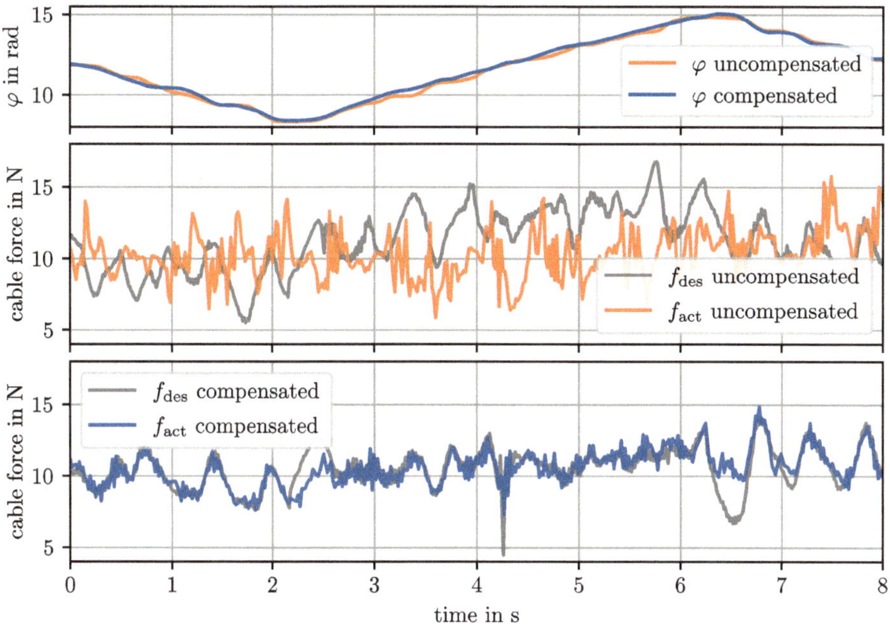

Fig. 6. Recorded motor angles φ and cable forces (desired f_{des} and measured f_{act}) without and with friction and cogging compensation for a single cable at an exemplary platform trajectory

5 Conclusion

The general possibility to set the cable forces via the motor currents is shown for cable-driven robots. The accuracy of the cable force prediction can be notably improved by modeling and compensating cogging and friction, which are the dominant disturbance effects.

For further improvement, the introduced model based feedforward control allows to compensate inertial forces and to predict changes of the cable motion direction improving the implemented Dahl model. Methods for determining and compensating viscous friction and the wrapping-angle-dependency of the pulley friction are also conceivable in the future.

Nevertheless, the achieved cable force accuracy of the presented approach can already be sufficient for many applications. It is shown that force based control is possible using a simple design approach without force sensors. A basic requirement is to use a gearless drive unit and a minimum number of pulleys for the cable guidance. The proposed design helps to reduce costs for cable-driven robots in general. Due to the determination of the cable forces, advanced control schemes are still deployable.

References

1. ODrive Firmware: ODrive Robotics, March 2021. https://github.com/odriverobotics/ODrive
2. Dahl, P.R.: A solid friction model. Technical Report. Defense Technical Information Center, Fort Belvoir, May 1968. https://doi.org/10.21236/ADA041920
3. Fang, S.: Design, Modeling and Motion Control of Tendon-Based Parallel Manipulators. No. 1076 in Fortschritt-Berichte VDI Reihe 8, Mess-, Steuerungs- Und Regelungstechnik, VDI-Verl, Düsseldorf (2005)
4. Hiller, M., Fang, S., Mielczarek, S., Verhoeven, R., Franitza, D.: Design, analysis and realization of tendon-based parallel manipulators. Mech. Mach. Theory **40**(4), 429–445 (2005). https://doi.org/10.1016/j.mechmachtheory.2004.08.002
5. Islam, M.S., Islam, R., Sebastian, T., Chandy, A., Ozsoylu, S.A.: Cogging torque minimization in PM motors using robust design approach. IEEE Trans. Ind. Appl. **47**(4), 1661–1669 (2011). https://doi.org/10.1109/TIA.2011.2154350
6. Kraus, W., Kessler, M., Pott, A.: Pulley friction compensation for winch-integrated cable force measurement and verification on a cable-driven parallel robot. In: 2015 IEEE International Conference on Robotics and Automation (ICRA), pp. 1627–1632, May 2015. https://doi.org/10.1109/ICRA.2015.7139406
7. Kraus, W.: Force control of cable-driven parallel robots. No. 49 in Stuttgarter Beiträge Zur Produktionsforschung, Fraunhofer Verlag, Stuttgart (2016)
8. Kraus, W., Schmidt, V., Rajendra, P., Pott, A.: System identification and cable force control for a cable-driven parallel robot with industrial servo drives. In: 2014 IEEE International Conference on Robotics and Automation (ICRA), Hong Kong, China, pp. 5921–5926. IEEE, May 2014. https://doi.org/10.1109/ICRA.2014.6907731
9. Lamaury, J., Gouttefarde, M.: Control of a large redundantly actuated cable-suspended parallel robot. In: 2013 IEEE International Conference on Robotics and Automation, pp. 4659–4664, May 2013. https://doi.org/10.1109/ICRA.2013.6631240
10. Lamaury, J., Gouttefarde, M., Chemori, A., Herve, P.E.: Dual-space adaptive control of redundantly actuated cable-driven parallel robots. In: Proceedings of the IEEE/RSJ International Conference on Intelligent Robots and Systems. IEEE/RSJ International Conference on Intelligent Robots and Systems, pp. 4879–4886, November 2013. https://doi.org/10.1109/IROS.2013.6697060
11. Lamaury, J., Gouttefarde, M., Michelin, M., Tempier, O.: Design and control of a redundant suspended cable-driven parallel robots. In: Lenarcic, J., Husty, M. (eds.) Latest Advances in Robot Kinematics, pp. 237–244. Springer, Dordrecht (2012). https://doi.org/10.1007/978-94-007-4620-6_30
12. Otis, M., Nguyen-Dang, T., Laliberté, T., Ouellet, D., Laurendeau, D., Gosselin, C.: Cable tension control and analysis of reel transparency for 6-DOF haptic foot platform on a cable-driven locomotion interface. Int. J. Electr. Comp. Syst. Eng. **3**, 16–29 (2009)
13. Piao, J., Kim, E.S., Choi, H., Moon, C.B., Choi, E., Park, J.O., Kim, C.S.: Indirect force control of a cable-driven parallel robot: tension estimation using artificial neural network trained by force sensor measurements. Sensors **19**(11), 2520 (2019). https://doi.org/10.3390/s19112520
14. Piccoli, M., Yim, M.: Cogging torque ripple minimization via position based characterization. In: Robotics: Science and Systems X. Robotics: Science and Systems Foundation, July 2014

15. Pott, A.: An improved force distribution algorithm for over-constrained cable-driven parallel robots. In: Computational Kinematics, vol. 15, pp. 139–146. Springer, Dordrecht (2014)
16. Pott, A.: Cable-Driven Parallel Robots: Theory and Application. Springer, Heidelberg (2018)
17. Pott, A., Mütherich, H., Kraus, W., Schmidt, V., Miermeister, P., Verl, A.: IPAnema: a family of cable-driven parallel robots for industrial applications. In: Mechanisms and Machine Science, vol. 12, pp. 119–134 (2013). https://doi.org/10.1007/978-3-642-31988-4_8

Control Stability Workspace for a Cable-Driven Parallel Robot Controlled by Visual Servoing

Zane Zake[1,2(✉)], François Chaumette[3], Nicolò Pedemonte[2], and Stéphane Caro[1,4]

[1] Laboratoire des Sciences du Numérique de Nantes, UMR CNRS 6004, 1, Rue de la Noë, 44321 Nantes, France
{Zane.Zake,stephane.caro}@ls2n.fr

[2] IRT Jules Verne, Chemin du Chaffault, 44340 Bouguenais, France
nicolo.pedemonte@irt-jules-verne.fr

[3] Inria, Univ Rennes, CNRS, IRISA, Rennes, France
Francois.Chaumette@inria.fr

[4] Centre National de la Recherche Scientifique (CNRS), 1, Rue de la Noë, 44321 Nantes, France

Abstract. A novel workspace, called Control Stability Workspace (CSW), is proposed and analyzed in this paper. It is a workspace from the control perspective. It allows us to determine whether the closed-loop controller will be able to guide a cable-driven parallel robot (CDPR) to a desired pose, given that there are perturbations present in the system. The effect of the different perturbations on the CSW is analyzed for two spatial CDPRs with a pose-based visual servoing (PBVS) controller. Indeed, this workspace is not only robot architecture-dependent, but also control-dependent. It was found that the model errors, that is, the errors in the cable exit and anchor point coordinates, have the most significant effect on the CSW. However, if the pulleys are small enough, then neglecting their geometry does not lead to a significant reduction of the workspace.

Keywords: Cable-driven parallel robots · Visual servoing · Stability · Workspace

1 Introduction

Cable-driven parallel robots (CDPRs) are characterized by a large workspace. However, knowing this workspace is not a trivial task. A workspace consists of all the CDPR moving-platform (MP) poses, in which certain conditions are fulfilled. Depending on the selected conditions multiple workspaces can be determined [1–3]. The simplest one is the Static Feasible Workspace (SFW), where the static equilibrium of the MP is used as the condition for defining the workspace [1,3]. The Wrench Feasible Workspace (WFW) is based on the static equilibrium along with the condition that the MP is able to withstand a set of external wrenches,

M. Gouttefarde et al. (Eds.): CableCon 2021, MMS 104, pp. 284–296, 2021.
https://doi.org/10.1007/978-3-030-75789-2_23

named Required Wrench Set (RWS), while staying in equilibrium. SFW is a special case of WFW when the external wrench is equal to zero and the static equilibrium is achieved by compensating only the MP weight.

Cable-cable interferences can affect the pose the MP reaches. Indeed, as the cable is no longer a straight line, but instead is bent at the interference point, the model no longer corresponds to the real CDPR and thus a different MP pose is obtained. To avoid such situations one can compute the Interference Free Workspace (IFW), which is the set of all MP poses without collisions between cables [4,5]. An extension of this workspace is the Collision Free Workspace (CFW), which is the set of all MP poses without collisions between the cables, the MP, and the environment.

For some tasks a CDPR will be required to attain high velocities and thus a Twist Feasible Workspace should be used [6]. When the dynamics of the robot need to be taken into account, the dynamic workspace needs to be considered. For example, the Dynamic Feasible Workspace (DFW) is the set of dynamic feasible MP poses. Here, a pose is dynamic feasible if the prescribed MP acceleration set is feasible, while cable tensions are kept within their bounds. For planar CDPRs the dynamic equations were solved analytically and thus the boundary of DFW could be found [7]. However, the same strategy cannot be applied to spatial CDPRs due to their increased complexity. For this reason a new workspace called Improved DFW is defined in [8].

All of the mentioned workspaces portray the physical ability of the MP to arrive at a certain pose given a set of conditions that need to be fulfilled. The control of the CDPR is not taken into account, meaning that there is no knowledge whether the controller will be able to guide the MP to the desired pose. For a closed-loop control, such as visual servoing, one can write the closed-loop equation and analyze the system stability with respect to perturbations [9–11]. Based on this analysis it was then possible to briefly introduce a novel workspace called Control Stability Workspace (CSW) [11]. It allows us to determine whether the controller will be able to guide the MP to the desired pose, given that there are perturbations present in the system. In this paper we make a thorough analysis of the sensitivity of this workspace with respect to perturbations in different parameters. Furthermore, the CSW is determined for two different CDPRs, and thus conclusions are made about the scalability of perturbation ranges as a function of the CDPR size.

2 Visual Servoing of a CDPR

A generic control scheme for visual servoing of a CDPR is shown in Fig. 1 [9–11]. Figure 2 illustrates the main components of a CDPR and the frames

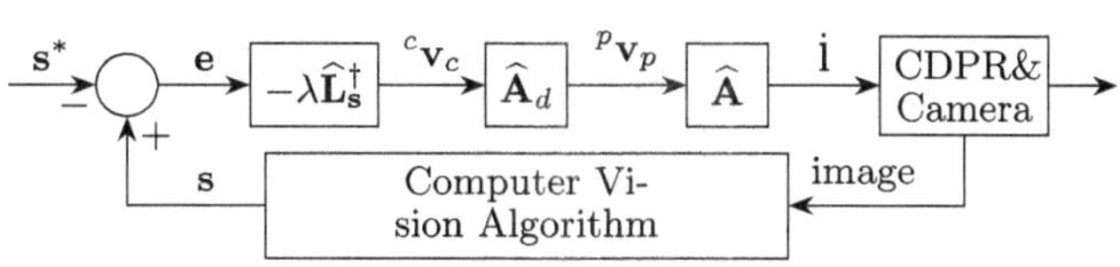

Fig. 1. Control scheme for visual servoing of a CDPR

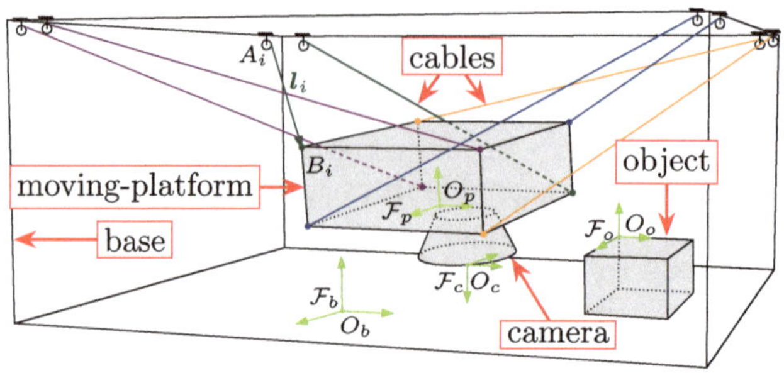

Fig. 2. Schematic of a spatial CDPR with eight cables, a camera mounted on the MP and an object in the workspace

of interest. The camera is mounted on the MP in the eye-in-hand configuration. Comparing the current feature vector $\mathbf{s}$, which is available from a computer vision algorithm, to the desired feature vector $\mathbf{s}^*$, which is previously defined, it is possible to compute the desired cable velocities that will try to make $\mathbf{s}$ converge to $\mathbf{s}^*$. Note that the content of the feature vector $\mathbf{s}$ depends on the chosen visual servoing approach. In case of Pose-Based Visual Servoing (PBVS), we can define $\mathbf{s} = ({}^c\mathbf{t}_o, \theta\mathbf{u})$, where ${}^c\mathbf{t}_o$ is the object position in the camera frame $\mathcal{F}_c$, and $\theta\mathbf{u}$ is the axis-angle representation of the rotation matrix ${}^{c^*}\mathbf{R}_c$ between the current and the desired camera frames [12].

An error $\mathbf{e}$ is computed by comparing $\mathbf{s}$ and $\mathbf{s}^*$. To ensure that this error converges to zero, we select as expected behavior:

$$\dot{\mathbf{e}} = -\lambda\mathbf{e} \tag{1}$$

where λ is a positive gain.

Upon derivation of $\mathbf{e} = \mathbf{s} - \mathbf{s}^*$ with respect to time, we get:

$$\dot{\mathbf{e}} = \mathbf{L_s}\,{}^c\mathbf{v}_c \tag{2}$$

where $\mathbf{L_s}$ is the interaction matrix and its composition depends on the chosen visual features [12]; ${}^c\mathbf{v}_c$ is the velocity of the camera with respect to the base frame $\mathcal{F}_b$, expressed in the camera frame $\mathcal{F}_c$.

It is then possible to express the camera velocity from (1) and (2):

$${}^c\mathbf{v}_c = -\lambda\,\widehat{\mathbf{L}}_{\mathbf{s}}^{\dagger}\,\mathbf{e} \tag{3}$$

where $\widehat{\mathbf{L}}_{\mathbf{s}}^{\dagger}$ is the pseudo-inverse of the interaction matrix estimation.

The relation between the camera velocity ${}^c\mathbf{v}_c$ and the MP velocity ${}^p\mathbf{v}_p$ is expressed as:

$${}^p\mathbf{v}_p = \mathbf{A}_d\,{}^c\mathbf{v}_c \tag{4}$$

where $\mathbf{A}_d$ is the Adjoint matrix [13]:

$$\mathbf{A}_d = \begin{bmatrix} {}^p\mathbf{R}_c & [{}^p\mathbf{t}_c]_\times {}^p\mathbf{R}_c \\ \mathbf{0}_3 & {}^p\mathbf{R}_c \end{bmatrix} \tag{5}$$

where ${}^{p}\mathbf{t}_{c}$ and ${}^{p}\mathbf{R}_{c}$ are the translation vector and rotation matrix that form the homogeneous transformation matrix ${}^{p}\mathbf{T}_{c}$ from $\mathcal{F}_p$ to $\mathcal{F}_c$.

The cable velocity vector $\dot{\mathbf{l}}$ is computed from MP twist ${}^{p}\mathbf{v}_{p}$ as follows:

$$\dot{\mathbf{l}} = \widehat{\mathbf{A}}\, {}^{p}\mathbf{v}_{p} \tag{6}$$

where $\widehat{\mathbf{A}}$ is the estimation of the Forward Jacobian matrix $\mathbf{A}$. For the modeling, the cables are supposed to be non-elastic and massless. The pulleys are assumed to be small, thus their radius is neglected. The components of matrix $\mathbf{A}$ can thus be expressed in the MP frame $\mathcal{F}_p$ as [14]:

$$\mathbf{A} = \begin{bmatrix} {}^{p}\mathbf{u}_1^T & ({}^{p}\mathbf{b}_1 \times {}^{p}\mathbf{u}_1)^T \\ \vdots & \vdots \\ {}^{p}\mathbf{u}_m^T & ({}^{p}\mathbf{b}_m \times {}^{p}\mathbf{u}_m)^T \end{bmatrix} \tag{7}$$

where ${}^{p}\mathbf{u}_i$ is the unit vector of ${}^{p}\overrightarrow{A_iB_i}$, pointing from the exit point A_i to the anchor point B_i of the ith cable, and ${}^{p}\mathbf{b}_i$ is the vector pointing from the origin of MP frame $\mathcal{F}_P$ to B_i.

The model of the system shown in Fig. 1 is written by injecting Eqs. (3), (4) and (6) into (2):

$$\dot{\mathbf{e}} = \mathbf{L_s}\, \mathbf{A}_d^{-1} \mathbf{A}^{\dagger}\, \dot{\mathbf{l}} \tag{8}$$

The output of the control scheme, that is, the cable velocity vector $\dot{\mathbf{l}}$, is expressed by injecting (3) and (4) into (6):

$$\dot{\mathbf{l}} = -\lambda\, \widehat{\mathbf{A}}\, \widehat{\mathbf{A}}_d\, \widehat{\mathbf{L}}_{\mathbf{s}}^{\dagger}\, \mathbf{e} \tag{9}$$

where $\widehat{\mathbf{A}}_d$ is the estimation of the Adjoint matrix $\mathbf{A}_d$.

The closed-loop equation of the system is expressed by injecting (9) into (8):

$$\dot{\mathbf{e}} = -\lambda\, \mathbf{L_s}\, \mathbf{A}_d^{-1}\, \mathbf{A}^{\dagger}\, \widehat{\mathbf{A}}\, \widehat{\mathbf{A}}_d\, \widehat{\mathbf{L}}_{\mathbf{s}}^{\dagger}\, \mathbf{e} \tag{10}$$

The stability criterion $\mathbf{\Pi}$ is defined as:

$$\mathbf{\Pi} = \mathbf{L_s}\, \mathbf{A}_d^{-1}\, \mathbf{A}^{\dagger}\, \widehat{\mathbf{A}}\, \widehat{\mathbf{A}}_d\, \widehat{\mathbf{L}}_{\mathbf{s}}^{\dagger} \tag{11}$$

According to Lyapunov stability analysis [15], $\mathbf{\Pi} > \mathbf{0}$ is a sufficient condition to obtain global asymptotic stability. Indeed, as long as the matrix $\mathbf{\Pi}$ is positive definite, it is clear from (10) that the error $\mathbf{e}$ will converge to $\mathbf{0}$.

3 Control Stability Workspace

Before using a CDPR, one needs to know its workspace. As recalled in the introduction, many workspace types have been defined [1,2]. For example, the Static-Feasible Workspace (SFW) is defined as [16]:

$$\mathcal{S} = \{\mathbf{p}_p \in SE(3) : \exists \boldsymbol{\tau} \in \mathcal{T},\ \mathbf{W}\boldsymbol{\tau} + \mathbf{w}_g = \mathbf{0}_6\} \tag{12}$$

Namely, the workspace $\mathcal{S}$ is the set of all MP poses $\mathbf{p}_p$ for which there exists a vector of cable tensions $\boldsymbol{\tau}$ within the cable tension space $\mathcal{T}$ such that the CDPR can balance the gravity wrench $\mathbf{w}_g$, and $\mathbf{W}\boldsymbol{\tau} + \mathbf{w}_g = \mathbf{0}_6$. The tension space $\mathcal{T}$ is an m-dimensional box of feasible tensions $\mathcal{T} = \{\boldsymbol{\tau} \in \mathbb{R}^m : \boldsymbol{\tau}_{lb} \leq \boldsymbol{\tau} \leq \boldsymbol{\tau}_{ub}\}$ where $\boldsymbol{\tau}_{lb}$ and $\boldsymbol{\tau}_{ub}$ are the lower and upper tension bounds, respectively [17].

SFW is a kineto-static workspace that contains all the poses that the MP is physically able to attain.

From (11) it can be seen that the stability of the system depends on the MP pose, because it shows up in the stability criterion $\mathbf{\Pi}$ through the Jacobian matrix $\mathbf{A}$, which is a function of the MP pose. Accordingly, a novel workspace, named Control Stability Workspace (CSW), can be defined as follows [11]:

$$\mathcal{Z} = \{\mathbf{p}_p \in SE(3) : \forall \mathbf{d} \in \mathcal{D}, \mathbf{\Pi} > \mathbf{0}\} \tag{13}$$

The workspace $\mathcal{Z}$ is the set of all MP poses $\mathbf{p}_p$, for which the stability criterion $\mathbf{\Pi}$ is positive definite for any vector of perturbations $\mathbf{d}$ that is within bounds $\mathcal{D}$. It means that for any MP pose within its CSW, the robot controller will be able to guide the MP to its goal.

3.1 The Definition of Perturbation Bounds $\mathcal{D}$

As was shown in [9,10], many parameters can induce perturbations. Furthermore, the perturbations affect the stability of the system simultaneously. To be able to analyze the effect of the perturbations on the CSW, we must first define the perturbation bounds.

Parameters can be either translational or rotational. For example, if we consider the camera pose in the MP frame $\mathcal{F}_p$, then the translational parameter is the camera position ${}^p\mathbf{t}_c$ and the rotational parameter is the camera orientation ${}^p\mathbf{R}_c$ expressed as the axis-angle.

Given a translational parameter T, the translational perturbation bound is defined as the radius r_t of the sphere shown in Fig. 3. It signifies that the perturbation $\boldsymbol{d}_t$ of parameter T that is within the range $\mathcal{D}_t$ can have a magnitude of up to r_t and can be in any direction. Thus, the perturbed value of parameter T can be anywhere within the sphere, e.g. it could take the value T_{d1} or T_{d2}.

For the rotational parameter the axis-angle representation is used. The rotational perturbation bound is defined as the angle $\Delta\theta$ for all axis directions $\mathbf{u}$.

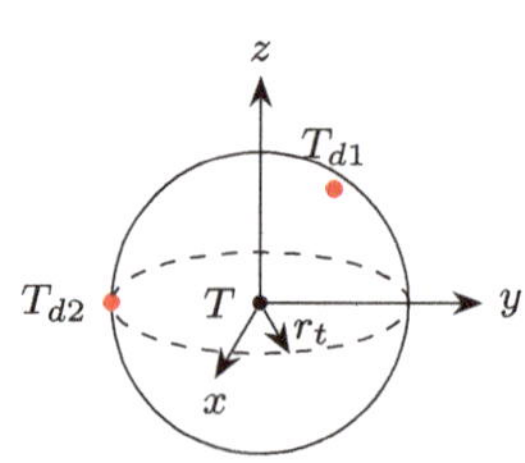

Fig. 3. Translational parameter T and its perturbation radius r_t

That is a perturbation bound $\mathcal{D}_\theta$ contains all rotational perturbations with the angle less than or equal to $\Delta\theta$ and any axis $\mathbf{u}$.

4 Case Study

4.1 CDPR Prototypes

As can be seen from (11), the stability criterion $\mathbf{\Pi}$ depends on the Jacobian matrix $\mathbf{A}$ and thus we need to consider the CDPR for whom the CSW will be obtained. Here, we use two suspended CDPRs with eight cables, named ACROBOT and CAROCA, shown in Fig. 4. ACROBOT is a $1.2\,\text{m} \times 1.2\,\text{m} \times 1.2\,\text{m}$ cube with a MP size $0.18\,\text{m} \times 0.17\,\text{m} \times 0.07\,\text{m}$ and its mass is 3.5 kg. The dimensions of CAROCA are $7\,\text{m} \times 4\,\text{m} \times 3\,\text{m}$, its MP is $0.42\,\text{m} \times 0.32\,\text{m} \times 0.23\,\text{m}$ and its mass is 150 kg. For these CDPRs we define the baseline perturbation bounds $\mathcal{D}_b = (r_{bp}, \Delta\theta_{bp}, r_{pc}, \Delta\theta_{pc}, r_{Ai}, r_{Bi})$ with the following values:

- MP position error range $r_{bp} = 0.03\,\text{m}$
- MP orientation error range $\Delta\theta_{bp} = 3°$
- camera position error range $r_{pc} = 0.01\,\text{m}$
- camera orientation error range $\Delta\theta_{pc} = 3°$
- cable exit point error range $r_{Ai} = 0.005\,\text{m}$
- cable anchor point error range $r_{Bi} = 0.005\,\text{m}$

These values are chosen based on the geometry of ACROBOT and taking into account some manufacturing errors that could be present on the robot. For example, there are pulley sheaves of 9 mm diameter on cable exit points that are not taken into account in the kinematic modeling of the CDPR. Similarly, cable anchor points are modeled as points, while they are actually located on a sphere

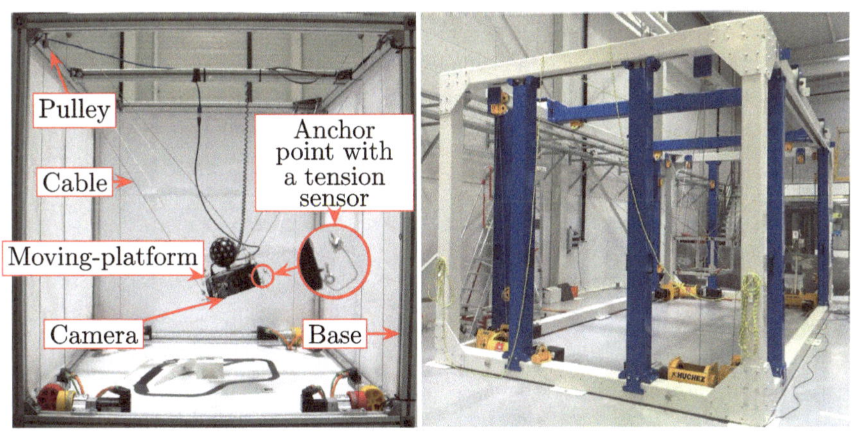

(a) (b)

Fig. 4. CDPR prototypes at IRT Jules Verne: **(a)** ACROBOT; **(b)** CAROCA of 0.008 m diameter about the modeled points. Camera position and orientation errors can also be amounted to manufacturing errors. Finally, the MP position and orientation errors illustrate the possible coarseness of estimation of the MP pose. Indeed, for many CDPRs the initial or *homing* MP pose is often measured by hand and thus having a small error on the pose estimation is clearly possible. $\mathcal{D}_b$ is kept the same for CAROCA to simplify the comparison of results. However, for the supplementary perturbations that will be added to $\mathcal{D}_b$, we will consider the particular size of each robot.

4.2 CSW for ACROBOT with PBVS

For ACROBOT, MP positions are assigned as follows. The potential workspace is discretized to a set of 12 planes that are parallel to the global $\boldsymbol{xy}$ plane. Each plane is then discretized to a matrix of 33 × 33 points with X and Y coordinates ranging from −0.48 m to 0.48 m and the step being 0.03 m. The Z coordinate of the planes ranges from 0 m to 1.1 m with the step being 0.1 m. Thus, the full workspace volume is 1.014 m^3. The evolution of the workspace volume is shown in Fig. 5. Note that the vertical axis denotes the ratio of the current volume over the maximum volume of 1.014 m^3. The CSW is defined in (13) for all MP poses. Thus, at each position multiple MP orientations $\theta\mathbf{u}$ are tested with a maximum angle of $\theta = 45°$ about any unit axis $\mathbf{u}$. If $\boldsymbol{\Pi} > \mathbf{0}$ for all MP orientations associated to a given position, then that MP position is added to the CSW.

First, the workspace corresponding to baseline perturbation bounds $\mathcal{D}_b$ is computed. It is shown in Fig. 6(a) and its volume is 0.88 m^3, namely, 87.0% of the full workspace volume. Then the perturbation ranges are incrementally increased one by one, leaving the rest of perturbations as defined in $\mathcal{D}_b$. Some examples are shown in Fig. 6. It can be seen that all workspace reduction occurs in a downward direction along $\boldsymbol{z}$ axis no matter the perturbation. This is because ACROBOT is a suspended CDPR and its most stable position is ${}^b\mathbf{t}_p = [0\ \ 0\ \ 0]^\top$. Indeed, in Fig. 6(i) the tiny workspace is centered around ${}^b\mathbf{t}_p = [0\ \ 0\ \ 0]^\top$ with about 0.1 m displacement possible in any direction.

It can be observed that some perturbations have a higher impact on the CSW size than others. Indeed, increasing camera position error range r_{pc} expressed

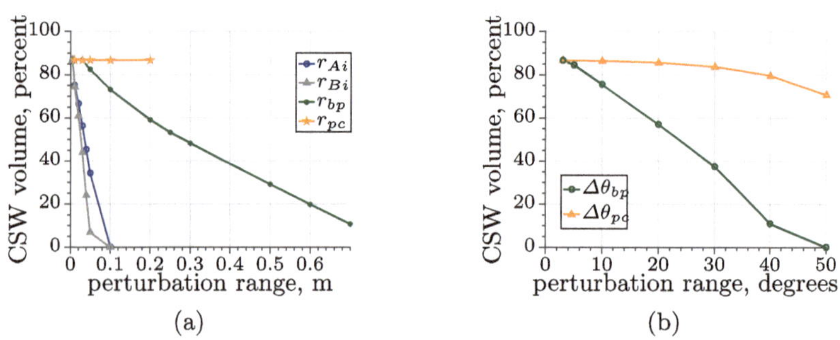

Fig. 5. CSW volume as a function of perturbation range, ACROBOT with PBVS

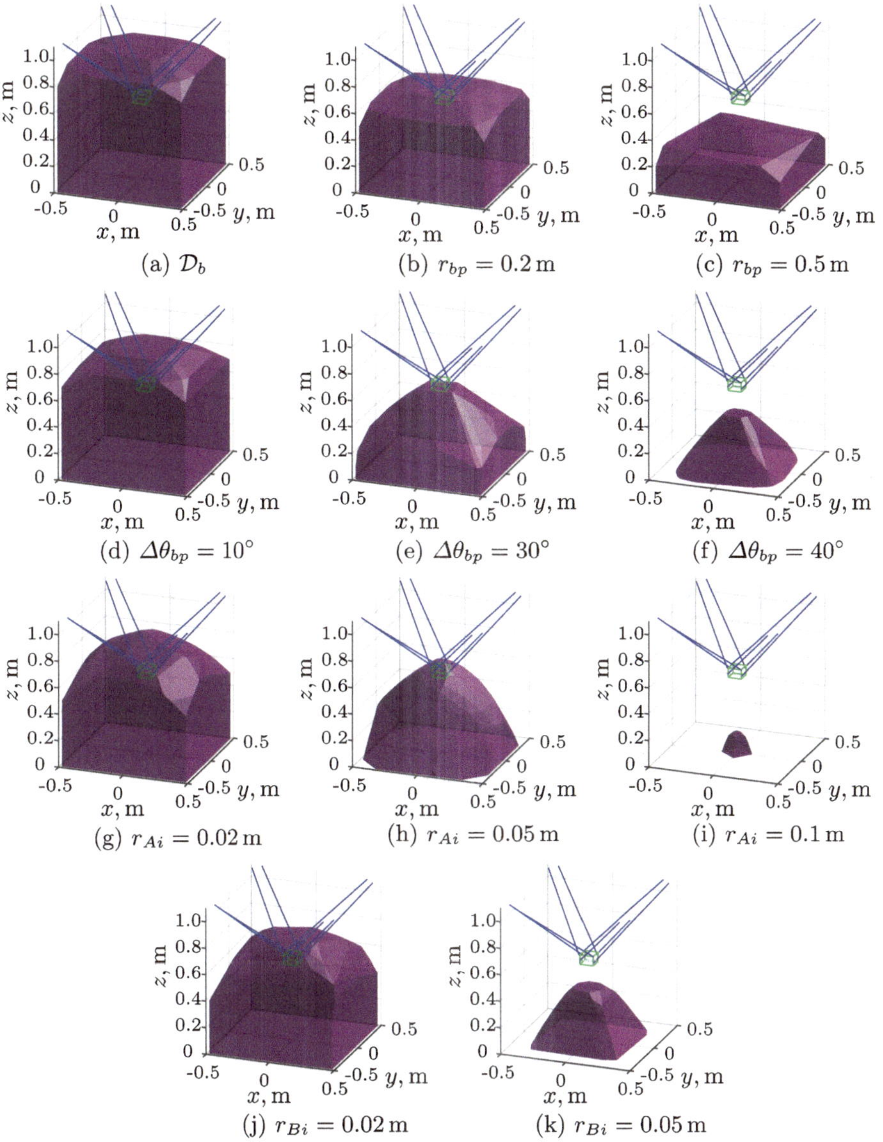

(a) $\mathcal{D}_b$ (b) $r_{bp} = 0.2\,\mathrm{m}$ (c) $r_{bp} = 0.5\,\mathrm{m}$

(d) $\Delta\theta_{bp} = 10°$ (e) $\Delta\theta_{bp} = 30°$ (f) $\Delta\theta_{bp} = 40°$

(g) $r_{Ai} = 0.02\,\mathrm{m}$ (h) $r_{Ai} = 0.05\,\mathrm{m}$ (i) $r_{Ai} = 0.1\,\mathrm{m}$

(j) $r_{Bi} = 0.02\,\mathrm{m}$ (k) $r_{Bi} = 0.05\,\mathrm{m}$

Fig. 6. Visualization of CSW for ACROBOT with PBVS

in $\mathcal{F}_p$ from 0.01 m to 0.20 m, only slightly decreases the workspace volume from 0.880 m^3 to 0.878 m^3, as shown in Fig. 5(a). Indeed, only 0.4% of workspace is lost by increasing r_{pc} twenty times. Furthermore, considering the MP size it means that the camera can be anywhere on the MP, without knowing where it is. On the contrary, coarse MP position estimation leads to a steady reduction of CSW, as can be seen in Fig. 5(a). For example, as shown in Fig. 6(b), with

$r_{bp} = 0.2\,\text{m}$, 59.1% of the workspace is still available. Of course this is only true if the other perturbations are kept within the limits defined by $\mathcal{D}_b$.

Regarding rotational errors, the results are similar. That is, increasing the MP rotational error $\Delta\theta_{bp}$ has a worse effect on the workspace area than the rotational error on camera pose $\Delta\theta_{pc}$, as can be seen in Fig. 5(b).

The errors in the CDPR model, namely in cable anchor and exit point coordinates have the highest influence on the CSW volume, as can be seen in Figs. 5(a) and 6. For example, for $r_{Bi} = 0.02\,\text{m}$, CSW volume is $0.618\,\text{m}^3$ (shown in Fig. 6(j)), while for $r_{Bi} = 0.05\,\text{m}$ the workspace reduces to $0.07\,\text{m}^3$ (shown in Fig. 6(k)). Similarly, when r_{Ai} is increased, the CSW volume decreases rapidly (shown in Figs. 6(g)) to 6(i)). However, this decrease is slower than for cable anchor point coordinate errors. Note that, if the error is no more than 0.01 m, then the workspace remains sufficiently large. For ACROBOT, the pulley sheave radius is 0.0045 m and thus is lower than $r_{Ai} = 0.005\,\text{m}$ defined for $\mathcal{D}_b$. Thus, it is clear that for ACROBOT, the pulleys are small enough to be negligible.

4.3 CSW for CAROCA with PBVS

For CAROCA, the potential workspace is discretized into 11 planes with 33×33 points per plane, just like for ACROBOT. However, here the step between the points and planes is larger. The full workspace volume is equal to $82.33\,\text{m}^3$.

The CSW computed for CAROCA with baseline perturbation bounds $\mathcal{D}_b$ has a volume of $63.06\,\text{m}^3$, which is 76.6% of the full workspace. The change of the CSW volume depending on perturbation range is shown in Fig. 7, while some examples are shown in Fig. 8. Note that in Fig. 7 the percentage shown is the obtained volume for a given perturbation set over the maximum volume of $82.33\,\text{m}^3$. In general, the shape of the curves in Fig. 7 is similar to those in Fig. 5 for ACROBOT. More precisely, perturbation in CDPR model has the highest effect on CSW volume, following by the MP pose estimation and finishing with hand-eye calibration errors. However, in Fig. 8(a) the green curve of the translational error r_{bp} has an exponential shape. Furthermore, with the increased CDPR size, the range of this perturbation has increased. Even having set $r_{bp} =$

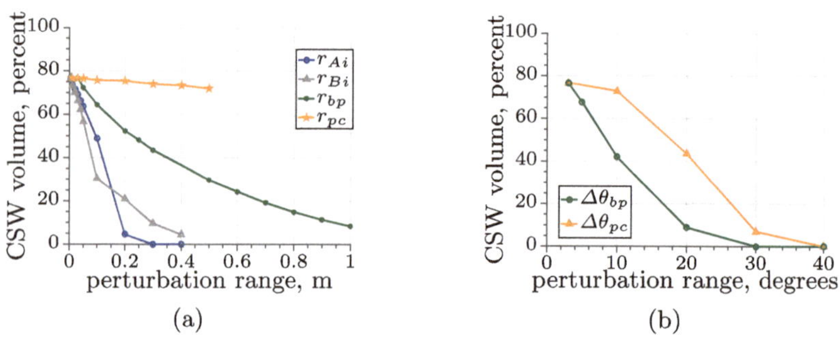

Fig. 7. CSW volume as a function of perturbation range, CAROCA with PBVS

1 m does not make the workspace null, instead it is equal to approximately 10% of the full workspace and appears to be almost flat, as can be seen in Fig. 8(d). Similarly, having $r_{Bi} = 0.1$ m or $r_{Ai} = 0.1$ m does not make the CSW vanish for CAROCA. Instead, approximately 30% and 50% of the full workspace remains available, respectively, as can be seen in Figs. 8(j) and 8(l). Furthermore, even setting the camera position error range $r_{pc} = 0.5$ m has little effect on the CSW size, which remains at 71.9%. On the other hand, rotational errors rapidly reduce the workspace. Especially those in the MP pose estimation, where at $\Delta\theta_{bp} = 10°$ the volume is at 42.1% (Fig. 8(e)) and at $\Delta\theta_{bp} = 30°$ the volume is only at 0.003% (Fig. 8(f)). While the rotational error $\Delta\theta_{pc}$ on the camera pose also has a significant effect on the CSW size, however having $\Delta\theta_{pc} = 30°$ allows us to use 43.4% of the workspace (Fig. 8(h)), which is significantly larger than for $\Delta\theta_{bp}$ of the same size.

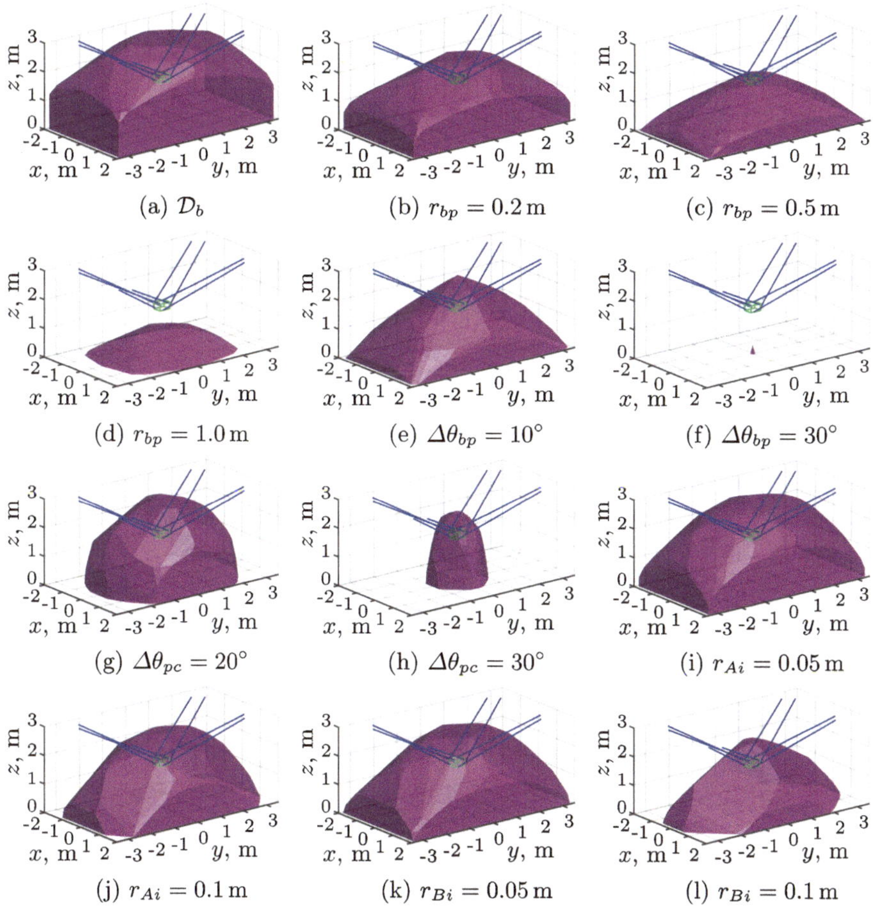

(a) $\mathcal{D}_b$ (b) $r_{bp} = 0.2$ m (c) $r_{bp} = 0.5$ m

(d) $r_{bp} = 1.0$ m (e) $\Delta\theta_{bp} = 10°$ (f) $\Delta\theta_{bp} = 30°$

(g) $\Delta\theta_{pc} = 20°$ (h) $\Delta\theta_{pc} = 30°$ (i) $r_{Ai} = 0.05$ m

(j) $r_{Ai} = 0.1$ m (k) $r_{Bi} = 0.05$ m (l) $r_{Bi} = 0.1$ m

Fig. 8. Visualization of CSW for CAROCA with PBVS

Having such a large CDPR also allows us to compare the change of CSW shape when some perturbations are increased. For example, the increase of r_{bp} is shown in Figs. 8(a)–8(d) and appears to flatten the workspace by greatly reducing the range along $\boldsymbol{z}$ axis before affecting the workspace range in $\boldsymbol{xy}$ plane. On the other hand, rotational errors in hand-eye calibration appear to affect the $\boldsymbol{xy}$ range more severely. This can be seen by comparing Fig. 8(a) with 8(g) and 8(h).

4.4 Discussion

Having computed the CSW for two different CDPRs we can conclude that:

- A perturbation on camera position in the MP frame has little to no effect on CSW volume and system stability;
- Similarly, a perturbation on camera orientation in the MP frame a smaller effect than the same perturbation on the MP orientation in the base frame;
- There can be a significant perturbation on the MP position in the base frame without making the system unstable;
- Perturbations on the CDPR model, that is on cable exit and anchor point coordinates, have the largest effect on CSW volume;
- A perturbation on cable anchor points leads to a smaller CSW than same perturbation on cable exit points;
- Translational perturbation bounds scale with CDPR geometry:
    - The larger the MP, the larger r_{Bi} without making the workspace vanish;
    - The larger the CDPR, the larger r_{Ai} and r_{bp} can be without making the workspace vanish;
- On the other hand, the rotational perturbations $\Delta\theta_{bp}$ and $\Delta\theta_{pc}$ do not scale with CDPR geometry. They remain approximately the same for ACROBOT and CAROCA.

Note that the obtained CSW is conservative because the perturbation ranges for each parameter were tested in many directions. Thus, the obtained CSW i s conservative, because it is the workspace where $\boldsymbol{\Pi} > \boldsymbol{0}$ for any perturbation direction. An additional study could be done with a fixed set of perturbation ranges, such as $\mathcal{D}_b$, to find perturbation direction combinations leading to the largest or to the smallest CSW size.

5 Conclusions

A novel workspace, named Control Stability Workspace (CSW), was analyzed in this paper. It is a workspace from the control perspective that allows us to determine whether the controller will be able to guide the MP of the CDPR to the desired pose in the presence of perturbations. The workspace was computed for two CDPRs of different sizes.

Many perturbations can exist in the system and all of them affect the system stability in a combined way. However, the sensitivity is a function of the perturbed parameter, as discussed in the previous section. Most importantly, the perturbation ranges within system stability scale with the CDPR geometry.

To conclude, CDPRs with visual servoing control are robust to many perturbations, including modeling errors and MP pose estimation errors. Computation of the CSW is a good approach to evaluate this robustness.

References

1. Stump, E., Kumar, V.: Workspaces of cable-actuated parallel manipulators. J. Mech. Des. **128**(1), 159–167 (2006)
2. Verhoeven, R.: Analysis of the Workspace of Tendon-Based, PhD thesis, Duisburg Gerhard Mercator University (2004)
3. Gagliardini, L.: Discrete Reconfigurations of Cable-Driven Parallel Robots, PhD thesis, Ecole Centrale de Nantes (2016)
4. Perreault, S., Cardou, P., Gosselin, C.M., Otis, M.J.-D.: Geometric determination of the interference-free constant-orientation workspace of parallel cable-driven mechanisms. J. Mech. Robot. **2**(3) (2010)
5. Arsenault, M.: Computation of the interference-free wrench feasible workspace of a 3-dof translational tensegrity robot. In: International Conference on Cable-Driven Parallel Robots, pp. 185–196 Springer (2019)
6. Lessanibahri, S., Gouttefarde, M., Caro, S., Cardou, P.: Twist feasibility analysis of cable-driven parallel robots. In: 3rd International Conference on Cable-Driven Parallel Robots, series Mechanisms and Machine Science, vol. 53, Québec, Canada, pp. 128–139, August 2017
7. Barrette, G., Gosselin, C.M.: Determination of the dynamic workspace of cable-driven planar parallel mechanisms. J. Mech. Des. Trans. ASME **127**(2), 242–248 (2005)
8. Gagliardini, L., Gouttefarde, M., Caro, S.: Determination of a dynamic feasible workspace for cable-driven parallel robots. In: Advances in Robot Kinematics. Springer **4**, 361–370 (2018)
9. Zake, Z., Chaumette, F., Pedemonte, N., Caro, S.: Vision-based control and stability analysis of a cable-driven parallel robot. IEEE Robot. Autom. Lett. **4**(2), 1029–1036 (2019)
10. Zake, Z., Caro, S., Suarez-Roos, A., Chaumette, F., Pedemonte, N.: Stability analysis of pose-based visual servoing control of cable-driven parallel robots. In: International Conference on Cable-Driven Parallel Robots, CableCon, pp. 73–84 (2019)
11. Zake, Z., Chaumette, F., Pedemonte, N., Caro, S.: Robust 21/2d visual servoing of a cable-driven parallel robot thanks to trajectory tracking. IEEE Robot. Autom. Lett. **5**(2), 660–667 (2020)
12. Chaumette, F., Hutchinson, S.: Visual servoing and visual tracking. In: Siciliano, B., Khatib, O. (eds.) Handbook of Robotics, pp. 563–583. Springer (2008)
13. Siciliano, B., Sciavicco, L., Villani, L., Oriolo, G.: Grimble, M., Johnson, M. (eds.) Robotics Modelling, planning, control (2010)
14. Pott, A.: Cable-Driven Parallel Robots: Theory and Application. Springer (2018)
15. Khalil, H.K.: Nonlinear Systems (2002)

16. Gagliardini, L., Caro, S., Gouttefarde, M., Girin, A.: Discrete reconfiguration planning for cable-driven parallel robots. Mech. Mach. Theory **100**, 313–337 (2016)
17. Ruiz, A.L.C., Caro, S., Cardou, P., Guay, F.: ARACHNIS : analysis of robots actuated by cables with handy and neat interface software. In: Cable-Driven Parallel Robots, pp. 293–305. Springer (2015)

A Simple and Efficient Non-model Based Cable Tension Control

João Cavalcanti Santos(✉) and Marc Gouttefarde

LIRMM, Univ Montpellier, CNRS, Montpellier, France
{cavalcanti,gouttefarde}@lirmm.fr

Abstract. Although the control of cable tensions plays a paramount role in the control of Cable-Driven Parallel Robots (CDPRs), studies addressing the subject are very few. The present paper proposes and experimentally validates a simple cable tension feedback control scheme. In order to avoid issues related to friction and saturation of the feedback correction, a velocity-controlled winch is considered. The closed-loop scheme consists of an ordinary PI applied with respect to the measured cable tensions. Thanks to its simplicity, the main advantage of the proposed scheme is its ease of implementation. The experimental validation is performed using two scenarios. First, a simplified case consisting of a single degree-of-freedom CDPR is considered. Finally, the performance is validated with a six degrees-of-freedom CDPR fully-constrained by eight cables.

Keywords: Cable tension control · Non-model based · Velocity-controlled motor

1 Introduction

Most often, motion control schemes of Cable-Driven Parallel Robots (CDPRs) define, at some point, a set of desired cable tensions. Therefore, a strategy should be defined to convert desired cable tensions to a variable that is meaningful to the motor drivers, typically motor torques (or currents), positions or velocities. Such control strategies are referred to as *cable tension control* schemes.

The majority of published experimental results were obtained with open-loop cable tension control. A common solution is the application of motor torques compensating the friction in the winches [2,7,15] which led to satisfying results in the context of position tracking control [6] and of force-based control [12]. Nevertheless, the application of a closed-loop approach should notably improve the robustness to the uncertainties present in the controlled system.

A closed-loop cable tension control can be seen as a simplified case of feedback force control which is a well established topic in robotics [16,17]. Its variants,

This work was supported by the European Union's H2020 Program (H2020/2014-2020) under the grant agreement No. 732513 (HEPHAESTUS project).

M. Gouttefarde et al. (Eds.): CableCon 2021, MMS 104, pp. 297–308, 2021.
https://doi.org/10.1007/978-3-030-75789-2_24

which can be exemplified by impedance control [3] and hybrid position/force control [1], were successfully implemented in both serial [10] and parallel robots [8]. Nevertheless, studies proposing closed-loop cable tension control strategies in the context of CDPR control are few.

Based on an elastic cable model, Kraus *et al.* propose a cable tension control scheme using motor positions as control inputs to the motor drivers [5]. This work presents details on the identification of the dynamic system and experimental results on position tracking control. Similar methods are used in [4] and in [11]. Nevertheless, the application of a position-controlled inner-loop may lead to the saturation of the feedback correction. Kraus *et al.* proposes the implementation of an anti-windup strategy in order to overcome this issue.

Besides, torque-controlled winches allow the motor torques to be directly used as the control inputs. An advanced technique was tested through numerical simulations in [9] using $\mathcal{H}_\infty$ robust control in order to compensate the influence of the cable sagging and elasticity. However, torque-based strategies requires a precise model of the friction present in the winches. This may be a significant issue when gear trains with a large reduction ratio are used.

The present paper proposes a rather simple control strategy considering velocity-controlled winches. This non-model based approach relies on a PI feedback correction and, therefore, presents simple implementation and tuning. Furthermore, the proposed scheme does not demand any identification of the controlled system. The control scheme can be seen as an application of a standard velocity-based admittance control, as discussed in [13]. Experimental validation is fulfilled using two scenarios. First, a simplified case of a one degree-of-freedom (DoF) CDPR is considered. Finally, the application on the CDPR motion control is validated on a 6-DoF CDPR fully-constrained by eight cables. In contrast to position-based control schemes, real-time experiments notably indicate that the proposed control scheme is not prone to saturation issues.

2 Set-Up Description

A simple 1-DoF CDPR is first considered in order to study the cable tension control of CDPRs. This set-up is illustrated in Fig. 1. It consists of two cables of a CDPR connected one to each other at point P. The influence of gravity is neglected under the assumption that the cables are kept sufficiently taut. Accordingly, the position of the point P is determined by means of the scalar ℓ as shown in Fig. 1. For this 1-DoF CDPR, the stretched cable length, *i.e.* the actual length of the cable under tension, is equal to ℓ plus a constant value, the latter corresponding to the stretched cable length when point P is in its reference position.

Cable tensions are measured with the aid of two load pins positioned in the pulley axes (as shown in Fig. 1). The cable tension measurements corresponding to each cable are denoted as τ_1 and τ_2. Initially, neglecting the pulleys inertia and friction, it is assumed that $\tau_1 = \tau_2 = \tau$. Consequences of this assumption are

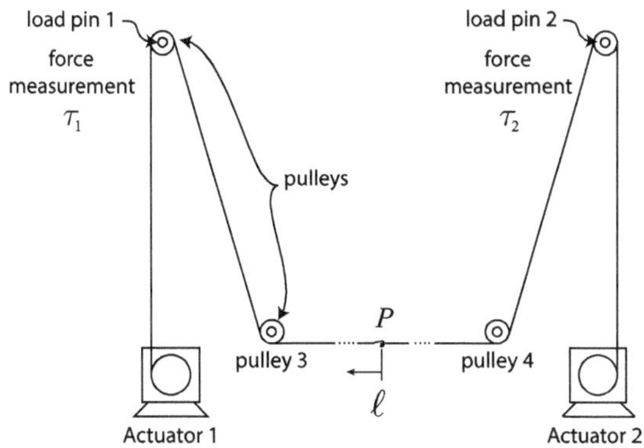

Fig. 1. An illustrative schematic view of the experimental set-up.

discussed in Sect. 4. Details on the experimental set-up components are presented in Sect. 4.

The synchronized movement of actuators 1 and 2 enables the control of the tension τ and the position ℓ. As a matter of fact, the essential aspects involved in the control of the cable tensions of a CDPR can be addressed with this system. Section 3 proposes a velocity-based feedback policy able to track a desired cable tension τ_d by means of controlling the velocity of winch 1, denoted as $\dot{q}$.

3 Cable Tension Control Scheme

The present section proposes a control scheme able to track a set of desired cable tensions by means of controlling the *motor velocities*. At first glance, a velocity-based strategy may seem a peculiar choice. A scheme controlling directly the *motor torques* may instead be considered as an appropriate option. Disregarding cable distributed mass, friction in the winches and in the routing pulleys, and dynamics forces, the torque applied by the drum is proportional to the cable tension. Moreover, independently of the chosen strategy, the innermost control loop regulates the motor current which is closely related to the motor torque.

Nevertheless, the torque applied on the drum is often transmitted from the corresponding motor by means of a gearbox which generally possesses non-negligible friction. As a result, the torque applied on the drum and the one generated by the motor are commonly related by a highly non-linear system. This issue is demonstrated in Fig. 2. The data displayed in this chart was obtained for a constant position ℓ of the 1-DoF CDPR shown Fig. 1 whose winches have gear trains with reduction ratio of 25—other technical details on the mechanical parts used in the experimental setup are described in the next section. One may note that a continuous variation of the motor torque results in significant discontinuities in the variation of the cable tension. This is due to the so-called *stick-slip* phenomenon.

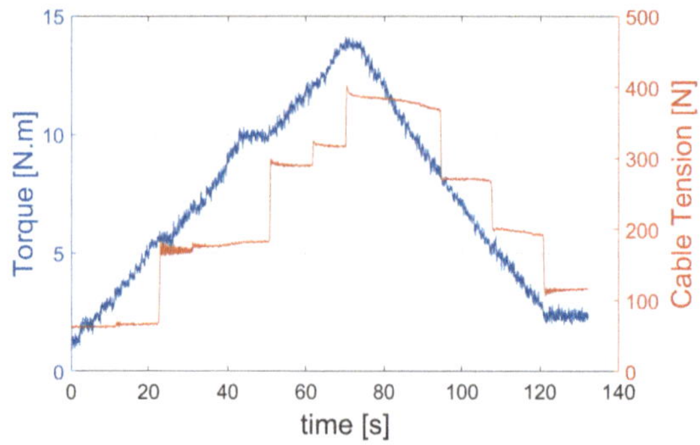

Fig. 2. Simultaneous plot of the applied torque and measured cable tension (with the cable attached to a fixed support).

In contrast, the relation between *motor positions* and cable tensions is much more tractable since it is mainly governed by the cable elasticity. Performing a test similar to the above mentioned experiment, the data depicted in Fig. 3 is obtained. In this case, a continuous variation of the motor position leads to a continuous variation of the cable tension. As a matter of fact, the results presented in Fig. 3 are rather obvious considering the elastic behavior of the cable. Since ℓ is constant in this experiment, the unstretched length variations presented in Fig. 3 takes place with constant stretched length. The comparison between Figs. 2 and 3 demonstrates that controlling the cable tension through the motor position should be more suitable than through the motor torque.

In general, a simple position control is able to perform a sufficiently precise position tracking independently of the external forces and torques. Hence, while positioning precision may be obtained independently of the cable tensions, the motor angular position may be controlled in order to generate a desired cable tension. Noting that the relation between the cable tension and the position is continuous and well behaved, good results with a *position-based* control scheme can be expected. This is indeed the approach proposed in [4,5] but some secondary complications come up with this method. Notably, the suppression of steady-state errors demands the application of an anti-windup feedback strategy.

For this reason, the present section proposes a scheme in which the motor velocity is taken as the control input. Similarly to a position-controlled winch, a simple velocity control is able to perform a sufficiently precise velocity tracking independently of the external forces and torques. Therefore, this approach is also able to reduce the influence of friction on the tracking of cable tensions. Furthermore, results presented in the remainder of the paper indicates that this velocity-based scheme is not prone to saturation issues.

As discussed earlier, a crucial advantage of the proposed control scheme lies in its simplicity of implementation. Indeed, as shown in Fig. 4, the proposed strategy consists of a simple PI feedback correction combined with a feedforward

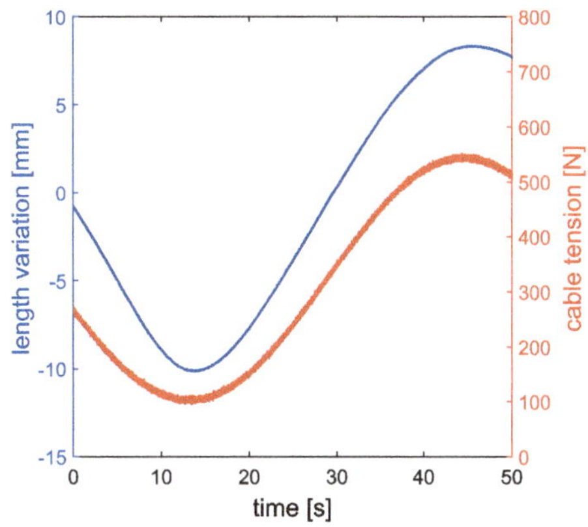

Fig. 3. Simultaneous plot of variations in the unstretched cable length and the measured cable tension (with the cable attached to a fixed support).

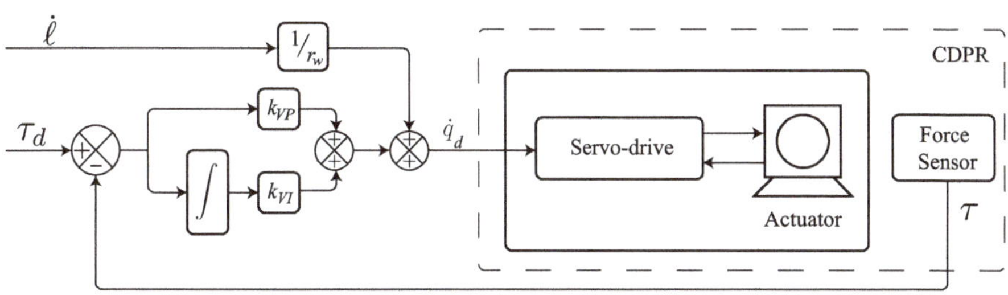

Fig. 4. Block diagram of the proposed velocity-based cable tension control scheme.

term proportional to the velocity $\dot{\ell}$. More precisely, the desired winch angular velocity $\dot{q}_d$ is computed according to:

$$\dot{q}_d = k_P \left(\tau_d - \tau\right) + k_I \int_{t_0}^{t_a} \left(\tau_d - \tau\right) dt + \frac{\dot{\ell}}{r_w}, \tag{1}$$

with positive scalars k_P and k_I representing the feedback gains. The limits of integration in (1) refer to the initial and actual time, t_0 and t_a, respectively. τ_d is the desired cable tension and τ is the measured one. Moreover, the positive scalar r_w denotes the radius of the winch drum. Consequently, the feedforward term $\dot{\ell}/r_w$ represents the motor angular velocity necessary to generate the tangential linear velocity $\dot{\ell}$ of the drum. Thanks to the simplicity of the control law (1), the gains k_P and k_I can be easily tuned by trial-and-error.

In practice, the value of the angular velocity $\dot{q}_d$ is used as a set-point in an inner-loop control responsible for tracking this velocity by means of controlling the motor currents. Typically, servo-drives are able to manage this inner-loop. While the proposed control scheme presents a remarkable simplicity, the next section shows that the resulting performance is satisfying.

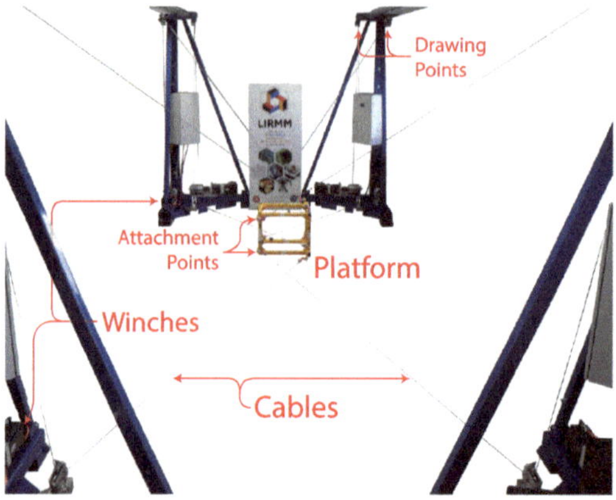

Fig. 5. HRPCable prototype.

4 Experimental Results

The experimental validation presented in this section was performed using the HRPCable prototype shown in Fig. 5. The mobile platform of HRPCable has six degrees of freedom and is driven by eight ⌀4 mm steel cables in a fully-constrained configuration. Each winch is driven by a motor Beckhoff AM8061 with nominal torque of 12.8 Nm and nominal power of 3.46 kW. The reduction is performed by a two stage gear train Beckhoff AG2210 with reduction ratio of 25:1. Motor servo drivers Beckhoff AX5112 are controlled by an IPC Beckhoff C6920. Control implemented in the IPC is developed in C++ (TwinCAT environment).

Load pins Sensy 5300 1T SL positioned in the axes of the pulleys measure the cable tensions. These sensors have linearity error of less than 1% and a nominal sensibility of 1 mV/V. The analogical amplifier of this sensor is integrated to the body of the load pin. Based on the authors' practical experience, this is a decisive requirement in order to obtain a reduced level of noise on the force measurements.

Section 4.1 presents the experimental validation of the control scheme proposed in Sect. 3. Moreover, supplementary results are discussed in Sect. 4.2. The influence of the cable hysteresis and the friction in the pulleys are quantified.

4.1 Main Validation

First, two kinematic chains (two winches and cables) of the HRPCable prototype were used to form the 1-DoF CDPR depicted in Fig. 1. Figure 6 shows the results obtained with the velocity-based control of Eq. (1) with a step input. The lower level is 190 N and the upper one is 440 N. The rise time to reach 440 N is 0.4 s. The position ℓ was kept constant.

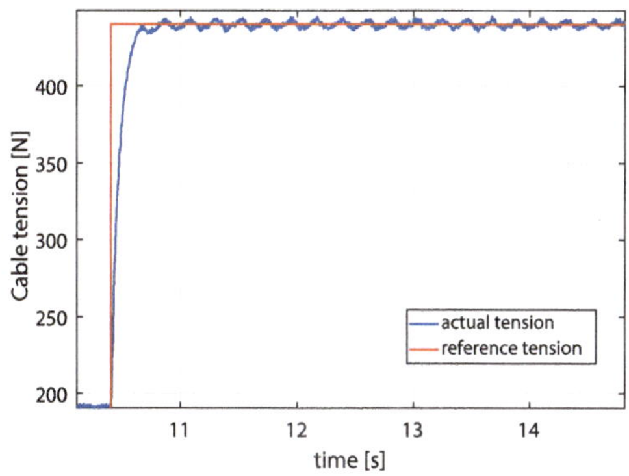

Fig. 6. Measured cable tension with a step input in the desired tension.

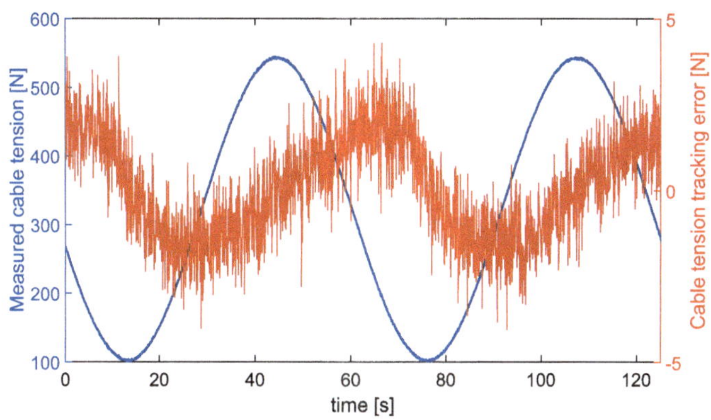

Fig. 7. Measured cable tensions and tracking errors. Displacement ℓ is constant and the desired cable tension is determined according $\tau_d = 320 + 220\cos(2\pi\, t/62.5)$ [N]. The RMS of the tracking error is equal to 1.51 N.

Figure 7 shows the results of another test where τ_d varies at low frequency. This figure shows that the obtained error is negligible so that desired and measured cable tensions are virtually equivalent. Moreover, the small error obtained with a slow variation of τ_d shows that the proposed strategy successfully avoids the stick-slip problem.

Moreover, Fig. 8 depicts the results for a sinusoidal ℓ and constant τ_d. This figure compares strategies with and without the feedforward term $\dot{\ell}/r_w$. It shows that a control scheme without the feedforward term would lead to large errors when variable ℓ is used. The small errors obtained when the feedforward term is applied indicates that the compensation of $\dot{\ell}$ is satisfying.

Figure 9 presents the results with sinusoidal ℓ and τ_d. A small delay may be observed when the maximum velocities are attained ($t \approx 12$ s and $t \approx 22$ s). Nevertheless, the obtained results with the proposed control strategy are con-

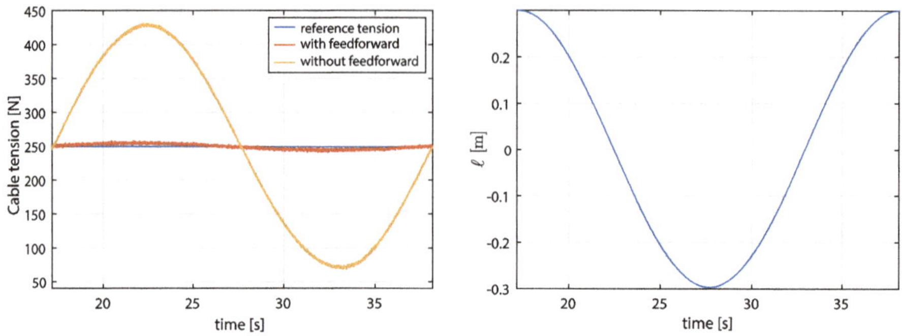

Fig. 8. Experimental results for a constant desired cable tension and a sinusoidal displacement ℓ. The chart on the left presents the desired and measured tensions comparing schemes with and without the feedforward term $\frac{\dot{\ell}}{r_w}$. The chart on the right presents the sinusoidal displacement ℓ.

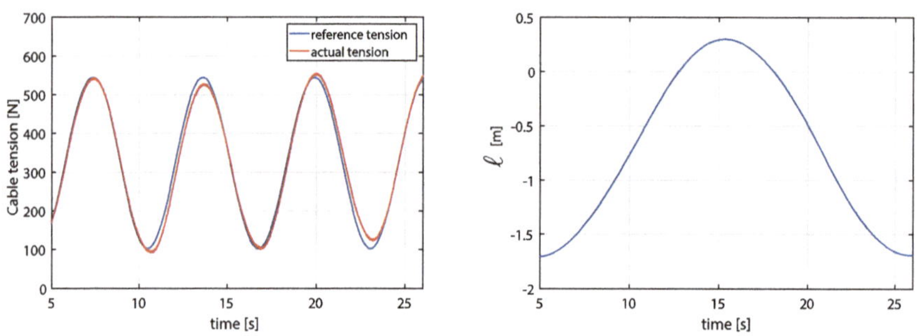

Fig. 9. Measured and desired cable tensions (on the left) and displacements ℓ (on the right).

sidered suitable. Accordingly, the proposed velocity-based cable tension control can be used in the CDPRs motion control.

Finally, the eight winches, cables and load pins of the HRPCable prototype were used and the proposed cable tension control scheme of Eq. (1) was implemented as part of a position tracking control strategy. More precisely, the Model Predictive Control (MPC) strategy proposed by the authors in [14] was implemented and used to perform the trajectory illustrated in Fig. 10. The resulting measured cable tensions and cable tension tracking errors are shown in Fig. 11. The tracking errors were mostly limited to values between $\pm$ 10 N, which are of the same order of magnitude of the measurement noise and allow an efficient closed-loop tracking of the desired cable tensions as can be seen in Fig. 11 (a).

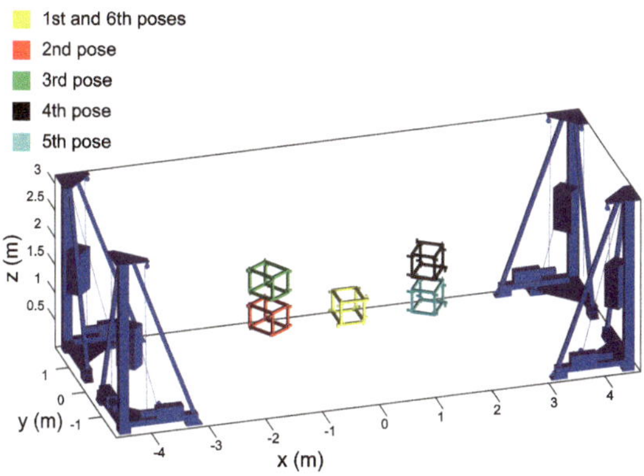

Fig. 10. Visited poses of the performed trajectory.

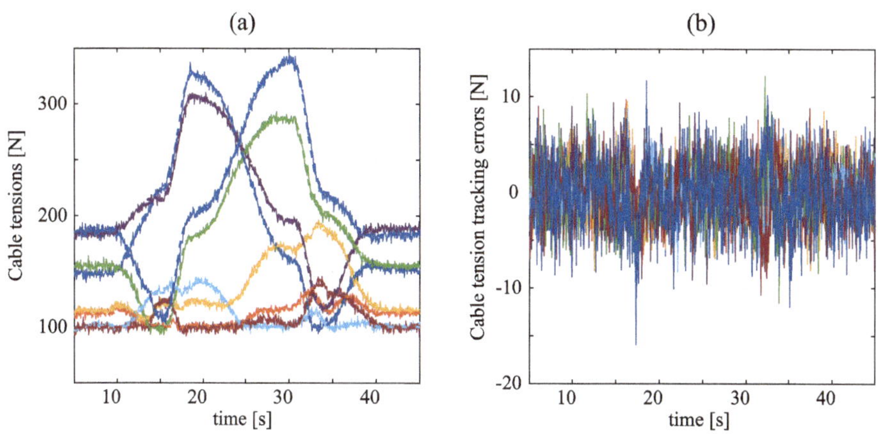

Fig. 11. (a) Graph of both measured (solid lines) and desired (dashed lines) cable tensions. (b) Tracking errors of the cable tensions.

4.2 Additional Results

Some additional results were obtained during the experiments described in the previous section. Figure 12 shows several repetitions of the experiment depicted in Fig. 7. One can see that the relation between the cable tension and the displacement of the winch $(q\, r_w)$ is repeatable. The plot displaying the elongation vs. the cable tension indicates that the relation between stress and strain for increasing cable tensions can be reasonably approximated by an affine function. Nevertheless, the values obtained for decreasing tensions shows that significant hysteresis is present.

As shown in Fig. 1, there is a second load pin in pulley 2 measuring the cable tension. Figure 13 presents the values obtained with the two load pins. One may see that the cable tension τ_1 measured by sensor 2 shifts when the direction of

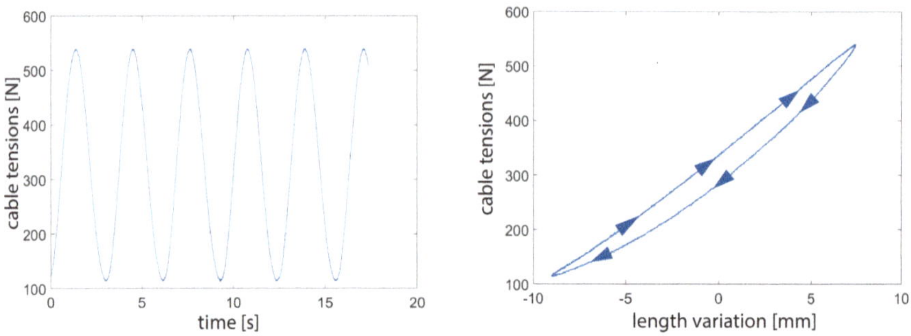

Fig. 12. Experimental results highlighting the hysteresis present in the cable elongation behavior.

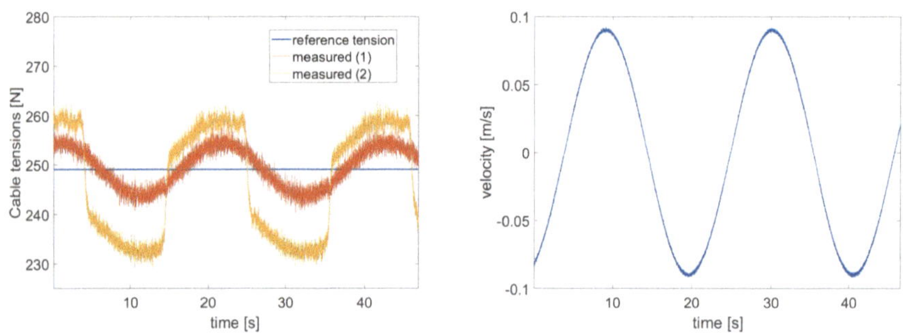

Fig. 13. Difference between cable tensions measured by load pins 1 and 2 (τ_1 and τ_2).

movement changes (velocity equals to zero). This shift is probably caused by the friction in pulleys 3 and 4 shown in Fig. 1. Since τ_1 is the value used in the control loop, it is continuous and close to the desired cable tension.

In order to clarify the influence of the friction in the pulleys, the following simplified model $\tilde{\tau}_2$ of the cable tension τ_2 can be considered:

$$\tilde{\tau}_2 = \tau_1 - 2\,\tau_f\,\mathrm{sign}(\dot{x}) + \tau_k, \tag{2}$$

where τ_f is the the dry friction force due to pulleys 3 and 4, the friction forces in these pulleys being considered equal. The scalar τ_k is a constant that depends on the initialization of the system. Parameters τ_f and τ_k were identified using data from Fig. 13. The fitted data is shown in Fig. 14 where it can be seen that the simple model in Eq. (2) is accurate. The pulley dry friction force obtained is $\tau_f = 3.9$ N.

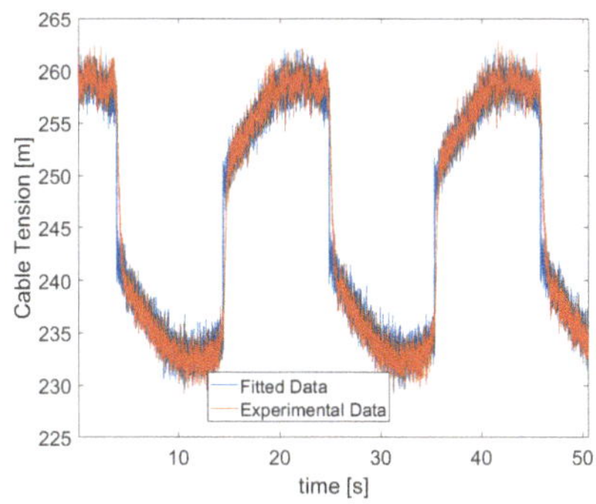

Fig. 14. Identification of the pulley friction: Measured cable tension τ_2 (experimental data) and predicted value $\tilde{\tau}_2$ given by Eq. (2) (fitted data).

5 Conclusions

This paper proposed and validated experimentally a non-model based cable tension control scheme. The desired cable tension is tracked by means of a velocity-controlled winch. The control scheme consists of an ordinary PI feedback correction combined with a feedforward term of the motor angular velocity. The advantage of the proposed approach in comparison to existing state-of-the-art strategies is its simplicity of implementation allied to satisfying performance. This approach did not led to saturation and proved to be barely sensitive to the friction in the winch. Future works may address the influence of friction in the pulleys and the cable elasticity (and hysteresis).

References

1. Chiaverini, S., Sciavicco, L.: The parallel approach to force/position control of robotic manipulators. IEEE Trans. Robot. Autom. **9**(4), 361–373 (1993). https://doi.org/10.1109/70.246048
2. Fang, S., Franitza, D., Torlo, M., Bekes, F., Hiller, M.: Motion control of a tendon-based parallel manipulator using optimal tension distribution. IEEE/ASME Trans. Mechatron. **9**(3), 561–568 (2004)
3. Hogan, N.: Impedance control: an approach to manipulation: Part I—theory. J. Dyn. Syst. Measure. Control **107**(1), 1–7 (1985). https://doi.org/10.1115/1.3140702
4. Kraus, W., Miermeister, P., Schmidt, V., Pott, A.: Hybrid position-force control of a cable-driven parallel robot with experimental evaluation. Mech. Sci. **6**, 119–125 (2015). https://doi.org/10.5194/ms-6-119-2015
5. Kraus, W., Schmidt, V., Rajendra, P., Pott, A.: System identification and cable force control for a cable-driven parallel robot with industrial servo drives. In: Proceedings - IEEE International Conference on Robotics and Automation, pp. 5921–5926. IEEE (2014)
6. Lamaury, J., Gouttefarde, M.: Control of a large redundantly actuated cable-suspended parallel robot. In: IEEE International Conference on Robotics and Automation, p. 211. IEEE (2013)

7. Lamaury, J., Gouttefarde, M., Chemori, A., Herve, P.E.: Dual-space adaptive control of redundantly actuated cable-driven parallel robots. In: IEEE International Conference on Intelligent Robots and Systems, pp. 4879–4886. IEEE (2013)
8. Merlet, J..: Force-feedback control of parallel manipulators. In: Proceedings. 1988 IEEE International Conference on Robotics and Automation, vol.3, pp. 1484–1489 (1988). https://doi.org/10.1109/ROBOT.1988.12277
9. Meunier, G., Boulet, B., Nahon, M.: Control of an overactuated cable-driven parallel mechanism for a radio telescope application. IEEE Trans. Control Syst. Technol. **17**(5), 1043–1054 (2009)
10. Ott, C., Mukherjee, R., Nakamura, Y.: Unified impedance and admittance control. In: 2010 IEEE International Conference on Robotics and Automation, pp. 554–561 (2010). https://doi.org/10.1109/ROBOT.2010.5509861
11. Picard, E.: Modeling and Robust Control of Cable-Driven Parallel Robots for Industrial Applications. Phd thesis, École centrale de Nantes, December 2019. https://tel.archives-ouvertes.fr/tel-02954030
12. Reichert, C., Müller, K., Bruckmann, T.: Robust internal force-based impedance control for cable-driven parallel robots. In: Pott, A., Bruckmann, T. (eds.) Cable-Driven Parallel Robots. Mechanisms and Machine Science, vol. 32, pp. 131–143. Springer, Cham (2015). https://doi.org/10.1007/978-3-319-09489-2_10
13. Roy, J., Whitcomb, L.L.: Adaptive force control of position/velocity controlled robots: theory and experiment. IEEE Trans Robot. Autom. **18**(2), 121–137 (2002). https://doi.org/10.1109/TRA.2002.999642
14. Santos, J.C., Chemori, A., Gouttefarde, M.: Redundancy resolution integrated model predictive control of cdprs : concept , implementation and experiments. In: Proceedings - IEEE International Conference on Robotics and Automation, pp. 3889–3895. Paris, France (2020)
15. Shang, W., Zhang, B., Zhang, B., Zhang, F., Cong, S.: Synchronization control in the cable space for cable-driven parallel robots. IEEE Trans. Ind. Electron. **66**(6), 4544–4554 (2019)
16. Siciliano, B., Villani, L.: Robot Force Control, 1st edn. Springer, US (1999)
17. Whitney, D.E.: Historical perspective and state of the art in robot force control. Int. J. Robot. Res. **6**(1), 3–14 (1987). https://doi.org/10.1177/027836498700600101

Modeling and Observer-Based Control of a Planar Two-Cable Crane Manipulator

Erik Hildebrandt[1,2], Tobias Handreg[2], Steffen Dryba[2], Pascal Froitzheim[2], Wilko Flügge[2], and Christoph Woernle[1(✉)]

[1] University of Rostock, Rostock, Germany
{erik.hildebrandt,woernle}@uni-rostock.de
[2] Fraunhofer Institute for Large Structures in Production Engineering IGP, Rostock, Germany
{erik.hildebrandt,tobias.handreg,steffen.dryba,pascal.froitzheim, wilko.fluegge}@igp.fraunhofer.de

Abstract. Two-cable crane manipulators move heavy payloads through large workspaces with computer-controlled winches mounted on trolleys. To make the payload track prescribed translational and rotational trajectories, a control method is presented based on a mechanical model. A nonlinear feedforward control that exploits the flatness property of the system model provides the control inputs for the undisturbed motion along reference trajectories. Trajectory deviations and sway motions are stabilised by a superimposed feedback controller. A nonlinear observer for the estimation of nonmeasurable states completes the control concept.

Keywords: Feedforward control · Linearisation · Nonlinear observer · Feedback control · Dual-hoist crane system · Cable-driven parallel robot

1 Introduction

The task of two-cable crane manipulators is to transport heavy payloads along prescribed translational and rotational trajectories in large working areas. As with conventional cranes the problem with automation is that the acceleration and deceleration of the trolleys as well as external disturbances cause undesired payload sway, which due to the low self-damping of the system only decay slowly and thus considerably complicate and delay positioning and orientation of the payload. Since economic crane operation depends significantly on the accuracy and speed of a working cycle, it is of interest to apply active control to prevent payload oscillations or at least reduce them to a reasonable level.

For the control of conventional cranes numerous model-based approaches have been proposed in the last decades. A comprehensive overview is documented in the papers by Abdel-Rahmen et al. [1] and Ramli et al. [12]. Besides control approaches by exact linearisation [7,8] and linear state feedback [13], energy-based as well as adaptive control methods [2,4] have proven to be efficient. Significantly less studies are available for the control of cranes with multi-cable

M. Gouttefarde et al. (Eds.): CableCon 2021, MMS 104, pp. 309–320, 2021.
https://doi.org/10.1007/978-3-030-75789-2_25

suspension. In-plane payload positioning with two-cable cranes is described in [9,11,15,16], while spatial positioning has been studied in [5,10].

In this contribution the control concept for a crane manipulator consisting of flatness-based feedforward and linear feedback control [5,17] is extended by a nonlinear observer for state estimation. As an example a two-cable configuration is considered. In Sect. 2 a mechanical model is derived in terms of the dependent and minimal coordinates. Based on the linearised equations of motion introduced in Sect. 3 the control concept is presented in Sect. 4. It includes feedforward control, feedback control and observer design. In Sect. 5 a simulation example shows the effectiveness of the proposed method.

2 Nonlinear Equations of Motion

2.1 Control Design Model

The mechanical model of a planar two-cable crane manipulator used for the derivation of the equations of motion is shown in Fig. 1. Along a crane runway two trolleys (masses m_{t}) move frictionless. Mounted on the trolleys are winches (radii r_{c}, moments of inertia J_{c}), whose massless and stiff cables guide a rigid payload platform (mass m_{p}, length $2l_{\mathrm{p}}$, height $2h_{\mathrm{p}}$, moment of inertia θ_{p} with respect to the centre of mass $\mathcal{S}_{\mathrm{p}}$) suspended at its corners.

The four independently controllable actuator coordinates $\mathbf{p} \in \mathbb{R}^4$ are the positions of the two trolleys $\mathbf{p}_{\mathrm{t}} \in \mathbb{R}^2$ and the cables $\mathbf{p}_{\mathrm{c}} \in \mathbb{R}^2$,

$$\mathbf{p}^{\mathrm{T}} = \begin{bmatrix} \mathbf{p}_{\mathrm{t}}^{\mathrm{T}} \ \mathbf{p}_{\mathrm{c}}^{\mathrm{T}} \end{bmatrix} \quad \text{with} \quad \mathbf{p}_{\mathrm{t}}^{\mathrm{T}} = \begin{bmatrix} p_{\mathrm{t}1} \ p_{\mathrm{t}2} \end{bmatrix}, \quad \mathbf{p}_{\mathrm{c}}^{\mathrm{T}} = \begin{bmatrix} p_{\mathrm{c}1} \ p_{\mathrm{c}2} \end{bmatrix}. \tag{1}$$

The four physical control inputs $\mathbf{u} \in \mathbb{R}^4$ include the drive forces of the trolleys $\mathbf{u}_{\mathrm{t}} \in \mathbb{R}^2$ and the drive torques of the winches $\mathbf{u}_{\mathrm{c}} \in \mathbb{R}^2$,

$$\mathbf{u}^{\mathrm{T}} = \begin{bmatrix} \mathbf{u}_{\mathrm{t}}^{\mathrm{T}} \ \mathbf{u}_{\mathrm{c}}^{\mathrm{T}} \end{bmatrix} \quad \text{with} \quad \mathbf{u}_{\mathrm{t}}^{\mathrm{T}} = \begin{bmatrix} u_{\mathrm{t}1} \ u_{\mathrm{t}2} \end{bmatrix}, \quad \mathbf{u}_{\mathrm{c}}^{\mathrm{T}} = \begin{bmatrix} u_{\mathrm{c}1} \ u_{\mathrm{c}2} \end{bmatrix}. \tag{2}$$

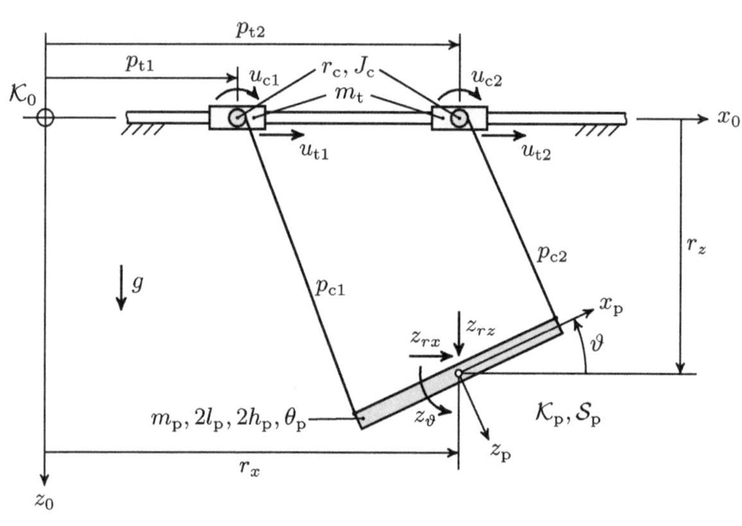

Fig. 1. Coordinates of the two-cable crane manipulator.

The payload coordinates $\mathbf{y} \in \mathbb{R}^3$ describe the position and orientation of the platform-fixed coordinate system $\mathcal{K}_\mathrm{p}$ relative to the inertial-fixed coordinate system $\mathcal{K}_0$ by the two Cartesian coordinates r_x, r_z of the origin of $\mathcal{K}_\mathrm{p}$ and the angle ϑ. For further considerations they are partitioned into the coordinate $y_\mathrm{d} \in \mathbb{R}^1$ suitable for describing the sway motion and the remaining dependent coordinates $\mathbf{y}_\mathrm{s} \in \mathbb{R}^2$,

$$\mathbf{y}^\mathrm{T} = \begin{bmatrix} y_\mathrm{d} & \mathbf{y}_\mathrm{s}^\mathrm{T} \end{bmatrix} \quad \text{with} \quad y_\mathrm{d} = r_x \,, \quad \mathbf{y}_\mathrm{s}^\mathrm{T} = \begin{bmatrix} r_z & \vartheta \end{bmatrix} \,. \tag{3}$$

Assumed disturbances $\mathbf{z} \in \mathbb{R}^3$ include unknown forces $\mathbf{z_r} \in \mathbb{R}^2$ on the origin of the platform-fixed coordinate system $\mathcal{K}_\mathrm{p}$ and a moment $z_\vartheta \in \mathbb{R}^1$,

$$\mathbf{z}^\mathrm{T} = \begin{bmatrix} \mathbf{z}_\mathbf{r}^\mathrm{T} & z_\vartheta \end{bmatrix} \quad \text{with} \quad \mathbf{z}_\mathbf{r}^\mathrm{T} = \begin{bmatrix} z_{rx} & z_{rz} \end{bmatrix} \,. \tag{4}$$

With the four actuator motions and the sway motion of the payload the system has five degrees of freedom, i.e. the motion of the system is described by five coordinates. Suitable minimal coordinates $\mathbf{q} \in \mathbb{R}^5$ are the actuator coordinates $\mathbf{p}$ and the coordinate $y_\mathrm{d} = r_x$,

$$\mathbf{q}^\mathrm{T} = \begin{bmatrix} \mathbf{p}^\mathrm{T} & y_\mathrm{d} \end{bmatrix} \,. \tag{5}$$

2.2 Equations of Motion in Dependent Coordinates

For feedforward control design and linearisation the equations of motion are formulated in terms of the actuator coordinates and the payload coordinates. The kinematic compatibility between the coordinates is ensured by the implicit constraints.

Implicit Constraints. Due to the stiff cables two implicit constraints hold between the actuator coordinates $\mathbf{p}$ and the payload coordinates $\mathbf{y}$,

$$\mathbf{g}(\mathbf{p},\mathbf{y}) \equiv \frac{1}{2} \begin{bmatrix} p_{c1}^2 - \mathbf{c}_1^\mathrm{T}\mathbf{c}_1 \\ p_{c2}^2 - \mathbf{c}_2^\mathrm{T}\mathbf{c}_2 \end{bmatrix} = \begin{bmatrix} 0 \\ 0 \end{bmatrix} \,. \tag{6}$$

The two cable vectors $\mathbf{c}_1, \mathbf{c}_2$ are expressed in the coordinate system $\mathcal{K}_0$ by

$$\begin{aligned} \mathbf{c}_1 &= r_x\mathbf{e}_{x0} + r_z\mathbf{e}_{z0} + \mathbf{d}_1 - p_{t1}\mathbf{e}_{x0} \quad \text{with} \quad \mathbf{d}_1 = {}^\mathrm{p}\mathbf{T}(\vartheta){}^\mathrm{p}\mathbf{d}_1 \,, \\ \mathbf{c}_2 &= r_x\mathbf{e}_{x0} + r_z\mathbf{e}_{z0} + \mathbf{d}_2 - p_{t2}\mathbf{e}_{x0} \quad \text{with} \quad \mathbf{d}_2 = {}^\mathrm{p}\mathbf{T}(\vartheta){}^\mathrm{p}\mathbf{d}_2 \,, \end{aligned} \tag{7}$$

with the canonical unit vectors $\mathbf{e}_{x0}^\mathrm{T} = \begin{bmatrix} 1 & 0 \end{bmatrix}, \mathbf{e}_{z0}^\mathrm{T} = \begin{bmatrix} 0 & 1 \end{bmatrix}$ from $\mathcal{K}_0$, the rotation matrix ${}^\mathrm{p}\mathbf{T}(\vartheta)$ from $\mathcal{K}_\mathrm{p}$ to $\mathcal{K}_0$ and the platform-fixed vectors ${}^\mathrm{p}\mathbf{d}_1, {}^\mathrm{p}\mathbf{d}_2$. If the centre of mass coincides with the centroid of the rectangular payload, the platform vectors are given by ${}^\mathrm{p}\mathbf{d}_1^\mathrm{T} = \begin{bmatrix} -l_\mathrm{p} & -h_\mathrm{p} \end{bmatrix}$ and ${}^\mathrm{p}\mathbf{d}_2^\mathrm{T} = \begin{bmatrix} l_\mathrm{p} & -h_\mathrm{p} \end{bmatrix}$.
The time derivative of (6) yields the implicit constraints at velocity level

$$\begin{aligned} \dot{\mathbf{g}}(\mathbf{p},\dot{\mathbf{p}},\mathbf{y},\dot{\mathbf{y}}) &\equiv \begin{bmatrix} \mathbf{G}_{\mathbf{p}_\mathrm{t}} & \mathbf{G}_{\mathbf{p}_\mathrm{c}} \end{bmatrix} \begin{bmatrix} \dot{\mathbf{p}}_\mathrm{t} \\ \dot{\mathbf{p}}_\mathrm{c} \end{bmatrix} + \begin{bmatrix} \mathbf{G}_{y_\mathrm{d}} & \mathbf{G}_{\mathbf{y}_\mathrm{s}} \end{bmatrix} \begin{bmatrix} \dot{y}_\mathrm{d} \\ \dot{\mathbf{y}}_\mathrm{s} \end{bmatrix} = \mathbf{0} \\ \dot{\mathbf{g}}(\mathbf{p},\dot{\mathbf{p}},\mathbf{y},\dot{\mathbf{y}}) &\equiv \mathbf{G}_\mathbf{p}(\mathbf{p},\mathbf{y}) \quad \dot{\mathbf{p}} \quad + \mathbf{G}_\mathbf{y}(\mathbf{p},\mathbf{y}) \quad \dot{\mathbf{y}} \quad = \mathbf{0} \,, \end{aligned} \tag{8}$$

with the analytically representable Jacobi matrices

$$\begin{aligned}\mathbf{G_p} &= \begin{bmatrix}\frac{\partial \mathbf{g}}{\partial \mathbf{p}_\mathrm{t}} & \frac{\partial \mathbf{g}}{\partial \mathbf{p}_\mathrm{c}}\end{bmatrix} = \left[\begin{array}{cc|cc}\mathbf{c}_1^\mathrm{T}\mathbf{e}_{x0} & 0 & p_{\mathrm{c}1} & 0\\ 0 & \mathbf{c}_2^\mathrm{T}\mathbf{e}_{x0} & 0 & p_{\mathrm{c}2}\end{array}\right] \in \mathbb{R}^{2\times 4}\,,\\ \mathbf{G_y} &= \begin{bmatrix}\frac{\partial \mathbf{g}}{\partial y_\mathrm{d}} & \frac{\partial \mathbf{g}}{\partial \mathbf{y}_\mathrm{s}}\end{bmatrix} = \left[\begin{array}{c|cc}-\mathbf{c}_1^\mathrm{T}\mathbf{e}_{x0} & -\mathbf{c}_1^\mathrm{T}\mathbf{e}_{z0} & -\mathbf{c}_1\times\mathbf{d}_1\\ -\mathbf{c}_2^\mathrm{T}\mathbf{e}_{x0} & -\mathbf{c}_2^\mathrm{T}\mathbf{e}_{z0} & -\mathbf{c}_2\times\mathbf{d}_2\end{array}\right] \in \mathbb{R}^{2\times 3}\,.\end{aligned} \tag{9}$$

The time derivative of (8) leads to the implicit constraints at acceleration level

$$\ddot{\mathbf{g}}(\mathbf{p},\dot{\mathbf{p}},\ddot{\mathbf{p}},\mathbf{y},\dot{\mathbf{y}},\ddot{\mathbf{y}}) \equiv \mathbf{G_p}\,\ddot{\mathbf{p}} + \mathbf{G_y}\,\ddot{\mathbf{y}} + \dot{\mathbf{G}}_\mathbf{p}\,\dot{\mathbf{p}} + \dot{\mathbf{G}}_\mathbf{y}\,\dot{\mathbf{y}} = \mathbf{0}\,. \tag{10}$$

Descriptor Form. The descriptor form of the equations of motion of the planar two-cable crane manipulator in terms of the actuator coordinates $\mathbf{p}$ and the payload coordinates $\mathbf{y}$, comprised in the vector $\mathbf{s} \in \mathbb{R}^7$, consists of the two implicit constraints (6) and seven kinetic differential equations, see Fig. 2,

$$\begin{aligned}\begin{bmatrix}\mathbf{M_p} & \mathbf{0}\\ \mathbf{0} & \mathbf{M_y}\end{bmatrix}\begin{bmatrix}\ddot{\mathbf{p}}\\ \ddot{\mathbf{y}}\end{bmatrix} &= \begin{bmatrix}\mathbf{0}\\ \mathbf{k_y}\end{bmatrix} + \begin{bmatrix}\mathbf{B_p}\\ \mathbf{0}\end{bmatrix}\mathbf{u} + \begin{bmatrix}\mathbf{0}\\ \mathbf{E_y}\end{bmatrix}\mathbf{z} + \begin{bmatrix}\mathbf{G}_\mathbf{p}^\mathrm{T}\\ \mathbf{G}_\mathbf{y}^\mathrm{T}\end{bmatrix}\begin{bmatrix}\lambda_1\\ \lambda_2\end{bmatrix}\\ \mathbf{M_s}\quad\ \ddot{\mathbf{s}} &= \ \ \mathbf{k_s} \ + \ \mathbf{B_s}\ \mathbf{u} + \ \mathbf{E_s}\ \mathbf{z} + \mathbf{G}_\mathbf{s}^\mathrm{T}(\mathbf{s})\ \boldsymbol{\lambda}\ \ ,\end{aligned} \tag{11}$$

with the constant mass matrices $\mathbf{M_p} = \mathrm{diag}\{m_\mathrm{t}\mathbf{I}_2, J_\mathrm{c}r_\mathrm{c}^{-2}\mathbf{I}_2\} \in \mathbb{R}^{4\times 4}$ and $\mathbf{M_y} = \mathrm{diag}\{m_\mathrm{p}\mathbf{I}_2, \theta_\mathrm{p}\} \in \mathbb{R}^{3\times 3}$, the vector of forces $\mathbf{k}_\mathbf{y}^\mathrm{T} = \begin{bmatrix}0 & m_\mathrm{p}g & 0\end{bmatrix} \in \mathbb{R}^3$, the input matrix $\mathbf{B_p} = \mathrm{diag}\{\mathbf{I}_2, r_\mathrm{c}^{-1}\mathbf{I}_2\} \in \mathbb{R}^{4\times 4}$, the disturbance matrix $\mathbf{E_y} = \mathbf{I}_3 \in \mathbb{R}^{3\times 3}$ and the minimal coordinates of the cable forces $\boldsymbol{\lambda} \in \mathbb{R}^2$. Here $\mathbf{I}_n \in \mathbb{R}^{n\times n}$ denotes the identity matrix of dimension n.

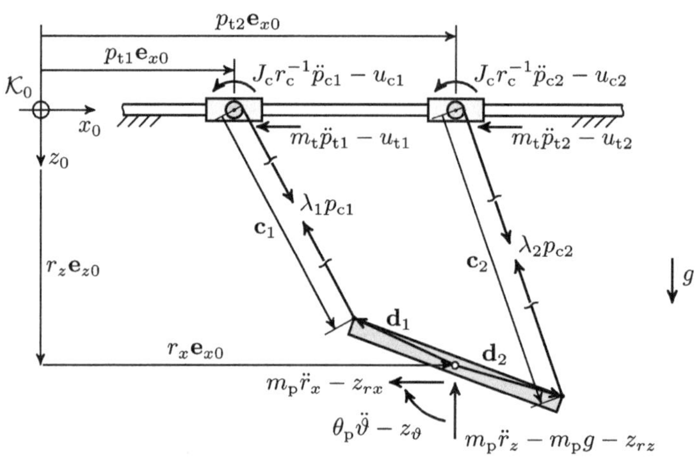

Fig. 2. Model of the two-cable crane manipulator.

2.3 Equations of Motion in Minimal Coordinates

The descriptor form in the dependent coordinates can be transferred into equations of motion in terms of the minimal coordinates. The coordinate transformation is accomplished by eliminating the cable force coordinates using explicit constraints. The equations of motion are then represented in state-space form, that is well suited for simulation and control design.

Explicit Constraints. For the coordinate transformation the explicit constraints associated with the implicit constraints (6) are required,

$$\begin{aligned} \begin{bmatrix} \mathbf{p} \\ \mathbf{y} \end{bmatrix} &= \begin{bmatrix} \mathbf{p} \\ \boldsymbol{\psi}_{\mathbf{y}} \end{bmatrix} \\ \mathbf{s} &= \boldsymbol{\Psi}(\mathbf{q}) \, , \end{aligned} \tag{12}$$

with the three explicit constraints $\boldsymbol{\psi}_{\mathbf{y}}$ implicitly defined by (6), that are not readily analytically solvable with respect to the payload coordinates $\mathbf{y}$. The numerical calculation can be carried out, for example, by the Newton-Raphson method.

The time derivative of (12) yields the explicit constraints at velocity level

$$\begin{aligned} \begin{bmatrix} \dot{\mathbf{p}} \\ \dot{\mathbf{y}} \end{bmatrix} &= \begin{bmatrix} \mathbf{I}_4 \\ \mathbf{J}_{\mathbf{y}} \end{bmatrix} \dot{\mathbf{q}} & \quad & \begin{bmatrix} \dot{y}_{\mathrm{d}} \\ \dot{\mathbf{y}}_{\mathrm{s}} \end{bmatrix} &= \begin{bmatrix} \mathbf{0} & 1 \\ -\mathbf{G}_{\mathbf{y}_{\mathrm{s}}}^{-1}\mathbf{G}_{\mathbf{p}} & -\mathbf{G}_{\mathbf{y}_{\mathrm{s}}}^{-1}\mathbf{G}_{y_{\mathrm{d}}} \end{bmatrix} \begin{bmatrix} \dot{\mathbf{p}} \\ \dot{y}_{\mathrm{d}} \end{bmatrix} \\ \dot{\mathbf{s}} &= \mathbf{J}_{\mathbf{s}}(\mathbf{q})\,\dot{\mathbf{q}} & \quad \text{with} \quad & \dot{\mathbf{y}} &= \mathbf{J}_{\mathbf{y}}(\mathbf{q}) \qquad \dot{\mathbf{q}} \, , \end{aligned} \tag{13}$$

where the Jacobi matrix $\mathbf{J}_{\mathbf{y}} \in \mathbb{R}^{3\times 5}$ is analytically obtained by solving the implicit constraints (8) according to the dependent payload coordinates $\mathbf{y}_{\mathrm{s}}$. The time derivative of (13) leads to the explicit constraints at acceleration level

$$\ddot{\mathbf{s}} = \mathbf{J}_{\mathbf{s}}(\mathbf{q})\,\ddot{\mathbf{q}} + \dot{\mathbf{J}}_{\mathbf{s}}(\mathbf{q},\dot{\mathbf{q}})\,\dot{\mathbf{q}} \, . \tag{14}$$

Minimal Form. The minimal form of the equations of motion (11) is obtained by expressing the dependent coordinates $\mathbf{s}$ in terms of the minimal coordinates $\mathbf{q}$ using the explicit constraints (12), (13) and (14) and by eliminating the cable force coordinates $\boldsymbol{\lambda}$ using the orthogonality property $\mathbf{J}_{\mathbf{s}}^{\mathrm{T}}\mathbf{G}_{\mathbf{s}}^{\mathrm{T}} = \mathbf{0}$,

$$\begin{aligned} \mathbf{J}_{\mathbf{s}}^{\mathrm{T}}\mathbf{M}_{\mathrm{s}}\mathbf{J}_{\mathrm{s}}\,\ddot{\mathbf{q}} + \mathbf{J}_{\mathbf{s}}^{\mathrm{T}}\mathbf{M}_{\mathrm{s}}\dot{\mathbf{J}}_{\mathbf{s}}\,\dot{\mathbf{q}} &= \mathbf{J}_{\mathbf{s}}^{\mathrm{T}}\mathbf{k}_{\mathrm{s}} + \mathbf{J}_{\mathbf{s}}^{\mathrm{T}}\mathbf{B}_{\mathrm{s}}\,\mathbf{u} + \mathbf{J}_{\mathbf{s}}^{\mathrm{T}}\mathbf{E}_{\mathrm{s}}\,\mathbf{z} \\ \mathbf{M}_{\mathbf{q}}(\mathbf{q})\,\ddot{\mathbf{q}} + \mathbf{k}_{\mathbf{q}}^{c}(\mathbf{q},\dot{\mathbf{q}}) &= \mathbf{k}_{\mathbf{q}}^{e}(\mathbf{q}) + \mathbf{B}_{\mathbf{q}}\,\mathbf{u} + \mathbf{E}_{\mathbf{q}}(\mathbf{q})\,\mathbf{z} \, , \end{aligned} \tag{15}$$

with the generalised mass matrix $\mathbf{M}_{\mathbf{q}} \in \mathbb{R}^{5\times 5}$, the vector of generalised centrifugal and Coriolis forces $\mathbf{k}_{\mathbf{q}}^{c} \in \mathbb{R}^{5}$, the vector of generalised applied forces $\mathbf{k}_{\mathbf{q}}^{e} \in \mathbb{R}^{5}$, the input matrix $\mathbf{B}_{\mathbf{q}} \in \mathbb{R}^{5\times 4}$ and the disturbance matrix $\mathbf{E}_{\mathbf{q}} \in \mathbb{R}^{5\times 3}$.

State-Space Form. With the state vector $\mathbf{x} \in \mathbb{R}^{10}$ the dynamic equations (15) can be written as a system of ten nonlinear state-space equations,

$$\begin{aligned} \begin{bmatrix} \dot{\mathbf{q}} \\ \ddot{\mathbf{q}} \end{bmatrix} &= \begin{bmatrix} \dot{\mathbf{q}} \\ \mathbf{M}_{\mathbf{q}}^{-1}\{\mathbf{k}_{\mathbf{q}}^{e} - \mathbf{k}_{\mathbf{q}}^{c}\} \end{bmatrix} + \begin{bmatrix} \mathbf{0} \\ \mathbf{M}_{\mathbf{q}}^{-1}\mathbf{B}_{\mathbf{q}} \end{bmatrix} \mathbf{u} + \begin{bmatrix} \mathbf{0} \\ \mathbf{M}_{\mathbf{q}}^{-1}\mathbf{E}_{\mathbf{q}} \end{bmatrix} \mathbf{z} \, , & \begin{bmatrix} \mathbf{q}(t_0) \\ \dot{\mathbf{q}}(t_0) \end{bmatrix} &= \begin{bmatrix} \mathbf{q}_0 \\ \dot{\mathbf{q}}_0 \end{bmatrix} \\ \dot{\mathbf{x}} &= \mathbf{a}(\mathbf{x}) + \mathbf{B}\,\mathbf{u} + \mathbf{E}(\mathbf{x})\,\mathbf{z} \, , & \mathbf{x}(t_0) &= \mathbf{x}_0 \, , \end{aligned} \tag{16}$$

with the state-dependent vector $\mathbf{a} \in \mathbb{R}^{10}$, the constant input matrix $\mathbf{B} \in \mathbb{R}^{10\times4}$ and the state-dependent disturbance matrix $\mathbf{E} \in \mathbb{R}^{10\times3}$. The minimal coordinates $\mathbf{q}$ are defined as output variables $\mathbf{w} \in \mathbb{R}^5$. Thus the output equation with the constant output matrix $\mathbf{C} \in \mathbb{R}^{5\times10}$ is linearly defined by

$$\begin{aligned} \mathbf{q} &= \begin{bmatrix}\mathbf{I}_5 & \mathbf{0}\end{bmatrix} \mathbf{x} \\ \mathbf{w} &= \quad \mathbf{C} \quad\;\; \mathbf{x} \,. \end{aligned} \tag{17}$$

3 Linearised Equations of Motion

A linearised model is derived for the control design, which describes the payload sway dynamics. By a suitable assumption the linearised equations of motion can be obtained in minimal form from the descriptor form. For control design the equations of motion are then presented in state-space form. Since the disturbances are not directly used in the control design, they are not taken into consideration.

Minimal Form. Since the sway dynamics of the two-cable crane manipulator is mainly described by the coordinate y_d, the coordinates $\mathbf{y}_\mathrm{s}$ and the cable force coordinates $\boldsymbol{\lambda}$ can be considered as slowly varying parameters. Under this assumption the Taylor series expansion of the five nonlinear equations of motion from (11) up to the first order terms yields the linearised minimal form

$$\mathbf{M}_{\Delta\mathbf{q}}\Delta\ddot{\mathbf{q}} = \mathbf{K}_{\Delta\mathbf{q}}(\mathbf{q})\Delta\mathbf{q} + \mathbf{B}_{\Delta\mathbf{q}}\Delta\mathbf{u} \,, \tag{18}$$

with the small deviations of the minimal coordinates $\Delta\mathbf{q} \in \mathbb{R}^5$ and the input variables $\Delta\mathbf{u} \in \mathbb{R}^4$ as well as the coefficient matrices

$$\mathbf{M}_{\Delta\mathbf{q}} = \begin{bmatrix} m_\mathrm{t}\mathbf{I}_2 & \mathbf{0} & \mathbf{0} \\ \mathbf{0} & J_\mathrm{c} r_\mathrm{c}^{-2}\mathbf{I}_2 & \mathbf{0} \\ \mathbf{0} & \mathbf{0} & m_\mathrm{p} \end{bmatrix} \in \mathbb{R}^{5\times5} \,, \qquad \mathbf{B}_{\Delta\mathbf{q}} = \begin{bmatrix} \mathbf{I}_2 & \mathbf{0} \\ \mathbf{0} & r_\mathrm{c}^{-1}\mathbf{I}_2 \\ \mathbf{0} & \mathbf{0} \end{bmatrix} \in \mathbb{R}^{5\times4}$$

$$\mathbf{K}_{\Delta\mathbf{q}} = \begin{bmatrix} \dfrac{\partial \mathbf{G}_\mathbf{p}^\mathrm{T}\boldsymbol{\lambda}}{\partial \mathbf{q}^\mathrm{T}} & \dfrac{\partial \mathbf{G}_{y_\mathrm{d}}^\mathrm{T}\boldsymbol{\lambda}}{\partial \mathbf{q}^\mathrm{T}} \end{bmatrix}^\mathrm{T} = \begin{bmatrix} -\operatorname{diag}\{\boldsymbol{\lambda}\} & \mathbf{0} & \boldsymbol{\lambda} \\ \mathbf{0} & \operatorname{diag}\{\boldsymbol{\lambda}\} & \mathbf{0} \\ \boldsymbol{\lambda}^\mathrm{T} & \mathbf{0} & -\operatorname{sum}\{\boldsymbol{\lambda}\} \end{bmatrix} \in \mathbb{R}^{5\times5} \,.$$

The cable force coordinates are calculated from the two remaining static equilibrium conditions from (11) using the explicit constraints (12),

$$\boldsymbol{\lambda} = -(\mathbf{G}_{\mathbf{y}_\mathrm{s}}^\mathrm{T})^{-1}\mathbf{k}_{\mathbf{y}_\mathrm{s}} \quad \text{with} \quad \mathbf{k}_{\mathbf{y}_\mathrm{s}}^\mathrm{T} = \begin{bmatrix} m_\mathrm{p} g \;\, 0 \end{bmatrix} . \tag{19}$$

Observation of the structure shows that the linearisation decouples the system into a subsystem with the coordinates $\Delta\mathbf{p}_\mathrm{t}, \Delta y_\mathrm{d}$ describing the sway dynamics and a subsystem with the coordinates $\Delta\mathbf{p}_\mathrm{c}$ describing the hoisting dynamics. The former subsystem corresponds to a three-mass oscillator with the spring stiffnesses $\boldsymbol{\lambda}$, see Fig. 3.

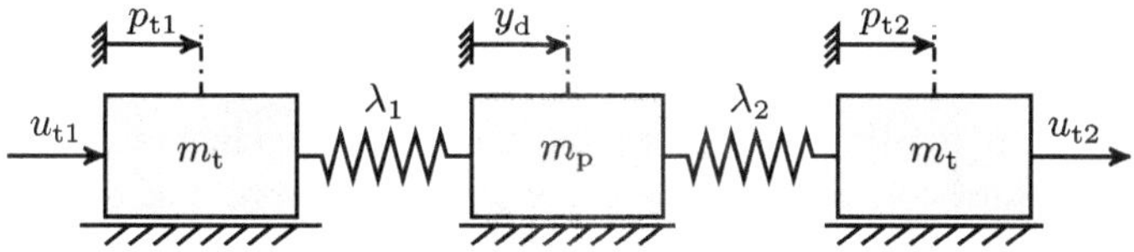

Fig. 3. Model representation of the linearised sway dynamics.

State-Space Form. With the small deviations of the state variables $\Delta\mathbf{x} \in \mathbb{R}^{10}$ the ordinary differential equations (18) can be transformed into the linear, time-variant state-space system

$$\begin{aligned} \begin{bmatrix}\Delta\dot{\mathbf{q}}\\ \Delta\ddot{\mathbf{q}}\end{bmatrix} &= \begin{bmatrix}\mathbf{0} & \mathbf{I}_5\\ \mathbf{M}_{\Delta\mathbf{q}}^{-1}\mathbf{K}_{\Delta\mathbf{q}} & \mathbf{0}\end{bmatrix}\begin{bmatrix}\Delta\mathbf{q}\\ \Delta\dot{\mathbf{q}}\end{bmatrix} + \begin{bmatrix}\mathbf{0}\\ \mathbf{M}_{\Delta\mathbf{q}}^{-1}\mathbf{B}_{\Delta\mathbf{q}}\end{bmatrix}\Delta\mathbf{u}\,, \quad \begin{bmatrix}\Delta\mathbf{q}(t_0)\\ \Delta\dot{\mathbf{q}}(t_0)\end{bmatrix} = \begin{bmatrix}\mathbf{0}\\ \mathbf{0}\end{bmatrix} \\ \Delta\dot{\mathbf{x}} &= \mathbf{A}_\Delta(\mathbf{x}) \quad \Delta\mathbf{x} + \quad \mathbf{B}_\Delta \quad \Delta\mathbf{u}\,, \quad \Delta\mathbf{x}(t_0) = \mathbf{0}\,, \end{aligned} \tag{20}$$

with the state-dependent system matrix $\mathbf{A}_\Delta \in \mathbb{R}^{10\times 10}$ and the constant input matrix $\mathbf{B}_\Delta \in \mathbb{R}^{10\times 4}$. As output variables $\Delta\mathbf{w} \in \mathbb{R}^5$ of the linearised model the small displacements of the minimal coordinates $\Delta\mathbf{q}$ are available. Thus, the linear output equation with the constant output matrix $\mathbf{C}_\Delta \in \mathbb{R}^{5\times 10}$ is

$$\begin{aligned} \Delta\mathbf{q} &= \begin{bmatrix}\mathbf{I}_5 & \mathbf{0}\end{bmatrix}\Delta\mathbf{x} \\ \Delta\mathbf{w} &= \quad \mathbf{C}_\Delta \quad \Delta\mathbf{x}\,. \end{aligned} \tag{21}$$

A requirement for the feasibility of designing a state controller and observer is that the system model (20)–(21) is controllable by the input variables $\Delta\mathbf{u}$ and observable by the output variables $\Delta\mathbf{w}$. Observation of the structure shows that for physical parameter values the system $(\mathbf{A}_\Delta, \mathbf{B}_\Delta, \mathbf{C}_\Delta)$ is completely controllable and observable according to Kalman.

4 Control Design

4.1 Nonlinear Feedforward Control

Feedforward control is applied to move the payload platform along desired translational and rotational trajectories. By exploiting the flatness property the required control inputs can be calculated in two successive steps, the so-called generalised inverse kinematics and inverse dynamics, see also [5,17].

Generalised Inverse Kinematics. In the first step the actuator coordinates $\check{\mathbf{p}} \in \mathbb{R}^4$ and cable force coordinates $\check{\boldsymbol{\lambda}} \in \mathbb{R}^2$ are expressed in terms of the payload coordinates $\check{\mathbf{y}} \in \mathbb{R}^3$. With the three kinetic differential equations of the payload (11) and the two implicit constraints (6) for the six unknowns, the problem is kinematically single redundant. The resolution can be done by task space augmentation [5] or optimisation [14]. The also required actuator speeds $\dot{\check{\mathbf{p}}}$ and accelerations $\ddot{\check{\mathbf{p}}}$ are calculated by time derivation. This leads to the calculations

$$\check{\mathbf{p}}^{(0:2)} = \boldsymbol{\Phi}_\mathbf{p}(\check{\mathbf{y}}, \dot{\check{\mathbf{y}}}, \ddot{\check{\mathbf{y}}}, \check{\mathbf{y}}^{(3)}, \check{\mathbf{y}}^{(4)})\,, \quad \check{\boldsymbol{\lambda}} = \boldsymbol{\Phi}_\lambda(\check{\mathbf{y}}, \dot{\check{\mathbf{y}}}, \ddot{\check{\mathbf{y}}})\,. \tag{22}$$

Inverse Dynamics. The second step is the calculation of the control inputs $\check{\mathbf{u}} \in \mathbb{R}^4$ in terms of $\check{\mathbf{y}}$. Solving the kinetic differential equations of the actuators (11) with respect to the inputs $\check{\mathbf{u}}$ and inserting the generalised inverse kinematics (22) yields a feedforward control law that calculates the inputs $\check{\mathbf{u}}$ as a function of the output variables $\check{\mathbf{y}}$ specified up to the fourth-order time derivatives

$$\check{\mathbf{u}} = \boldsymbol{\Phi}_{\mathbf{u}}(\check{\mathbf{y}}, \dot{\check{\mathbf{y}}}, \ddot{\check{\mathbf{y}}}, \check{\mathbf{y}}^{(3)}, \check{\mathbf{y}}^{(4)}) \,. \tag{23}$$

4.2 Linear Feedback Control

If the equations of motion (11) perfectly represent the dynamics of the two-cable crane manipulator, the feedforward control (22)–(23) moves the payload platform exactly along desired trajectories. However due to model uncertainties and external disturbances a controller is required to stabilise the real system along the prescribed trajectories and actively dampen payload sway [6].

The controller design is based on the linear, time-variant state-space equation (20). For the model a linear feedback of the complete state vector $\Delta\mathbf{x}$ with the feedback matrix $\mathbf{L}_\mathrm{C} \in \mathbb{R}^{4\times 10}$ is used,

$$\Delta\mathbf{u} = -\mathbf{L}_\mathrm{C}(\hat{\mathbf{x}})\Delta\mathbf{x} \quad \text{with} \quad \Delta\mathbf{x} = \hat{\mathbf{x}} - \check{\mathbf{x}} \,, \tag{24}$$

so that the feedback system $\mathbf{A}_\Delta - \mathbf{B}_\Delta\mathbf{L}_\mathrm{C}$ asymptotically stabilises deviations of the actual values $\hat{\mathbf{x}}$ from the desired values $\check{\mathbf{x}}$. A procedure to stabilise linear systems by a complete state feedback represents the linear-quadratic controller design according to Riccati. Here the feedback matrix $\mathbf{L}_\mathrm{C}$ is determined by minimising a quadratic cost function

$$J = \frac{1}{2}\int_0^\infty \left(\Delta\mathbf{x}^\mathrm{T}\mathbf{Q}_\mathrm{C}\Delta\mathbf{x} + \Delta\mathbf{u}^\mathrm{T}\mathbf{R}_\mathrm{C}\Delta\mathbf{u}\right) \mathrm{d}t \tag{25}$$

with the weighting matrix $\mathbf{Q}_\mathrm{C} \in \mathbb{R}^{10\times 10}$ of the state variables $\Delta\mathbf{x}$ and the weighting matrix $\mathbf{R}_\mathrm{C} \in \mathbb{R}^{4\times 4}$ of the input variables $\Delta\mathbf{u}$. The weighting matrices $\mathbf{Q}_\mathrm{C}$ and $\mathbf{R}_\mathrm{C}$ represent positive definite, constant diagonal matrices, which allow a systematic penalty of deviations of individual state variables $\Delta\mathbf{x}$ or input variables $\Delta\mathbf{u}$ by a suitable choice of elements. The steady-state solution of the cost function (25) leads to the algebraic Riccati equation

$$\mathbf{A}_\Delta^\mathrm{T}\mathbf{P}_\mathrm{C} + \mathbf{P}_\mathrm{C}\mathbf{A}_\Delta - \mathbf{P}_\mathrm{C}\mathbf{B}_\Delta\mathbf{R}_\mathrm{C}^{-1}\mathbf{B}_\Delta^\mathrm{T}\mathbf{P}_\mathrm{C} + \mathbf{Q}_\mathrm{C} = \mathbf{0} \,, \tag{26}$$

which determines the optimal feedback

$$\mathbf{L}_\mathrm{C} = \mathbf{R}_\mathrm{C}^{-1}\mathbf{B}_\Delta^\mathrm{T}\mathbf{P}_\mathrm{C} \,. \tag{27}$$

Equation (26) is a system of nonlinear, algebraic equations for the unknown elements of the positive definite matrix $\mathbf{P}_\mathrm{C} = \mathbf{P}_\mathrm{C}^\mathrm{T} \in \mathbb{R}^{10\times 10}$, whose solution succeeds numerically at each time step using the matrix sign function method [3].

4.3 Nonlinear Observer

The feedback controller (24) requires the actual values of all system states with high temporal resolution. Typically for multi-cable crane manipulators only the actuator coordinates $\mathbf{p}$ and the payload coordinates $\mathbf{y}$ are available as measurements [6,17]. The task of the nonlinear observer is therefore to estimate from the partially disturbed measurement variables $\mathbf{w}$ a smoothed time curve of the state variables $\hat{\mathbf{x}}$ for the controller.

On the basis of the nonlinear state-space equation (16) and the linear output equation (17) the following estimation approach is used for the observer,

$$\begin{cases} \dot{\hat{\mathbf{x}}} = \mathbf{a}(\hat{\mathbf{x}}) + \mathbf{B}\mathbf{u} + \mathbf{L}_\mathrm{O}(\hat{\mathbf{x}})\,(\mathbf{w} - \hat{\mathbf{w}}) \;, \quad \hat{\mathbf{x}}(t_0) = \hat{\mathbf{x}}_0 \;, \\ \hat{\mathbf{w}} = \mathbf{C}\hat{\mathbf{x}} \;, \end{cases} \tag{28}$$

where $\hat{\mathbf{x}} \in \mathbb{R}^{10}$ and $\hat{\mathbf{w}} \in \mathbb{R}^5$ denote the estimated state and output variables, respectively. In analogy to the feedback controller design using the linearised equations of motions (20)–(21) a stabilising observer gain $\mathbf{L}_\mathrm{O} \in \mathbb{R}^{10\times 5}$ is obtained by minimising a cost function

$$J = \frac{1}{2}\int_0^\infty \left(\tilde{\mathbf{x}}^\mathrm{T}\mathbf{Q}_\mathrm{O}\tilde{\mathbf{x}} + \tilde{\mathbf{w}}^\mathrm{T}\mathbf{R}_\mathrm{O}\tilde{\mathbf{w}}\right) dt \;, \tag{29}$$

with the weighting matrix $\mathbf{Q}_\mathrm{O} \in \mathbb{R}^{10\times 10}$ of the estimation error of the state variables $\tilde{\mathbf{x}} = \mathbf{x} - \hat{\mathbf{x}}$ and the weighting matrix $\mathbf{R}_\mathrm{O} \in \mathbb{R}^{5\times 5}$ of the estimation error of the output variables $\tilde{\mathbf{w}} = \mathbf{w} - \hat{\mathbf{w}}$. The matrices $\mathbf{Q}_\mathrm{O}$ and $\mathbf{R}_\mathrm{O}$ represent constant, positive definite diagonal matrices, which allow the systematic penalty of state noise and measurement noise by a suitable choice of elements. The steady-state solution of the cost function (29) leads to the algebraic Riccati equation

$$\mathbf{A}_\Delta\mathbf{P}_\mathrm{O} + \mathbf{P}_\mathrm{O}\mathbf{A}_\Delta^\mathrm{T} - \mathbf{P}_\mathrm{O}\mathbf{C}_\Delta^\mathrm{T}\mathbf{R}_\mathrm{O}^{-1}\mathbf{C}_\Delta\mathbf{P}_\mathrm{O} + \mathbf{Q}_\mathrm{O} = \mathbf{0} \;, \tag{30}$$

which determines the optimal observer gain

$$\mathbf{L}_\mathrm{O} = \mathbf{P}_\mathrm{O}\mathbf{C}_\Delta^\mathrm{T}\mathbf{R}_\mathrm{O}^{-1} \;. \tag{31}$$

By exploiting the duality between controller and observer design the same matrix sign function method as for (26) can be used to solve the system of nonlinear, algebraic equations (30) for the unknown elements of the positive definite matrix $\mathbf{P}_\mathrm{O} = \mathbf{P}_\mathrm{O}^\mathrm{T} \in \mathbb{R}^{10\times 10}$ numerically at each time step.

4.4 Control Structure

The overall structure of the control method including nonlinear feedforward control, linear feedback control and nonlinear state observation is shown in Fig. 4. Note that all subsystems are adapted to the time-varying control design model. To simulate the forward dynamic system behaviour, the nonlinear equations of motion are used.

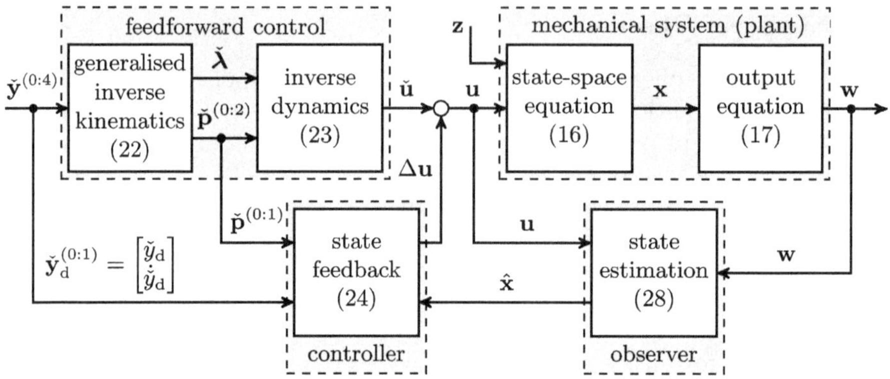

Fig. 4. Modular structure of the control method.

5 Simulation Results

As an example a rectangular payload with $m_\mathrm{p} = 12.5\,\mathrm{kg}$, $l_\mathrm{p} = 1\,\mathrm{m}$, $h_\mathrm{p} = 0.25\,\mathrm{m}$, $\theta_\mathrm{p} = 1\,\mathrm{kg\,m^2}$ is considered. The physical parameters of the actuators are given by $m_\mathrm{t} = 34\,\mathrm{kg}$, $r_\mathrm{c} = 0.075\,\mathrm{m}$, $J_\mathrm{c} = 0.0873\,\mathrm{kg\,m^2}$. The numerical calculations were done in Matlab. According to Fig. 5 the payload is moved from an initial rest position at $t_0 = 0\,\mathrm{s}$ to a final rest position at $t_1 \approx 4.2\,\mathrm{s}$ with the initial and final conditions of the actuator and payload coordinates

$$\begin{aligned}\mathbf{p}^\mathrm{T}(t_0) &= \left[1.00\,\mathrm{m}\ 3.00\,\mathrm{m}\ 3.25\,\mathrm{m}\ 3.25\,\mathrm{m}\right], \mathbf{y}^\mathrm{T}(t_0) = \left[2.00\,\mathrm{m}\ 3.50\,\mathrm{m}\ 0.00\,\mathrm{rad}\right],\\ \mathbf{p}^\mathrm{T}(t_1) &= \left[5.01\,\mathrm{m}\ 6.74\,\mathrm{m}\ 5.28\,\mathrm{m}\ 4.28\,\mathrm{m}\right], \mathbf{y}^\mathrm{T}(t_1) = \left[6.00\,\mathrm{m}\ 5.00\,\mathrm{m}\ 0.52\,\mathrm{rad}\right].\end{aligned} \tag{32}$$

The actuator coordinates belonging to the given payload coordinates are calculated by the generalised inverse kinematics (22) using minimal chain lengths as optimisation criterion. Also in the experiment the payload is displaced about 20° by an external disturbance z_{rx} starting at time $t = 10\,\mathrm{s}$.

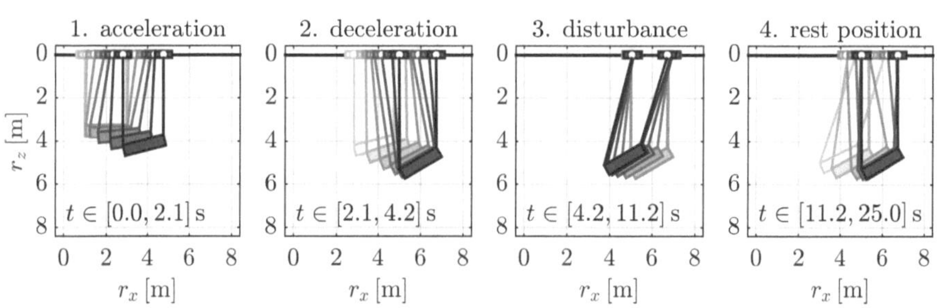

Fig. 5. Snapshots of the two-cable crane manipulator for a controlled motion with external disturbance depicted in four snapshots with increasing darkness of the snapshots as time increases during respective sequence.

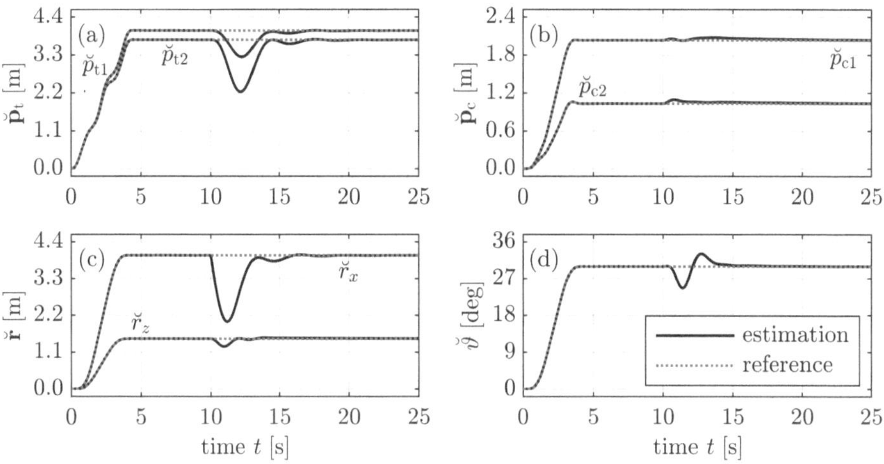

Fig. 6. Translational motion of the payload with rotation around the lateral axis with external disturbance.

The corresponding time histories are illustrated in Fig. 6. Subfigures 6a-b show the trolley displacements $\breve{\mathbf{p}}_t(t) = \mathbf{p}_t(t) - \mathbf{p}_t(t_0)$ and cable length displacements $\breve{\mathbf{p}}_c(t) = \mathbf{p}_c(t) - \mathbf{p}_c(t_0)$ calculated from the generalised inverse kinematics and estimated from the state observer. Subfigure 6c-d represent the desired and observed time histories of the payload coordinates $\breve{\mathbf{y}}(t) = \mathbf{y}(t) - \mathbf{y}(t_0)$. The time trajectories of the desired payload coordinates are interpolated by functions being steady up to the fourth-order time derivatives. The results show how the developed control concept can move the payload along desired trajectories and compensate undesired payload sway.

6 Conclusion

A control concept for a two-cable crane manipulator is presented. Based on the derived equations of motion a nonlinear flatness-based feedforward control is designed to move the payload along desired translational and rotational trajectories. A superimposed, linear feedback compensates for trajectory deviations and payload sway caused by external disturbances. The required actual states are estimated by a nonlinear observer. To guarantee the global asymptotic stability of the closed loop system despite the time-varying control design model, the algebraic Riccati equation is solved at each time step for the sought control and observer gain. The efficiency of the proposed control method is shown by simulation results involving tracking of desired trajectories and compensating external disturbances. Ongoing research is devoted to the extension of the control concept to more complex three-cable and four-cable configurations as well as experimental validation and verification.

References

1. Abdel-Rahman, E.M., Nayfeh, A.H., Masoud, Z.N.: Dynamics and control of cranes: a review. J. Vibr. Control **9**(7), 863–908 (2003)
2. Aschemann, H.: Passivity-based trajectory control of an overhead crane by interconnection and damping assignment. In: Motion and Vibration Control, pp. 21–30. Springer (2009)
3. Bini, D.A., Iannazzo, B., Meini, B.: Numerical solution of algebraic riccati equations. SIAM (2011)
4. Fang, Y., Ma, B., Wang, P., Zhang, X.: A motion planning-based adaptive control method for an underactuated crane system. IEEE Trans. Control Syst. Technol. **20**(1), 241–248 (2011)
5. Heyden, T., Woernle, C.: Dynamics and flatness-based control of a kinematically undetermined cable suspension manipulator. Multibody Syst. Dyn. **16**(2), 155–177 (2006)
6. Hildebrandt, E., Handreg, T., Froitzheim, P., Flügge, W., Woernle, C.: Validation of flatness-based feedforward control for a four-chain crane manipulator. In: European Conference on Mechanism Science, pp. 86–94. Springer (2020)
7. Knierim, K.L., Krieger, K., Sawodny, O.: Flatness based control of a 3-DOF overhead crane with velocity controlled drives. IFAC Proc. Volumes **43**(18), 363–368 (2010)
8. Kolar, B., Rams, H., Schlacher, K.: Time-optimal flatness based control of a gantry crane. Control Eng. Pract. **60**, 18–27 (2017)
9. Lu, B., Fang, Y., Sun, N.: Adaptive output-feedback control for dual overhead crane system with enhanced anti-swing performance. IEEE Trans. Control Syst. Technol. **28**(6), 2235–2248 (2019)
10. Miller, A.S., Sarvepalli, P., Singhose, W.: Dynamics and control of dual-hoist cranes moving triangular payloads. In: Dynamic Systems and Control Conference. American Society of Mechanical Engineers (2014)
11. Otto, S., Rückwald, T., Seifried, R.: Control of multibody systems in DAE form applied to a cable robot. In: DINAME 2019 - Proceedings of the XVIII International Symposium on Dynamic Problems of Mechanics (2019)
12. Ramli, L., Mohamed, Z., Abdullahi, A.M., Jaafar, H.I., Lazim, I.M.: Control strategies for crane systems: a comprehensive review. Mech. Syst. Sign. Process. **95**, 1–23 (2017)
13. Sawodny, O., Aschemann, H., Lahres, S.: An automated gantry crane as a large workspace robot. Control Eng. Pract. **10**(12), 1323–1338 (2002)
14. Stoltmann, M., Froitzheim, P., Fuchs, N., Woernle, C.: Flatness-based feedforward control of a crane manipulator with four load chains. In: Springer (ed.) European Conference on Mechanism Science, pp. 61–68 (2018)
15. Sun, N., Fu, Y., Yang, T., Zhang, J., Fang, Y., Xin, X.: Nonlinear motion control of complicated dual rotary crane systems without velocity feedback: design, analysis, and hardware experiments. IEEE Trans. Autom. Sci. Eng. **17**(2), 1017–1029 (2020)
16. Vaughan, J., Yoo, J., Knight, N., Singhose, W.: Multi-input shaping control for multi-hoist cranes. In: American Control Conference (2013)
17. Woernle, C.: Trajectory tracking for a three-cable suspension manipulator by nonlinear feedforward and linear feedback control. In: Cable-Driven Parallel Robots, pp. 371–386. Springer (2013)

Calibration and Performance Improvement

Path Following Demonstration of a Hybrid Cable-Driven Parallel Robot

Micaël Michelin[1], Pierre Elie Hervé[1], Olivier Tempier[2], Jean Baptiste Izard[1], and Marc Gouttefarde[2(✉)]

[1] TECNALIA, Basque Research and Technology Alliance, Montpellier, France
[2] LIRMM, Univ Montpellier, CNRS, Montpellier, France
marc.gouttefarde@lirmm.fr

Abstract. In the context of applications requiring large workspace such as painting of airliners, a cable-driven parallel robot (CDPR) having a manipulator (robot arm) on-board its mobile platform possesses an interesting potential. Such a robotic system has typically more than six degrees of freedom (DOFs) and, hence, more than the minimum required to accomplish a task in Cartesian space. In particular, a robotic system consisting of a six-DOF cable-driven parallel robot whose mobile platform is equipped with a seven-DOF manipulator is considered in this paper. This robotic system being kinematically redundant, a redundancy resolution scheme is used to allow its end-effector to follow a relatively complex path in Cartesian space while avoiding cable collisions and satisfying some constraints on the CDPR workspace.

Keywords: Hybrid cable-driven parallel robot · Kinematic redundancy

1 Introduction

Cable-Driven Parallel Robots (CDPRs) are known for their potentially very large workspace and heavy payload capabilities, e.g. [1–6]. Manufacturing, manipulation, and inspection in various industries (e.g. aeronautic, naval, nuclear, construction, renewable energy) may require a local dexterity that the CDPR mobile platform alone cannot provide. Examples of such applications are painting or sandblasting [7] and assembly [8] of large parts. Therefore, in order to accomplish such tasks, it can be relevant to equip the mobile platform of a large-dimension CDPR with a manipulator (robot arm).

Placing an actuated mechanical device on-board the mobile platform of a CDPR is not a novel idea, e.g. [9–11], but only few previous works deal with it. A kinematic analysis of a 3 degree-of-freedom (DOF) manipulator suspended on one cable is presented in [10]. In [12], a pendulum-like underactuated robot, consisting of a 1-DOF actuated pendulum suspended on a single cable, is introduced. Moreover, the works presented in [13, 14] deal with underconstrained CDPRs having a multiple DOF manipulator placed on their mobile platform. In [14], the static analysis of a planar 2-cable suspended CDPR having an on-board 4-DOF planar manipulator is presented. In [13], a 2-DOF planar manipulator is placed on a planar 2-cable suspended CDPR in order to improve

M. Gouttefarde et al. (Eds.): CableCon 2021, MMS 104, pp. 323–335, 2021.
https://doi.org/10.1007/978-3-030-75789-2_26

the range of possible orientations along a trajectory. Hence, in these previous works, the CDPR is underconstrained which means here that the mobile platform is driven by less cables than its number of DOFs. In the present paper, the case of redundantly actuated 6-DOF CDPRs, i.e., CDPRs driven by more than six cables, is considered. A few previous works deal with redundantly actuated CDPRs equipped with a manipulator. A static analysis of planar 3-DOF suspended CDPRs with a manipulator on-board the mobile platform is proposed in [15]. Recently, in [16], the decoupled modeling, state estimation, and model-predictive vibration control of a kinematically-constrained redundant planar CDPR equipped with two 1-DOF pendulums are discussed.

When a 6-DOF CDPR mobile platform is equipped with a multiple DOF manipulator, the resulting Hybrid Cable-Driven Parallel Robot (HCDPR) is kinematically redundant. Hence, in order to generate feasible generalized joint coordinates allowing the HCDPR end-effector to follow a path, standard kinematic redundancy resolution or path planning techniques can be used. In the recent works [17, 18], continuous path validation methods integrated into the path planning software HPP [19] were proposed for the HCDPR consisting of the six-DOF CDPR prototype CoGiRo [3, 20] equipped with a seven-DOF manipulator Yaskawa SIA20F. In the present paper, the same HCDPR is considered but standard kinematic redundancy resolution techniques based on local optimizations are used to generate feasible generalized joint coordinates which allow the HCDPR end-effector to follow a path, while avoiding cable collisions and satisfying some workspace constraints. The results are validated in simulations and in experimental demonstrations [21, 22] which form the contribution of this work since, to the best of our knowledge, such practical experimental results were never reported before.

This paper is organized as follows. Section 2 describes the HCDPR considered in this work. Section 3 presents the kinematic model while Sect. 4 introduces the kinematic redundancy resolution method used to solve the inverse kinematics. Finally, Sect. 5 presents the simulation and experimental results.

2 HCDPR Description

One concern when installing a manipulator on a large-dimension CDPR mobile platform is the electrical cable(s) required to bring power to the CDPR mobile platform. In the case of the CDPR prototype CoGiRo [3], whose global dimensions are 15 m $\times$ 11 m $\times$ 6 m (L $\times$ l $\times$ h), it is critical to integrate both the serial manipulator and its controller on-board the mobile platform so that only one data link (Ethernet cable) and one power cable are required. The prototype CoGiRo is a suspended CDPR (all its cable drawing points are located on top of this working volume) and its mobile platform is a cube with a side length of 1 m. The on-board manipulator should thus be placed below the mobile platform and should have a working range of around 1 m. The maximum payload capability of CoGiRo across a large part of its wrench-feasible workspace being 500 kg, the desired net mass of the on-board manipulator and its controller should not be larger than 300 kg. Moreover, the manipulator payload capability should not be too limiting and its controller should have communication capabilities with external systems (here, the CDPR control system). We therefore set the main requirements on the choice of the manipulator to be placed on-board the CDPR mobile platform as follows:

- Manipulator payload capability: 20 kg;
- Working range: 1 m;
- Controller box fitting inside a 1 m side cube;
- Target point on-the-fly request capabilities with relatively small latency;
- Total mass (manipulator + controller box) lower than 300 kg.

According to the requirements summarized above, the manipulator Yaskawa SIA20F was found to be a good choice. Its controller FS100 has a relatively low mass and can remotely receive on-the-fly target points, e.g. such as demonstrated by the capability of the available ROS compatible FS100 drivers to move the end-effector in Cartesian space using a gamepad [23]. The SIA20F is a 7-DOF anthropomorphic manipulator (S-R-S manipulator) with a rated payload of 20 kg, a relatively low net mass (120 kg for the manipulator itself and 26 kg for the FS100 controller) and a working range of 1118 mm from the elbow central joint to the wrist flange. It also features a high torque capacity at the wrist allowing operations such as drilling, which is one of the objectives of the DexterWide project [24]. With the SIA20F manipulator and its controller on-board the mobile platform of the CDPR CoGiRo (Fig. 1), the platform of the resulting robotic system weighs a total of 296 kg (without payload at the end-effector of the SIA20F).

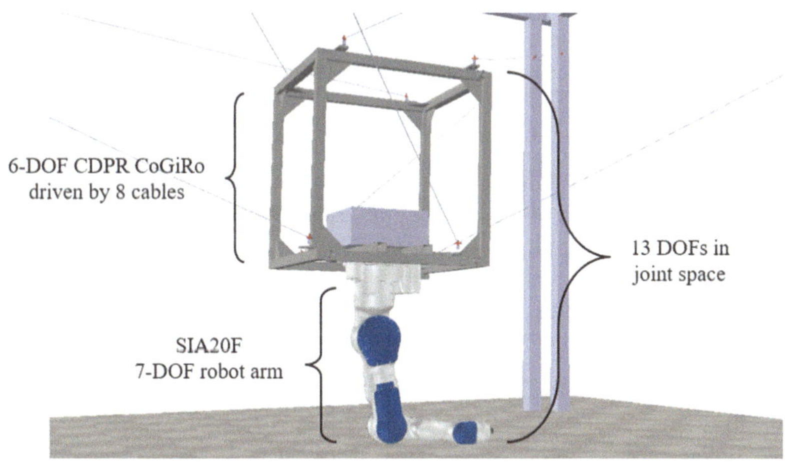

Fig. 1. The HCDPR consisting of the CDPR CoGiRo equipped with a Yaskawa SIA20F manipulator (shown as implemented in a V-REP simulator of the system).

3 Kinematic Model

3.1 Kinematic and Actuation Redundancy

The robotic system shown in Fig. 1 is an HCDPR (a hybrid parallel/serial robot system) having two levels of redundancy. The CoGiRo CDPR has an actuation redundancy of degree two with eight driving cables for a six-DOF mobile platform. The manipulator

SIA20F has one degree of kinematic redundancy with seven joints for six DOFs at its end-effector. The HCDPR has then fifteen actuated DOFs for six DOFs at the end-effector and possesses both actuation and kinematic redundancies. Methods allowing to manage these two types of redundancy are widely presented in the literature. For serial manipulators, a well-known method is to use an iterative solution of the inverse geometric problem based on the pseudo-inverse of manipulator Jacobian matrix. The generalization of this method with projection of joint velocities in the Jacobian nullspace allows to perform joint motions without any influence on the end-effector Cartesian pose. These "internal" joint motions can be used to achieve secondary tasks such as obstacle and joint limit avoidance, or maximizing manipulability, e.g. [25, 26]. Similar methods allow one to manage actuation redundancy of parallel robots using the transpose of the Jacobian matrix to control internal forces in the mechanical structure.

3.2 Selection of Generalized Joint Coordinates

In fact, the methods presented in [27] are currently used to solve the actuation redundancy of the CDPR CoGiRo and to control the motions of its mobile platform, the motion control being based on a controlled winding of the cables. Therefore, in the present paper, it is considered that the six DOFs of the mobile platform of the CDPR CoGiRo are properly controlled. Hence, the CDPR mobile platform DOFs can be considered directly in a purely kinematic redundancy resolution problem where the eight cable lengths are replaced by six variables defining the mobile platform pose:

$$[l_1 \ l_2 \ l_3 \ l_4 \ l_5 \ l_6 \ l_7 \ l_8]^T \rightarrow \left[x \ y \ z \ \phi \ \theta \ \psi\right]^T \tag{1}$$

The X-Y-Z Euler angle convention is used to define the three angles ϕ, θ, and ψ which define the orientation of the mobile platform, while x, y and z define the position of the reference point of the mobile platform. Then, for the overall system consisting of the CDPR mobile platform and its on-board manipulator (with joint coordinates q_i), the following generalized (hybrid) coordinate vector $\boldsymbol{Q}$ is defined by concatenating the six mobile platform pose variables and the seven SIA20F joint coordinates:

$$\boldsymbol{Q} = \left[z \ y \ x \ \phi \ \theta \ \psi \ q_1 \ q_2 \ q_3 \ q_4 \ q_5 \ q_6 \ q_7\right]^T \tag{2}$$

The thirteen components of vector $\boldsymbol{Q}$ are considered as joint coordinates. The mobile platform of the CDPR can thus be viewed as a floating base allowing to position and orient the on-board manipulator base arbitrarily in Cartesian space (within the limits of the CDPR position and orientation workspaces).

3.3 Serial Kinematic Chain and Denavit–Hartenberg Parameters

Based on Sect. 3.2, the overall system can be considered as a serial kinematic chain consisting of three prismatic joints and three revolute joints (platform of the CDPR CoGiRo), followed by seven revolute joints (manipulator SIA20F). The corresponding kinematic modeling and the Denavit-Hartenberg parameters are shown in Fig. 2 ($\sigma_j = 1$ for a prismatic joint and $\sigma_j = 0$ for a revolute joint). Frame R_0 is a fixed reference frame and frame R_{14} is the mobile frame attached to the HCDPR end-effector.

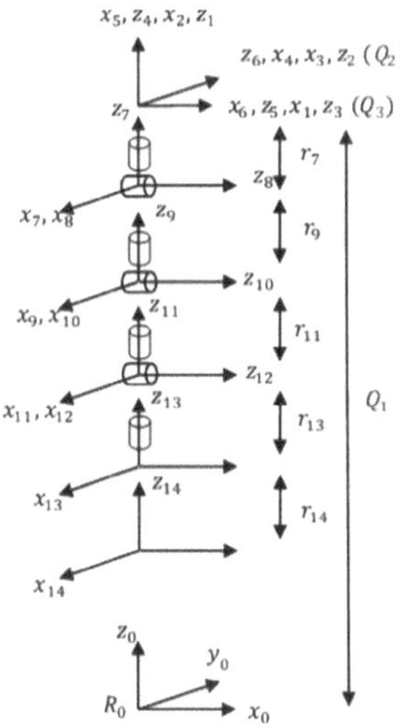

j	σ_j	α_j	d_j	θ_j	r_j
1	1	0	0	0	$Q_1\ (z)$
2	1	$-\pi/2$	0	$-\pi/2$	$Q_2\ (y)$
3	1	$-\pi/2$	0	$-\pi/2$	$Q_3\ (x)$
4	0	$-\pi/2$	0	$Q_4\ (\psi)$	0
5	0	$\pi/2$	0	$Q_5\ (\phi) + \pi/2$	0
6	0	$\pi/2$	0	$Q_6\ (\theta) + \pi/2$	0
7	0	$\pi/2$	0	$Q_7\ (q_1) - \pi/2$	$-r_7$
8	0	$-\pi/2$	0	$Q_8\ (q_2)$	0
9	0	$\pi/2$	0	$Q_9\ (q_3)$	$-r_9$
10	0	$-\pi/2$	0	$Q_{10}\ (q_4)$	0
11	0	$\pi/2$	0	$Q_{11}\ (q_5)$	$-r_{11}$
12	0	$-\pi/2$	0	$Q_{12}\ (q_6)$	0
13	0	$\pi/2$	0	$Q_{13}\ (q_7)$	$-r_{13}$
14	0	0	0	0	$-r_{14}$

Fig. 2. Kinematic modeling as an equivalent serial chain (left) and corresponding Denavit-Hartenberg parameters (right).

4 Kinematic Redundancy Resolution

For a kinematically redundant robotic systems such as the one under study, there exist infinitely many joint positions for a given end-effector pose. To deal with this redundancy, following the guidelines provided in [25, 26], an iterative solution of the inverse geometric problem of the serial kinematic chain introduced in Sect. 3 is presented in this section.

Iteratively solving the inverse geometric problem consists in computing the joint coordinate increments $d\boldsymbol{Q}$ corresponding to an end-effector pose increment $d\boldsymbol{X}$. Here, $\boldsymbol{X}$ is the manipulator end-effector pose vector with respect to the base frame R_0 and with Euler angles used to define the orientation:

$$\boldsymbol{X} = \begin{bmatrix} x_e & y_e & z_e & \phi_e & \theta_e & \psi_e \end{bmatrix}^T \tag{3}$$

The increment $d\boldsymbol{X}$ is computed from the current end-effector pose $\boldsymbol{X_t}$ and the desired one at the next sample period $\boldsymbol{X}_{t+1}$. These poses are first converted into the corresponding homogeneous transformation matrices relative to base frame R_0:

$${}^0\mathbf{T}_{\mathrm{t}} = \begin{bmatrix} \boldsymbol{s}_t\ \boldsymbol{n}_t\ \boldsymbol{a}_t\ \boldsymbol{P}_t \\ 0\ 0\ 0\ 1 \end{bmatrix}, \quad {}^0\boldsymbol{T}_{t+1} = \begin{bmatrix} \boldsymbol{s}_{t+1} & \boldsymbol{n}_{t+1} & \boldsymbol{a}_{t+1} & \boldsymbol{P}_{t+1} \\ 0 & 0 & 0 & 1 \end{bmatrix} \tag{4}$$

where $\boldsymbol{s}$, $\boldsymbol{n}$ and $\boldsymbol{a}$ are the columns of the corresponding rotation matrix and $\boldsymbol{P}$ is the position vector of the end effector in R_0. Then, the end-effector increment dX is given by:

$$dX = \begin{bmatrix} \boldsymbol{P}_{t+1} - \boldsymbol{P}_t \\ \frac{1}{2}(\boldsymbol{s}_t \times \boldsymbol{s}_{t+1} + \boldsymbol{n}_t \times \boldsymbol{n}_{t+1} + \boldsymbol{a}_t \times \boldsymbol{a}_{t+1}) \end{bmatrix} \tag{5}$$

The increment dQ can be obtained by means of the Jacobian matrix pseudo-inverse: $dQ = J^{+}dX$ where J is the Jacobian matrix of the serial kinematic chain. This method brings forth the particular solution minimizing $\dot{Q}^2$. Computing other solutions for dQ can classically be done by adding a term to $dQ = J^{+}dX$:

$$dQ = J^{+}dX + (I_n - J^{+}J)Z \tag{6}$$

This second term allows the projection of any joint velocity vector Z into the Jacobian matrix null space, these joint velocities thus generate no displacement of the end-effector. This second term can be used to optimize a cost function$H(Q)$, for instance by computing $Z = k_H \nabla H$ where H is a scalar function depending on the joint variables, $H(Q) = F(Q_1, \ldots, Q_{13})$ and ∇H is the gradient of$H(Q)$, $\nabla H = \left[\frac{\partial H}{\partial Q_1} \cdots \frac{\partial H}{\partial Q_{13}}\right]^T$. Equation (6) can then be written as the following well-known equation:

$$dQ = J^{+}dX + k_H(I_n - J^{+}J)\nabla H \tag{7}$$

When the scalar k_H is positive, the cost function H is maximized and when k_H is negative, it is minimized. Function H can express various objectives, for example the distance between a robot part and an environment object or the distance to joint limits. These functions may thus be used to avoid collisions and joint limits. Moreover, several optimization functions H_i and associated scalars k_{Hi} can be considered in order to include several objectives into the problem:

$$dQ = J^{+}dX + \sum k_{Hi}(I_n - J^{+}J)\nabla H_i \tag{8}$$

In the case of the CDPR CoGiRo with the manipulator SIA20F on-board its mobile platform, the Jacobian matrix J is computed from the thirteen homogeneous transformation matrices defining the position and orientation of joint frame j relative to frame R_0:

$$^{0}T_j = \begin{bmatrix} s_j\ n_j\ a_j\ P_j \\ 0\ \ 0\ \ 0\ \ 1 \end{bmatrix} \tag{9}$$

Then, this Jacobian matrix can be written as:

$$J = \begin{bmatrix} J_{p1}J_{p2} \cdots J_{p13} \\ J_{o1}J_{o2} \cdots J_{o13} \end{bmatrix} \tag{10}$$

with $J_{pj} = \sigma_j a_j + \overline{\sigma_j}(a_j \times (P_{14} - P_j))$ and $J_{oj} = \overline{\sigma_j} a_j$, where $\sigma_j = 1$ for a prismatic joint and $\sigma_j = 0$ for a revolute joint, and where P_{14} is the position vector of the frame attached to the end-effector.

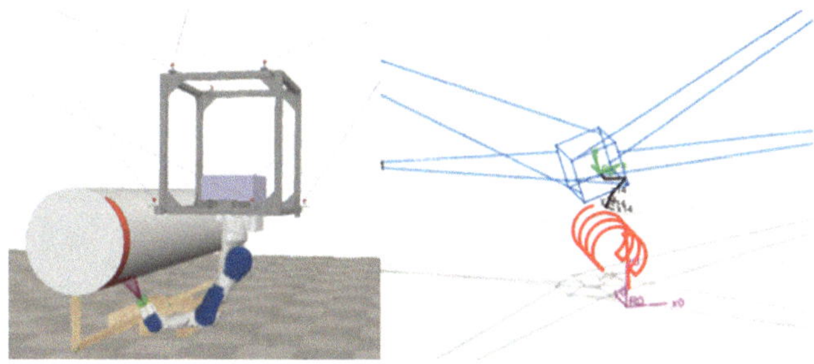

Fig. 3. Painting scenario (left); Painting path in red, CoGiRo in blue and SIA20F in black (right).

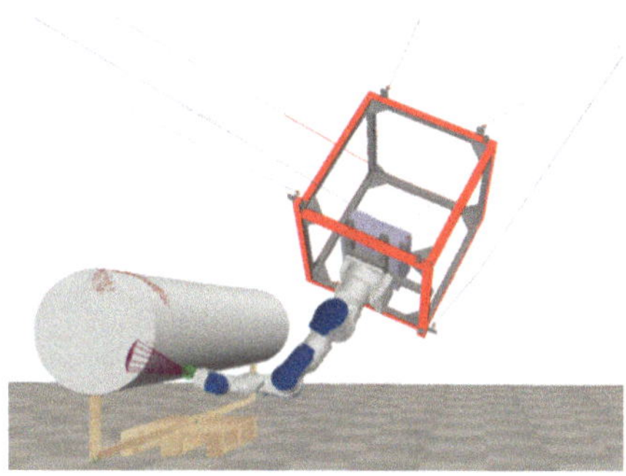

Fig. 4. Motion without optimization term: A cable-platform collision occurs.

5 Simulation and Experimental Results

In order to test the kinematic redundancy resolution method of Sect. 4, we consider a scenario where the manipulator on-board the CDPR paints a large cylinder, the latter representing for example a plane fuselage (Fig. 3).

In a first step, the equation $d\boldsymbol{Q} = \boldsymbol{J}^{+}d\boldsymbol{X}$ (without any optimization term) is used to find values for all joint coordinates to follow the desired path around the cylinder. The CoGiRo mobile platform orientation reaches large values which are not feasible on the real robot due to a cable colliding with the platform (Fig. 4). Equation (7) is then used, adding a first optimization term aiming at maintaining a platform orientation close to zero (Fig. 5). The cost functions H_{oi} used for that purpose are described in the appendix.

With this optimization of the platform orientation, the CDPR platform reaches high positions along the vertical z axis at some points along the path (Fig. 5). These positions are not feasible because of the limitations on the cable tensions, the latter being too large for positions high in the workspace. Therefore, a second cost function H_z (see the second paragraph of the appendix) is added to the local optimization in order to avoid overshooting a given limit on the CDPR platform vertical position (Fig. 5).

At this point however, some cables collide with the cylinder when painting the underside of the cylinder (Fig. 6). Adding a third set of cost functions H_{di} dealing with the distance between the cables and the cylinder (third paragraph of the appendix) allows

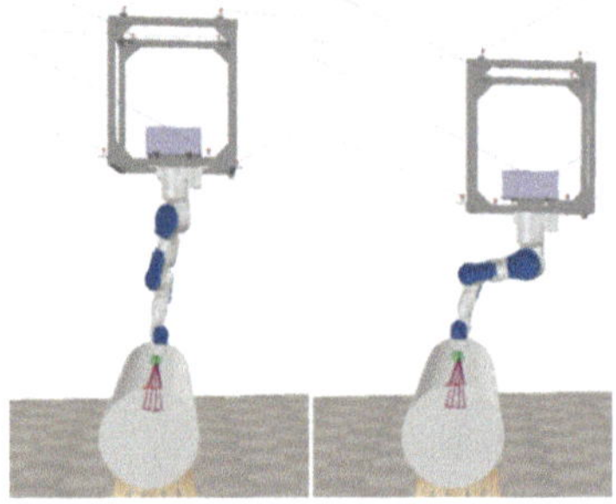

Fig. 5. Motion with an optimization of the platform orientation (left) and a further optimization to limit the platform height (right).

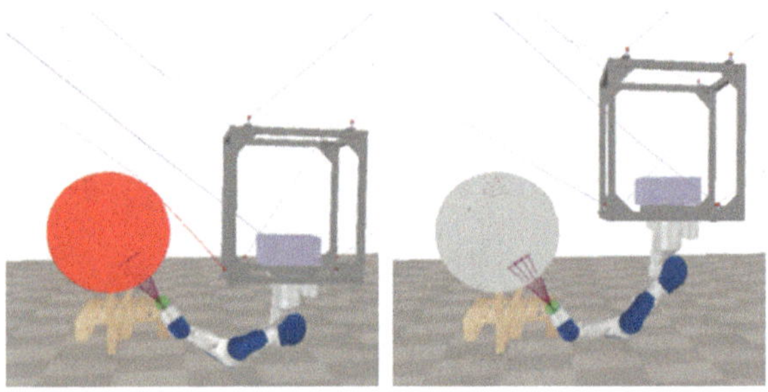

Fig. 6. Motion without (left)/with (right) optimization cost functions avoiding cable-cylinder collisions.

the method to generate joint coordinate trajectories which avoid cable-cylinder collisions (Fig. 6).

To avoid singularities, a manipulability optimization term is finally introduced as presented in the fourth paragraph of the appendix. This term significantly changes the HCDPR behavior as can be seen in Fig. 7. The aim is to maximize the cost function $H_m = \det\left[\boldsymbol{J}\boldsymbol{J}^T\right]$ (Fig. 8). Being given the hybrid nature of the considered robotic system (serial/parallel), the interpretation and relevance of using this manipulability optimization term deserves further investigations in future works.

A complete motion avoiding collisions, with feasible platform orientation, limited platform height, and manipulability optimization avoiding singularities while remaining into the limits of the manipulator joint angles is finally obtained. This final motion has been tested on the physical HCDPR testbed which consists of the CDPR CoGiRo equipped with the SIA20F manipulator. For this purpose, a motion synchronization strategy has been set up. This synchronization is necessary since the two robots operate on different PLCs. It is based on sending on-the-fly target points to both systems (CDPR and manipulator), which was one of the requirements for choosing the manipulator (see in Sect. 2). A discretized path is used on both robots with common indexes on the points composing the path. The SIA20F goes through the points, reporting the actual index to the CDPR CoGiRo, which speeds up and slows down every step by reducing (respectively increasing) the time shift before the next step in order to keep up with the

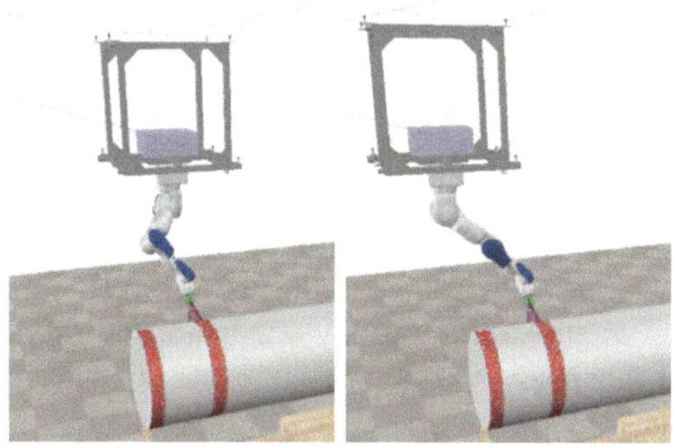

Fig. 7. Motion without (left)/with (right) manipulability optimization.

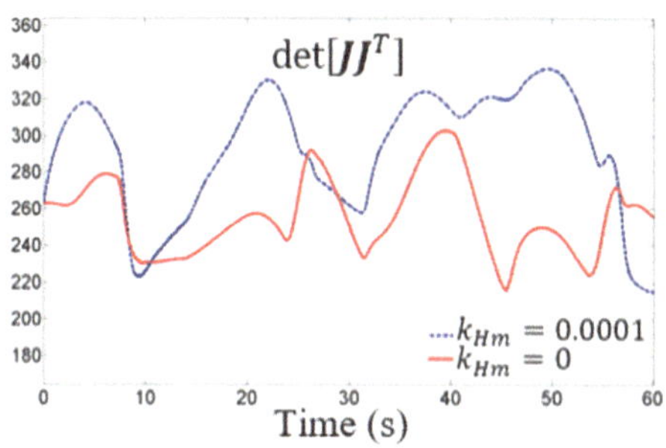

Fig. 8. Cost function $H_m = det\left[JJ^T\right]$ along the desired trajectory.

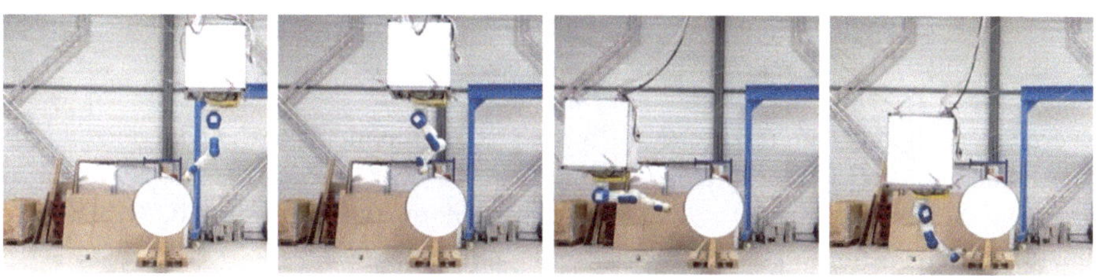

Fig. 9. CDPR CoGiRo and manipulator SIA20F following the painting path around a cylinder.

SIA20F. In the end, with a 2 ms discretized trajectory in input for the SIA20F and 20 ms for CoGiRo, the maximum latency between the two robots is 60 ms. Some HCDPR configurations attained during the motion around the cylinder are shown in Fig. 9. The full demonstration video of this motion is provided in [21] and, with further explanations, in [22].

6 Conclusion

This paper reported the use of standard kinematic redundancy resolution techniques based on local optimizations on a HCDPR. The six-DOF CDPR CoGiRo equipped with a seven-DOF industrial manipulator was used to demonstrate the capability of such a hybrid parallel/serial robotic system to follow a path defined along the surface of a large-dimension cylinder, imitating the painting of a large object such as an airplane fuselage part. In order to successfully follow this path, several null space projections were used to achieve secondary tasks related to cable collision avoidance, manipulability optimization, and workspace limitations. Among several possible future works, the inclusion of wrench-feasibility constraints into the kinematic redundancy resolution and the consideration of the HCDPR dynamics could be relevant. Methods ensuring that problem constraints remain verified, which is not the case with null space projection techniques, would also be relevant.

Acknowledgment. This work was supported by the ANR under grant ANR-15-CE10-0006, project DexterWide.

Appendices

1. *Cost functions related to the platform orientation*

To maintain the platform orientation close to zero, the following cost functions are considered:

$$H_{o1} = Q_4^2 \; H_{o2} = Q_5^2 \; H_{o3} = Q_6^2 \tag{11}$$

where Q_4, Q_5 and Q_6 are the fourth, fifth and sixth components of the generalized joint coordinate vector $\boldsymbol{Q}$ (Euler angles). Minimizing the cost functions H_{oi} is achieved by assigning the same value $k_{Ho} = -1$ to the corresponding three terms in Eq. (8). The cost functions H_{oi} are minimized when orientation angles are null.

2. *Platform height cost function*

To avoid too high vertical positions of the CDPR mobile platform, the following cost function is used, where Q_{1d} is the desired platform height:

$$H_z = (Q_1 - Q_{1d} - 3)^2 \tag{12}$$

With a positive coefficient k_{Hz}, the gradient $\nabla \boldsymbol{H}_z$ keeps the platform away from an upper limit situated at 3 m above the ground (base frame R_0). The coefficient k_{Hz} is calculated as $k_{Hz} = e^{(-20d)}$ where d is the distance between the current position of the platform reference point and the upper limit at 3 m above the ground. Coefficient k_{Hz} increases exponentially from zero to one as d decreases. The repulsive effect of the ceiling become noticeable when $d < 0.2\ m$.

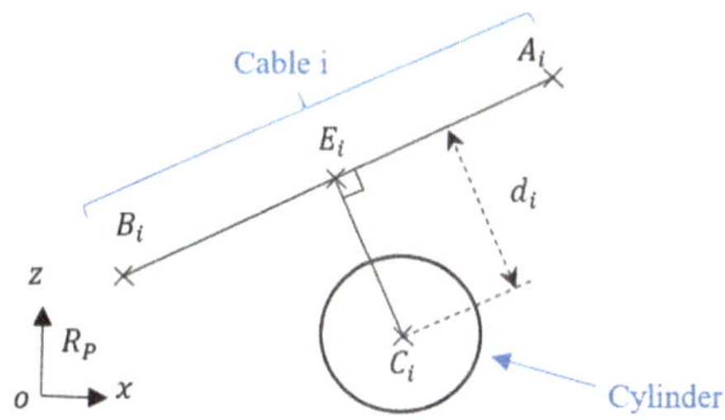

Fig. 10. Cable-cylinder distance

3. *Cable distance cost function*

The computation of the gradient $\nabla \boldsymbol{H}_{di}$ of the cable–cylinder distance cost function H_{di} is done as follows. A cable is considered to be a straight-line segment A_iB_i, A_i being the drawing point of the cable and B_i its anchor point on the platform (Fig. 10). Sagging is thus neglected. All coordinates are expressed with respect to the frame R_P whose y axis is parallel to the cylinder axis. Then, the squared distance d_i^2 is used as the cost function, $H_{di} = d_i^2$, where d_i is given by:

$$d_i = \overrightarrow{E_iC_i} = \sqrt{(x_{Ci} - x_{Ei})^2 + (z_{Ci} - z_{Ei})^2} \tag{13}$$

As illustrated in Fig. 10, the location of point C_i is known and the location of point E_i is computed as follows:

$$\begin{aligned} x_{Ei} &= x_{Bi} + \gamma_i(x_{Ai} - x_{Bi}) \\ z_{Ei} &= z_{Bi} + \gamma_i(z_{Ai} - z_{Bi}) \end{aligned} \tag{14}$$

Finding the location of E_i is equivalent to finding γ_i such as d^2 is minimal which is the case when:

$$\frac{\partial d_i^2}{\partial \gamma_i} = 0 \tag{15}$$

By using Eq. (13), (14) and (15), γ_i can be computed which leads to d_i^2 and then to the gradient $\nabla \boldsymbol{H}_{di}$. With a positive coefficient k_{Hdi}, the cable–cylinder distance cost function keeps away each cable from the cylinder. The coefficient k_{Hdi} is calculated in the same way as k_{Hz}, but relatively to d_i.

4. *Manipulability cost function*

To increase the manipulability and avoid singular configurations of the on-board manipulator, the following cost function is considered:

$$H_m = \det\left[\boldsymbol{J}\boldsymbol{J}^T\right] \tag{16}$$

In this case, the gradient is obtained by numerical derivation. Thus, the gradient:

$$\nabla \boldsymbol{H_m} = \left[\frac{\partial \det[\boldsymbol{J}\boldsymbol{J}^T]}{\partial Q_1} \cdots \frac{\partial \det[\boldsymbol{J}\boldsymbol{J}^T]}{\partial Q_{13}} \right] \tag{17}$$

is replaced by:

$$\nabla \boldsymbol{H_m} = \left[\frac{\det[\boldsymbol{J}_{\varepsilon 1}\boldsymbol{J}_{\varepsilon 1}^T] - \det[\boldsymbol{J}\boldsymbol{J}^T]}{\varepsilon_1} \cdots \frac{\det[\boldsymbol{J}_{\varepsilon 13}\boldsymbol{J}_{\varepsilon 13}^T] - \det[\boldsymbol{J}\boldsymbol{J}^T]}{\varepsilon_{13}} \right] \tag{18}$$

where:

$$\begin{gathered} \boldsymbol{J}_{\varepsilon 1} = \boldsymbol{J}(\boldsymbol{Q_c} + \boldsymbol{\varepsilon}_1) \\ \vdots \\ \boldsymbol{J}_{\varepsilon 13} = \boldsymbol{J}(\boldsymbol{Q_c} + \boldsymbol{\varepsilon}_{13}) \end{gathered} \tag{19}$$

and:

$$\begin{gathered} \boldsymbol{\varepsilon}_1 = [\varepsilon 000000000000] \\ \vdots \\ \boldsymbol{\varepsilon}_{13} = [000000000000\varepsilon] \end{gathered} \tag{20}$$

ε is a small increment in the joint coordinate which has been tuned at 1.10^{-6}. With a positive coefficient $k_{Hm} = 0.0001$, when used in Eq. (8), this gradient allows the manipulator to stay away from singular configurations.

References

1. Albus, J.S., Bostelman, R.V., Dagalakis, N.G.: The NIST ROBOCRANE. J. Robot. Syst. **10**(15), 709–724 (1993)
2. Duan, B., Qiu, Y.-Y., Zhang, F., Zi, B.: On design and experiment of the feed cable-suspended structure for super antenna. Mechatronics **19**(14), 503–509 (2009)
3. Gouttefarde, M., Collard, J.-F., Riehl, N., Baradat, C.: Geometry selection of a redundantly actuated cable-suspended parallel robot. IEEE Trans. Robot. **31**(12), 501–510 (2015)
4. Lambert, C., Nahon, M., Chalmers, D.: Implementation of an aerostat positioning system with cable control. IEEE/ASME Trans. Mechatron. **12**(11), 32–40 (2007)
5. Merlet, J.-P., Daney, D.: A portable, modular parallel wire crane for rescue operations. In: IEEE International Conference on Robotics and Automation, Anchorage, CA; USA (2010)
6. Pott, A., Mütherich, H., Kraus, W., Schmidt, V., Miermeister, P., Verl, A.: IPAnema: a family of cable-driven parallel robots for industrial applications. In: Bruckmann, T., Pott, A. (eds.) Cable-Driven Parallel Robots, pp. 119–134. Springer, Heidelberg (2013)
7. Tempier, O., Hervé, P.-E., Gouttefarde, M.: Cable-driven parallel robot CoGiRo moving a robot manipulator (2015). https://youtu.be/-CmjXwXEZJA
8. IRT Jules Verne, Cable-driven parallel robot equipped with a UR10 manipulator (2018). https://youtu.be/J3R01IgwUi8
9. Albus, J.: Modular Suspended Platform. US Patent 6,566,834 (2003)

10. Arai, T., Matsumura, S., Yoshimura, Y., Osumi, H.: A proposal for a wire suspended manipulator. Robotica **17**, 3–9 (1999)
11. Pott, A., et al.: Cable-Driven Parallel Robot IPAnema 2 at Automatica 2010 (2010). https://youtu.be/9-qQYv90_Og
12. Lefrançois, S., Gosselin, C.: Point-to-point motion control of a pendulum-like 3-Dof underactuated. In: IEEE International Conference on Robotics and Automation, Anchorage, USA (2010)
13. Bamdad, M., Taheri, F., Abtahi, N.: Dynamic analysis of a hybrid cable-suspended planar. In: IEEE International Conference on Robotics and Automation, Seattle, Washington (2015)
14. Osumi, H., Utsugi, Y.K.M.: Development of a manipulator suspended by parallel wire structure. In: IEEE/RSJ International Conference Intelligent Robotic Systems, Takamatsu, Japan (2000)
15. Gouttefarde, M.: Static analysis of planar 3-DOF cable-suspended parallel robots carrying a serial manipulator. In: 6th European Conference on Mechanism Science (EUCOMES), Nantes, France (2016)
16. Qi, R., Rushton, M., Khajepour, A., Melek, W.W.: Decoupled modeling and model predictive control of a hybrid cable-driven robot. Robot. Auton. Syst. **118**, 1–12 (2019)
17. Bury, D., Izard, J.-B., Gouttefarde, M., Lamiraux, F.: Continuous collision detection for a robotic arm mounted on a cable-driven parallel robot. In: IEEE/RSJ International Conference on Intelligent Robots and Systems (IROS), Macau, China (2019)
18. Bury, D., Izard, J.-B., Gouttefarde, M., Lamiraux, F.: Continuous tension validation for cable-driven parallel robots. In: IEEE/RSJ International Conference on Intelligent Robots and Systems (IROS), Las Vegas, United States (2020)
19. Mirabel, J., et al.: HPP: a new software for constrained motion planning. In: IEEE/RSJ Intelligent Robots and Systems (IROS), Daejeon, Korea (2016)
20. Gouttefarde, M., et al.: CoGiRo: A large-dimension cable-driven parallel robot (2012). https://youtu.be/2b4YwFZhtIE
21. Tempier, O., Michelin, M., Hervé, P.-E., Gouttefarde, M., Izard, J.-B.: Cable robot CoGiRo and Yaskawa SIA20F: Path Following – ANR project DexterWide (2019). https://youtu.be/g0_OqK-ZWdU
22. Michelin, M., Tempier, O., Hervé, P.-E., Gouttefarde, M., Izard, J.-B.: Cable-driven parallel robot CoGiRo equipped with a robot arm Yaskawa SIA20F (2021). https://youtu.be/UKbbWN43UcQ
23. Xamla, 2017-07-07 SDA10D Joystick Control (2018). https://youtu.be/KgzYKmDy8jk
24. ANR, Project DexterWide (2019). https://www.lirmm.fr/dexterwide
25. Khalil, W., Dombre, E.: Modeling, Identification and Control of Robots. Hermes Penton, London (2002)
26. Siciliano, B., Khatib, O. (eds.): Springer Handbook of Robotics. Springer, Heidelberg (2008)
27. Gouttefarde, M., Lamaury, J., Reichert, C., Bruckmann, T.: A versatile tension distribution algorithm for n-DOF parallel robots driven by n+2 cables. IEEE Trans. Rob. **31**(6), 1444–1457 (2015)

Development of an Active End Effector for Cable Robot Calibration

Pauline Marie Nüsse(✉), Marcus Hamann, David Lukas Winter, and Christoph Ament

Faculty of Applied Computer Science, University of Augsburg, 86159 Augsburg, Germany
{pauline.nuesse,marcus.hamann,david.winter,christoph.ament}@uni-a.de
https://www.uni-augsburg.de/de/fakultaet/fai/informatik/prof/rt/

Abstract. This paper presents the development of an active end effector for cable robots. Together with a laser tracker, it is possible to calibrate any cable robot. In this case, a planar cable robot with three degrees of freedom is used as a demonstrator. As part of the calibration, four winch positions are calculated. The task of the active end effector is to determine the translational and rotational degrees of freedom of the platform. For this purpose, Kalman filters are used. The pose of the platform is then used to determine the cable exit points with the help of a model-based approach in combination with optimization.

Keywords: Cable-driven parallel robot · Calibration · Kalman filter

1 Introduction

Cable robots have a high potential for many applications due to their flexibility and high payload. The flexibility provides advantages. Especially when the robot has to be adapted to a new and large operating environment, cable robots have an advantage over robot arms. This reconfiguration must be carried out quickly and precisely. One way of calibrating cable robots using a non-contact method is to use a laser tracker. The University of Augsburg is using such a laser tracker. It is equipped with relatively measuring interferometers and is very fast in tracking a target. The fast determination of a target position is thus possible very precisely. However, a challenge that always arises when using a laser tracker is the determination of the rotation of the platform. One way is the use of multiple targets or reflectors, respectively. Another alternative will be presented in this paper.

A planar cable robot with three degrees of freedom is used as a demonstrator in this work. However, other configurations of cable robots are also conceivable. The aim of this work is the determination of the winch positions with the help of the end effector extension. For the calibration, an active end effector is attached to the platform. The extended platform can be seen in Fig. 1. In the figure, the rotating reflector, which revolves around the platform with a defined radius and

M. Gouttefarde et al. (Eds.): CableCon 2021, MMS 104, pp. 336–347, 2021.
https://doi.org/10.1007/978-3-030-75789-2_27

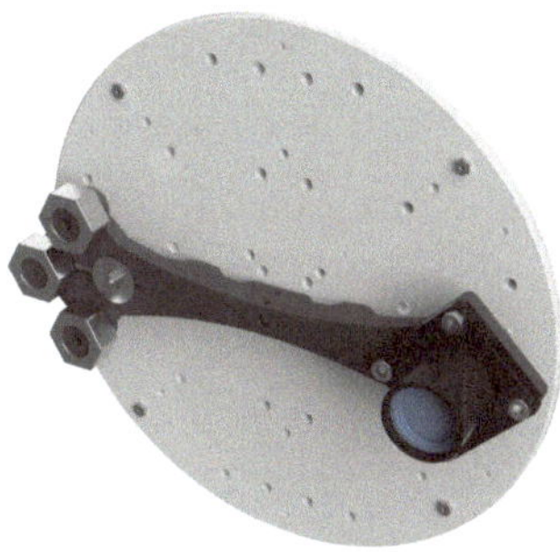

(a) Front view

(b) Back view

Fig. 1. CAD of the platform extension

a defined angular velocity, can be seen. Besides, a microcontroller, the power supply and a stepper motor can also be seen in the back view. This setup offers two main advantages: On the one hand, it allows multiple positions on the platform to be measured dynamically, and on the other hand, the defined radius provides a reference length, which is required for the laser tracker. Laser-based measurement systems have already been used in the past to calibrate cable robots. In [6] and [5], a laser distance sensor is attached to the platform of the robot. This sensor determines the distance to a reflector during the calibration. In [3] and [4], the calibration of a spatial cable robot with the laser tracker that is used in this paper is presented. In [8], a large cable robot for rescue operations is presented. The positions of the cable exit points are determined there as well. In [7], a method for determining initial cable lengths is presented. The method is based on the measurement of relative cable lengths by encoders on the drive side. [1] presents two methods for calibrating a cable robot based on cable force measurements. In [2], the prototype *CoGiRo* is calibrated using external and internal sensors. Many of the publications mentioned above use inverse kinematics and optimization to determine the geometric system parameters. This approach is also followed here.

For this purpose, the mathematical basics are presented in Sect. 2. In Sect. 3, the calibration is presented in order to present experimental results in Sect. 4. In Sect. 5, a conclusion is drawn and further work is presented.

2 Mathematical Foundations

2.1 Cable Robot Modeling

For the calibration, we need the inverse kinematics of the cable robot, the inverse kinematics of the laser tracker and the path followed by the reflector on the platform. All three aspects are described in the following section.

Inverse Kinematics of the Cable Robot. Figure 2 shows a sketch of the cable robot.

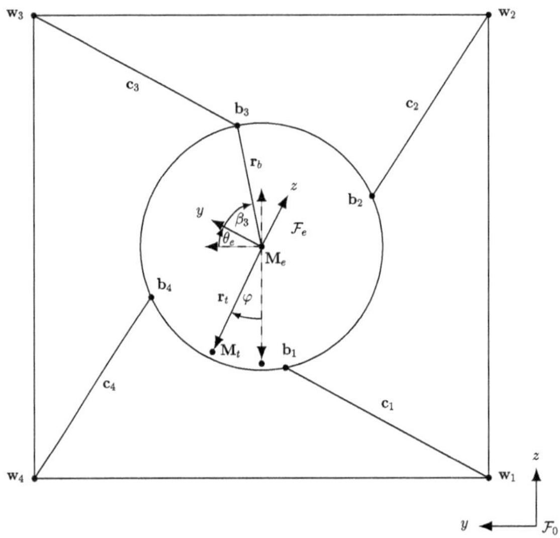

Fig. 2. Sketch of the platform with all geometric parameters

The center of the platform $\mathbf{M}_e$ ist given by

$$\mathbf{M}_e = \begin{bmatrix} y_e \\ z_e \end{bmatrix} . \tag{1}$$

The orientation of the platform in the reference frame $\mathcal{F}_0$ is given by θ_e. Together with the orientation β_i of connection point $\mathbf{b}_i$ the connection point $\mathbf{b}_i$ in the end effector frame $\mathcal{F}_e$ is written as

$$\mathbf{b}_i = \mathbf{R}(\beta_i) \cdot \begin{bmatrix} r_b \\ 0 \end{bmatrix} = r_b \cdot \begin{bmatrix} \cos\beta_i \\ \sin\beta_i \end{bmatrix} \tag{2}$$

with rotation matrix

$$\mathbf{R}(\beta_i) = \begin{bmatrix} \cos\beta_i & -\sin\beta_i \\ \sin\beta_i & \cos\beta_i \end{bmatrix} \tag{3}$$

and distance r_b. The connection point ${}^0\mathbf{b}_i$ in the reference frame $\mathcal{F}_0$ is given as

$$\begin{aligned} {}^0\mathbf{b}_i &= \begin{bmatrix} y_e \\ z_e \end{bmatrix} + \mathbf{R}(\theta_e) \cdot \mathbf{b}_i \\ &= \begin{bmatrix} y_e \\ z_e \end{bmatrix} + r_b \cdot \begin{bmatrix} \cos\beta_i \cos\theta_e - \sin\beta_i \sin\theta_e \\ \cos\beta_i \sin\theta_e + \sin\beta_i \cos\theta_e \end{bmatrix} \\ &= \begin{bmatrix} y_e \\ z_e \end{bmatrix} + r_b \cdot \begin{bmatrix} \cos(\beta_i + \theta_e) \\ \sin(\beta_i + \theta_e) \end{bmatrix} \end{aligned} \tag{4}$$

with $i = 1 \dots 4$ and rotation matrix

$$\mathbf{R}(\theta_e) = \begin{bmatrix} \cos\theta_e & -\sin\theta_e \\ \sin\theta_e & \cos\theta_e \end{bmatrix} . \tag{5}$$

The cable vector $\mathbf{c}_i$ is composed of the detachment point $\mathbf{w}_i = [w_{i,y}, w_{i,z}]^T$ and the connection point $\mathbf{b}_i$ and written as

$$\begin{aligned} \mathbf{c}_i &= \mathbf{w}_i - {}^0\mathbf{b}_i \\ &= \mathbf{w}_i - \mathbf{R}(\theta_e) \cdot \mathbf{b}_i \\ &= \begin{bmatrix} w_{i,y} - y_e - r_b \cdot \cos(\beta_i + \theta_e) \\ w_{i,z} - z_e - r_b \cdot \sin(\beta_i + \theta_e) \end{bmatrix} . \end{aligned} \tag{6}$$

The inverse kinematics of the cable robot denoted for the i-th cable thus results in

$$\varphi_{\text{inv},i}^{\text{cr}} = |\mathbf{c}_i| . \tag{7}$$

Inverse Kinematics of the Laser Tracker. The inverse kinematics of the laser tracker describes the vector from the mirror surfaces of the laser tracker to the center of the reflector attached to the platform. This vector is given by

$$\varphi_{\text{inv},i}^{\text{lt}} = \rho_{0,i} + \Delta\rho_i = |\mathbf{M}_l - \mathbf{M}_t| \tag{8}$$

where $\rho_{0,i}$ is the initial laser beam path length of the i-th laser tracker, $\Delta\rho_i$ is the laser beam path length variation for the current measurement, $\mathbf{M}_l$ is the laser exit point and $\mathbf{M}_t$ is the current target position.

Reflector Path. The vector to the center of the reflector is defined as

$$\begin{aligned} \mathbf{M}_t &= \mathbf{M}_e + \mathbf{R}(\varphi)\mathbf{r}_t \\ &= \begin{bmatrix} x_e \\ y_e \\ z_e \end{bmatrix} + \mathbf{R}(\varphi) \begin{bmatrix} r_{t,x} \\ r_{t,y} \\ 0 \end{bmatrix} . \end{aligned} \tag{9}$$

It is composed of the vector to the center of the platform and the vector of the reflector in the end effector frame. Now, an auxiliary vector $\mathbf{r}'$ can be defined, which has an arbitrary length in direction x and a normalized length in direction y. The scalar product between this vector and the normal vector of the platform is defined as

$$\mathbf{r}' \circ \mathbf{n} = \begin{bmatrix} x' \\ 1 \\ 0 \end{bmatrix} \circ \begin{bmatrix} n_x \\ n_y \\ n_z \end{bmatrix} = 0 \tag{10}$$

and is equal to zero. After rearranging and normalizing Eq. (10) we obtain

$$\mathbf{r}_t = \begin{bmatrix} r_{t,x} \\ r_{t,y} \\ 0 \end{bmatrix} = \begin{bmatrix} -n_y \\ n_x \\ 0 \end{bmatrix} \cdot \frac{|\mathbf{r}_t|}{\sqrt{n_y^2 + n_x^2}} . \tag{11}$$

Using the rotation matrix

$$\mathbf{R}(\varphi) = \begin{bmatrix} n_x^2(1-c\varphi)+c\varphi & n_xn_y(1-c\varphi)-n_zs\varphi & n_xn_z(1-c\varphi)+n_ys\varphi \\ n_yn_x(1-c\varphi)+n_zs\varphi & n_y^2(1-c\varphi)+c\varphi & n_yn_z(1-c\varphi)-n_xs\varphi \\ n_zn_x(1-c\varphi)-n_ys\varphi & n_zn_y(1-c\varphi)+n_xs\varphi & n_z^2(1-c\varphi)+c\varphi \end{bmatrix} \tag{12}$$

around the axis $\mathbf{n}$, we derive

$$\mathbf{M}_t = \begin{bmatrix} x_e + \frac{|\mathbf{r}_t|}{\sqrt{n_y^2+n_x^2}}\left(-n_yc\varphi - n_xn_zs\varphi\right) \\ y_e + \frac{|\mathbf{r}_t|}{\sqrt{n_y^2+n_x^2}}\left(n_xc\varphi - n_yn_zs\varphi\right) \\ z_e + \frac{|\mathbf{r}_t|}{\sqrt{n_y^2+n_x^2}}\left(n_x^2s\varphi + n_y^2s\varphi\right) \end{bmatrix}, \tag{13}$$

where $c\varphi = \cos\varphi$ and $s\varphi = \sin\varphi$.

2.2 Extended Kalman Filter

For the calibration, an Extended Kalman filter is applied. For this purpose, the following state space model is used.

$$\mathbf{x}(k+1) = \mathbf{A}(k)\mathbf{x}(k) + \mathbf{w}(k) \tag{14}$$

$$\mathbf{y}(k) = \mathbf{g}\left(\mathbf{x}(k)\right) + \mathbf{v}(k) \tag{15}$$

In this model, $\mathbf{x}$ corresponds to the state vector, $\mathbf{A}$ to the system matrix, $\mathbf{g}$ to the nonlinear output matrix and $\mathbf{y}$ to the noisy measurement. The system is disturbed by a process noise $\mathbf{w}$ and a measurement noise $\mathbf{v}$. The Kalman filter is a state estimator. It is based on a prediction step with

$$\mathbf{P}^-(k) = \mathbf{A}(k-1)\mathbf{P}^+(k-1)\mathbf{A}^T(k-1) + \mathbf{Q}(k-1) \tag{16}$$

$$\hat{\mathbf{x}}^-(k) = \mathbf{A}(k-1)\hat{\mathbf{x}}^+(k-1) \tag{17}$$

and a correction step with

$$\mathbf{K}(k) = \mathbf{P}^-(k)\mathbf{C}^T(k)\left[\mathbf{R}(k) + \mathbf{C}(k)\mathbf{P}^-(k)\mathbf{C}^T(k)\right]^{-1} \tag{18}$$

$$\hat{\mathbf{x}}^+(k) = \hat{\mathbf{x}}^-(k) + \mathbf{K}(k)\left[\mathbf{y}(k) - \mathbf{g}\left(\hat{\mathbf{x}}^-(k)\right)\right] \tag{19}$$

$$\mathbf{P}^+(k) = (\mathbf{I} - \mathbf{K}(k)\mathbf{C}(k))\mathbf{P}^-(k)(\mathbf{I} - \mathbf{K}(k)\mathbf{C}(k))^T + \mathbf{K}(k)\mathbf{R}(k)\mathbf{K}^T(k) \tag{20}$$

where $\mathbf{P}$ is the covariance of the estimation error, $\mathbf{Q}$ and $\mathbf{R}$ are known covariance matrices of the noise processes $\mathbf{w}$ and $\mathbf{v}$ respectively, $\mathbf{K}$ is the Kalman filter gain, $\mathbf{I}$ is the identity matrix and $\mathbf{C}$ is the linearized output matrix with

$$\mathbf{C} = \left.\frac{\partial \mathbf{g}\left(\mathbf{x}\right)}{\partial \mathbf{x}}\right|_{\mathbf{x}=\hat{\mathbf{x}}^-}. \tag{21}$$

The superscript "+" denotes a posteriori and the superscript "−" denotes a priori estimation. Further information about Kalman filters can be found in [9].

3 Calibration

The calibration aims to identify the winch positions. Therefore, pose measurements are needed. To obtain these from the interferometer signals three Extended Kalman filters are used. Subsequently, the winch positions are determined by optimization. The overall calibration procedure is depicted in Fig. 3.

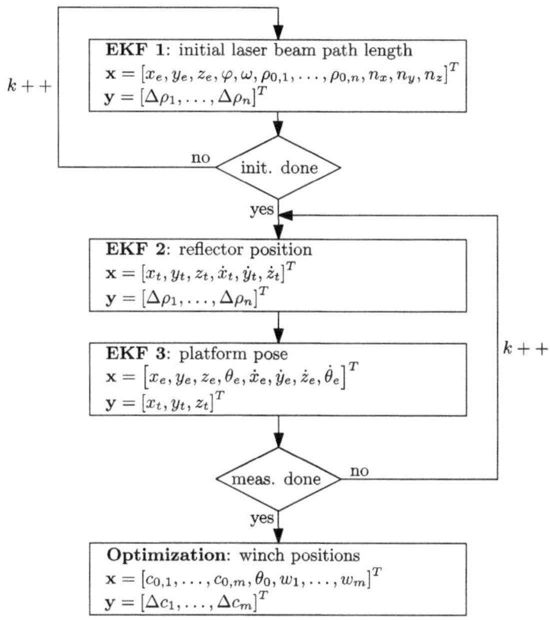

Fig. 3. Calibration procedure

3.1 EKF 1

Since the laser tracker used only provides changes in the laser beam path lengths, the initial laser beam path lengths must first be estimated. In order to calculate the rotation of the reflector around the proper axis, the normal vector of the plane spanned by the cable robot is also determined. The state vector used is given by

$$\mathbf{x}_{EKF1} = [x_e, y_e, z_e, \varphi, \omega, \rho_{0,1}, \ldots, \rho_{0,n}, n_x, n_y, n_z]^T . \tag{22}$$

For this step, it is sufficient to leave the platform in a fixed position and only let the reflector spin around the center of the platform. So all states except φ are constant. The platform orientation θ is assumed to be zero in this point. All further estimates are therefore only changes to this inital orientation. The system matrix thus results in

$$\mathbf{A}_{EKF1} = \begin{bmatrix} 1 & 0 & 0 & \cdots & & & 0 \\ 0 & 1 & 0 & \ddots & & & \vdots \\ \vdots & \ddots & 1 & 0 & \ddots & & \vdots \\ \vdots & & \ddots & 1 & T & \ddots & \vdots \\ \vdots & & & \ddots & 1 & 0 & \ddots \\ \vdots & & & & \ddots & \ddots & 0 \\ 0 & \cdots & & & 0 & 1 & 0 \end{bmatrix}, \tag{23}$$

where T is the sampling time. The measurements y_{EKF1} used in the filter are the laser beam path length changes $\Delta\rho_i$. The measurement equations are derived from Eq. (8) as

$$\mathbf{g}_{EKF1} = \varphi^{\mathrm{lt}}_{\mathrm{inv},i} - \rho_{0,i} \,. \tag{24}$$

The output matrix $\mathbf{C}_{EKF1}$ can consequently be calculated according to Eq. (21).

3.2 EKF 2

By knowing the initial values, the platform can be moved along a trajectory from now on. With the second filter, the position of the reflector is estimated. The same measurement expression as in EKF 1 can be used, but with known initial laser beam path lengths $\rho_{0,i}$. The state vector changes to

$$\mathbf{x}_{EKF2} = [x_t, y_t, z_t, \dot{x}_t, \dot{y}_t, \dot{z}_t]^T \,. \tag{25}$$

The system matrix is given by

$$\mathbf{A}_{EKF2} = \begin{bmatrix} 1 & 0 & 0 & T & 0 & 0 \\ 0 & 1 & 0 & 0 & T & 0 \\ \vdots & \ddots & 1 & 0 & 0 & T \\ \vdots & & \ddots & 1 & 0 & 0 \\ \vdots & & & \ddots & 1 & 0 \\ 0 & \cdots & & & 0 & 1 \end{bmatrix} \tag{26}$$

and C_{EKF2} can be calculated as before.

3.3 EKF 3

The next step is to obtain the pose of the moving platform from the reflector position. As the phase of φ and ω are known from EKF 1, the current φ can be calculated in every step. This helps to distinguish the rotation of the platform and the reflector. The state vector is given by

$$\mathbf{x}_{EKF3} = \left[x_e, y_e, z_e, \theta_e, \dot{x}_e, \dot{y}_e, \dot{z}_e, \dot{\theta}_e\right]^T . \tag{27}$$

And the system matrix results in

$$\mathbf{A}_{EKF3} = \begin{bmatrix} 1 & 0 & 0 & 0 & T & 0 & \cdots & 0 \\ 0 & 1 & 0 & 0 & 0 & T & \ddots & \vdots \\ \vdots & \ddots & 1 & 0 & 0 & 0 & T & 0 \\ \vdots & & \ddots & 1 & 0 & 0 & 0 & T \\ \vdots & & & \ddots & 1 & 0 & 0 & 0 \\ \vdots & & & & \ddots & 1 & 0 & 0 \\ \vdots & & & & & \ddots & 1 & 0 \\ 0 & \cdots & & & & \cdots & 0 & 1 \end{bmatrix} \tag{28}$$

The measurement used is the position of the reflector, hence the measurement equation is given by Eq. 13 and C_{EKF3} is calculated as before.

3.4 Optimization

Using EKF 2 and 3 in every time step, the platform can be moved to several poses along a trajectory, where the current cable lengths and the platform pose are estimated by EKF 3. These states are used to calculate the winch positions by optimization. Since the cable lengths and the platform rotation are only given as changes in the variable, the initial cable lengths and platform rotation are determined as well. As the normal vector of the plane of the cable robot is known, it is sufficient to only optimize the components of the position of the first winch and the components y and z of the remaining winches. The parameter vector for optimization results in

$$x_{opt} = \left[c_{0,1}, \ldots, c_{0,m}, w_{1,x}, w_{1,y}, w_{1,z}, w_{2,y}, w_{2,z}, \ldots, w_{m,y}, w_{m,z}\right] . \tag{29}$$

The residuals used are obtained from the inverse kinematics of the cable robot given by Eq. (7). They are given by

$$r_{j,i} = \varphi^{cr}_{inv,j,i} - (c_{0,i} + \Delta c_{j,i}) \tag{30}$$

for all measurement positions j and cables i. A non-linear least squares method is used as regression algorithm, which is implemented as `lsqnonlin` in Matlab.

4 Results

In the following section, the calibration procedure presented is evaluated. For **R** and **Q** of all filters used, the standard deviation of the measurement noise σ_v is determined by measurement in steady state. Then **R** can be derived directly from this, while **Q** is chosen in the order of magnitude between σ_v and **R**.

4.1 Simulation

At the initial position, EKF 1 is used to determine the initial laser beam path lengths. The convergence of $l_{0,i}$ and the normal vector $\mathbf{n}$ can be seen in Fig. 4.

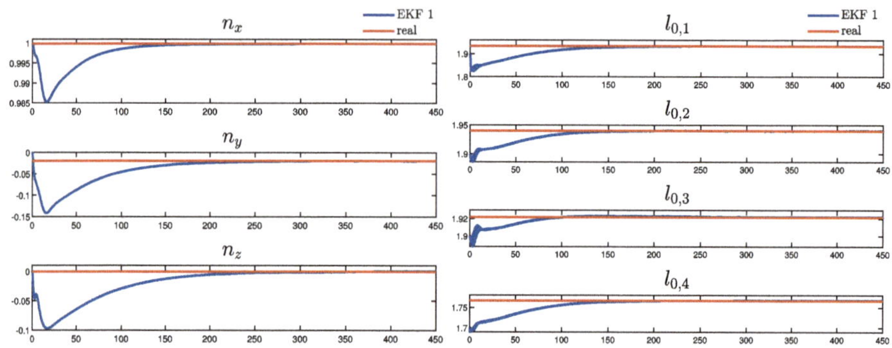

(a) normal vector $\mathbf{n}$ (b) inital laser beam path lenghts $l_{0,i}$

Fig. 4. Simulation results EKF 1

Then, the platform can start moving to dedicated measurement positions. For this purpose a grid of 16 points was chosen. The platform pauses for 20 s at every point to assure convergence. The path describing the reflector movement is shown in Fig. 5(a). The red markers depict the average TCP position estimated by EKF 3 during the 20 s measurement. The results from EKF 3 can be seen in Fig. 5(b). The part of the error in θ that is linear with time results from a slightly incorrect estimation of ω in EKF 1 and can easily be eliminated before optimization.

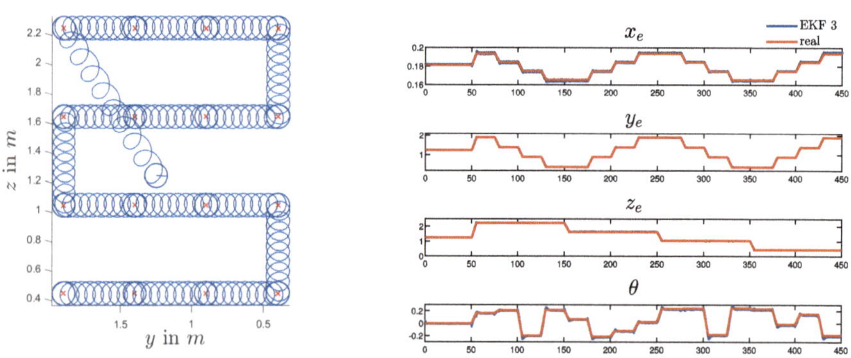

(a) Trajectory of reflector (b) EKF 3

Fig. 5. Simulation results EKF 2 + 3

To reduce the error due to slightly oscillating values, the mean value of the measurements in every measurement pose is calculated. The optimized winch positions compared to the real positions are presented in Table 1.

Table 1. Optimized winch positions vs. simulation in m

w_i	Optimization	Simulation
w_1	$[0.1690\ 0.1207\ 0.2576]^T$	$[0.1600\ 0.1200\ 0.2550]^T$
w_2	$[0.1681\ 0.1186\ 2.4403]^T$	$[0.1600\ 0.1200\ 2.4400]^T$
w_3	$[0.2091\ 2.2220\ 2.4420]^T$	$[0.2000\ 2.2230\ 2.4400]^T$
w_4	$[0.2100\ 2.2236\ 0.2543]^T$	$[0.2000\ 2.2230\ 0.2550]^T$

4.2 Experiment

In the experiment, the same procedure is used. The platform is moved using simplified kinematics and roughly measured winch positions. Only the upper two cables are used to move to positions while the lower cables are just kept under tension. Looking at the first filter in Fig. 6 it can be seen, that it converges like in simulation. The path described by the reflector is depicted in Fig. 7(a).

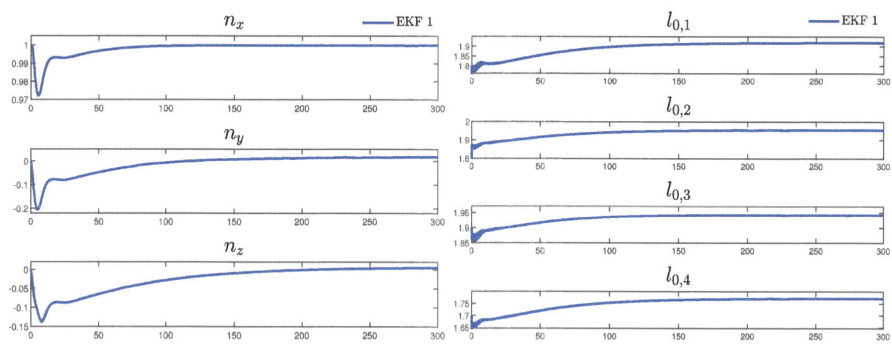

(a) normal vector $\mathbf{n}$ (b) inital laser beam path lenghts $l_{0,i}$

Fig. 6. Experimental results EKF 1

The platform is oscillating out of its plane and the revolving reflector leads to a wobbling movement of the platform. This is noticeable in the results of EKF 3 in Fig. 7(b). The estimation of x_e is quite noisy. Furthermore, θ drifts and no clear plateaus are recognisable. In this case, it is difficult to eliminate the linear error. The oscillation of the platform might cause a slightly incorrect estimation of the initial values in EKF 1, which leads to a worsening of the result of EKF 3. At

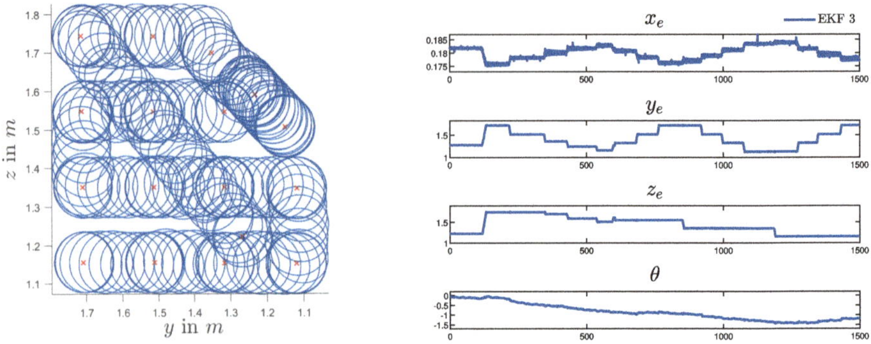

(a) Trajectory of reflector (b) EKF 3

Fig. 7. Experimental results EKF 2 + 3

least the noise of signal x_e can be reduced for optimization by taking the mean value. So the optimization results in the winch positions given in Table 2. For comparison, the roughly measured winch positions used for moving the platform are given as well.

Table 2. Optimized winch positions - experiment in m

w_i	Optimization	Measured
w_1	$[0.2100\ 0.0308\ 0.1653]^T$	$[0.1600\ 0.1200\ 0.2550]^T$
w_2	$[0.1962\ 0.1833\ 2.4482]^T$	$[0.1600\ 0.1200\ 2.4400]^T$
w_3	$[0.1621\ 2.1890\ 2.4447]^T$	$[0.1600\ 2.2230\ 2.4400]^T$
w_4	$[0.1722\ 2.2236\ 0.2597]^T$	$[0.1600\ 2.2230\ 0.2550]^T$

5 Conclusion

The presented paper introduced a novel platform extension for cable robots. The resulting active end effector makes it possible to determine the pose of the platform and the position of the winches with the help of a laser tracker. Results were shown in simulation and an experiment. In the experiment, deficits were identified concerning the convergence of the orientation of the platform and noisy state estimation. Furthermore, it turned out that the platform oscillates due to the rotation of the reflector. This influence can be reduced by improved construction or higher cable tensions.

The further mechanical and methodical improvement of the presented concept is one of the goals for the future. It is conceivable that the convergence will be improved by using a different filter concept. Besides, a validation by an external, absolute measuring laser tracker is planned for the future.

References

1. Borgstrom, P.H., Jordan, B.L., Borgstrom, B.J., Stealey, M.J., Sukhatme, G.S., Batalin, M.A., Kaiser, W.J.: Nims-pl: a cable-driven robot with self-calibration capabilities. IEEE Trans. Robot. **25**(5), 1005–1015 (2009). https://doi.org/10.1109/TRO.2009.2024792
2. dit Sandretto, J.A., Daney, D., Gouttefarde, M.: Calibration of a fully-constrained parallel cable-driven robot. In: Padois, V., Bidaud, P., Khatib, O. (eds.) Romansy 19 - Robot Design, Dynamics and Control, CISM International Centre for Mechanical Sciences, vol. 544, pp. 77–84. Springer, Vienna (2013). https://doi.org/10.1007/978-3-7091-1379-0_10
3. Hamann, M.: Calibration procedure for a geometrically reconfigurable 3-dof cable-driven parallel robot. In: Modelling, Simulation and Identification (2018). https://doi.org/10.2316/P.2018.857-013
4. Hamann, M., Nüsse, P.M., Ament, C.: Towards a precise cable-driven parallel robot - a model-driven parameter identification enhanced by data-driven position correction. In: Pott, A., Bruckmann, T. (eds.) Cable-Driven Parallel Robots, Mechanisms and Machine Science, vol. 74, pp. 367–376. Springer, [S.l.] (2019). https://doi.org/10.1007/978-3-030-20751-9_31
5. Jin, X., Jung, J., Ko, S.Y., Choi, E., Park, J.O., Kim, C.S.: Geometric parameter calibration for a cable-driven parallel robot based on a single one-dimensional laser distance sensor measurement and experimental modeling. Sensors (Basel, Switzerland) **18**(7), 2392 (2018). https://doi.org/10.3390/s18072392
6. Jin, X., Jung, J., Park, S., Park, J.O., Ko, S.Y.: Geometric parameter calibration using a low cost laser distance sensor for a planar cable robot: matlab simulation. In: 2016 13th International Conference on Ubiquitous Robots and Ambient Intelligence (URAI), pp. 534–537. IEEE, Piscataway, NJ (2016). https://doi.org/10.1109/URAI.2016.7625772
7. Lau, D.: Initial length and pose calibration for cable-driven parallel robots with relative length feedback. In: Gosselin, C., Cardou, P., Bruckmann, T., Pott, A. (eds.) Cable-Driven Parallel Robots, Mechanisms and Machine Science, vol. 53, pp. 140–151. Springer International Publishing, Cham (2018). https://doi.org/10.1007/978-3-319-61431-1_13
8. Merlet, J.P., Daney, D.: A portable, modular parallel wire crane for rescue operations. In: IEEE International Conference on Robotics and Automation (ICRA), 2010, pp. 2834–2839. IEEE, Piscataway, NJ (2010). https://doi.org/10.1109/ROBOT.2010.5509299
9. Simon, D.: Optimal state estimation: Kalman, H [infinity], and nonlinear approaches. Wiley-Interscience, Hoboken, NJ (2006). https://doi.org/10.1002/0470045345, https://site.ebrary.com/lib/alltitles/docDetail.action?docID=10308111

Accuracy Improvement for CDPRs Based on Direct Cable Length Measurement Sensors

Christoph Martin[1(✉)], Marc Fabritius[1], Johannes T. Stoll[1], and Andreas Pott[2]

[1] Robot and Assistive Systems Department, Fraunhofer Institute for Manufacturing Engineering and Automation IPA, 70569 Stuttgart, Germany
christoph.martin@ipa.fraunhofer.de

[2] Institute for Control Engineering of Machine Tools and Manufacturing Units (ISW) University of Stuttgart, 70049 Stuttgart, Germany

Abstract. In many applications of cable-driven parallel robots (CDPRs), accuracy is an important requirement. The accuracy of CDPRs with a controller based only on the standard kinematic model is limited because of effects like cable elongation. Carrying payload on the platform, i.e. applying an external wrench, increases the influence of this effect. To address this problem, we present a cable length correction method based on direct cable length measurement sensors (DCLM-Sensors). With this method, effects like cable elongation can be compensated. In experiments, the position accuracy of the cable robot IPAnema 3 could be improved by 61.49% without additional payload, and 86.31% with additional payload. We present the integration of the sensor feedback in the cable robot controller and the results of an experimental evaluation on the cable robot IPAnema 3.

Keywords: Cable-driven parallel robot · CDPR · Direct cable length measurement · DCLM-Sensor · Laser sensor · Accuracy · Elongation · Creep

1 Introduction

Cable-driven parallel robots (CDPRs) are a class of parallel robots which use cables to manipulate a mobile platform relative to a fixed frame. The use of cables brings several advantages, such as a large workspace, scalability, a good payload-to-weight ratio, and high dynamics.

Many applications of CDPRs that exploit these properties like 3D-printing or pick-and-place also require a high robot accuracy. Satisfying this requirement can be critical for the suitability of CDPRs [1]. Therefore, improving the accuracy of CDPRs is an important and active field of CDPR research [7,16]. The accuracy of CDPRs is influenced by various factors such as:

M. Gouttefarde et al. (Eds.): CableCon 2021, MMS 104, pp. 348–359, 2021.
https://doi.org/10.1007/978-3-030-75789-2_28

- The CDPR model of the controller and the effects which it considers or neglects (e.g. cable elasticity, cable sagging due to the cable's weight, pulley kinematic)
- The accuracy of the calibration of the parameters of the CDPR model
- Cable properties (e.g. elasticity, creep, hysteresis, wear)
- The behavior of the cables on the winches (e.g. cable flattening, coiling errors, non-linear cable length-to-drive ratio)
- The mechanical properties of the remaining CDPR components: frame, platform, pulleys, drive trains (e.g. manufacturing accuracy, mechanical play, compliance, friction)

A more detailed list can be found in [14, page 26 ff].

Many CDPR controllers in practice are based on simple CDPR models which are able to run in real-time and only require few parameters to be calibrated. They move the platform by translating a desired platform pose to cable lengths through an inverse kinematic code. These lengths are then set with the motor encoders in the winches. A common assumption is that the cable lengths are a linear function of the encoder values. For CDPRs that use coiling winches like the IPAnema [12] or CoGiRo [4], the encoders measure the revolutions of the winch drums. In CDPRs with linear actuators like the MARIONET-VR [7] or IPAnema 2 planar [10, page 357], the encoders measure the linear displacement. Such CDPR controllers neglect the effects listed above. Among these effects, the elongation of the cables due to elasticity plays a major role for the accuracy of CDPRs, especially for CDPRs with plastic fiber cables [16].

In the literature, the following approaches can be found to combat the negative impact of cable elongation. The issue of cable creep which changes the relationship between home pose and cable lengths over time can be mitigated through frequent recalibration of the home pose. Calibration methods for CDPRs can be found in [13] and an automated one in [8]. Both methods take measurements at different poses in the workspace, which may not be possible depending on the application. Sandretto et al. use measurements from a laser tracker [13], while Miermeister uses the internal position and cable force sensors [8]. Another approach is to use a more complex cable model that can account for creep [9]. The application of such models is often hindered by their dependency on parameters that are difficult to measure and may change due to environmental influences.

Focussing on elasticity, in [16], a simple elongation model is used to increase the accuracy of CDPRs. This results in a decrease of the average positioning error by almost 40%. Hereby, it is assumed that the external wrench acting on the platform is known and constant. Any deviation in this parameter, such as payload, decreases the accuracy of this approach.

Another possibility is to employ feedback control using external sensors. In [1], multiple cameras are used to determine the pose of the platform and the cable directions at the pulleys. This requires that their view on the platform must not be obstructed. Fortin-Côté et al. use additional information from angular sensors measuring the angle of the cables at the pulleys of a suspended cable robot with two cables to increase the accuracy in [3]. Merlet proposes two approaches to directly measure the cable length in [6,7]. Both are using markers attached to the cable in known intervals. Measuring the actual distance between the markers, the current cable length can be determined. In [7], the idea is to use small strips of magnetic tape or colored markers which can be detected by a hall sensor or IR optical sensor, respectively. In [6], the method based on colored markers is further investigated. The resolution of this setup depends on the distance of the color sensors and markers, the number of the color sensors and the cable length. With this measurement method, a resolution of 0.1 m is achieved for a cable of 60 m length with 84 markers and 15 color sensors in the given examples.

This paper presents a new correction method for CDPRs based on direct cable length measurement sensors (DCLM-Sensors) attached to each cable. With this method, it is possible to compensate inaccuracies due to cable creep, elasticity, coiling errors, flattening, and non-linear cable length-to-drive ratio. Errors due to other effects like inaccuracies in the calibrated parameters of the CDPR model remain unaffected.

The mechanical design of the DCLM-Sensors used in this paper is presented in [5]. Furthermore, an experimental evaluation of the DCLM-Sensors, comparing the cable lengths measured with the DCLM-Sensors and the cable lengths set with the motor encoders on the IPAnema 3, can be found in [5].

The structure of this paper is as follows: In Sect. 2, the standard kinematic model of a CDPR and the kinematic of the DCLM-Sensor is defined. In Sect. 3, the standard CDPR controller and the implementation of the integration of the DCLM-Sensor feedback are explained. Section 4 presents the results of the experimental evaluation, after which the paper closes with the conclusion.

2 Kinematic

In the following, the standard kinematic model of a CDPR and the kinematic of the DCLM-Sensor are introduced. Vectors and matrices are labeled in bold.

2.1 Standard Kinematic Model of a CDPR

The standard kinematic model of a CDPR is visualized in Fig. 1. Two coordinate systems are used. The global coordinate system $\mathcal{K}_0$ which is fixed to the cable robot frame and the moving coordinate system $\mathcal{K}_\mathrm{p}$ which is attached to the moving platform of the CDPR. The CDPR has m cables. The vectors $\mathbf{a}_i \in \mathrm{SE}_3$, $i \in \{1, ..., m\}$ describe the coordinates of the proximal anchor points A_i with respect to the coordinate system $\mathcal{K}_0$, where the cables are connected to the cable robot frame. The vectors $\mathbf{b}_i \in \mathrm{SE}_3$, $i \in \{1, ..., m\}$ describe the distal anchor points B_i of the cables with respect to the coordinate system $\mathcal{K}_\mathrm{p}$, where the cables are connected to the cable robot platform. The pose of the platform $\mathcal{K}_\mathrm{p}$ with respect to $\mathcal{K}_0$ is described by the position vector $\mathbf{r} \in \mathrm{SE}_3$ and the rotation matrix $\mathbf{R} \in \mathrm{SO}_3$. The cables are assumed to be straight lines. Thus, the vectors $\mathbf{l}_i \in \mathrm{SE}_3$, $i \in \{1, ..., m\}$, describing the cables, are defined as follows:

$$\mathbf{l}_i = \mathbf{a}_i - \mathbf{r} - \mathbf{R}\mathbf{b}_i \quad \text{for} \quad i \in \{1, ..., m\} \tag{1}$$

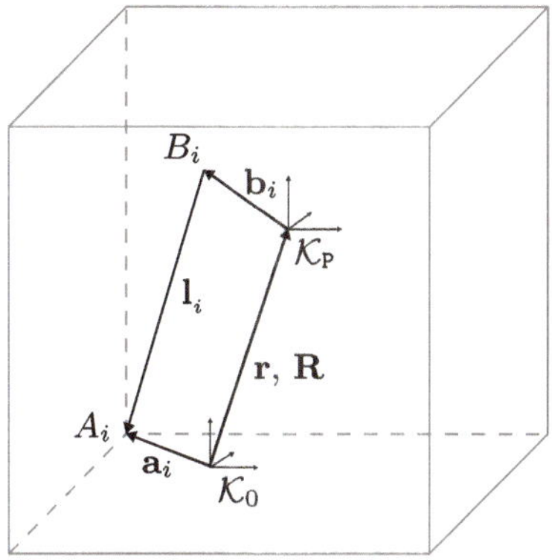

Fig. 1. Kinematic model of a CDPR (modified from [2])

Furthermore, at the proximal anchor points A_i the cables are redirected by pulleys. To take the pulleys into account, the standard kinematic model is developed further in [11]. In this paper, the inverse kinematic (IK) is based on the inverse pulley kinematic model from [11].

2.2 Kinematic of the DCLM-Sensor

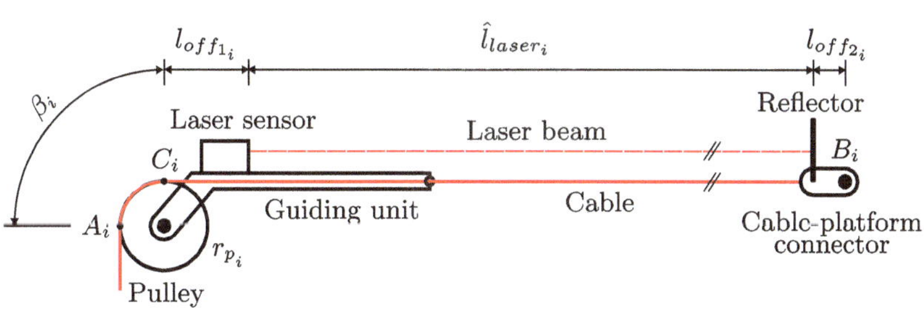

Fig. 2. Schematic sketch of a DCLM-Sensor (modified from [5])

The kinematic of the DCLM-Sensor is explained based on Fig. 2. The aim is to determine the cable length from the proximal anchor point A_i to the distal anchor point B_i. The third point C_i is defined as the point where the cable leaves the pulley tangentially. The cable length wound up on the pulley l_{AC_i} is equal to the angle β_i times the pulley radius r_{p_i}. The angle β_i is obtained from the IK and depends on the commanded platform pose. For the straight part of the cable between C_i and B_i, the laser beam is assumed to be aligned parallel to the cable. The straight length is composed of the two offsets l_{off1_i}, l_{off2_i} and the measurement of the laser distance sensor $\hat{l}_{laser_i}$. The offsets are set in the home pose as constant values. As a result, we receive the following equation for the cable length $\hat{l}_i$:

$$\hat{l}_i = \beta_i r_{p_i} + l_{off1_i} + \hat{l}_{laser_i} + l_{off2_i} \quad (2)$$

3 Implementation

The control scheme of the implementation of the cable length correction controller with the DCLM-Sensors is visualized in Fig. 3. The controller is implemented on the IPAnema 3, a redundantly constrained cable robot with eight cables. Each cable of the CDPR is equipped with a DCLM-Sensor. The controller is implemented in TwinCAT 3 from BECKHOFF except for the forward kinematic (FK). The FK is implemented in WiPy[1], an open-source CDPR simulation library developed at Fraunhofer IPA. The algorithm of the FK is based on [15]. It is a proof of concept implementation for static poses to show that the correction method works.

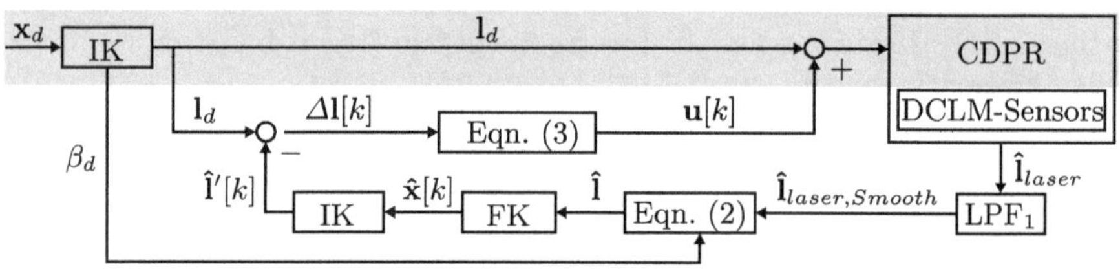

Fig. 3. Control scheme of the DCLM correction controller

3.1 Standard CDPR Controller

The standard CDPR controller without any correction is shown in the grey-shaded area in Fig. 3 and works as follows:

The commanded desired pose $\mathbf{x}_d$ is given to the IK. The output of the IK, the cable length $\mathbf{l}_d$, are commanded to the drive amplifiers of the CDPR. With the standard controller, there is no feedback whether the real cable lengths and the commanded cable lengths are the same.

3.2 DCLM Correction Controller

The approach in this paper augments the standard controller with the cable length feedback of the DCLM-Sensors. The controller is developed for the static case, where the pose $\mathbf{x}_d$ does not change. To get the feedback cable lengths from the DCLM-Sensors, the measured lengths of the DCLM-Sensors $\hat{\mathbf{l}}_{laser}$ are first filtered with a first-order low pass filter LPF_1 for smoothing. With the filtered measurements $\hat{\mathbf{l}}_{laser,Smooth}$, the cable lengths $\hat{\mathbf{l}}$ are calculated according to Eq. 2.

[1] https://gitlab.cc-asp.fraunhofer.de/wek/wirex.git.

The DCLM-Sensor measurements $\hat{\mathbf{l}}_{laser}$ are updated with the laser sensor output rate of 2.5 ms (see Table 2). The values $\hat{\mathbf{l}}_{laser,Smooth}$ and $\hat{\mathbf{l}}$ are calculated each millisecond. The cable lengths based on the DCLM-Sensor measurements $\hat{\mathbf{l}}$ are given to the FK as an input, which results in the pose $\hat{\mathbf{x}}[k]$. Since the FK is implemented in WiPy, the execution is not real-time capable. Therefore, the index $[k]$ is introduced to denote each execution cycle. With the pose $\hat{\mathbf{x}}[k]$ and the IK, we receive the feedback cable lengths $\hat{\mathbf{l}}'[k]$. To get the control deviations $\Delta\mathbf{l}[k]$, the feedback cable lengths $\hat{\mathbf{l}}'[k]$ are subtracted from the desired cable lengths $\mathbf{l}_d$. The desired cable lengths $\mathbf{l}_d$ and also the angles $\boldsymbol{\beta}_d$ are received from the IK with the commanded desired pose $\mathbf{x}_d$ as input. Since the pose is static, $\mathbf{l}_d$, $\boldsymbol{\beta}_d$, and $\mathbf{x}_d$ do not change. The cable lengths have to be corrected about the cable length deviations $\Delta\mathbf{l}[k]$. Therefore, the deviations are fed to an integration block. The integration block can be described as follows:

$$\begin{aligned} \mathbf{u}[k] &= \mathbf{u}[k-1] + K_I \cdot \Delta\mathbf{l}[k] \\ K_I &= 0.4 \text{ (heuristically chosen value)} \\ k &\text{ ... incremented about 1 each 1.7 s, WiPy processing time} \end{aligned} \tag{3}$$

If new correction values $\Delta\mathbf{l}[k]$ are available, which means k is incremented by 1, it is multiplied with the factor K_I and smoothly added to the output value of the last cycle $\mathbf{u}[k-1]$ within 1 s. Finally, the output $\mathbf{u}[k]$ of the integration block is added to the cable lengths $\mathbf{l}_d$ to correct the cable length deviations.

In a first approach, the cable length corrections were directly applied to their related cable. By using DCLM-Sensors to measure and control the cable lengths, the cables can be regarded as non-elastic connections. Due to errors in the calibration of the cable robot (positions of A_i and B_i), the mathematical model has errors compared to the real cable robot. Furthermore, there are errors in the measurements of the laser distance sensors. Since the CDPR is redundantly constrained, these errors can lead to set points in the joint space that do not correspond to any platform pose for non-elastic cables. Thus, the first approach could not reach a stationary state. The remedy is to ensure that all points in the joint space correspond to a platform pose. This is done by concatenating an FK and an IK block. The FK is computed by a least-square optimization, which can be interpreted as using pretensioned springs instead of cables minimizing their potential energy [15]. The cable lengths $\hat{\mathbf{l}}'[k]$, which are the output of the IK block, are guaranteed to have a corresponding platform pose under the assumption of non-elastic cables.

4 Experimental Evaluation

The DCLM correction controller is evaluated on the cable robot demonstrator IPAnema 3 at the Fraunhofer IPA. Its geometrical parameters are listed in Table 1. As laser distance sensors for the DCLM-Sensors, DL50-P2228 from SICK[2] are used. The technical specification of these laser distance sensors can be found in Table 2. Figure 4 shows the overall experimental setup.

[2] https://www.sick.com/de/en/p/p346664 [Accessed: 04-February-2021].

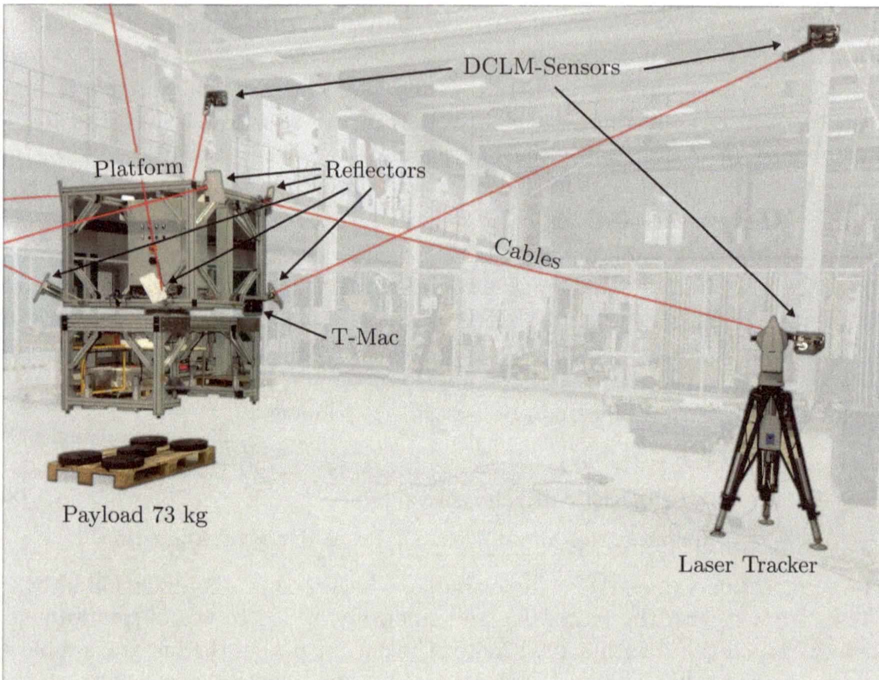

Fig. 4. Experimental evaluation on the IPAnema 3

Table 1. Geometrical parameters of the cable robot IPAnema 3

Cable index i	Proximal anchor points $\mathbf{a}_i$ [m]	Distal anchor points $\mathbf{b}_i$ [m]	Pulley radii r_{P_i} [m]
1	$[7.6029, 5.5952, 3.5665]^T$	$[0.5488, 0.46, -0.48]^T$	0.057
2	$[7.5238, -5.5789, 3.5745]^T$	$[0.5488, -0.46, -0.48]^T$	0.057
3	$[-7.4614, -5.5621, 3.7536]^T$	$[-0.5488, -0.46, -0.48]^T$	0.057
4	$[-7.3114, 5.6113, 3.7408]^T$	$[-0.5488, 0.46, -0.48]^T$	0.057
5	$[7.6126, 5.5812, -0.7817]^T$	$[0.36, 0.7488, 0.48]^T$	0.057
6	$[7.5427, -5.5851, -0.8525]^T$	$[0.36, -0.7488, 0.48]^T$	0.057
7	$[-7.7641, -5.5694, -0.9384]^T$	$[-0.36, -0.7488, 0.48]^T$	0.057
8	$[-7.6959, 5.6008, -0.9036]^T$	$[-0.36, 0.7488, 0.48]^T$	0.057

In the experiments, the accuracy of the cable robot with the standard CDPR controller based on the IK is compared to the accuracy of the DCLM correction controller (see Sect. 3.2). For the accuracy measurement, the exact pose of the platform of the CDPR is measured with a Leica laser tracker AT960[3] with a

[3] https://www.hexagonmi.com/de-de/products/laser-tracker-systems/leica-absolute-tracker-at960 [Accessed: 04-February-2021].

Table 2. Technical specification of the SICK laser distance sensor DL50-P2228

Measuring range	200 – 50 000 mm (with reflector foil)
Typ. repeatability (1 σ)	0.25 mm (moving average set to slow)
Accuracy	Distance $\leq$ 4 m: $\pm$ 5 mm; Distance > 4 m: $\pm$ 3 mm
Output rate	2.5 ms

T-Mac attached to the platform. With the T-Mac, all six degrees of freedom can be measured. The measurement accuracy is specified in Table 3.

Table 3. Technical specification of the Leica AT960

Position accuracy	$\pm$ (15 µm + 6 µm/m)
Typical rotation accuracy	$\pm$ 0.01°

For the evaluation of the DCLM correction controller, a grid of 25 static poses with constant orientation of the CDPR-platform in an x-y-plane with a constant z-coordinate (z = 0 m) is chosen. The grid is visualized in Fig. 5 with blue dots. The proximal anchor points $A_1, ..., A_8$ of the cable robot are displayed with orange dots. The evaluated grid is quite small compared to the cable robot frame but due to the setup of the cable robot with the heavy end-effector the workspace is limited. The grid is evaluated twice in the order following the red arrow. The first time directly after recalibration of the home pose and without payload. The second time, a payload of 73 kg is added. Each run takes around 40 min. At each pose, first the accuracy of the IK is measured. Afterwards, the DCLM correction controller is enabled and after a waiting time of 25 s to

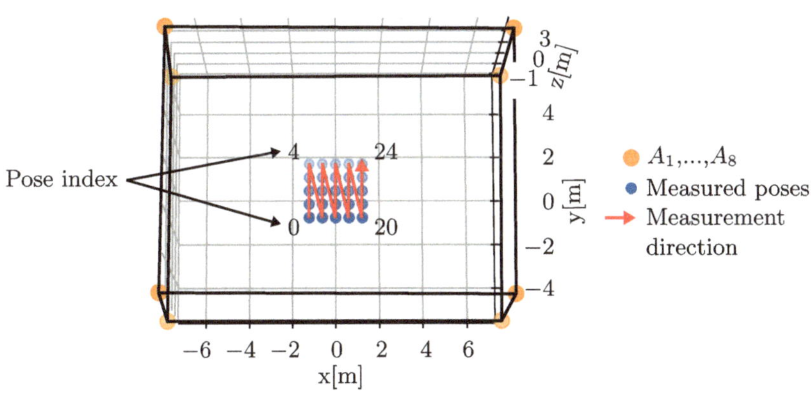

Fig. 5. Measured poses to evaluate the DCLM correction controller [5]

let the controller reach a stationary state, the accuracy of the DCLM correction controller is measured. Each measurement is taken and averaged over one second.

4.1 Accuracy Results of the DCLM Correction Controller

The results of the evaluation are shown in Fig. 6 and Fig. 7. Heatmaps are chosen to visualize the error. The evaluated poses are marked with black dots, the color is a measure for the error. The relation between color and error value is given by the color bar on the right for each line, respectively. Between the measured poses, bilinear interpolation is chosen. The heatmaps in the left column show the results of the error with the standard controller based on the IK, the ones in the right column the error with the DCLM correction controller. Figure 6a to Fig. 6d show the position error, Fig. 7a to Fig. 7d the orientation error. The displayed position error is calculated by the 2-norm between the commanded position and the measured position with the laser tracker. The orientation error is calculated by the angle of the rotating phasor representation.

As can be seen from the graphs, the mean position and orientation accuracy can be significantly improved with the DCLM correction controller. Without payload, the mean position accuracy is improved by 61.49%, the mean orientation accuracy by 62.5%. For the experiments with a payload of 73 kg, the position accuracy improvement accounts to 86.31%, the orientation accuracy increases about 61.9%. The absolute numbers of the experimental results can be found in Table 4.

Table 4. Results of the experimental evaluation

	Position error [mm]				Orientation error [°]			
	Mean	Min.	Max.	Std. Dev.	Mean	Min.	Max.	Std. Dev.
IK	3.09	1.43	6.57	1.3314	0.16	0.04	0.31	0.0661
DCLM-Sensor	1.19	0.42	2.60	0.6064	0.06	0.02	0.10	0.0217
IK, 73 kg	13.51	8.58	17.00	1.8928	0.21	0.03	0.35	0.0807
DCLM-Sensor, 73 kg	1.85	0.79	3.77	0.7601	0.08	0.04	0.13	0.0250

The orientation improvement of the experiments with payload is in the same magnitude as without payload. Regarding the position accuracy, the improvement for the experiments with payload is much higher compared to the experiments without payload. This is due to the high position deviation of the standard CDPR controller with payload caused by the elasticity of the cables. Comparing the mean position accuracy of the DCLM correction controller with and without payload, the accuracy with payload is about 0.66 mm worse. The assumption for this deterioration is that the cable force distribution changed due to the payload. The level of the forces of the lower cables decreases. Due to cable sagging, the laser does not measure exactly parallel to the cable which leads to cable length measurement errors.

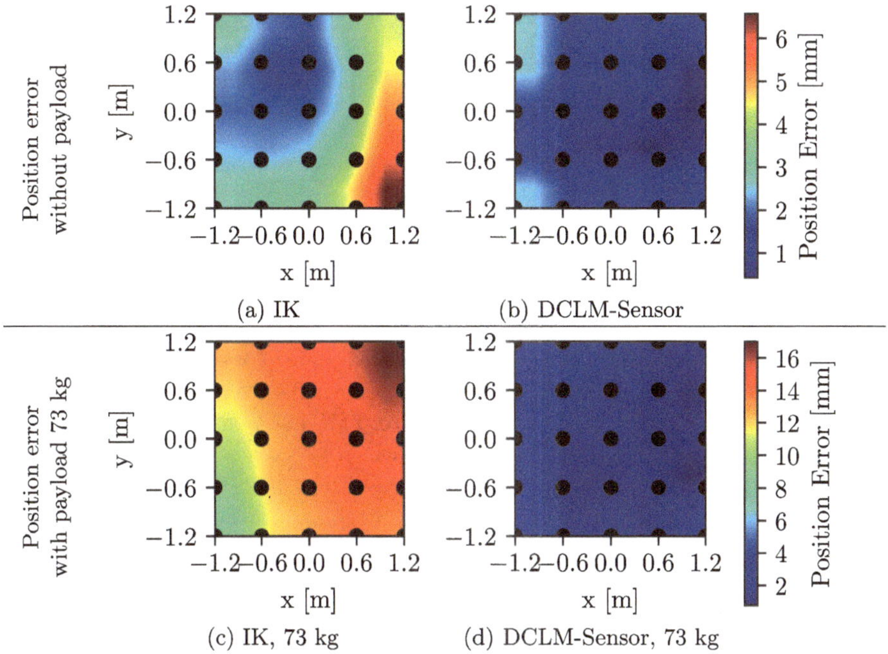

Fig. 6. Position accuracy of IK versus DCLM-Sensor, x-y-plane (z = 0 m)

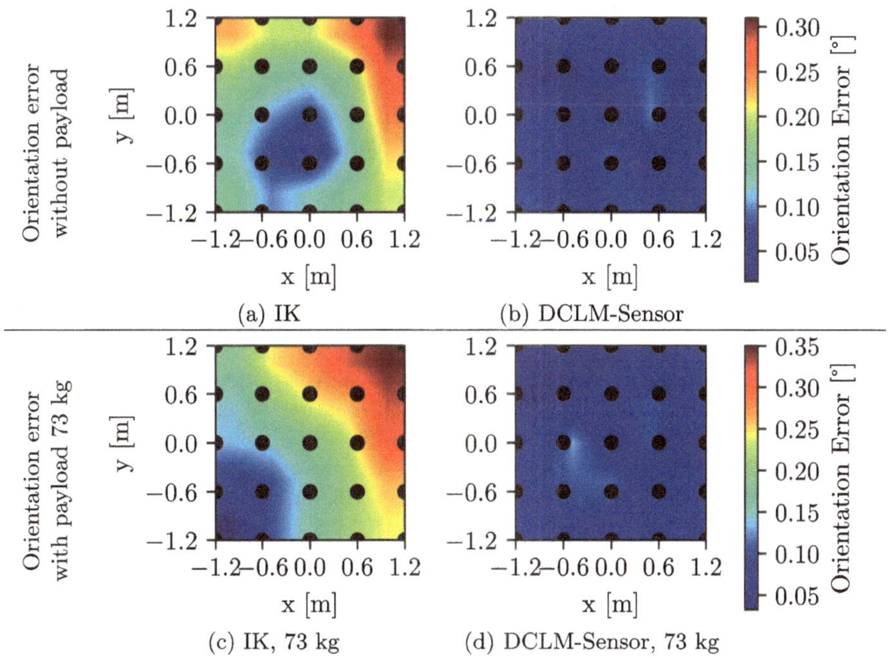

Fig. 7. Orientation accuracy of IK versus DCLM-Sensor, x-y-plane (z = 0 m)

Furthermore, in Fig. 6c and Fig. 7c the cable creep effect due to the payload can be seen. The accuracy becomes worse the longer the experiment takes. Both effects, elasticity and creep, can be compensated with the DCLM correction controller.

5 Conclusion

In this paper, we present a new cable length correction method based on a laser sensor system for direct cable length measurement to increase the accuracy of CDPRs. Thus, it is possible to compensate effects on the accuracy, such as cable elongation due to elasticity or creep. An experimental comparison of the accuracy between a standard CDPR controller based on the inverse pulley kinematic model [11] and a controller with feedback from DCLM-Sensors shows a significant improvement of the cable robot accuracy.

In future research, we plan to evaluate the direct cable length measurement correction controller with another cable robot platform and thus in a bigger workspace. Also, different orientations are planned to be evaluated. Furthermore, we are working towards a control concept to integrate the feedback of the DCLM-Sensors in a real-time capable way, since the presented controller is not real-time capable. Another possible application of the direct cable length measurement system is the auto calibration of the home pose of CDPRs.

References

1. Dallej, T., et al.: Modeling and vision-based control of large-dimension cable-driven parallel robots using a multiple-camera setup. Mechatronics **61**, 20–36 (2019)
2. Fabritius, M., Martin, C., Pott, A.: Calculation of the collision-free printing workspace for fully-constrained cable-driven parallel robots. In: ASME 2018 International Design Engineering Technical Conferences and Computers and Information in Engineering Conference. American Society of Mechanical Engineers Digital Collection (2018)
3. Fortin-Côté, A., Cardou, P., Campeau-Lecours, A.: Improving cable driven parallel robot accuracy through angular position sensors. In: 2016 IEEE/RSJ International Conference on Intelligent Robots and Systems (IROS), pp. 4350–4355. IEEE (2016)
4. Lamaury, J., Gouttefarde, M.: Control of a large redundantly actuated cable-suspended parallel robot. In: 2013 IEEE International Conference on Robotics and Automation, pp. 4659–4664. IEEE (2013)
5. Martin, C., et al.: A laser-based direct cable length measurement sensor for CDPRs. In: Robotics 10.2 (2021). ISSN: 2218-6581. https://doi.org/10.3390/robotics10020060. https://www.mdpi.com/2218-6581/10/2/60
6. Merlet, J.-P.: Improving cable length measurements for large CDPR using the vernier principle. In: Pott, A., Bruckmann, T. (ed.) Cable-Driven Parallel Robots, vol. 74, pp. 47–58. Springer (2019)
7. Merlet, J.-P.: MARIONET, a family of modular wire-driven parallel robots. In: Advances in Robot Kinematics (ARK), pp. 53–61. Springer (2010)

8. Miermeister, P., Pott, A.: Auto calibration method for cable-driven parallel robots using force sensors. In: Lenarcic, J., Husty, M. (ed.) Latest Advances in Robot Kinematics, pp. 269–276. Springer (2012)
9. Piao, J., et al.: A polymer cable creep modeling for a cable-driven parallel robot in a heavy payload application. In: Gosselin, C., et al. (ed.) Cable-Driven Parallel Robots, vol. 53, pp. 62–72. Springer (2018)
10. Pott, A.: Cable-Driven Parallel Robots: Theory and Application, vol. 120. Springer, Berlin (2018)
11. Pott, A.: Inuence of pulley kinematics on cable-driven parallel robots. In: Lenarcic, J., Husty, M. (ed.) Latest Advances in Robot Kinematics, pp. 197–204. Springer (2012)
12. Pott, A., et al.: IPAnema: a family of cable-driven parallel robots for industrial applications. In: Bruckmann, T., Pott, A. (ed.) Cable-Driven Parallel Robots, vol. 12, pp. 119–134. Springer (2013)
13. Dit Sandretto, J.A., et al.: Certifited calibration of a cable-driven robot using interval contractor programming. In: Thomas, F., Perez Gracia, A. (ed.) Computational Kinematics, vol. 15, pp. 209–217. Springer (2014)
14. Schmidt, V.: Modeling Techniques and Reliable Real-Time Imple- mentation of Kinematics for Cable-Driven Parallel Robots using Polymer Fiber Cables. PhD thesis. University of Stuttgart (2017)
15. Schmidt, V., Pott, A.: Implementing extended kinemat- ICS of a cable-driven parallel robot in real-time. In: Cable-Driven Parallel Robots, pp. 287–298. Springer, Berlin (2013)
16. Schmidt, V., Pott, A.: Increase of position accuracy for cable-driven parallel robots using a model for elongation of plastic fiber ropes. In: Wenger, P., Flores, P. (ed.) New Trends in Mechanism and Machine Science, vol. 43, pp. 335–343. Springer (2017)

Development of Safety Concepts for Cable-Driven Parallel Robots

David Lukas Winter(✉) and Christoph Ament

Faculty of Applied Computer Science, University of Augsburg,
86159 Augsburg, Germany
{david.winter,christoph.ament}@uni-a.de
https://www.uni-augsburg.de/de/fakultaet/fai/informatik/prof/rt/

Abstract. This work presents a set of safety concepts for cable-driven parallel robots (CDPR) based on the kinematic variables position, velocity and acceleration, as well as applicable forces. A strategy for shortening the stopping distance during emergency braking is also presented. The approaches are first described for a single motor with a holding brake and subsequently extended to monitor the robot workspace using multiple winches. This is demonstrated for a planar two DoF CDPR with four motor-driven winches as an experimental setup. It is shown that the end effector can be stopped before reaching defined limits, regardless of the current velocity.

Keywords: Safety systems · Collision prevention · Cable-driven parallel robot

1 Introduction

In robotics, safety means protecting the environment from the robot, while security means protecting the robot from malicious attacks [8]. Since safety cannot be guaranteed without security, these aspects go hand in hand. However, security is mainly achieved through physical and digital access restrictions. The cybersecurity required for this is a domain not unique to robotics and will not be considered further in this paper.

Due to the physical connection of robots with their environment, safety plays a particularly important role. Cable robots in particular offer some application advantages over ordinary robots, due to their easily scalable workspace and high dynamics. However, the advantages from an application perspective lead to increased risks from a safety perspective. As already shown in [16], the safety of cable robots is significantly affected by cable malfunctions, since the positioning of the end effector cannot be guaranteed. In the work of [6], the possible mechanical failures of CDPRs are also analyzed and a concept using auxiliary cables to maintain safety in case of cable breakage is presented. Furthermore, failure depended strategies to ensure safety are introduced. These strategies consist of

M. Gouttefarde et al. (Eds.): CableCon 2021, MMS 104, pp. 360–371, 2021.
https://doi.org/10.1007/978-3-030-75789-2_29

immobilizing the robot or moving toward a safe pose. Such recovery strategies to achieve a safe pose after a cable break have been analyzed in further works. In [9], a method is presented in which the tension in the cables is minimized during recovery. A recovery strategy consisting of multiple phases is presented in [2]. In the first phase after a cable break, the end effector is moved into the remaining working space. Following this, the end effector is brought to a standstill in the new working space using the remaining cables. An elliptical trajectory towards a safe pose is presented in [14]. In this case, the recovery time is minimized. Also, the strategy shown in [3] aims at moving the end effector to defined safe poses. Thereby, the remaining cables are kept taut at all times. A strategy developed in [4] consists of minimizing the kinetic energy of the end effector, while also considering a reasonable force distribution in the cables.

While in the previously mentioned literature a cable break is assumed, the following strategies assume a fully functional system. An approach considering cable forces to ensure the safety of humans when using a rehabilitation robot is presented in [7]. Other safety aspects to be considered when operating rehabilitation robots are presented in [5]. Accordingly, the end effector position and velocity, as well as cable forces or motor torques must stay within defined limits. This requires safety mechanisms that prevent the limits from being exceeded. However, these have not been considered in the previous reference. In the work of [12], velocity, acceleration and even jerk are limited in the trajectory planning. Furthermore, a stop trajectory with a feasibility check of the cable forces at the halt point is presented. However, the limitations are only considered at the planning level and deviations during movement due to external influences are neglected. The force that can actually be applied by the motors for braking is not taken into account. In the present work, safety mechanisms for the working range and velocity are presented for general cable robots and implemented for a planar cable robot. Thereby the concept of immobilization is used while considering the forces applicable for deceleration.

2 Safety Systems

For the safety systems now presented, motors with a holding brake are used. The AM-8131F motors from Beckhoff used in the experiments have a rated torque of 1.35 Nm and a peak torque of 6.07 Nm [1]. During operation, the maximum torque can be limited by software, which will be referred to as $M_{\max}$. With the given radius $r = 0.05$ m of the motor-driven winch, the motor torque and speed can be converted to forces and cable velocities. Moreover, the holding brake used in the safety concepts has an application delay time of 8 ms, which is taken into account in the following experiments. The holding torque of the mechanical brake is 2.0 Nm which leads to a maximum detainable weight of about 4 kg per cable. However, due to some electrical limitations, the maximum applicable motor torque is 0.8 Nm. This results in a maximum sustainable weight of approximately 1.6 kg. The real-time system used to control the motors in all experiments works with a cycle time of 0.0005 s.

2.1 Joint Space

In the following, a system similar to a lifting crane with a single winch and pulley is considered. The weight of the end effector is 0.15 kg and the motor torque is limited to $M_{\text{max}} = 0.5\,\text{Nm}$.

To avoid damage to the hardware, the limits of the cable lengths $L_{\text{crit,min}}$ and $L_{\text{crit,max}}$ that can be approached safely must be known. Assuming that the end effector does not swing transverse to the cable direction, these cable lengths determine the usable working space $\mathcal{W}_{\text{crit}}$ without causing damage. In addition, two smaller workspaces are defined. In the smallest workspace $\mathcal{W}$ (see Fig. 1), the target cable lengths may be located and can be approached without hesitation. The second area $\mathcal{W}_{\text{tol}}$ serves as a buffer zone in which the end effector may only be located in the event of a slight overshoot. When the buffer zone is exceeded, a triggered holding brake only engages after the motor has braked the load. For this reason, a stopping distance between $\mathcal{W}_{\text{crit}}$ and $\mathcal{W}_{\text{tol}}$ is necessary.

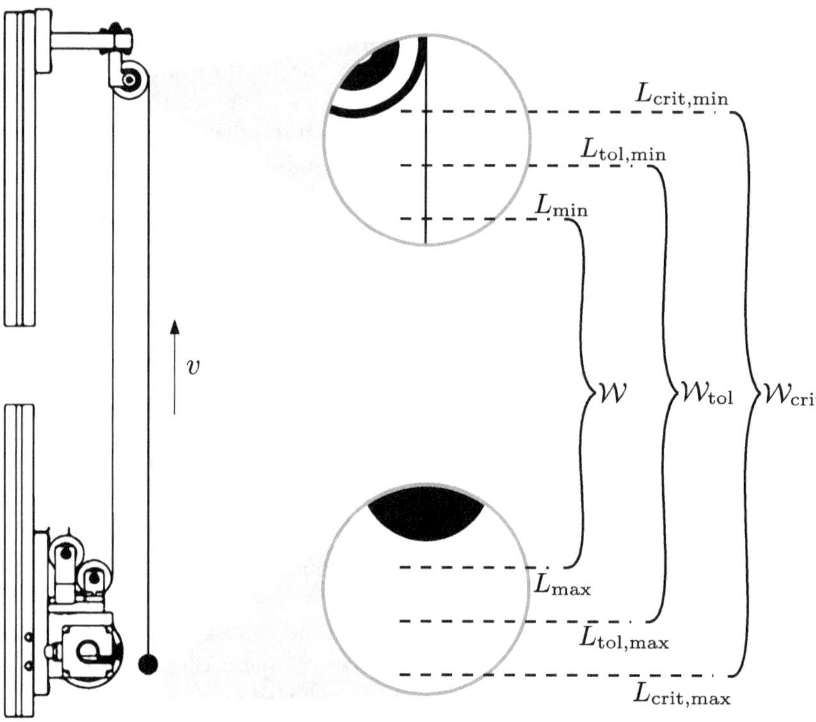

Fig. 1. Example for cable length limits resulting in the workspaces $\mathcal{W}, \mathcal{W}_{\text{tol}}$ and $\mathcal{W}_{\text{crit}}$.

With the current cable length L, the end effector is located within the allowed working range as long as the condition

$$L_{\text{tol,min}} \leq L \leq L_{\text{tol,max}} \tag{1}$$

is satisfied. Otherwise, the braking area is reached and the holding brake of the motor must be triggered. Before manually deactivating the automatic brake, the motor torque should be reduced and the end effector held by hand. After releasing the brake, the end effector can be moved manually back into the working area. To reduce the speed-dependent braking distance, the cable speed can be limited. Analogous to the position-based safety system, there are two limits for the speed. The target velocity v_{max} is considered as the highest specified speed for the motor controller and the trajectory planning. The brake needs to be triggered in case a defined maximum speed v_{brake} is exceeded. Therefore, the maximum speed must fulfill the condition

$$0 < v_{\text{max}} < v_{\text{brake}} \quad . \tag{2}$$

To prevent the safety system from an automatic release of the motor after a braking process has led to a stop, a state machine as in Fig. 2 is used. If the speed exceeds v_{brake}, the system switches to the state *emergency brake*. This state can only be left by a manual trigger after which the speed is monitored in the state *observe velocity* again.

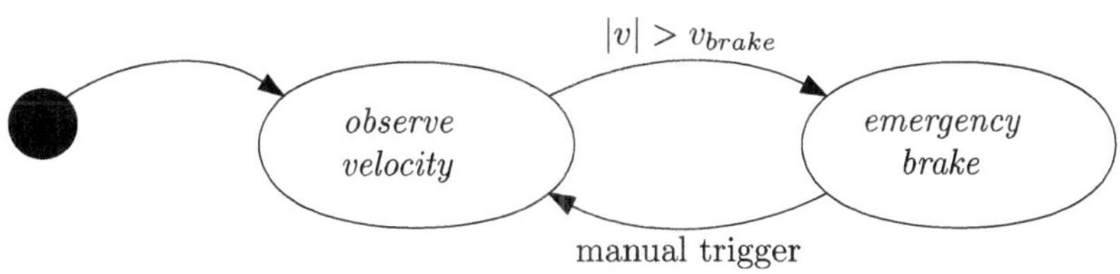

Fig. 2. State machine for velocity observation.

In the following, the behavior of the emergency brake when exceeding the maximum velocity of $v_{\text{brake}} = 0.4\,\text{ms}^{-1}$ is investigated. For this purpose, the motor torque i s et to zero, which causes the end effector to drop. In Fig. 3, the actual measured variables motor torque M, cable length L and velocity v are shown. When the speed limit is reached, the brake is triggered (see BREAK flag). After that, a delay of 0.5 s is visible until the motor begins to brake the load and velocity decreases. It takes a total of 530 ms until the holding brake is activated and the motor is no longer enabled (see ENABLED flag). Taking the actual braking distance into account, the delay results in an overshoot of $L_{\text{ovs}} = 33.9\,\text{cm}$ after the brake trigger position. The exceeding is not critical as long as the end effector is in the permitted working range during the entire deceleration process. However, if the distance covered in this 0.5 s delay is bigger than the

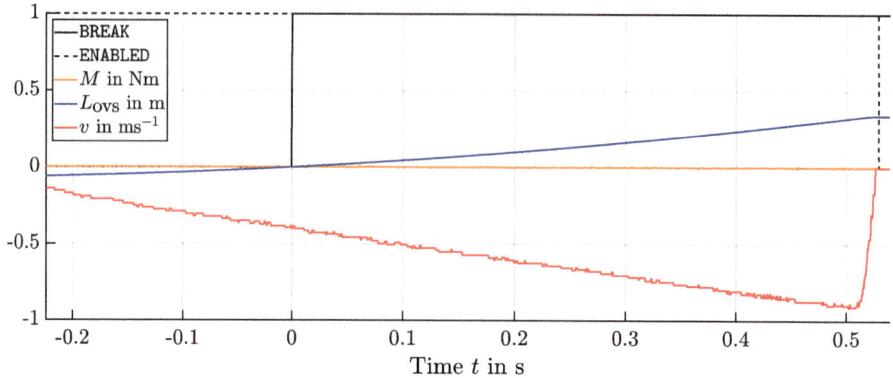

Fig. 3. Behavior of motor torque M, overshooting cable length L_{ovs} and velocity v with delay of 0.5 s until deceleration after exceeding the maximum speed.

remaining distance, a collision occurs before the motor has even started the actual deceleration process.

To ensure an earlier activation of the holding brake to minimize damage, the *emergency brake* state is extended to include active braking. Consequently, not only the holding brake of the motor is triggered, but also decelerated with the maximum torque set in form of a step contrary to the current movement:

$$M_{\text{set}} = -\text{sign}(v) \cdot M_{\text{max}}. \quad (3)$$

With $M_{\text{max}} = 0.5\,\text{Nm}$, as chosen in this experiment, it takes less than 0.05 s from $v_{\text{brake}} = 0.4\,\text{ms}^{-1}$ to standstill, while the overshoot is reduced to $L_{\text{ovs}} = 5.5\,\text{mm}$ as shown in Fig. 4. The peaks in motor torque of up to 0.6 Nm are a result of the motor-internal torque controller.

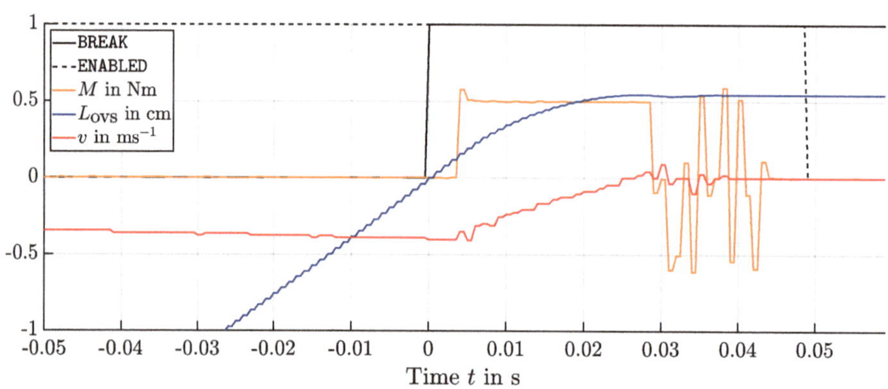

Fig. 4. Behavior of motor torque M, overshooting cable length L_{ovs} and velocity v with active deceleration after exceeding the maximum speed.

As a final extension of the safety systems, the speed-dependent braking distance can be calculated for a known maximum motor torque and end effector weight m. For this purpose,

$$F_{\text{max}} = -\text{sign}(v) \cdot M_{\text{max}}/r \tag{4}$$

is used to calculate the maximum cable force which counteracts the current movement. A minimal force to keep the cable under tension is not taken into account. Since cables cannot push for physical reasons, for positive velocities only gravity is used to decelerate the end effector. The resulting brake force F_{res} at the end effector is

$$F_{\text{res}} = \max\left(0, F_{\text{max}}\right) + \text{sign}\left(v\right) \cdot mg. \tag{5}$$

If $F_{\text{res}} > 0$, the end effector can thus be braked with a maximum acceleration of

$$a_{\text{max}} = \frac{F_{\text{res}}}{m} \quad . \tag{6}$$

Otherwise, the gravitational force outweighs the braking cable force and the emergency brake must get activated immediately.

Looking closely at Fig. 4, a delay of 5 ms between the time the brake gets triggered and the time where the motor is using the desired torque can be observed. This delay arises due to the motor-internal torque controller. In addition, the holding brake has an application delay time of 8 ms. The current acceleration a is assumed to be constant during the total delay time of $t_d = 13\,\text{ms}$. This gives the distance covered during t_d as

$$L_{\text{d}} = v \cdot t_d + \frac{1}{2} \cdot a \cdot {t_d}^2 \tag{7}$$

and the speed after the time t_d as

$$v_d = v + a \cdot t_d \quad . \tag{8}$$

Using *Linear Segments with Parabolic Blends (LSPB)* as shown in [17], this results in a braking distance

$$L^*_{\text{brake}} = \frac{1}{2} \cdot a_{\text{max}} \cdot {t_{\text{brake}}}^2 \quad . \tag{9}$$

With given v_d, the duration of the actual braking process t_{brake} is calculated using

$$t_{\text{brake}} = \frac{|v_d|}{a_{\text{max}}} \quad . \tag{10}$$

Depending on the movement direction, the remaining distance L_{rem} until the braking zone is reached, can be calculated using

$$L_{\text{rem}} = L - L_{\text{tol,min}} \quad \text{or} \quad L_{\text{rem}} = L_{\text{tol,max}} - L \quad . \tag{11}$$

Taking the distance covered during the delay time into account, this results in the braking distance of

$$L_{\text{brake}} = L^*_{\text{brake}} + L_d \tag{12}$$

and the emergency brake is activated, if the condition

$$L_{\text{rem}} <= L_{\text{brake}} \tag{13}$$

is satisfied. In the experiment, the end effector is dropped at a 20 cm distance from $L_{\text{tol,max}} = 0.6\,\text{m}$ with the velocity-based safety system being disabled. Figure 5 shows the application of the emergency brake before reaching the limit. Although the absolute value of the velocity is larger than $0.6\,\text{ms}^{-1}$, the end effector is able to stop 1.5 mm before reaching $L_{\text{tol,max}}$. Thus, triggering the brake at a certain velocity is no longer necessary when considering a static workspace.

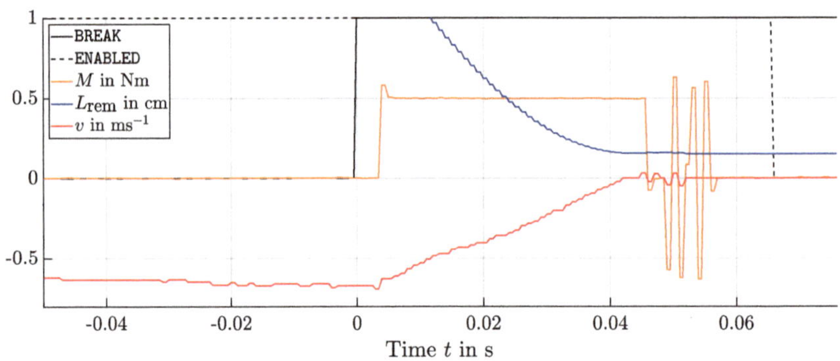

Fig. 5. Motor torque M, remaining cable Length L_{rem} and velocity v during emergency brake considering brake force with $L_{\text{tol,max}} = 0.6\,\text{m}$.

2.2 Operational Space

To apply the presented safety systems in the global coordinate system with multiple cables, the position of the end effector must be determined. Since the analytical calculation of the direct kinematics is a complex problem for parallel robots – especially considering pulleys – the position $\mathbf{x} = \begin{bmatrix} x_e \; y_e \; z_e \end{bmatrix}^{\text{T}}$, velocity $\dot{\mathbf{x}}$ and acceleration $\ddot{\mathbf{x}}$ of the end effector can be determined using a Kalman filter or an external measuring system as used in [11].

Also, the workspace $\mathcal{W}_{\text{crit}}$ in which the end effector may be located must be known. There are several possibilities to determine this, as shown in [15]. Using a given workspace $\mathcal{W}_{\text{crit}}$, the smaller workspace $\mathcal{W}_{\text{tol}}$ with a braking zone to $\mathcal{W}_{\text{crit}}$ must be calculated. Generally, this can be done by scaling down the boundary points around the center point of the workspace. For the position-based safety system, the emergency brake gets triggered if

$$\mathbf{x} \notin \mathcal{W}_{\text{tol}} \quad . \tag{14}$$

To determine this for a planar cable robot, it must be verified that the end effector position is within the workspace polygon. The crossings test as shown in [10] is suitable for this. In this case, $\mathbf{x} \in \mathcal{W}_{\mathrm{tol}}$ applies if a straight line – starting from the end effector position – into an arbitrary direction intersects the edges of the workspace in an odd number.

For three-dimensional workspaces, the principle of the crossings test is applied twice. In the first step, the intersection points of an arbitrary straight line starting from the end effector position with the planes spanned by the corner points of the workspace are determined. In the second step, it is examined whether the intersections with the planes are inside the spanning polygon, for which the crossings test is applied as described for the planar case. If the straight line intersects the plane within the polygon, it is a real intersection point, which is counted. For an odd number of intersections, $\mathbf{x} \in \mathcal{W}_{\mathrm{tol}}$ applies analogously to the two-dimensional case.

The velocity-based safety system for operational space works analogously to joint space, using

$$v = \|\dot{\mathbf{x}}\|_2 \quad . \tag{15}$$

For the extension of the brake-force-based safety concept to the to the operational space, the intersection point $\mathbf{x}_{\mathrm{tol}}$ between the line of movement $\mathbf{x}(t)$ described by

$$\mathbf{x}(t) = \mathbf{x} + t \cdot \dot{\mathbf{x}} \quad t \geq 0 \tag{16}$$

and the workspace hull of $\mathcal{W}_{\mathrm{tol}}$ is needed. This can be calculated by intersecting the line with all outer workspace planes and checking if the resulting point is actually part of the workspace hull of $\mathcal{W}_{\mathrm{tol}}$, using the crossings test as described before. Knowing the first intersection point $\mathbf{x}_{\mathrm{tol}}$, the remaining distance results in

$$d_{\mathrm{rem}} = \|\mathbf{x} - \mathbf{x}_{\mathrm{tol}}\|_2 \quad . \tag{17}$$

If $\mathbf{x} \notin \mathcal{W}_{\mathrm{tol}}$, the remaining distance gets negated. For the maximum force counteracting the current movement, only the motors (partially) pulling into the opposite direction are considered. If there is no movement, the braking distance d_{brake} is zero. Using n motors, for motor $i \in [1 \cdots n]$ with pulley position $\mathbf{P}_i = \left[x_{p,i}\ y_{p,i}\ z_{p,i}\right]^{\mathrm{T}}$, the pulling direction

$$\mathbf{p}_i = \mathbf{P}_i - \mathbf{x} \tag{18}$$

is calculated. The counteracting cable force results in

$$F_{\mathrm{max}} = \sum_{i=0}^{n} \max\left(0, -\frac{\dot{\mathbf{x}} \circ \mathbf{p}_i}{\|\dot{\mathbf{x}}\|_2 \cdot \|\mathbf{p}_i\|_2}\right) \frac{M_{\mathrm{max},i}}{r_i} \tag{19}$$

where $\circ$ denotes the scalar product. Considering gravity force

$$\mathbf{F}_{\mathbf{g}} = [0 \quad 0 \quad -m \cdot g]^{\mathrm{T}} \quad , \tag{20}$$

the resulting brake force at the end effector is

$$F_{\text{res}} = F_{\max} - \frac{\dot{\mathbf{x}} \circ \mathbf{F_g}}{\|\dot{\mathbf{x}}\|_2 \cdot \|\mathbf{F_g}\|_2} \|\mathbf{F_g}\|_2 \quad . \tag{21}$$

The resulting stopping distance d_{brake} is calculated analogously to Eqs. (6-12), using the velocity as shown in Eq. (15) and the acceleration as calculated by

$$a = \frac{\dot{\mathbf{x}} \circ \ddot{\mathbf{x}}}{\|\dot{\mathbf{x}}\|_2 \cdot \|\ddot{\mathbf{x}}\|_2} \|\ddot{\mathbf{x}}\|_2 \quad . \tag{22}$$

With d_{rem} from Eq. (17), the emergency brake is activated for

$$d_{\text{rem}} <= d_{\text{brake}} \quad . \tag{23}$$

In the experiment, a planar 2-DoF cable robot with four pulleys at positions $\mathbf{P}_i$ is used. With the upper cables, the point mass end effector is held 10 cm above the lower limit of the working range $\mathcal{W}_{\text{tol}}$. The end effector position $\mathbf{x}$ and velocity $\dot{\mathbf{x}}$ are measured with a laser tracker as shown in Fig. 6.

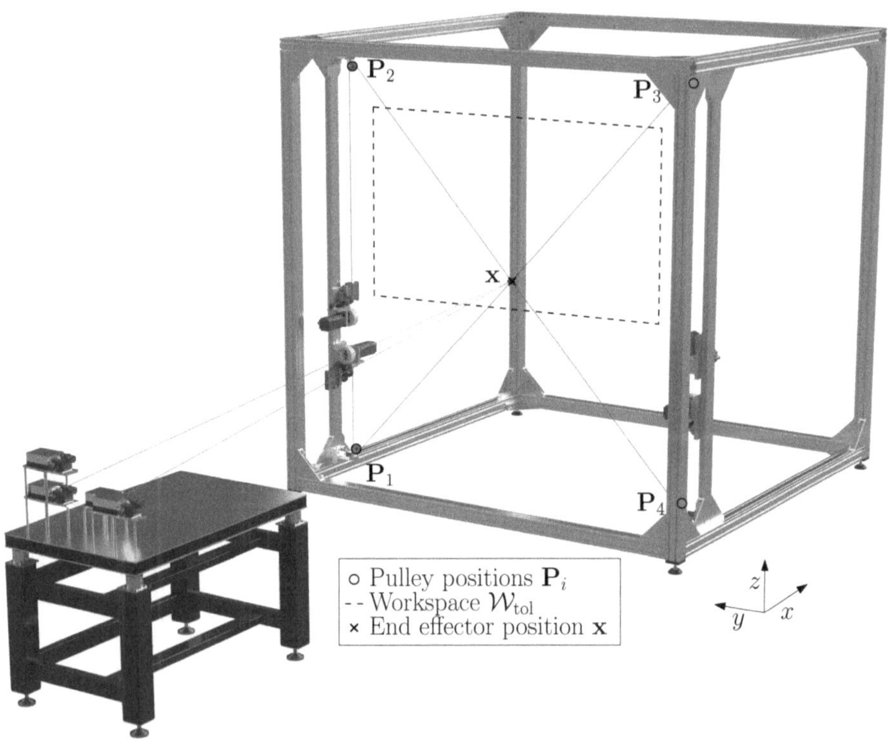

Fig. 6. Setup of planar cable robot with pulley positions $\mathbf{P}_i$, workspace $\mathcal{W}_{\text{tol}}$ and end effector position $\mathbf{x}$.

The end effector weight is increased to 0.240 g whereas $M_{\max}$ is staying at 0.5 Nm because of the higher number of motors. To keep the lower cables under tension during normal operation, the lower motors pull with 0.12 Nm. Then the torque of the motors actuating the cables towards the upper pulleys $\mathbf{P}_2$ and $\mathbf{P}_3$ is set to zero so that the end effector falls while the lower cables additionally pull towards the ground. In Fig. 7, the braking behavior is shown. Although the end effector speed in z-direction reaches 0.85 $\frac{\text{m}}{\text{s}}$ at the moment the brake is triggered, the motion can be stopped at a distance of 0.04 mm before reaching the z-axis limit. The stepwise descent of the `ENABLED` flag shows that not all motors have activated the holding brake at the same time. Nevertheless, the braking process is completed after 67 ms. To increase clarity, the motor torques are not shown in the graph. The behavior of the individual motors during braking is analogous to joint space.

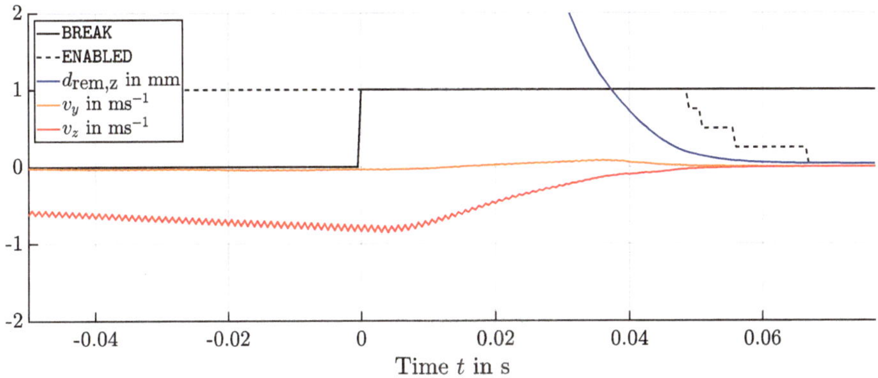

Fig. 7. Remaining distance $d_{\text{rem,z}}$ and axial velocities v_y, v_z during emergency brake for a planar cable robot considering collision with boundary of $\mathcal{W}_{\text{tol}}$.

3 Conclusion

In the present work, we have three safety concepts for preventing damage to the hardware of a CDPR and the environment. The first safety system is based on the specification of position limits which must not be exceeded. The speed-based safety concept assumes a maximum velocity of the end effector. If this is exceeded, the emergency brake is activated. The third concept calculates the model-based braking distance for a given mass of the end effector and given maximum motor torques. The braking process is triggered exactly when the braking distance is equal to the remaining distance to the beginning of the braking zone. This prevents braking unnecessarily early as well as braking too late. For all mentioned safety systems, the emergency brake has also been extended to include active braking with the maximum permissible motor torques. Hereby, the jerk was neglected due to the use of direct drives. Considering gears, a limitation of the jerk is recommended. For this purpose, the present safety system can be combined with the method for stop trajectory calculation presented in [12]. The experimental functional verification for a cable robot with three translational degrees of freedom is pending. Furthermore, an extension of the safety system to rigid bodies with rotational degrees of freedom is planned.

The safety concepts presented assume a static workspace without possible collisions with the cables. To extend the safety systems to a dynamic workspace and protect people inside, suitable measuring equipment would first be necessary. Also, it must be taken into account that people in the working area can move toward the end effector and thus the brake must be triggered earlier. To keep the braking distance as short a s possible and minimize personal injury in the event of unavoidable collisions, the speed, as well as the force, can be reduced if people are detected in the work area. This procedure is shown in [13] for ordinary industrial robots and can also be applied to cable robots. Alternatively, the obstacles can be bypassed to increase safety. A way to do this is shown in [18]. Here, possible collisions with the cables are also taken into account during obstacle avoidance. However, to ensure the safety requirements in terms of personal protection of ISO 13849, the holding brake of the motors is not sufficient and further arrangements for braking must be made [1].

References

1. Beckhoff Automation GmbH & Co. KG: Operating instructions synchronous servomotors for compact drive technology. https://download.beckhoff.com/download/document/motion/am8100_ba_en.pdf
2. Berti, A., Gouttefarde, M., Carricato, M.: Dynamic recovery of cable-suspended parallel robots after a cable failure. In: Lenarčič, J., Khatib, O. (eds.) Advances in Robot Kinematics, Springer Proceedings in a Advanced Roborics, vol. 4, pp. 331–339. Springer, Cham [etc.] (2018). https://doi.org/10.1007/978-3-319-56802-7_35
3. Boschetti, G., Minto, R., Trevisani, A.: Experimental investigation of a cable robot recovery strategy. Robotics **10**(1), 35 (2021). https://doi.org/10.3390/robotics10010035
4. Boumann, R., Bruckmann, T.: Development of emergency strategies for cable-driven parallel robots after a cable break. In: Bruckmann, T., Pott, A. (eds.) Cable-Driven Parallel Robots, Mechanisms and Machine Science, vol. 74, pp. 269–280. Springer International Publishing, Cham (2019). https://doi.org/10.1007/978-3-030-20751-9_23
5. Carbone, G., Gherman, B., Ulinici, I., Vaida, C., Pisla, D.: Design issues for an inherently safe robotic rehabilitation device. In: Ferraresi, C., Quaglia, G. (eds.) Advances in Service and Industrial Robotics, Mechanisms and Machine Science, vol. 49, pp. 1025–1032. Springer International Publishing, Cham and s.l. (2018). https://doi.org/10.1007/978-3-319-61276-8_110
6. Caro, S., Merlet, J.P.: Failure analysis of a collaborative 4-1 cable-driven parallel robot. In: Pisla, D., Corves, B., Vaida, C. (eds.) New Trends in Mechanism and Machine Science, vol. 89, pp. 440–447. Springer, Cham (2020). https://doi.org/10.1007/978-3-030-55061-5_50

7. Chen, Q., Zi, B., Sun, Z., Li, Y., Xu, Q.: Design and development of a new cable-driven parallel robot for waist rehabilitation. IEEE/ASME Trans. Mechatron. **24**(4), 1497–1507 (2019). https://doi.org/10.1109/TMECH.2019.2917294
8. Dieber, B., Schlotzhauer, A., Brandstötter, M.: Safety & security – erfolgsfaktoren von sensitiven robotertechnologien. e i Elektrotech. Informationstechnik **134**(6), 299–303 (2017). https://doi.org/10.1007/s00502-017-0512-4
9. Ghaffar, A., Hassan, M.: Failure analysis of cable based parallel manipulators. Appl. Mech. Mater. **736**, 203–210 (2015). https://doi.org/10.4028/www.scientific.net/AMM.736.203
10. Haines, E.: Point in polygon strategies. In: Heckbert, P.S. (ed.) Graphics Gems, pp. 24–46. The Graphics Gems Series, Kaufmann, San Diego (1994). https://doi.org/10.1016/B978-0-12-336156-1.50013-6
11. Hamann, M., Nüsse, P.M., Winter, D., Ament, C.: Towards a precise cable-driven parallel robot - a model-driven parameter identification enhanced by data-driven position correction. In: Bruckmann, T., Pott, A. (eds.) Cable-Driven Parallel Robots, Mechanisms and Machine Science, vol. 74, pp. 367–376. Springer International Publishing, Cham (2019). https://doi.org/10.1007/978-3-030-20751-9_31
12. Lemmen, P., Heidel, R., Bruckmann, T.: A conditional stop capable trajectory planner for cable-driven parallel robots. In: Bruckmann, T., Pott, A. (eds.) Cable-Driven Parallel Robots, Mechanisms and Machine Science, vol. 74, pp. 281–292. Springer International Publishing, Cham (2019). https://doi.org/10.1007/978-3-030-20751-9_24
13. Malm, T., Salmi, T., Marstio, I., Montonen, J.: Dynamic safety system for collaboration of operators and industrial robots. Open Eng. **9**(1), 61–71 (2019). https://doi.org/10.1515/eng-2019-0011
14. Passarini, C., Zanotto, D., Boschetti, G.: Dynamic trajectory planning for failure recovery in cable-suspended camera systems. Journal of Mechanisms and Robotics, vol. 11, no. 2 (2019). https://doi.org/10.1115/1.4041942
15. Pott, A. (ed.): Cable-Driven Parallel Robots. Springer Tracts in Advanced Robotics, Springer International Publishing, Cham (2018). https://doi.org/10.1007/978-3-319-76138-1
16. Roberts, R.G., Graham, T., Lippitt, T.: On the inverse kinematics, statics, and fault tolerance of cable-suspended robots. J. Robot. Syst. **15**(10), 581–597 (1998). https://doi.org/10.1002/(SICI)1097-4563(199810)15:10⟨581::AID-ROB4⟩3.0.CO;2-P
17. Spong, M.W., Hutchinson, S., Vidyasagar, M.: Robot Modeling and Control. Wiley, Hoboken, NJ (2006)
18. Xu, J., Park, K.S.: Moving obstacle avoidance for cable-driven parallel robots using improved RRT. Microsystem Technologies (2020). https://doi.org/10.1007/s00542-020-05100-4

Applications

Evaluation of a Cable-Driven Parallel Robot: Accuracy, Repeatability and Long-Term Running

Marceau Métillon[1,2], Nicolò Pedemonte[3], and Stéphane Caro[1,2(✉)]

[1] Laboratoire des Sciences du Numérique de Nantes (LS2N), UMR CNRS 6004, 1, rue de la Noë, 44321 Nantes, France
{marceau.metillon,stephane.caro}@ls2n.fr

[2] Centre National de la Recherche Scientifique (CNRS), 1, rue de la Noë, 44321 Nantes, France

[3] IRT Jules Verne, Chemin du Chaffault, 44340 Bouguenais, France
nicolo.pedemonte@irt-jules-verne.fr

Abstract. The performance evaluation of Cable-Driven Parallel Robots in terms of accuracy, repeatability and long term capacities is a key for their industrial deployment. This paper presents and analyses corresponding experimental results for the PACE prototype, a fully-constrained, redundantly restrained, eight cables and six degrees of freedom cable-driven parallel robot. First, a long-term running test is carried out to evaluate the deviation of the moving-platform pose along time of use. A second experiment consists into a full design of experiments to evaluate the effect of some factors specific to this type of robot on its absolute accuracy and repeatability. Experimental results are presented and discussed.

Keywords: Parallel robot · Cable · Accuracy · Repeatability · Long-term running

1 Introduction

Several industrial sectors, e.g. the naval and renewable energy industries, are facing with the necessity to manufacture and manipulate large and complex shape products. Cable-Driven Parallel Robots (CDPRs) are suitable for this type of application thanks to their high payload-to-weight ratio, large workspace and reconfiguration capabilities.

CDPRs are a particular class of parallel robots whose moving-platform (MP) is connected to the robot fixed base frame by a number of cables. The cables are coiled on motorized winches which control the cable lengths. The cables are routed from the winches to the MP through pulleys. The connection points between the cables and the base frame are referred to as exit points and are located on the exit pulleys and the connection points between the cables and the MP are referred to as anchor points.

M. Gouttefarde et al. (Eds.): CableCon 2021, MMS 104, pp. 375–388, 2021.
https://doi.org/10.1007/978-3-030-75789-2_30

The knowledge of the accuracy and repeatability of CDPRs is a key feature to guarantee their correct fitting to the industrial task they are assigned to. Various approaches are proposed to improve the accuracy of CDPRs such as the use of additional sensors and data fusion. In [5,6], cable angle measurement sensors are used in a sensor fusion to improve the MP pose accuracy. In [8], a fusion of data from a camera and three laser distance sensors is used as a feedback in a closed-loop Cartesian pose controller. In [10], various combinations of sensor measurements such as IMU and cable angle are studied while in [12], cable tension sensors are used to simplify and solve the analytical Direct Kinematic (DK) problem of CDPRs. In [13], the estimation of the MP mass and payload using cable tension is studied. In [18] various algorithms and filters are considered in IMU-sensors fusion for solving the DK problem of a Gough-Stewart platform. In [11], three lidars are embedded on the MP during long-term running of a CDPR and are used to correct the Cartesian MP pose.

The use of elastic cable models also increases the robot accuracy. In [17], the use of a simple elastic cable model is implemented and experimentations exhibit an improved position accuracy. In [2], the use of an elasto-dynamic model of a non-redundant CDPR is studied in term of position accuracy improvement. In [14], the stiffness of the robot is studied and a Stiffness Oriented Tension Distribution Algorithm reduces the deviation of the MP under external wrenches. Exteroceptive vision-based control algorithms also contribute to the improvement of CDPR accuracy. In [4], a 3D pose kinematic visual servoing and a vision-based computed torque control were implemented for the control of a CDPR. In [3], a vision is used to track the robot position using tracked visual features placed on the MP and an external fixed camera. In [20], a 2 $\frac{1}{2}$D visual servoing is studied and implemented experimentally on a CDPR to increase its robustness to manufacturing and calibration errors as well as to improve its accuracy.

Once designed and manufactured, an experimental evaluation allows one to determine the prototype accuracy and repeatability performances. In [15] the long-term running of a CDPR printer was analyzed. After a week of continuous running the joint lag was measured exhibiting the cable creep. In [7] the position accuracy of a three-dof CDPR performing dynamic trajectories was evaluated using an external vision tracking system. In [16], the vertical position error of a cable-based robotic crane were determined, a distribution of position errors along horizontal axes was also described and analysed. Others techniques can be used to increase CDPR accuracy. In [1], a Neural Network uses IMU sensor data to control the robot and images taken by an embedded camera are used to close the control loop. In [9], the cable elasticity is taken into account in the fusion of data from embedded IMU and encoder information.

In this paper, the accuracy and repeatability of a CDPR prototype, named PACE and located at IRT Jules Verne, are analyzed experimentally through long-term running. Furthermore a full factorial design of experiments is run to evaluate the effect of factors specific to CDPRs on the MP pose accuracy and repeatability. The paper is organized as follows: Sect. 2 presents the experimental setup considered in this work. Section 3 describes the method and shows the results obtained with the long-term experiment. Section 4 presents the full-factorial design of experiments and the obtained results. Conclusions and future work are drawn in Sect. 5.

2 Experimental Setup

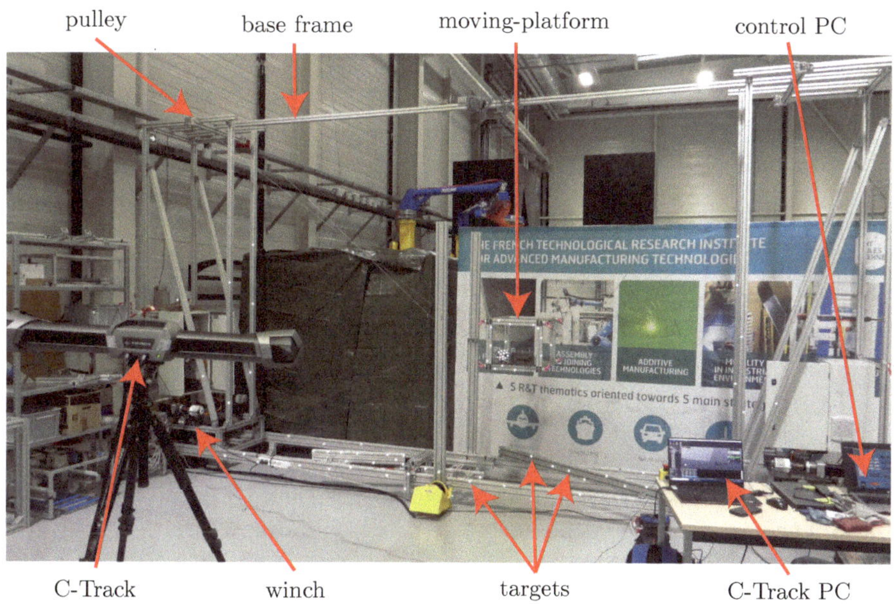

Fig. 1. PACE prototype and measurement device

2.1 Robot Prototype

Figure 1 shows the fully-constrained CDPR under study named PACE and located at IRT Jules Verne, France. It is composed of eight cables and the size of its frame is: $L = 4.6$ m $\times$ $l = 0.9$ m $\times$ $H = 2.8$ m. The cables are synthetic Dyneema® which reference is DYNE02SE99[1] with a nominal diameter of 2 mm. Eight motorized winches are mounted on the frame. The coiling drum of the winches have grooves so that a single layer is coiled around the nominal diameter of the drums equal to 90 mm. The cable length measurement estimation is obtained using the motor encoder with a resolution of 0.001°. Eight force sensors are located between the base frame and the respective eight winches in order to measure the cable tensions. Cables are routed through pulleys and connected to

[1] Acquired from Corderie Lancelin.

the vertex of a parallelepiped aluminium profile MP. Two control cabinets are mounted on each side of the frame and connect all the motors and needed equipment such as motors drivers, sensors, power supply and communication devices. A control PC is located in the switching cabinets. Using a remote connexion, it is possible to control the robot and download recorded data from the robot controller PC.

2.2 Exteroceptive Measurement

A dynamic tracking module VXtrack along with a C-Track™ measuring device from Creaform are used to measure the MP pose. The device is composed of a set of two cameras mounted on a support and connected to a controller communicating with a computer for data acquisition. Self-adhesive circular dots, of diameter equal to 20 mm, called targets are placed on the rigid main frame of the robot and identified as reference targets. The latter are used to define the global frame. Targets are also placed on the MP and on a faceted 3D-printed sphere fixed to the MP. They are identified as tracked targets and are used to measure the MP pose in the global frame of the robot at a rate 30 Hz. The measurement repeatability and accuracy are equal to 0.02 mm and 0.1 mm, respectively within a 16 m^3 measurement volume.

3 Long-Term Running Experiments

The long-term running evaluation of the robot refers to the evaluation of the deviation of the MP pose during long-term on-duty and off-duty periods of the robot. The MP deviation amounts to the MP absolute accuracy, namely the difference between the desired MP pose and the measured MP pose.

3.1 Method

In this experiment, the robot is controlled using a classical closed-loop position-based joint-space control. The desired trajectory motion described in operational space is converted into a set of joint coordinates using the Inverse Geometric Model (IGM) of the robot. A Feed-Forward model determines the needed joint torques to fulfil the static and dynamic equilibrium of the MP along the trajectory. In addition to this open-loop control architecture a feedback controller uses joint position information to compute a correcting torque to ensure the trajectory tracking without error between the desired trajectory and the trajectory followed by the robot in the joint space. The difference between joint position measurement and desired joint position (the joint error) is used in a proportional-integral-derivative (PID) feedback controller.

A path[2] alternating translation and orientation movements of the MP was specifically defined for the long-term running experiments and is shown in Fig. 2.

[2] Here is a link to the trajectory followed by the CDPR MP during long-term experiments: https://metillon.net/CableCon2021.

Fig. 2. Test path, targets and MP home poses

The path is composed of four segments passing through the waypoints P_{E1}, P_{E2}, P_{E3} and P_{E4}. The Cartesian coordinates of these four points expressed in the global frame $\mathcal{F}_b$ are the following:

$$\mathbf{p}_{E1} = [1.000, 0.346, 0.400]^T \ m \tag{1}$$
$$\mathbf{p}_{E2} = [1.500, 0.346, 1.200]^T \ m \tag{2}$$
$$\mathbf{p}_{E3} = [3.200, 0.346, 1.400]^T \ m \tag{3}$$
$$\mathbf{p}_{E4} = [2.800, 0.346, 0.800]^T \ m \tag{4}$$

The orientation of the MP remains constant once the MP translates. The MP performs pure rotations at points P_{E2} and P_{E4}. Accordingly, the test-trajectory is as follows:

1. Starting at point P_{E1}, translation to point P_{E2}
2. Rotation of the MP of $+5°$ then $-3°$ around axis $\boldsymbol{x}_b$, $+5°$ then $-5°$ around axis $\boldsymbol{y}_b$ and $+5°$ then $-5°$ around axis $\boldsymbol{z}_b$ successively while staying at position P_{E2}
3. Translation from P_{E2} to P_{E4} via P_{E3}
4. Rotation of the MP of $+5°$ then $-3°$ around axis $\boldsymbol{x}_b$, $+5°$ then $-5°$ around axis $\boldsymbol{y}_b$ and $+5°$ then $-5°$ around axis $\boldsymbol{z}_b$ successively while staying at position P_{E4}
5. Translation from P_{E4} to P_{E1}
6. Pause of 10 s at point P_{E1}

This trajectory is running on a loop all along the experiment. The MP velocity follows a trapezoidal profile along each segment and the maximum velocity is 1 m/s. One loop takes 40 s to complete.

At the beginning of the experiment, the robot is calibrated at its home pose once and is then controlled to reach the initial point of the specific trajectory. At this point a first measurement of the MP pose is performed and is used as a reference for further measurements. From this point, the robot is expected to work three 8-h work sessions on three different days without re-calibration. During a session, the robot repeats the trajectory on a loop during the whole session. At each iteration of the loop, the robot does stop during ten seconds at the initial point. This allows the exteroceptive measurement of the MP pose after settling of dynamic vibrations.

During sessions, three measurements are performed namely at the beginning, at the middle and at the end of the session. At the end of the session, once the measurement is made, the trajectory is stopped and the MP remains at the initial pose of the trajectory. The robot is then safely shut down using the brakes and standard procedure awaiting next session. At the beginning of the next session, the robot is powered on. The MP pose is measured in order to determine the static deviation occurring during the off-duty period between the two sessions. Then the same procedure is repeated as per the first session for the second and third sessions.

3.2 Long-Term Running Experimental Results

At the end of the experiment, nine pose measurements of the MP were recorded corresponding to the beginning, the middle and the end of all three sessions as shown in Fig. 3. The boxes represent the working session (namely on-duty) and the part of the graph outside the boxes represents the periods when the MP was held in position at initial point and not moving (namely off-duty).

By analyzing the different MP pose measurements it is possible to distinguish two different deviations. One is referred to as *dynamic deviation* which is accumulated during running hours of the robot, the other is referred to as *static deviation* and consists into the deviation added up during off-duty periods. Figure 3 shows the static and dynamic position and orientation deviations of the MP along time on the overall experiment. Figure 4 describes the part of the deviation obtained during on-duty and off-duty periods. The results exhibit a creep phenomenon of the cable during off-duty periods. As far as we understand, this is mainly due the MP mass exerting cable tensions and thus a permanent elongation of the cables. The opposite sign in the deviations comes from the routine used to stop the robot at the end of each running period. The trajectory is stopped when the robot comes back to the initial pose. The measurement at the end of the day is performed then the brakes are activated and the drivers stopped. Once the off-duty period is over, in the beginning of the next running period, the robot is powered on by activating drivers and then releasing the brakes while servoing the motors on the current position.

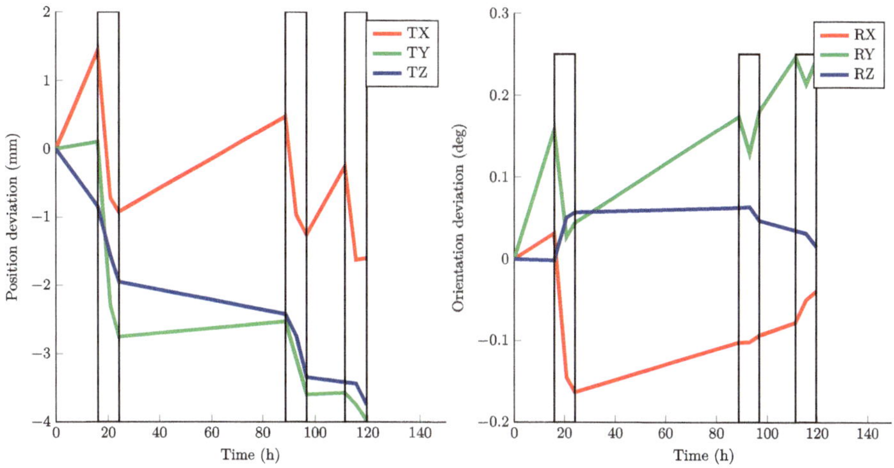

(a) Translational displacement (b) Rotational displacement

Fig. 3. MP pose deviation along the three eight-hour sessions and during rest periods (Total time = 120 h = 5 days)

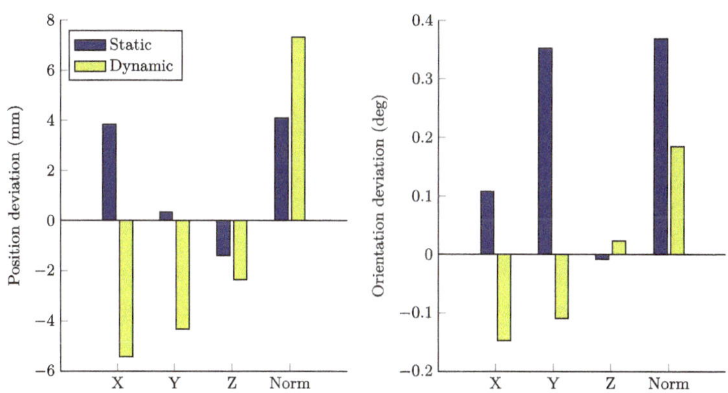

Fig. 4. Static and dynamic MP position and orientation deviations

4 Accuracy and Repeatability

This section aims to evaluate the absolute accuracy and repeatability of the prototype. The absolute accuracy is the capability of the robot to position its MP at a desired pose. The repeatability is the ability of the robot to repeatedly return to the same location under the same program with the same conditions [19]. Moreover, this section aims to determine the influence of factors specific to CDPR on the accuracy and repeatability of the MP pose.

4.1 Method

In order to identify the specific influence of each factor, a full factorial design of experiments is considered over the two-levels of each factor. The considered factors are the following : *cable tension*, *home MP pose*, *elasticity* and *payload*. Each factor has two-levels as described in Table 1.

Table 1. Full factorial design factors

Factors	Low level (−1)	High level (1)
Cable tension	50 N	28 N
Home MP pose	$\mathbf{p}_1 = [2.177, 0.349, 1.083]^T$ m	$\mathbf{p}_2 = [2.127, 0.423, 0.017]^T$ m
Elasticity	Without	With
Payload	Without ($0\,kg$)	With (8 kg)

Considering the four factors, the full factorial design of experiment has $2^4 = 16$ different experiments. In order to take into account the repeatability, all experiments are repeated five times. Therefore 80 experiments are carried out in a randomised order.

4.2 Factors

The *home MP pose* factor defines the pose where the MP is calibrated. Two different calibration poses are considered, the first pose, named P_1, is located around the centre of the robot workspace while the second, named P_2 is located at the bottom of the robot workspace. The *cable tension* factor corresponds to the initial cable tension set. The initial cable tension is the parameter defining the amount of tension given to the cable during the calibration phase. During the calibration procedure, the MP is located at its home pose then the initial cable tension is gradually applied simultaneously to all cables. This ensures a certain pre-tension in the cable. Two sets are defined, each yielding the same initial cable tension for all cables. The *elasticity* factor defines whether the cable elasticity compensation algorithm of the prototype is used for the experiment or not. The cable elasticity compensation algorithm aims to compensate the cable elongation under the cable tension due to its elasticity. Using the cable tension and length information, a linear elastic cable model is used to determine the cable elongation, which is then used in the robot geometric model so that the desired cable length accounts for the elongation. The elasticity of the cable was identified experimentally such that $ES = 121\,775.96$ N. The *payload* factor corresponds to the presence or the absence of the payload on the MP.

4.3 Experimental Procedure

Two points of interest, known as target points, called P_A and P_B whose coordinates expressed in the base frame $\mathcal{F}_b$ are $\mathbf{p}_A = [1.000, 0.346, 0.400]^T$ m and $\mathbf{p}_B = [3.200, 0.346, 1.400]^T$ m are defined in the robot workspace. The pose of the MP is to be evaluated at these two points and compared to the desired pose. The accuracy can then be expressed as the mean of the error between the desired and measured poses over the five repetitions of a given experiment. The repeatability is the standard deviation of the error over the repetitions. For each experiment, the procedure is the following:

1. The platform is placed on the calibration point corresponding to the *home MP pose* factor (P_1 or P_2).
2. The load corresponding to the *payload* factor is mounted on the platform (with or without).
3. The elasticity algorithm is set corresponding to the *elasticity* factor (with or without).
4. The initial cable tension set corresponding to the *cable tension* factor (high or low) is gradually applied to the cables in increments of 25% of the final value.
5. The robot reaches the target P_A and waits.
6. The MP pose is measured using the C-Track™ measurement system.
7. The robot reaches the target P_B and waits.
8. The MP pose is measured using the C-Track™ measurement system.
9. The robot measurements are recorded.
10. The MP returns to the calibration point.

In this experiment, the control architecture is close to the one used in Sect. 3. The only difference lies in the cable elasticity compensation algorithm used, which is a function of the state of the *cable elasticity* factor.

4.4 Results

As per the exteroceptive measurements, the pose error is obtained as the difference between the desired pose and the measured pose of the MP. For a given experiment, the error is evaluated for each of the five repetitions at the two target positions P_A and P_B. Figure 5 represents the MP pose error in terms of position and orientation error at P_A for the first experiment configuration. Similarly, the same information is obtained concerning the pose P_B but is not depicted in this paper for the sake of shortness. In this case the configuration is: *home MP pose* factor is set to P_1, the *cable tension* factor is set to high, the *payload* factor is set to without and the *elasticity* factor is set to without. Position and orientation error at points P_A and P_B constitute the responses of the full-factorial design and are collected for the 80 experiments constituting the full-factorial design.

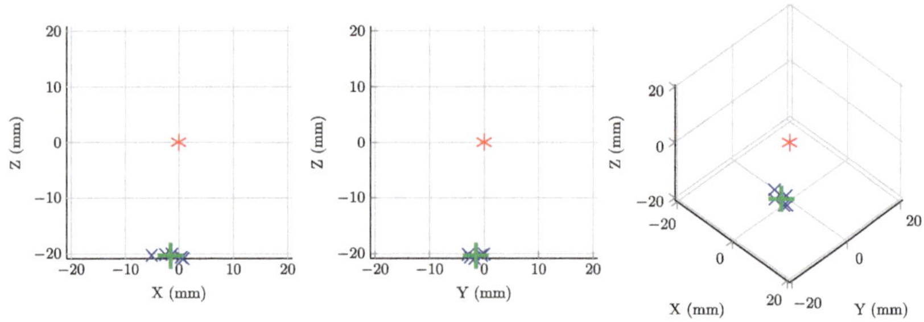

(a) Measured positions at point A

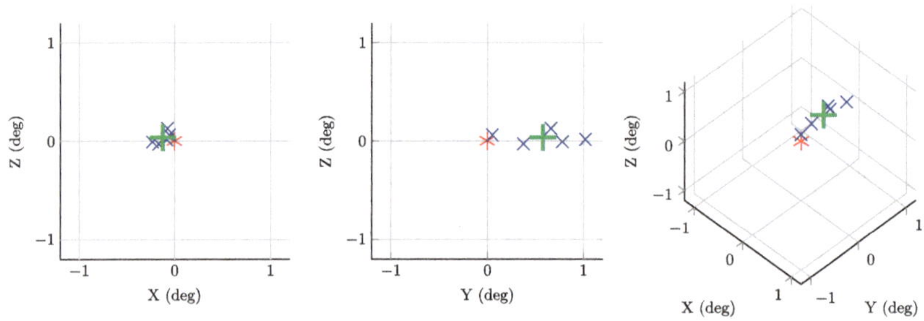

(b) Measured orientations at point A

Fig. 5. MP accuracy and repeatability at Experiment 1: the desired target in red; the measured poses in blue; the mean error in green

Figure 6 shows the results in terms of accuracy and repeatability of the MP position at both points P_A and P_B for each of the sixteen experiments of the full factorial design. Likewise, the information on accuracy and repeatability in terms of MP orientation is obtained but not given in the paper for conciseness. For a good understanding of the effect of each factor and as well as the effect of interactions between factors on the robot accuracy and repeatability a factorial regression analysis was performed using the Minitab® software. Figure 7 shows the effects of each factor on the MP accuracy and repeatability. The factors influencing the accuracy are evaluated using a factorial regression on the mean of the error of the repetitions of the experiments while the factors influencing the repeatability are evaluated using a variability analysis of the error over the repetitions. It should be noted that the most influential factors on the robot accuracy are the *home MP pose* factor and the *payload* ones as well as their interaction with other factors.

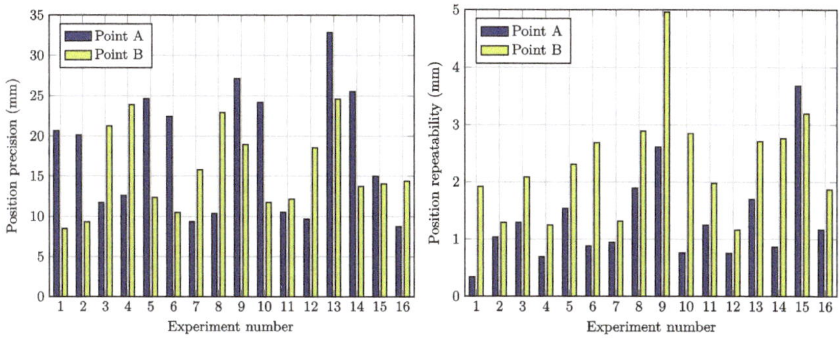

(a) Accuracy (b) Repeatability

Fig. 6. Position accuracy and repeatability for each experiment

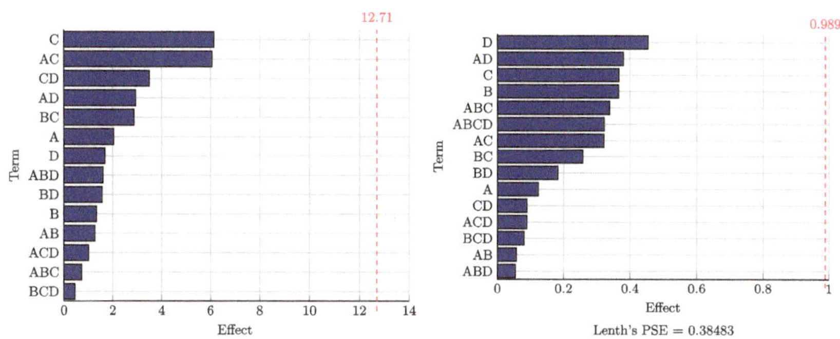

(a) Position accuracy (b) Position repeatability

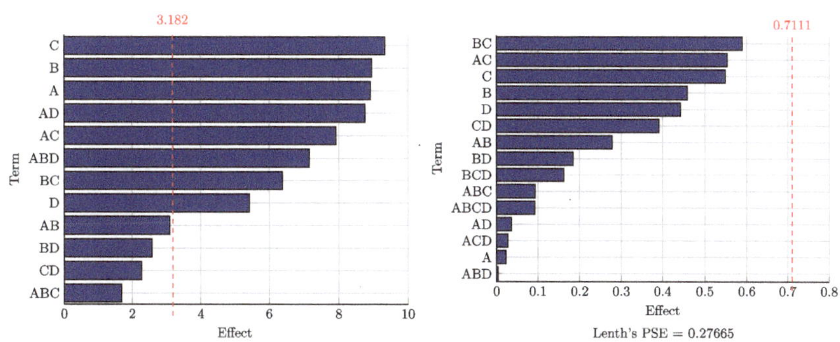

(c) Orientation accuracy (d) Orientation repeatability

Fig. 7. Pareto effects of factors on responses: A, B, C and D stand for *payload*, *cable tension*, *home MP pose* and *elasticity*, respectively.

Figure 8 shows the factor levels that optimize the MP accuracy and repeatability responses in term of position and orientation where the responses are equally weighted. The effect of each factor level on each response is also displayed. The composite desirability D shows how well the proposed level set optimizes the four responses where individual desirability d assesses how the factor set optimizes each response. The optimized choice of factor level is as followed: *payload* is set to "with", *cable tension* is set to "high", *home MP pose* is set to P_2 and *elasticity* is set to "with". The proposed set yields a good composite desirability of 0.8614. The targeted values of the responses are 1.18 mm and 13.7 mm for position repeatability and accuracy, respectively, and 0.06° and 0.62° for orientation repeatability and accuracy, respectively.

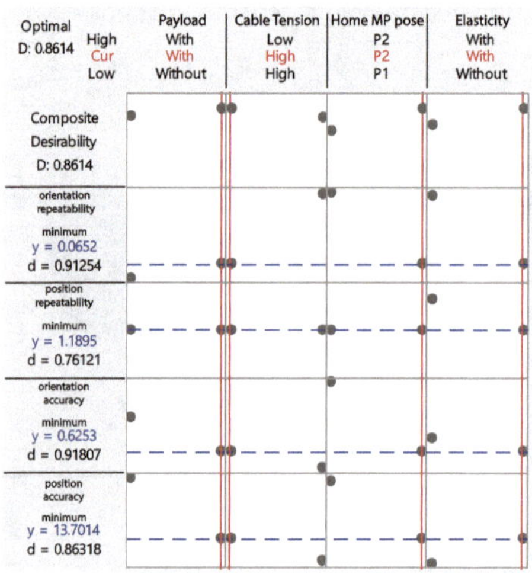

Fig. 8. Response optimization

5 Conclusion

This paper proposed an experimental method to evaluate the long-term running, accuracy and repeatability of a CDPR using exteroceptive measurements. In addition an investigation of the influence of factors was led using a full-factorial design of experiments. The long-term running evaluation of the prototype yields a static deviation less than 5 mm in terms of translation and 0.5° in terms of orientation while the dynamic deviation is less than 10 mm in terms of translation and 0.2° in terms of orientation.

Acknowledgements. This research work is part of the PACE project managed by IRT Jules Verne (French Institute in Research and Technology in Advanced Manufacturing Technologies for Composite, Metallic and Hybrid Structures). The authors wish to associate the industrial and academic partners of this project, namely, Naval

Group, University of Nantes, IMT-Atlantique and CNRS. Tanguy Moro from IRT JV and the anonymous reviewers are dutifully acknowledged for their great help in using Minitab® software and improving the paper, respectively. This work is also supported by the ANR CRAFT project, grant ANR-18-CE10-0004 https://anr.fr/Project-ANR-18-CE10-0004.

References

1. Aflakian, A., Safaryazdi, A., Tale Masouleh, M., Kalhor, A.: Experimental study on the kinematic control of a cable suspended parallel robot for object tracking purpose. Mechatronics **50**, 160–176 (2018). https://doi.org/10.1016/j.mechatronics.2018.02.005
2. Baklouti, S., Caro, S., Courteille, E.: Elasto-dynamic model-based control of non-redundant cable-driven parallel robots. In: Arakelian, V., Wenger, P. (eds.) ROMANSY 22: Robot Design, Dynamics and Control, CISM Courses and Lectures, vol. 584, pp. 238–246. Springer, Cham (2019). https://doi.org/10.1007/978-3-319-78963-7_31
3. Bayani, H., Masouleh, M.T., Kalhor, A.: An experimental study on the vision-based control and identification of planar cable-driven parallel robots. Robot. Auton. Syst. **75**, 187–202 (2016). https://doi.org/10.1016/j.robot.2015.10.002
4. Dallej, T., Gouttefarde, M., Andreff, N., Michelin, M., Martinet, P.: Towards vision-based control of cable-driven parallel robots. In: 2011 IEEE/RSJ International Conference on Intelligent Robots and Systems, pp. 2855–2860. IEEE, Piscataway, September 2011. https://doi.org/10.1109/IROS.2011.6094591
5. Fortin-Cote, A., Cardou, P., Campeau-Lecours, A.: Improving cable driven parallel robot accuracy through angular position sensors. In: Staff, I. (ed.) 2016 IEEE RSJ International Conference on Intelligent Robots and Systems (IROS), pp. 4350–4355. IEEE, Piscataway, October 2016. https://doi.org/10.1109/IROS.2016.7759640
6. Garant, X., Campeau-Lecours, A., Cardou, P., Gosselin, C.: Improving the forward kinematics of cable-driven parallel robots through cable angle sensors. In: Gosselin, C., Cardou, P., Bruckmann, T., Pott, A. (eds.) Cable-Driven Parallel Robots, Mechanisms and Machine Science, vol. 53, pp. 167–179. Springer, New York (2017). https://doi.org/10.1007/978-3-319-61431-1_15
7. Gosselin, C., Foucault, S.: Experimental determination of the accuracy of a three-DOF cable-suspended parallel robot performing dynamic trajectories. In: Pott, A., Bruckmann, T. (eds.) Cable-Driven Parallel Robots, Mechanisms and Machine Science, vol. 32, pp. 101–112. Springer, Cham (2015). https://doi.org/10.1007/978-3-319-09489-2_8
8. Korayem, M.H., Tourajizadeh, H., Taherifar, M., Khayatzadeh, S., Maddah, M., Imanian, A., Tajik, A.: A novel method for recording the position and orientation of the end effector of a spatial cable-suspended robot and using for closed-loop control. Int. J. Adv. Manuf. Technol. **72**(5–8), 739–755 (2014). https://doi.org/10.1007/s00170-014-5681-2
9. Korayem, M.H., Yousefzadeh, M., Kian, S.: Precise end-effector pose estimation in spatial cable-driven parallel robots with elastic cables using a data fusion method. Measurement **130**, 177–190 (2018). https://doi.org/10.1016/j.measurement.2018.08.009

10. Merlet, J.P.: Direct kinematics of CDPR with extra cable orientation sensors: the 2 and 3 cables case with perfect measurement and sagging cables. In: 2017 IEEE/RSJ International Conference on Intelligent Robots and Systems (IROS), pp. 6973–6978 (2017). https://doi.org/10.1109/IROS.2017.8206622
11. Merlet, J.P., Papegay, Y., Gasc, A.V.: The prince's tears, a large cable-driven parallel robot for an artistic exhibition. In: 2020 IEEE International Conference on Robotics and Automation (ICRA), pp. 10378–10383. IEEE, Piscataway (2020). https://doi.org/10.1109/ICRA40945.2020.9197011
12. Oftadeh, R., Aref, M.M., Taghirad, H.D.: Forward kinematic analysis of a planar cable driven redundant parallel manipulator using force sensors. In: 2010 IEEE/RSJ International Conference on Intelligent Robots and Systems (IROS), pp. 2295–2300. IEEE (2010). https://doi.org/10.1109/IROS.2010.5649471
13. Picard, E., Caro, S., Claveau, F., Plestan, F.: Pulleys and force sensors influence on payload estimation of cable-driven parallel robots. IROS **2018**, 1429–1436 (2018). https://doi.org/10.1109/IROS.2018.8594171
14. Picard, E., Caro, S., Plestan, F., Claveau, F.: Stiffness oriented tension distribution algorithm for cable-driven parallel robots. In: Lenarčič, J., Siciliano, B. (eds.) Advances in Robot Kinematics 2020, Springer Proceedings in Advanced Robotics, vol. 15, pp. 209–217. Springer, Cham (2020). https://doi.org/10.1007/978-3-030-50975-0_26
15. Pott, A., Tempel, P., Verl, A., Wulle, F.: Design, implementation and long-term running experiences of the cable-driven parallel robot CaRo printer. In: Pott, A., Bruckmann, T. (eds.) Cable-Driven Parallel Robots, Mechanisms and Machine Science, vol. 74, pp. 379–390. Springer, Cham (2019). https://doi.org/10.1007/978-3-030-20751-9_32
16. Scalera, L., Gallina, P., Seriani, S., Gasparetto, A.: Cable-based robotic crane (CBRC): design and implementation of overhead traveling cranes based on variable radius drums. IEEE Trans. Robot. **34**(2), 474–485 (2018). https://doi.org/10.1109/TRO.2018.2791593
17. Schmidt, V., Pott, A.: Increase of position accuracy for cable-driven parallel robots using a model for elongation of plastic fiber ropes. In: Wenger, P., Flores, P. (eds.) New Trends in Mechanism and Machine Science, Mechanisms and Machine Science, vol. 43, pp. 335–343. Springer, New York (2016). https://doi.org/10.1007/978-3-319-44156-6_34
18. Schulz, S., Seibel, A., Schlattmann, J.: Performance of an IMU-based sensor concept for solving the direct kinematics problem of the Stewart-Gough platform. IROS **2018**, 5055–5062 (2018). https://doi.org/10.1109/IROS.2018.8594039
19. Siciliano, B., Khatib, O. (eds.): Springer Handbook of Robotics, 2nd edn. Springer, Cham (2016)
20. Zake, Z., Chaumette, F., Pedemonte, N., Caro, S.: Robust $2\frac{1}{2}$d visual servoing of a cable-driven parallel robot thanks to trajectory tracking. IEEE Robot. Autom. Lett. **5**(2), 660–667 (2020). https://doi.org/10.1109/LRA.2020.2965033

Under Constrained Cable-Driven Parallel Robot for Vertical Green Maintenance

Sebastian Schröder(✉)

TU Berlin, Institute for Machinery Design and System Technology, Chair Machinery System Design, Berlin, Germany
s.schroeder@tu-berlin.de

Abstract. Façade greening has great potential to combat problems in modern cities, such as urban heat islands. However, it is not used extensively inter alia due to its high maintenance efforts. In this paper a prototype of an under constrained parallel driven Robot for Vertical Green maintenance is presented. First the concept of cable driven robots is compared with other possible solutions. Afterwards the concept of a robot held by four cables and driven by two motors is explained for the given task with a large and flat work space. This configuration leads to two uncontrolled degrees of freedom. The rolling movement of the robot should be measured and counteracted by the swivel head for the tools. The distance perpendicular to the façade is planned to be measured with a laser level detector and controlled by a pair of actuated propellers. The work in progress on the robot and the results of the first field tests are shown. The robotic concept seems to be suitable for vertical green maintenance and other tasks with a comparable workspace.

Keywords: Vertical green maintenance · Façade greening · Under constrained cable driven robot

1 Introduction

1.1 Vertical Green (VG) for Cities

Facade greening and living walls have a great potential to encounter modern city challenges. They can cool buildings, recycle waste-water and produce food or bio-energy. In addition, green spaces have a positive impact on people's wellbeing and are increasing the local biodiversity. However, façade greening is remarkably seldom considered in building projects. The aim of the project Vertical Green 2.0 is to show and numerate the potential of VG and find answers for existing challenges. In the Project the Chair Machinery System Design at TU Berlin is working on solutions to simplify the time consuming and costly maintenance of vertical green. This drawback has been named in co-creation workshops by stakeholders again and again [1].

M. Gouttefarde et al. (Eds.): CableCon 2021, MMS 104, pp. 389–400, 2021.
https://doi.org/10.1007/978-3-030-75789-2_31

1.2 Maintenance of Vertical Green

Nowadays green facades are maintained manually with the help of cherry pickers. Detailed steps and technical requirements for VG maintenance are not well documented in literature. Robotic solutions on the market could not be found. Parallel to our Project the Fraunhofer IPA and the Institute for Energy Efficiency in Production (EEP) of University of Stuttgart started the project "Green Wall Robot – automated, smart green facades". The aim, to get rid of time-consuming and expensive manual maintenance, is the same. On the one hand, the team presents a design study of a cable driven robot (Fig. 1). On the other side they describe ideas for a rail driven system that can exchange and maintain greening modules [2].

Fig. 1. Design study of cable-driven maintenance robot for façade greening [2]

According to [3] the amount and type of construction material is crucial to get a positive CO_2 balance of a green façade. Due to the needed extensive structure, the approach of having interchangeable modules is not seen as an environmentally friendly solution from our point of view. Our focus is the maintenance of stationary plants, including ground-based vegetation like climbers and trellis plants as well as green walls.

The need of a system for cutting the plants was named most often by Stakeholders. Cutting can be divided into simple flattening of the façade and cutting back at windows or the eaves. Other tasks are the precise application of fertilizer and if need pesticides as well as replanting and harvesting. A 1 cm accuracy of the robot is assumed to be sufficient by the author. Keeping the costs per square meter low is important to enforce new this maintenance technology. As the desired frequency for VG maintenance is not very high, the focus lays on the development of a portable technology which can be used at various facades and thus helps to keep the costs per square meter low.

In order to fulfill complex tasks like harvesting specific fruits the robot is supposed to handle the tool with 6 DOF.

1.3 Justification of a Cable Driven Robotic Concept

For the described task with its large working space four main concepts could be identified. These are a serial motion structure, a parallel motion structure, a "walking" structure with changing base connections and an unguided, free moving drone. These concepts and examples of real implementations are shown in Table 1.

Table 1. Different approaches to move a tool relative to the ground

Serial motion structure	Parallel motion structure	Structure with changing base connections	Unguided, free moving structure
Motion portal	Cable driven Robot	Canadarm2	Agricultural Drone

The concepts were then evaluated with a Benefit Analysis following the VDI 2225 guideline. The Drone performed worst due to safety and efficiency reasons. The motion portal is a state-of-the-art solution, but the structure needs to be very massive and thus expensive for the required workspaces. The concept of having a structure with changing base connections walking along the façade sounds very promising and works successful for the pictured application in space without gravity. An automatic, stiff and safe coupling to connectors on the greened façade is assumed to be critical by the author. In addition, either many connectors in a dense arrangement or very strong ones are needed. Although this concept couldn't prevail, it is worth to be further assessed. The concept with the highest score in the benefit analysis was a parallel driven cable robot. The pictured system with a comparable working space has been set up by researchers at the University of Duisburg-Essen during the "LEAN" project. In the benefit analysis this concept could score with its potentially large working space, its robustness, its safety and energy efficiency.

1.4 Analysis of Degrees of Freedom

The moving platform of cable driven robots can be fully constrained with a minimum of 7 cables. For robots with 6 cables the stability of a positions depends on applied forces and gravity. This mode of operation is also called suspended or crane configuration [8].

The working space for vertical green maintenance can be described as very flat. For cutting the surface of the greening the movement orthogonal to the façade is minimal. For other maintenance operations tools may need to be moved into the greening, but in comparison to its height and width these travels are relatively small. In addition, robotic structure and anchor point should not protrude too far from the façade. Further outstanding anchor points could possibly reach over streets or pavements and are also harder to connect to building structures. Figure 2 shows in green an estimation of a possible greening size and working area of the tool center point (TCP).

Estimations for setting up a fully constrained Robot in a flat working space lead to high cable tension when forces orthogonal to the façade occur. Thus, high power

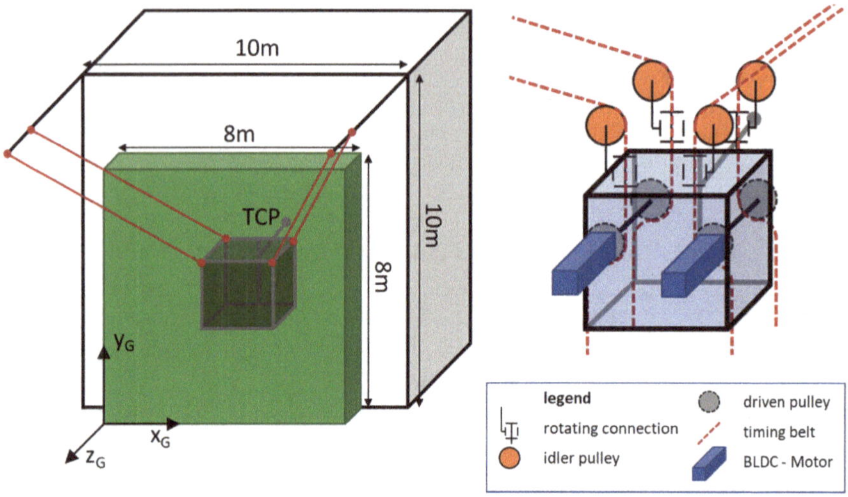

Fig. 2. Description of the working space and the parallel suspension concept

winches need to be installed. As described in Sect. 1.2 a portable robot with winches located in the motion platform is aspired. A number of 6 or even more winches with at least partly high power would lead to high weight and power consumption. A fully constrained robot is not seen as a feasible solution for this task. Reducing the number of cables leads to an increasing number of uncontrolled degrees of freedom.

An incompletely restrained positioning mechanism with 4 transmission elements driven by 2 brushless DC motors will be set up for this task (Fig. 2). The transmission elements on each side run parallel and are assumed to be stiff. In this configuration the height (y_G) and the horizontal motion parallel to the façade (x_G) can be controlled by the winch motors. The orientation around the axis x_G and y_G is constant for all poses.

One uncontrolled degree of freedom is a swinging movement orthogonal to the façade. Forces in this direction will not affect the cable tension. This allows a downsizing of the winch motors. Techniques to control this degree of freedom are described in Sect. 2.4. The second remaining degree of freedom is a rotation movement around the axis orthogonal to the façade. This degree of freedom would be avoidable with more cables, but the orientation is stabilized and limited by the gravity. Small deviations of the robot around this axis can be accepted, as long as the position of the tool head can be held in the wanted position. This can be achieved by counteracting the rotation of the robot with the installed turning axis for the tool head. More information is provided in Sect. 2.3.

2 Description of the Cable Driven Robot

Figure 3 shows the actual prototype. 400 W BLDC Motors from Nanotec® in combination with worm gears are used to drive the pulleys of the belts. Two rotos generate the thrust to control the distance to the façade and are running in contrary direction to avoid torque on the robot. The 3 turning axes in the swivel head allow the tool to be turned relative to the base. They are realized with a worm gear and a 2 DOF Robolink W cable driven joint both from the company Igus®. All 3 turning axis are driven by stepper motors. The system is supplied by a custom made 44,4 V and 20 Ah Li-Ion battery with a nominal current of 100 A. The control is running on a Raspberry Pi 3 with CODESYS. Inputs and outputs are made industry compatible with a PiXtend extension. The final mass will be around 50 kg.

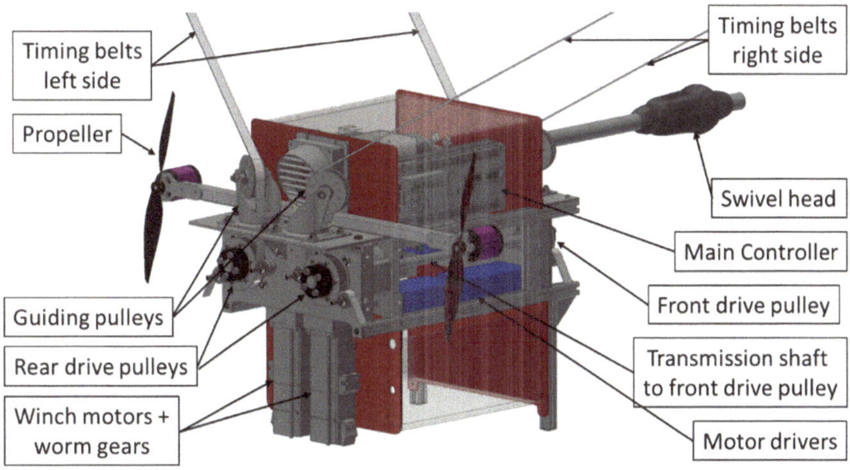

Fig. 3. Virtual prototype of the maintenance robot

2.1 Tolerated Working Space

Figure 5 shows the workspace of the robot for the use case shown in Fig. 3 assuming a force of both left winches respectively both right winches in-between 50 and 700 N. The sum of the forces in both left belts is calculated based on Eq. 1.

$$F_L = F_{L\_V} \cdot \sqrt{1 + \left(\frac{x}{H - y}\right)^2} = F_G \cdot \left(1 - \frac{x}{B}\right) \cdot \sqrt{1 + \left(\frac{x}{H - y}\right)^2} \qquad (1)$$

This equation is developed based on Fig. 4.

The equation is then evaluated for a map with 10 cm resolution. Afterwards the map was mirrored to get a new map of robot positions with tolerated forces for the right winches. All valid positions of both maps were marked in green. All other positions are marked in red.

At the area on the outside edge the robot is not in a stable configuration because of to low belt tension. The top area cannot be reached because the endurable forces in the belts and winches are limited. This means that the beams holding the robot need to be a bit higher and further to the side than the greened area.

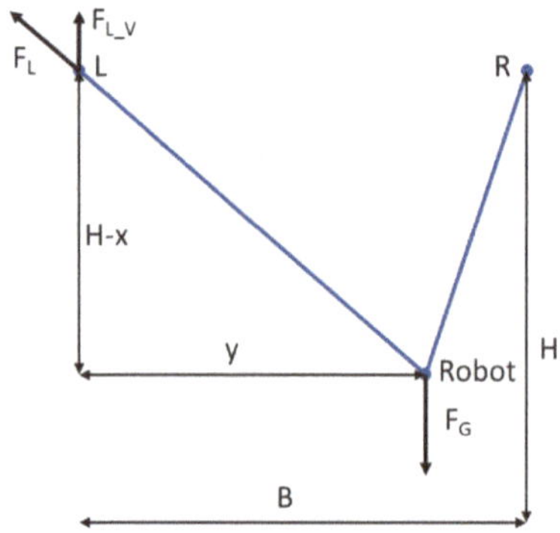

Fig. 4. Simplified diagram of the system forces

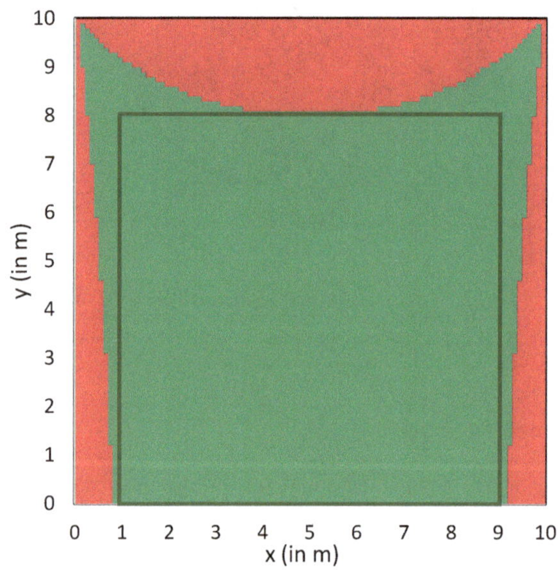

Fig. 5. Working area for the cable driven robot with anchor points in the upper corners

2.2 Selection of Traction Device

Cables, Chains and timing belts are evaluated for the use as a transmission element. If using cables, the drum needs to be winded only in one layer with a cable guidance system to allow an exact estimation of the cable length. Another problem with cables can be its strain under load. Heavy tools for example lead to high forces in the cables facing to the greening. Problems are expected when the winches roll up most of the cable in order to get to the top position. The free, unrolled part of the cable facing to the greening will be longer than the rear cable on the same side. This will turn the robot and the arm for holding the tool head will no longer be orthogonal to the façade.

Tensioning elements without slip like chains and timing belts can guarantee that both unrolled parts sitting on the same shaft have the same length.

First round link chains where have been considered due to their approval as a load carrying element and the simple way of storing them. Tests however showed that the slagging of the chains due to their high tare weight lead to a low stiffness of the robot.

Timing belts were chosen for the traction device. One challenge remaining potentially uneven force distribution in the timing belts due to angular misalignments of the pulley when the robot is swinging toward the facade. These angular deviations have to be avoided. With an additional pulley rotation around the vertical tangent of the main drive pulley the deviations can be transformed into a twist of the belt. This twisting is within the specification of the belt. The setup is shown in Fig. 4.

Fig. 6. Main drive pulleys and rotating guiding pulley above

Another drawback of the timing belt solution is the elaborately storing of the unneeded belt. First attempts with hanging down belts work but are not seen as a robust solution for real operation. A system for rolling up the unused part of the timing belt is under construction.

2.3 Compensation of Rolling Motions of the Robot

Rolling motions around its axis orthogonal to the façade lead to a turning of the tool head and in addition to a minor deflection parallel to the façade, because both belts can not reach at the robot in the tool rotation axis. The tool axis is located about 11 mm under the axes of the rotating pulleys. The deflections parallel to the façade for a total of nine positions (the edges, the midpoints of the borders and the center position of the work space) are shown in Fig. 6.

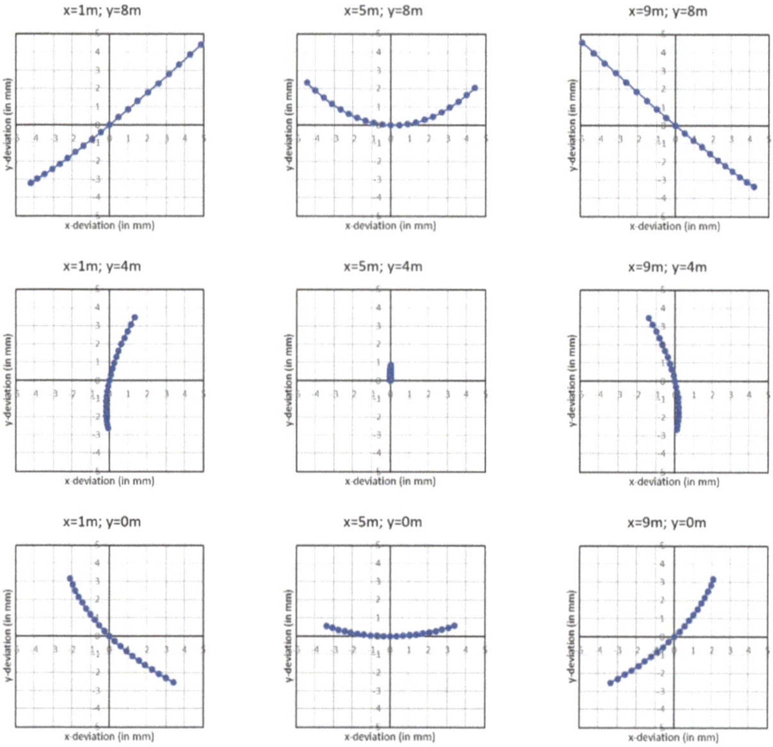

Fig. 7. Deflection of the tool axis for rolling motions of the robot up to 10° in both directions

Deviations for rotations up to an angle of 10° are within a range of 1 mm at the center position and increasing at the boundaries of the working area. With an expected total repeatability of 1 cm these deviations can just be accepted.

The turning of the tool head due to rolling motions of the robot will be compensated by counter rotation of the swivel head. Therefore the rolling motions will be calculated with a complementary filter based on data from a gyroscope and an accelerometer. The controller design is still under work but this part is not expected to be critical.

2.4 Compensation of Swinging Motions of the Robot

As described in Sect. 1.4 the movement orthogonal to the façade will not be controlled by the belt drive system. The robot can swing back and forth. These movements are planned to be controlled by a sensor system and a quadcopter rotor. The sensor system will consist of a rotary laser, setting up a laser level parallel to the façade, and a detector on the robot. The 1 m long detector is installed orthogonal to the laser level and will detect the crossing point of the robot and the laser level. With this information the distance to the façade can be calculated. Unfortunately, no commercial detectors with a range of about 1 m and an accuracy of 2–3 mm are currently available on the market. A custom-made system has been set up. The systems consist of 384 photodiodes with transimpedance amplifiers. The output is filtered with a high pass filter and compared with a threshold voltage in another operational amplifier. The signal is then captured with a interrupt and can be read out via an SPI bus. With a diode distance of 2,54 mm the resolution varies between 1,27 mm and 2,54 mm. Figure 7 shows one of the 12 detector circuit boards.

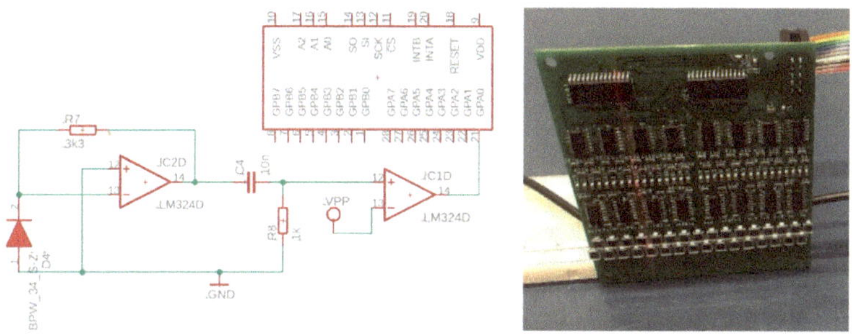

Fig. 8. circuit (left) and PCB (right) of the detector

The use of a quadcopter rotor as the actor is the most critical part of this robotic concept. In previous tests quadcopter rotors and collective pitch helicopter rotors have been compared. The quadcopter rotors control the thrust by changing the speed while the helicopter rotor runs with constant speed and is changing the pitch of its blades. Figure 8 shows the used test stands.

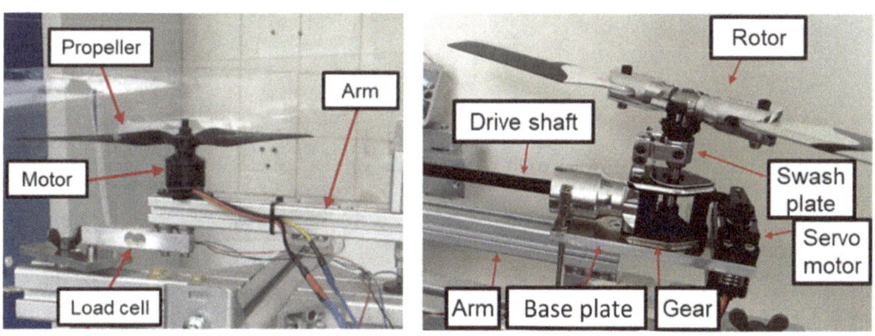

Fig. 9. Test stands for the quadcopter and the collective pitch rotor [9]

Figure 9 shows the generated thrust for both systems for a step from 0 to 100% power at the timestep 0 s. As expected higher dynamics of the helicopter system can be identified. Full thrust could be built up within 0.3 s. The quadcopter in our setup needed 0.6 s plus a delay time of 0.3 s. With a different speed controller this delay time can probably be reduced significantly. In the end the quadcopter will be twice slower, but therefore considerably more efficient and a lot cheaper, simpler and quieter.

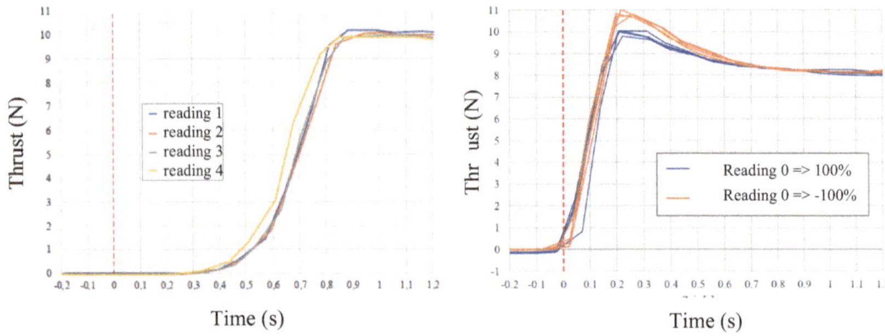

Fig. 10. Generated thrust for the quadcopter rotor (left) and the collective pitch rotor (right) [9]

3 Demonstration of the Actual Prototype

3.1 Current Status of the Robot

Figure 10 shows the actual Prototype on the façade. The winches with the belts drive the system as expected. Velocities up to 0.5 m/s can be reached. The tool can be orientated with the swivel head around all three rotating axis. The robot can be controlled by a browser via a web interface (Fig. 11).

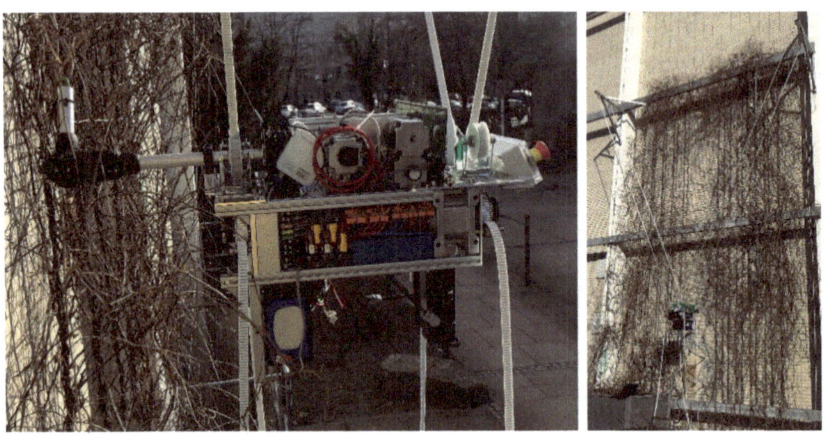

Fig. 11. Close-up view of the robot (left) and Photo of the robot installed at the façade greening (right)

3.2 Further Development Tasks

The robot is planned to be completed in summer 2021. The crucial and critical missing module is the rotor system for controlling the distance to the wall. Several tools for cutting have been build and need to be adapted to the robot. All other missing tasks and modules listed in Table 2 are seen uncritical and can be finished soon.

Table 2. Target deadlines for open tasks and modules

Task/module	Target deadline
Laser detector	March 2021
Belt storage system	April 2021
Rotor system	May 2021
Tool changer	June 2021
Adapting tools for cutting	June 2021
Final test and benchmarking	September 2021
Documentation	January 2022

4 Conclusion

Missing maintenance solutions hinder the extensive use of vertical green. The developed cable driven robot concept prevailed against all other motion structures. It is portable and can be used on multiple facades. The calculations and first tests show a sufficient precision. Multiple tools can be adapted to fulfill all needed task for vertical green maintenance. The robot is planned to be completed until summer 2021.

Acknowledgements. Many thanks go to all the students who supported this project with their master and bachelor theses. In addition a big thank you to the review team for the high number of qualitative advice.

The project "Vertical Green 2.0" is funded as a part of the program Sustainable Urban Global Initiative (http://www.sugi-nexus.org/) of the Joint Programming Initiative "Urban Europe" and the Belmont Forum.

References

1. Kühle, L.: An analysis of obstacles and elaboration of measures for an integrative implementation of façade greening in German Cities. Master thesis, Technical University of Berlin
2. Bregler, K., Kraus, N., Reisinger, M.: Handout green wall robot. https://www.ipa.fraunhofer.de/content/dam/ipa/de/documents/Projekte/Forschungsprojekte/2019_Handout_Green_Wall_Robot.pdf. Accessed 2 Mar 2021
3. Schmal, P.: Facade greening for the production of biomass –an energy balance. Master thesis, Technical University of Berlin
4. Scheugenpflug GmbH. https://commons.wikimedia.org/wiki/File:2_komponenten_einfachdosierer_auf_portal.jpg. Accessed 2 Mar 2021
5. Bruckmann, T., Sturm, C., Fehlberg, L., Reichert, C.: An energy-efficient wire-based storage and retrieval system. In: 2013 IEEE/ASME International Conference on Advanced Intelligent Mechatronics, Wollongong, NSW, Australia, pp. 631–636 (2013)
6. Canadian Space Agency/NASA. https://www.asc-csa.gc.ca/images/iss/canadarm2/fiche-comparative-canadarm-canadarm2-canadarm3.jpg. Accessed 2 Mar 2021

7. Savin, A.: WikiCommons. https://www.commons.wikimedia.org/wiki/File:DJI_Phantom_4Pro 04-2017_img3_in_flight.jpg. Accessed 12 Mar 2021
8. Pott, A.: Cable-Driven Parallel Robots. Springer, Heidelberg (2018). https://doi.org/10.1007/978-3-642-31988-4
9. Welling, M.: Konzeption und Bau einer Rotoreinheit zur gezielten Beeinflussung des Abstandes zur Wand von seilgeführten Bewirtschaftungsvorrichtungen für Gebäudebegrünung. Master thesis, Technical University of Berlin

Real-World Development of a Cleaning CDPR for Primary Lamella Sedimentation Tanks

Chun Ming Leung, Wai Yi Lam, Chun Keung Kwok, and Darwin Lau(✉)

The Chinese University of Hong Kong, Sha Tin, Hong Kong
cmleung@mae.cuhk.edu.hk, {waiyilam,chunkeungkwok,darwinlau}@cuhk.edu.hk

Abstract. Lamella sedimentation tanks are more efficient than conventional sedimentation tanks in treating sewage water because of their low footprint and high efficiency by the inclusion of inclined plates. However, regular cleaning is required for maintaining the efficiency of preventing clogging resulting from sludge accumulation. As the task is repetitive and dangerous, human cleaning is less preferable while no automatic solution is ready in the market. As a result, an automated solution is developed using a Cable Driven Parallel Robot (CDPR) to carry cleaning tools within the tank, due to its advantages of large workspace capability and simple mechanical set-up. In this paper, we introduce the application of lamella tank cleaning using CDPR, including the robot design, setup, operation and cleaning results. The development and results show the potential and maturity of using the developed CDPR system for real-world and longer-term uses.

1 Introduction

A primary lamella sedimentation tank is used to clarify raw sewage, the influent is first fed underneath the inclined plates then up through the plates, suspended particles settle on the plates while effluent exit the system via overflow box. However, sludges would accumulate overtime on the plates reducing the clarification efficiency and the quality of the effluent. Eventually, clogging may happen if the tank is not maintained and cleaned regularly. It is discovered that sludge has uneven distribution across the tank. One of the reasons for heavier accumulation in specific spots is the uneven water flow resulting in thicker accumulation on some plates. In the worst case, the saturated sludge would inevitably damage the plates because of overload [1]. Moreover, irregular accumulation results in more turbulence which also causes uneven flow and worsen the accumulation [2].

In order to prevent the light influence to drastic consequence from sludge accumulation, regular manual cleaning is essential. Currently, the tank is cleaned either by physical contact with a brush or water jet. Manual cleaning is a labour intensive and dangerous job for human workers. First, the worker is exposed to acidic gas for prolong period and at risk of falling into sewage. Second, the plates usually deep in water for one to two meters which implies the tools for cleaning

M. Gouttefarde et al. (Eds.): CableCon 2021, MMS 104, pp. 401–412, 2021.
https://doi.org/10.1007/978-3-030-75789-2_32

are hard to manoeuvre. In term of cleaning quality, there are several drawbacks of the existing methods. The facilities usually do not provide workers full access to every plate, temporary scaffold board is used to access the plate in the middle of the tank, which results in partial cleaning coverage. On the other hand, water jet could not penetrate the sewage, a thorough cleaning requires suspension of the tank operation.

Since the difficulties of human cleaning and needs for cleaning, a robotic solution to replace the tedious labour work is recommended. CDPRs is considered to be a solution for their large workspace capability and simple mechanical setup. In this particular application, the proposed solution involves the setup of a CDPR above the sedimentation tank. Manipulating and submerging the cleaning tool into sewage and clean the plates are the keys. Two different types of configuration has been considered, under-actuated suspended cable robots [3,4] and fully constrained cable robots [5–7]. The FAST [8,9] CDPR is the world's largest single-dish radio telescope used in surveying outer space is known for its huge workspace with a 500 m span. SpiderCam [10] and SkyCam [11] are other examples that demonstrate excellent workspace span, which is widely used in the stadium for dolly shots. The above examples are under-actuated suspended cable robot in class 2T and class 1R2T [12] which is vulnerable to external wrench. The Cable Robot Simulator (CRS) [13] is the world's first cable robot for human transport used for VR flight and driving simulation. The storage-retrieval machine [14,15] used in intralogistics for warehouse automated storage and retrieval are examples of fully constrained cable robot which provide better orientation capability. The capability to resist rotation is important in this particular application in which the end-effector has a long tubular tool extension. Hence, 8-6 configuration CDPR i s chosen while the workspace analysis has utilized CASPR [16] an open CDPRs research platform to study the system requirement of the robot.

In this paper, the cleaning CDPR for primary lamella sedimentation tank proposing an application scenario is introduced. To the best of the authors' knowledge, this is the first attempt to use a CDPR for lamella sedimentation tank cleaning. The proposed solution is able to save excessive manpower from tedious, while hazardous routine works by a streamlined workflow. Moreover, it tackled the problem of sludge accumulation which keep the tank efficiency at optimum. Our approach is capable to keep most part of the plates fresh, as verified by empirical results. Evaluation of the cleaning end tools is provided to optimize the cleaning performance and provide guidelines for researchers for further exploitation. This work demonstrates the maturity and feasibility of the solution with six months testing period, which overcomes the harsh weather condition.

2 Design

2.1 Supporting Frame

Figure 1 shows the original structures of the lamella sedimentation tank, the tank level is 5.5 m from the ground and no mounting points for the robot. As a result, a supporting frame is necessary. The frame has maximised the span while not colliding with the existing structure to maximize the workspace while minimizing the actuating power.

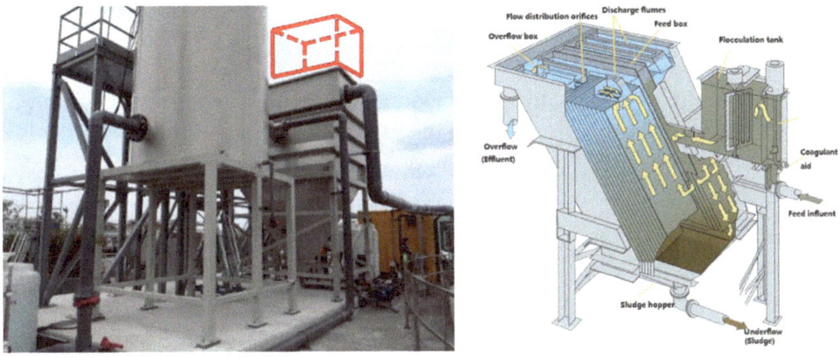

(a) Primary lamella sedimentation tank, red box showing the desired workspace

(b) Working principal of lamella sedimentation tank [17]

Fig. 1. Primary lamella sedimentation tank [17]

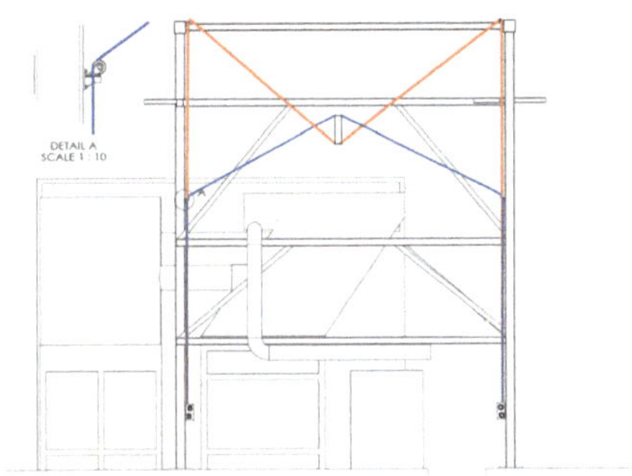

Fig. 2. Schematic of cleaning robot

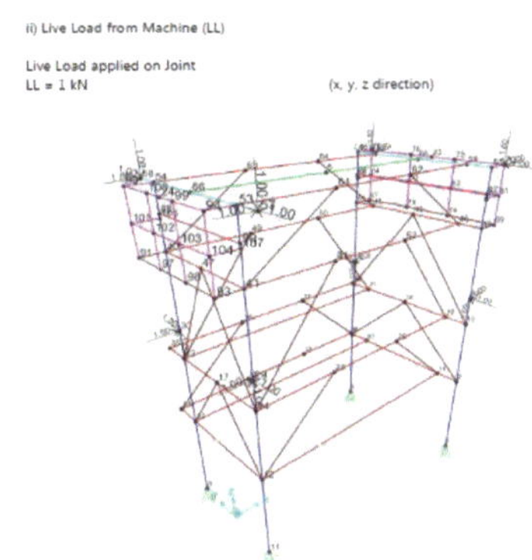

Fig. 3. Frame simulation with robot loading

For convenient maintenance and wiring of power and communication, all actuator winches are located at ground level. As such, all cables are therefore routed Fig. 2 from the ground level actuator winches to the bottom and top base cable attachment locations on the support structure. To prevent excessive friction, leading to wear and tear, on the cables, all attachment locations on the support structure are two degrees of freedom pulleys Fig. 5. The structural calculation of the frame has taken the robot loading acting on the frame into consideration Fig. 3. As the load direction varies during the robot operation, we took the most conservative assumption in the calculation. Each attachment point on the frame is assumed to have a live load of 1 kN in all x, y, z-direction, while the max cable force in normal operating condition is 880 N.

2.2 Analysis

Based on the tank and frame dimensions and thus attachment points, the required cable forces are computed using static workspace analysis.

$$\begin{aligned} \mathbf{f}^*(\mathbf{q}) &= \arg\min \quad \mathbf{f}^T\mathbf{f} \\ \text{s.t.} \quad \mathbf{g}(\mathbf{q}) &= -L(\mathbf{q})^T\mathbf{f} \\ \mathbf{f}_{min} &\leq \mathbf{f} \leq \mathbf{f}_{cont} \end{aligned} \tag{1}$$

where $\mathbf{f} = [f_1, f_2, f_3, f_4, f_5, f_6, f_7, f_8]^T$ represent the forces of each cable, and $\mathbf{g}(\mathbf{q})$ and $\mathrm{L}(\mathbf{q})$ are the gravity vector and the Jacobian matrix, respectively. At the pose $\mathbf{q}$ with fixed orientation (x y z 0 0 0), the existence of a feasible solution $\mathbf{f}^*$ implies that $\mathbf{q}$ is within the robot's workspace. Since the cleaning force is negligible, external forces are not considered in this analysis. For different selections of the continuous cable forces cont f, the resulting workspace allows the verification of whether the robot satisfies the operational requirements.

In Fig. 4, each column shows the particular workspace of desired cable force (300 N, 600 N & 900 N) at a different elevation (1.0 m, 1.75 m & 2.5 m). From the resulting workspace shown, a selection of continuous cable force of 900 N was deemed suitable for the system requirements. Hence, the required motor actuating torque is determined.

2.3 Winches and Pulleys

Figure 2, the distance from the caster pulleys to winches is 4.2 m and 7.8 m, resulting in lead angles β of 0.75° and 0.4° correspondingly. This provides an advantage to adopt a simple mechanism on the winch. For the system to physically operate within the required workspace of (6.5 m length × 3.3 m width × 3.6 m height), the maximum allowable change in cable length for the winch was designed to be 12.5 m.

To prevent the spools from back driving when the motors are powered off. A power-off brake (label 2 in Fig. 5) is selected to chain up the motor (label 1 in Fig. 5) and reducer (label 3 in Fig. 5) for the four upper attachment points. Figure 5 shows the CAD drawing of the winch components. As the setup of 8 cables is arranged such that each vertical pole would mount one actuation housing. Each housing house two motors and spools for convenience in assembly and mounting. And the winches are offset in-depth, so no cable interference along with cable routing. Most fabricated mechanical components are stainless steel 316 to overcome the adverse weather and acidic environment.

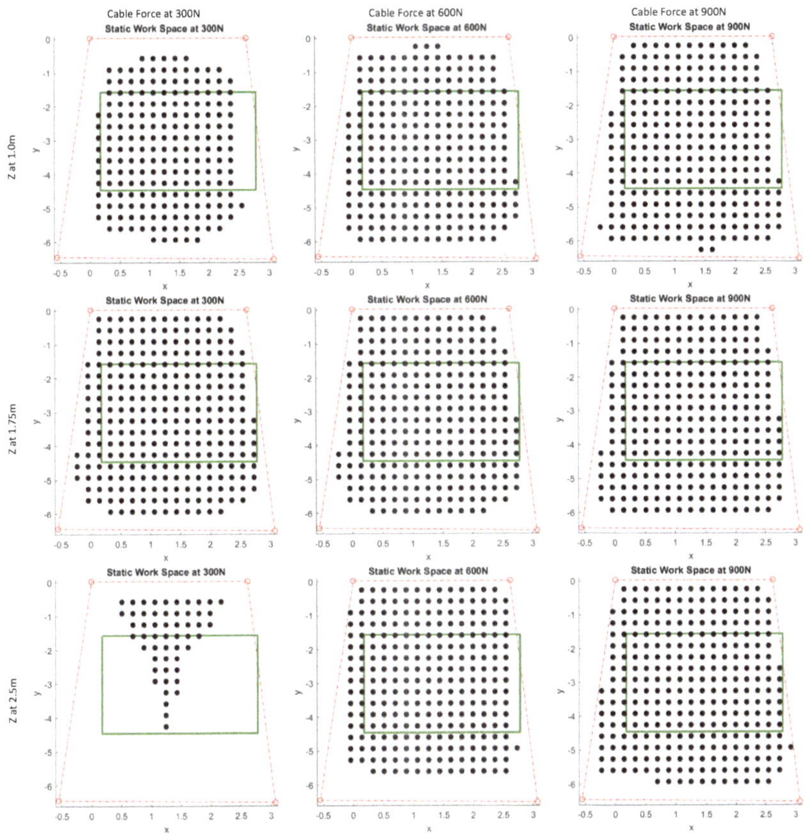

Fig. 4. Top view static work space analysis with the red boundary corresponding to the convex hull of the attachment locations and green boundary representing the lamella plate region to clean.

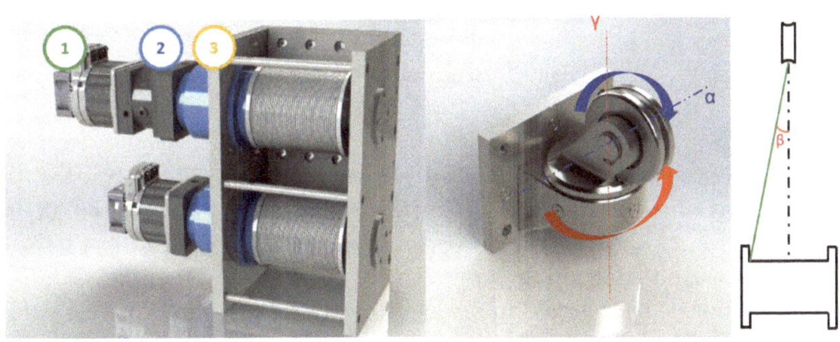

Fig. 5. CAD assembly of winch and caster pulley

2.4 Cleaning Tools

The cleaning tools have adopted a modular design, which consists of three components including the universal I-frame, the tools and the Bump-stop. The universal I-frame is where the cables attached. It has a holder at 55° and aligns with the inclination of the lamella plate. The tools are one meter long, and the tool head (brush and jet) is fixed at the end while the Bump-Stop is attached to the tube from 75 mm of the tip.

Brush. The nominal spacing dimension between lamella plates is 100 mm. The viscosity and adhesion strength of the sludge was tested by manually attempting to remove the sludge. From this test, it was found that the sludge can be removed by applying light force onto lamella plates. By these observations, the brush cleaning tool Fig. 6 was designed to use longer bristles. The length of the bristles was selected to be 140 mm. During operation, the centre of the cleaning tool is commanded to operate in the centre of two lamella plates. As such, the longer bristle length (20 mm) than the midway distance to the lamella plates (50 mm) ensures that the bristles would be bent when performing cleaning Fig. 7. The bent bristles provide both elasticities to absorb any slight positional inaccuracies of the cable-driven parallel robot and some stiffness to apply pressure to clean the plates.

Fig. 6. Brush cleaning tool

Jet. A non-contact approach is tested. A water jet nozzle is attached to the end-effector base mount Fig. 8. The entire setup consists of a water jet nozzle on the end-effector, and a compressor pump at the base platform of the tank (fixed location). As the cleaning is performed while the sedimentation tank is in operation, jet nozzle submerged in sewage, the nozzle pressure must be sufficient to clean the lamella plates at a distance of 50 mm. From tests, a cleaning pressure of 5 MPa is selected and provide a rough coverage of 75 mm.

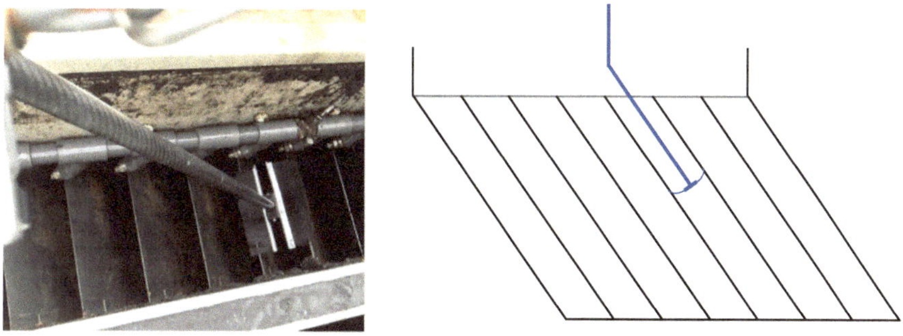

(a) Brush bristle over the lamella plate

(b) Schematic of brush bristle when within the lamella plate

Fig. 7. Brush bristle

Fig. 8. Water jet cleaning tool

Bump-Stop. For the protection of the robot system, limit switches were mounted to the cleaning end-effector. For convenience, a contact bumper module is developed to be attached to the end-effector. The sensor module comprises several limit switches (label 1 in Fig. 9) so that contact in front and back could be detected. The contact sensors connect to a wireless communication module (label 2 in Fig. 9), responsible for relaying any contact trigger signals to the processing PC to stop the robot.

3 Model Acquisition and Operation

3.1 Model Acquisition

Despite the carefully designed cable attachment pulley locations and corresponding mounting locations on the fabricated supporting frame, given the large scale and size of the structure, the actual attachment points are likely to vary from

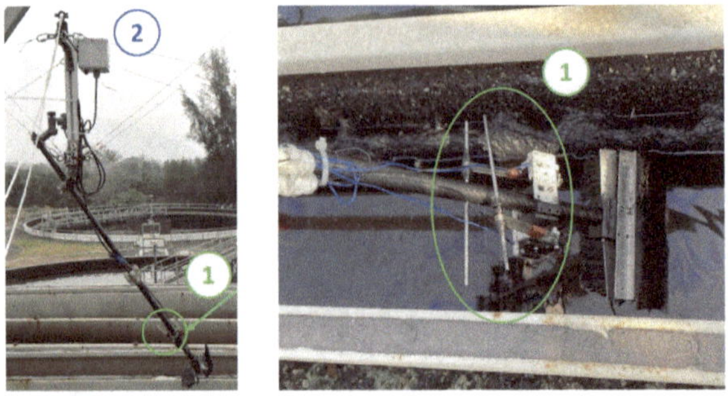

Fig. 9. Contact bumper sensor module

(a) Base Frame Configuration measurement

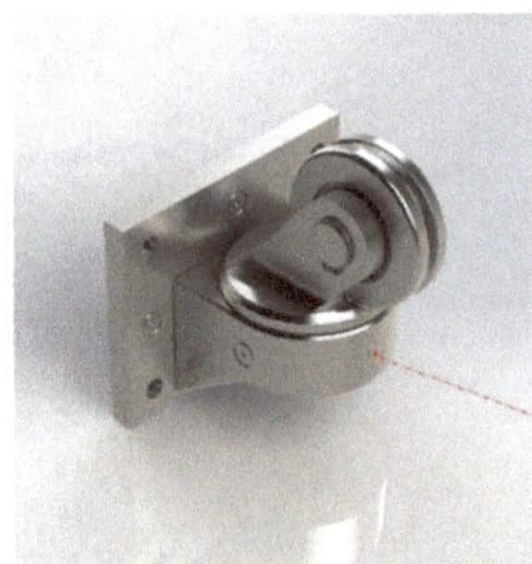

(b) Marker for calibration

Fig. 10. Calibration method

the designed position. As such, after the installation of the support structure and robot, the measurement of the caster pulley location is required.

The model acquisition is performed with a Leica DISTO s910, fixed at an arbitrary location that could cover the whole workspace. Measurements of marker locations Fig. 10 and the attachment points of the end-effector Fig. 11 are taken with the laser distance meter. The data acquired has an accuracy within ±5 mm in this convex hull range (Table 1). Retrieving the data and compensating the marker offset, the data is used to calculate the inverse kinematics with the software.

Apart from initial model acquisition, a vision-based calibration method is attempted for quick and effective auto-calibration (position and orientation) of the end-effector. In this application, an eye-to-hand approach is adopted because of a simpler mechanical setup. Basler acA2500-14um camera with Computar H0514-MP2 lens is situated on the base frame while placing the marker onto the end-effector. When the robot start-up, the actual pose and commanded home pose are compared, the robot will correct its posture accordingly.

(a) Attachments of base and end-effector

(b) Attachment of EE

Fig. 11. Attachment points

3.2 Operation

As the tool aligns with the plates, the orientation of the end-effector could assume to be constant and zero during cleaning. Such that the pose definition can be simplified as $\mathbf{q}(t) = (\mathbf{p}(t), 0, 0, 0)$. At a given time instant of a specific pose, the execution of the robot requires inverse kinematics for each cable length $\mathbf{l} = [l_1, l_2, l_3, l_4, l_5, l_6, l_7, l_8]^T$, where $\mathbf{l} = \mathbf{g}(\mathbf{q})$.

$$l_i = \|-\mathbf{a}_i + \mathbf{p} + \mathbf{b}_i\| \tag{2}$$

where $\mathbf{a_i}$ is the position vector from the base frame to the attachments on the frame. $\mathbf{b_i}$ is the position vector from the end-effector frame to attachments on the end-effector Fig. 11.

Table 1. Attachment points of the robot, a_i b_i.

i	Base a_i	End-effector b_i
0	$[2.683, 0.0505, 3.372]^T$	$[0.080, 0.050, -0.1775]^T$
1	$[-0.175, 0.0795, 3.375]^T$	$[-0.080, 0.050, -0.1775]^T$
2	$[2.614, 0.0325, 0.050]^T$	$[0.165, 0.045, 0.1775]^T$
3	$[0.000, 0.0325, 0.050]^T$	$[-0.165, 0.045, 0.1775]^T$
4	$[3.287, -6.5255, 3.555]^T$	$[0.080, -0.050, -0.1775]^T$
5	$[-0.680, -6.4305, 3.610]^T$	$[-0.080, -0.050, -0.1775]^T$
6	$[3.0680, -6.5075, 0.088]^T$	$[0.165, -0.045, 0.1775]^T$
7	$[-0.553, -6.4445, 0.075]^T$	$[-0.165, -0.045, 0.1775]^T$

The robot could run in manual mode and auto mode (predefined trajectory). The operation frequency is around 28 Hz and runs smoothly in position control. In normal cleaning operation, the nominal speed is set as $v = 50$ mm/s. The cleaning sequence is defined as Fig. 12.

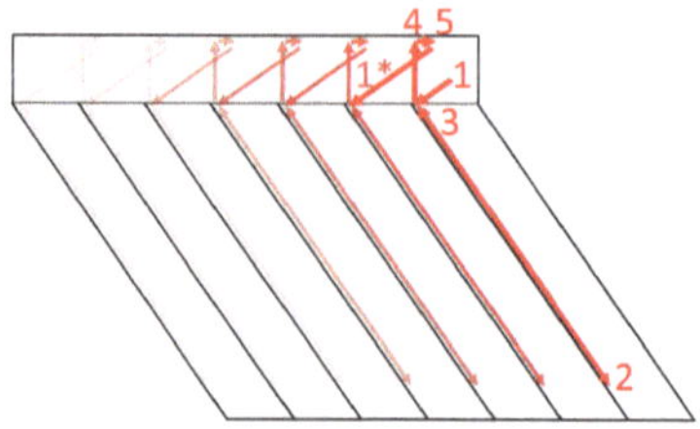

Fig. 12. Sequence of cleaning motion

4 Results

After testing, the robot is capable to jog at 100 mm/s with 12 kg load in the path specified in Fig. 13 with no significant cable slack. While the robot has tested in 300 mm/s with no load. The missing corner of the path is due to cable-frame interference. The robot could operate and perform effective cleaning during the tank operation. Both cleaning tools show satisfactory result and the bump-stop system works in both static and dynamic condition. In power-off condition, the robot remains rigid when subject to an external force.

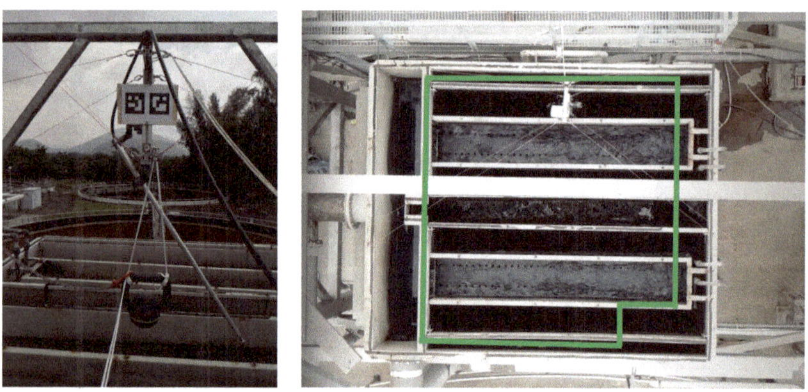

(a) Load with 12kg weight (b) Jogging path wit 12kg load

Fig. 13. Result with 12 kg load

4.1 Cleaning Effect

The tests show both types of cleaning tool have sufficient capability to clean the sludge accumulated on the lamella plate with appropriate contact strength. Different cleaning sequence and duration were tried. It is discovered that the sludge is relatively soft when the tank in normal operation while less soft and dry if the tank stops operating for a couple of days. However, in both cases, the robot could provide an equivalent cleaning effect. Above 90% of sludge is removed after one reciprocating motion by visual inspection.

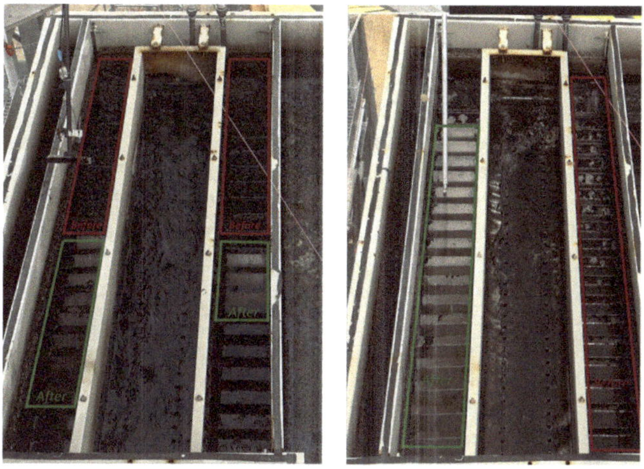

(a) Result of Brush (b) Result of Jet

Fig. 14. Cleaning Result

Figure 14 shows the cleaning effect of brush and jet, respectively. It clearly shows that the cleaning was able to significantly remove the black sludge to show the metallic surface of the plates. In conclusion, from both sets of cleaning tests, it is observed that both types of cleaning tools are sufficient in removing the accumulated sludge on the lamella plates. Although satisfying performance is obtained even with only a single cleaning run, multiple cleaning iterations would achieve better complete cleaning coverage.

5 Conclusion and Future Work

In this paper, a mature CDPR cleaning system is developed for primary lamella sedimentation tank cleaning. This system eliminates the risk of clogging the lamella tank and automates the cleaning operation that is dangerous and undesired to human workers. Future works will look into a long-term running with minimal human intervention to confirm the robustness of the system and the implementation of a CDPR for the cleaning of 30 m × 10 m full-size lamella sedimentation tanks.

References

1. Barquet, D.: Lamellar Modules: Maintenance Manual, July 2019. https://www.tecnoconverting.com/technical-articles/lamellar-modules-maintenancemanual/
2. McKean, T., Bourke, B., Mitchell, W., Caplygin, L., McKeown, R.: Novel application of a lamella clarifier for improved primary treatment of domestic wastewater. In: 73rd Annual Victorian Water Industry Engineers and Operators Conference, Exhibition Centre–Bendigo (2010)

3. Carricato, M., Abbasnejad, G.: Direct geometrico-static analysis of underconstrained cable-driven parallel robots with 4 cables. In: Cable-Driven Parallel Robots, pp. 269–285. Springer (2013)
4. Abbasnejad, G., Carricato, M.: Real solutions of the direct geometrico-static problem of under-constrained cable-driven parallel robots with 3 cables: a numerical investigation. Meccanica **47**(7), 1761–1773 (2012)
5. Gosselin, C.: Cable-driven parallel mechanisms: state of the art and perspectives. Mech. Eng. Rev. **1**(1), DSM0004–DSM0004 (2014)
6. Wu, Y., Cheng, H.H., Fingrut, A., Crolla, K., Yam, Y., Lau, D.: CU-brick cable-driven robot for automated construction of complex brick structures: from simulation to hardware realisation. In: 2018 IEEE International Conference on Simulation, Modeling, and Programming for Autonomous Robots (SIMPAR), pp. 166–173. IEEE (2018)
7. Lau, D., Oetomo, D., Halgamuge, S.K.: Generalized modeling of multilink cable-driven manipulators with arbitrary routing using the cable-routing matrix. IEEE Trans. Robot. **29**(5), 1102–1113 (2013)
8. Duan, B., Qiu, Y., Zhang, F., Zi, B.: Analysis and experiment of the feed cable-suspended structure for super antenna. In: 2008 IEEE/ASME International Conference on Advanced Intelligent Mechatronics, pp. 329–334. IEEE (2008)
9. Nan, R., et al.: The five-hundred-meter aperture spherical radio telescope (FAST) project. Int. J. Mod. Phys. D **20**(06), 989–1024 (2011)
10. Spdiercam, September 2018. https://www.spidercam.tv/
11. Skycam: Advanced technology, September 2018. http://www.skycam.tv/
12. Bruckmann, T., Mikelsons, L., Brandt, T., Hiller, M., Schramm, D.: Wire robots part I: kinematics, analysis & design (2008)
13. Miermeister, P., Lächele, M., Boss, R., Masone, C., Schenk, C., Tesch, J., Kerger, M., Teufel, H., Pott, A., Bültho , H.H.: The cablerobot simulator large scale motion platform based on cable robot technology. In: 2016 IEEE/RSJ International Conference on Intelligent Robots and Systems (IROS), pp. 3024–3029. IEEE (2016)
14. Bruckmann, T., Lalo, W., Nguyen, K., Salah, B.: Development of a storage retrieval machine for high racks using a wire robot. In: International Design Engineering Technical Conferences and Computers and Information in Engineering Conference, Vol. 45035, pp. 771–780. American Society of Mechanical Engineers (2012)
15. Hassan, M., Khajepour, A.: Analysis of a large-workspace cable-actuated manipulator for warehousing applications. In: International Design Engineering Technical Conferences and Computers and Information in Engineering Conference, Vol. 49040, pp. 45–53 (2009)
16. Lau, D., Eden, J., Tan, Y., Oetomo, D.: CASPR: a comprehensive cablerobot analysis and simulation platform for the research of cable-driven parallel robots. In: 2016 IEEE/RSJ International Conference on Intelligent Robots and Systems (IROS), pp. 3004–3011. IEEE (2016)
17. Controlling Sludge in Lamella Clarifiers, February 2017. https://www.partech.co.uk/controlling-sludge-in-lamella-clariers/

Cable-Driven Robot to Simulate the Buoyancy Force for Improving the Performance of Underwater Robots

Alejandro Rodriguez-Barroso(✉) and Roque Saltaren

Polytechnic University of Madrid, Madrid, Spain
alejandro.rbarroso@upm.es

Abstract. The underwater environment presents conditions very different from those found on the ground so the control algorithms and the design of submarine robots have to consider this circumstance. Usually, the experiments are performed in real pools after the computational simulations, imposing a high leap in the testing process that requires to implement various requirements and subsystems for the first real world tasks. For this reason, an on-ground simulator of the underwater environment is developed based on a cable-driven parallel robot. This technology allows us to create a scenario where the aquatic robots can test their algorithms in a secure, controlled and customizable environment to guarantee a good performance before the experiments performed with real immersions in swimming pools. The Buoyancy force is simulated on a rigid end-effector in different application points while its accuracy is measured with the reaction forces applied by the body to the ground and tracking the zero-moment point displacement. A method to select the motors and the mechanical components for this task is shown.

Keywords: Parallel robots · Robot kinematics · Underwater technology · Force measurement

1 Introduction

Making experiments in underwater environments with robots for seabed operations or tasks close to solid obstacles like the "Girona 500" [1], "DiverBot" [2], "Little Crabster" [3], "CR200" [4], "UX-1" [5], soft underwater walking robots like [6] among others, presents a series of difficulties that have to be considered before the experimental stage of development in real environments. For instance, the physical properties of the ground inside the pools where the initial tests are performed used to be slippery which difficult the crawling control algorithms because they use to depend on the friction between the floor and the limbs, as analyzed in [7]. The dimensions of those pools use to be too small so the motion of the robot is constrained to that region. If more space is needed and the tests are developed in environments like oceans, lakes, or rivers, the terrain is uneven and has obstacles like rocks, algae, or rifts that will impose additional requirements for the control algorithms of the robot to consider and handle them like [8]. Basic control algorithms developed at the first stages would find difficulties to achieve

M. Gouttefarde et al. (Eds.): CableCon 2021, MMS 104, pp. 413–425, 2021.
https://doi.org/10.1007/978-3-030-75789-2_33

their objectives in those real environments. The inherent characteristics of the aquatic environment also present some additional difficulties that have to be considered like the underwater currents, turbidity, or viscosity.

When a robot is submerged, the most common sources of failure are related to water infiltration or hydraulic oil leaks. Those failures could produce great damage to the electronics of the robot and the surrounding environment. Besides, the access to the submerged robot is more difficult while performing the experiments by technicians that would require making in-situ repairs, modifications of the robot's mechanisms, or making brief tests or measurements. Swimming suits and autonomous breathing devices should be needed to access the emplacement of the robot during the tasks.

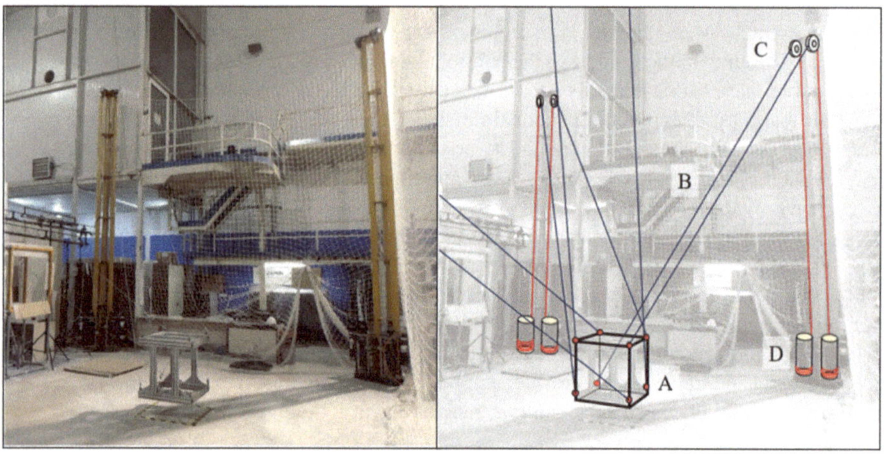

Fig. 1. *Left:* CDPR used in this analysis. *Right:* scheme of the CDPR with the end-effector (A), the cables (B), the last pulleys (C) and the winches (D).

The proposed solution is the novel implementation of a cable-driven parallel robot (CDPR) [9] to simulate the underwater conditions of the robots before submerging them, performing the desired experiments on-ground. Multi-body modeling considering hydrodynamic effects (added inertia, buoyancy, lift, and drag) is made in [10–12]. This new stage of the experimental development of the underwater robots would be performed before the tests in the real emplacement and after the computer simulations and virtual experiments [13] that validate the algorithms and the design of the prototype. The main contribution of this work is the validation of the buoyancy forces applied in a real prototype according to the analyzed buoyancy forces as well as the actuator selection according to this application.

A CDPR is composed of a moving end-effector, where the motion and wrenches are imposed; some cables attached to it; and the pulleys where the cables are coiled through until the winches, in which the motors apply the power to the system. A scheme of a CDPR is shown in Fig. 1. Attending to the advantages of the CDPRs related to their huge workspaces, the high wrench-to-weight ratio, and low inertia, the underwater dynamics and conditions are intended to be replied by using this kind of robot. The feasibility of the CDPRs was already demonstrated in previous prototypes where specific controlled wrenches were intended to be applied to human subjects like in rehabilitation as seen in [14, 15] or to act as haptic devices like in sports simulators [16], walking simulators [17], among others.

2 Buoyancy Force

The influence of the hydrodynamic forces is fundamental for underwater robots with fast movements; however, those that crawl on the seabed [18], due to their low speed, use to be focused on the contact forces with the environment and the static forces like the gravity and buoyancy applied in their center of mass (CoM) and center of buoyancy (CoB) respectively. These two conditions are also considered in lightweight terrestrial robots like BALLU [19]. The buoyancy force is defined as:

$$\boldsymbol{F_B} = -\rho V_f \boldsymbol{g} \tag{1}$$

Where $\boldsymbol{F_B}$ is the buoyancy force, ρ is the density of the fluid, V_f is the volume of the submerged body, and $\boldsymbol{g}$ is the vector corresponding to the gravity force. The buoyancy force has an upper vertical direction. In submerged rigid solids, this force can be modeled as applied in the CoB, which corresponds to the CoM of the displaced fluid. Assuming that a constant density of the water, the CoB will correspond to the center of volume or centroid of the rigid body.

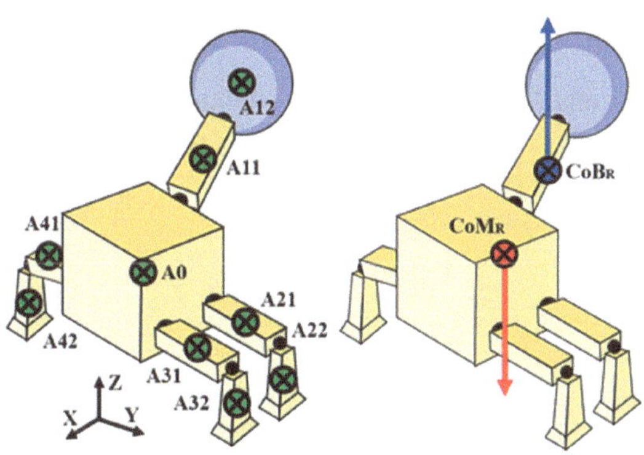

Fig. 2. *Left*: Underwater multi-body robot. Each limb has the CoM and CoB coincident (**Aij**). *Right:* resultant CoM and CoB of the entire robot. The limb 12 has lower density than the rest.

2.1 Resultant Buoyancy

A mechanism or model composed of several rigid bodies has one CoB for each limb. Despite some prototypes where more than one end-effector is used like [20], normally, the CDPR has the ability to impose a wrench to one rigid body so that, the body with higher volume can be chosen to be anchored to the cables of the robot and act as the end-effector.

The application of the wrench to one body by the cables requires the knowledge of the resultant buoyancy force and its point of application. In Fig. 2, the CoB of each limb is reduced to one force that acts over the entire robot considered as one body. The position of the resultant COB (**COB$_{\mathbf{R}}$**) is defined by the following coordinates:

$$CoB_{RX} = \frac{\sum_i \sum_j V_{ij} X_{ij}}{V_{ij}}; \tag{2}$$

$$CoB_{RY} = \frac{\sum_{ii} \sum_{jj} V_{ij} Y_{ij}}{V_{ij}}; \tag{3}$$

$$CoB_{RZ} = \frac{\sum_i \sum_j V_{ij} Z_{ij}}{V_{ij}} \tag{4}$$

Where the i-index corresponds to the number of arm or leg and the j-index to each limb. Equations (2)–(4) can be either used to obtain a CoB for each arm or to remove the influence of some limbs from the analysis.

The resultant buoyancy force exerted is the sum from all the limbs:

$$\boldsymbol{F_B} = \sum_i \sum_j \boldsymbol{F_{Bij}} \tag{5}$$

2.2 Wrench Exertion

The CDPR has to simulate just the underwater buoyancy so the wrench needed to exert by the cables corresponds to the resultant buoyancy force ($\boldsymbol{F_B}$) applied in the point **CoB$_{\mathbf{R}}$** = [XR, YR, ZR]. This resultant force is calculated for each new limb position as well as the Jacobian matrix. This wrench is:

$$\boldsymbol{W} = [\boldsymbol{F_B}(\boldsymbol{CoB_R} \times \boldsymbol{F_B})]^T \tag{6}$$

CDPR sometimes have more cables than DoF in the end-effector. This redundant constraint, analyzed in [21], allows us to handle the inability of the cables to push the load, increase the Wrench Feasible Workspace (WFW) from and apply wider ranges of wrenches. The relationship between this required wrench and the force exerted by the cable tension is:

$$\boldsymbol{W} = \boldsymbol{J}^T \boldsymbol{\tau} \tag{7}$$

Being $\boldsymbol{\tau}$ the tension of each cable and $\boldsymbol{J}$ the Jacobian Matrix, which is defined as:

$$\boldsymbol{J} = [\boldsymbol{u_k}(\boldsymbol{p_k} \times \boldsymbol{u_i})] \tag{8}$$

Being $\boldsymbol{u_k}$ the unitary vector parallel to each cable, from the end-effector to the last pulley and $\boldsymbol{p_k}$ the vector from the origin of coordinates of the end-effector and the $\boldsymbol{k}$ anchor of the $\boldsymbol{k}$ cable.

Due to the redundancy configuration, the value of the cable tension for each desired wrench has infinite solutions, which can be solved in the way from [22]:

$$\tau = J^{+}W + N\left(J^{T}\right)\lambda \tag{9}$$

Being (J^{+}) the pseudoinverse of the transposed Jacobian Matrix, $N(J^{T})$ the nullspace of $\mathbf{J}^{\mathrm{T}}$ and λ the independent values from the basis of the nullspace.

3 Reaction Force and Equilibrium

One way to verify that the CDPR is exerting a wrench equal to the expected buoyancy is to measure the zero-moment point (ZMP), whose geometric interpretation can be found in [23], obtained from the robot contacts with the ground.

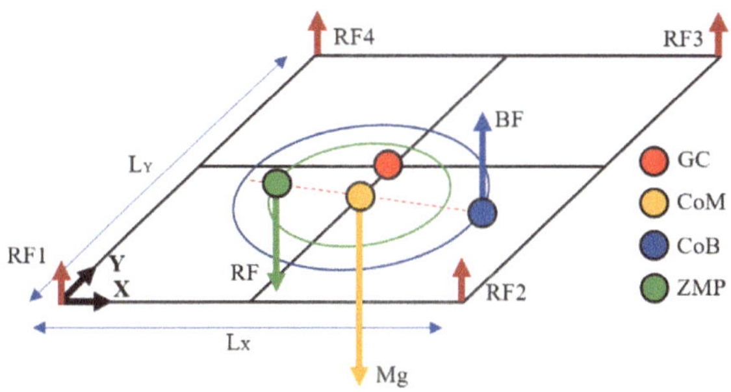

Fig. 3. Forces considered. BF: Buoyancy Force applied in the CoB; Mg: Weight applied in the COM; RF: Total reaction force applied. RFx: Reaction force; GC: Geometric center.

The reaction forces are measured by load cells situated in the legs of the robot or in its body to verify if the buoyancy is exerted with precision. Those reaction forces provide the ZMP and the total force exerted by the robot to the ground (RF), as observed in Fig. 3. The expressions to obtain this ZMP from the ground measures are obtained from the following equalities:

$$(RF_1 + RF_4)ZMP_X = (RF_3 + RF_2)(L_X - ZMP_X) \tag{10}$$

$$(RF_1 + RF_2)ZMP_Y = (RF_3 + RF_4)(L_Y - ZMP_Y) \tag{11}$$

Where the ZMP is:

$$ZMP_X = \frac{(RF_2 + RF_3)L_X}{(RF)}; \tag{12}$$

$$ZMP_Y = \frac{(RF_3 + RF_4)L_Y}{(RF)} \tag{13}$$

With:

$$\boldsymbol{RF} = \sum_{l} \boldsymbol{RF}_l \tag{14}$$

The ZMP can also be obtained from the forces applied to the robot. For this application, those are the gravity forces inherent to the bodies and the resultant buoyancy force, as observed in Fig. 3. The value of the RF from these forces is:

$$\boldsymbol{F_B} - M\boldsymbol{g} = \boldsymbol{RF} \tag{15}$$

And the ZMP from the inertial and buoyancy force:

$$\|\boldsymbol{CoB_R}\|\|\boldsymbol{F_B}\| - \|\boldsymbol{CoM_R}\|\|(M\boldsymbol{g})\| = \|\boldsymbol{RF}\|\|\boldsymbol{ZMP}\| \tag{16}$$

With these two definitions of the ZMP from the forces applied to the body and those applied to the ground, the validation of the performance of the buoyancy simulator can be performed.

4 Robot Description

In this section, the CDPR built in the Center of Automation and Robotics in Madrid for this purpose is described. It has an area of 9 $\times$ 7.5 m and 6 m in height. The experiments are performed with a rigid end-effector situated in a random and non-centered or symmetric position.

4.1 Mechanical Components

The end-effector in this application is the main limb of the underwater robot. This can be considered as that with the bigger volume or the limb that is in contact with the floor or the surrounding environment. For the tests considered in this article, the end-effector is a rigid body of 152 kg shown in Fig. 4. The anchor points are situated in the 8 corners of the prism with dimensions 0.63 $\times$ 0.77 $\times$ 0.65 m. The connection pattern followed is the same as the obtained in [24]. The suspended configuration allows to exert higher vertical forces.

The last pulleys are situated in the upper part of the robot so that 4 pillars are intended to support 2 pulleys in the top part of each one. The pillars are designed to be modular to have the possibility to choose the height of each one.

The selection of the characteristics of the cable are based on the maximum tension required and also the minimum bending radius. Higher tension requirements will need wider cables while smaller pulleys will require more flexible cables.

The main purposes o f the winches are:

- Impose the required tension to the cable coiled in it.
- Measure the tension value and the turns performed by the drum.
- Avoid that one section of the coiled cable could leap over other section.

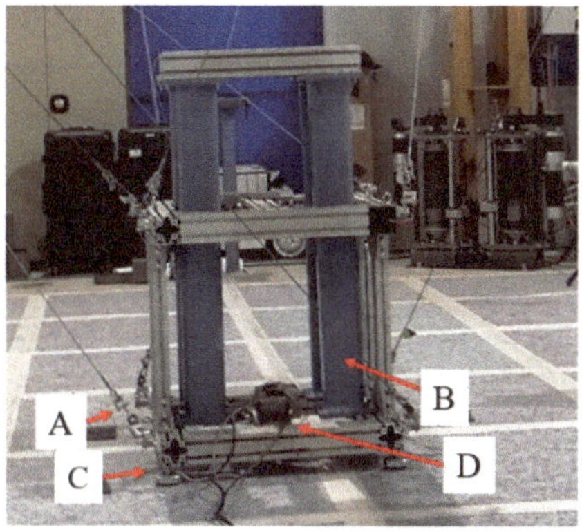

Fig. 4. Rigid end effector where the buoyancy force is simulated. A: Universal joints. B: Added weights. C: Four load cells in the corners of the base. D: Controller CompaqtRIO from *NI*.

Fig. 5. *Left:* A: motor. B: Drum to coil the cable. C: Toothed-belt. D: Aligning platform. E: Tension measurement mechanism. F: Worm to produce linear movement. G: Final alignment plate. H: Toothed-belt. *Center:* Two winches in their final location in the base of one pillar. *Right:* Tension measurement mechanism. A: Pulley that receive the horizontal cable from the drum and deploy it vertically. B: Load cell. C: Alignment 45° piece. D: Aligning platform. E: Bearing. F: Linear cam

The tension of the cable can be measured in different points: in the motor by measuring the current, in the winch, in the last pulley, or the attachment with the end-effector. Each location has some advantages and disadvantages: Measuring in the motor does not require additional load cells, but all the guiding elements that impose friction and inertia are not considered in the measurement. Measuring in the winch allows having the sensors in an accessible and compact place.

The last pulley could provide more accurate measures, despite the mass and inertia of the cable from it until the end-effector; however, a complex aligning mechanism is needed as well as a long distance to cover by the signaling cable to the acquisition device. Finally, the measurement in the end-effector provides the most accurate tension value;

nevertheless, the direction of its corresponding vector requires complex mechanisms able to measure without disturb cable direction and tension and the end-effector will need a data cable connected to it. In the proposed robot, the measurement is performed in the winch by using an alignment system detailed in Fig. 5.

4.2 Motor Selection

The selection of the motor of the winch, the gearbox, and the diameter of the drum depends on the application of the CDPR and the required ranges of forces and velocities. For this underwater simulator, low speeds are expected for the bodies submerged as observed in [25], where a maximum speed of 1.5 m/s can be accepted. The maximum tension value in the cables is established as 800 N attending to a previous study of the required wrenches in the end-effector along with the workspace by using (9). The relationship between the deploy velocity of the cable and the motor speed is:

$$\dot{l} = \frac{r_D \dot{\theta}_D}{i} \tag{17}$$

With l the cable length deployed, r_D the drum radius, i the gearbox reduction, and θ_D the angle rotated by the motor. On the other hand, the tension imposed on the cable related to the motor torque is:

$$\frac{F_D r_D}{i} = T_D \tag{18}$$

With F_D the force imposed to the cable by the drum and T_D the torque of the motor. The tension value of the cable close to the end-effector (τ) do not have to be the same of the be the same as F_D because there are some pulleys between both cable sections whose friction and inertias could modify the tension value.

As the cable is made of steel, the minimum radius of the drum (r_D) is defined by its flexibility. As the maximum required tension is 800N, the selected cable is the stainless steel 7×19 with a diameter of 4 mm. This cable has a Working Load Limit (WLL) of 1.81 kN and a Minimum Breaking Load (MBL) of 9 kN. These characteristics provide a security factor for the cable of $9/0.8 = 11.25$ for a normal working mode of the CDPR.

The desirable minimum diameter for this cable is 95 mm and its critical minimum is 73 mm. These magnitudes impose a constraint in the drum diameter. Assuming a motor without gearbox ($i = 0$) and the thinner drum possible of $r_{Dmin} = 47.5$ mm, the required speed and torque of the motor are:

$$\theta_D \geq \frac{1.5}{r_{Dmin}} = 402\text{RPM}; \tag{19}$$

$$T_D \geq 800 r_{Dmin} = 38\text{Nm} \tag{20}$$

The common values provided by the catalogs are the stall torque, the speed and the rated power. The power required by the motor attending to Eq. (19) and (20) is:

$$T_{DStall}\dot{\theta} = \frac{1280 r_{Dmin}}{i} \frac{1.5i}{r_{Dmin}} = 1920\text{W} \tag{21}$$

This required power is represented in Fig. 6, where the stall torque and the nominal motor speed of different motors from "Unimotor". The motor selected is the one with 6.3 Nm of stall torque, 3000 RPM and 1.76 Kw.

The ratio r_D/i expressed in (17) and (18) is 0.0048 and 0.0079 respectively. Taking the lower one, the boundary of the exerted power is applied at the desired speed while the tension of the cable can be increased until 1300 N. This ratio provides the values $r_D = 76.2$ mm and $i = 16$.

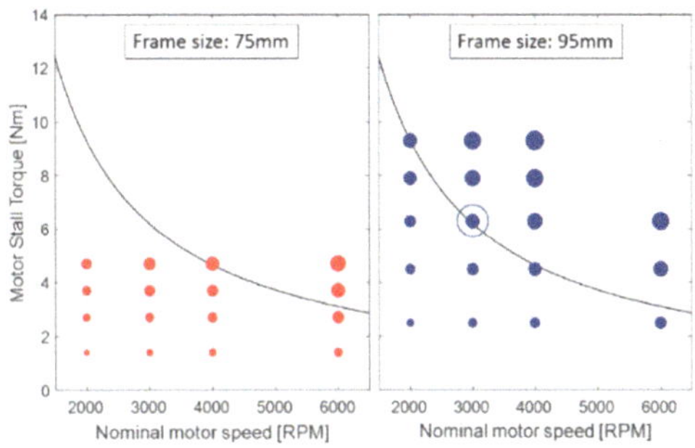

Fig. 6. Power of different DC motors from "Unimotor". The black line represents the required power, being the upper part the region with motors that fulfil the requirement.

5 Application

The experiments performed in this section allow to understand the response of heavy loads under the tensions that mimic the buoyancy force and also to measure the accuracy of the underwater simulation by tracking the ZMP measured by the load cells of the platform. The force distribution when a buoyancy force of 500 N i s applied at 0.2 m from the projection of the center of mass is shown in Fig. 3. This upward constant force is applied with the cables displacing the CoB_R along the blue circle centered in the CoM. To verify the accuracy of this applied force the resultant force is measured with the load cells situated in the corners of the base and computed the ZMP that should follow the green circle of radius 0.1 m.

In order to obtain the desired wrenches in the platform, a joint space control has been performed following a force reference and feedback of the tension values read from the winch mechanism described in Fig. 5.

The region of the trajectories followed by the ZMP in one test of the experiment, as well as its desired path is shown in Fig. 7. It corresponds to a circumference of 0.1 m of radius, according to Eq. (16). The point with a maximum radius value is 0.16 m while the minimum is 0.08 m. Considering the magnitude of the reaction force, the buoyancy force is desired to maintain its magnitude of 500 N while moving the application point along the green circle. The end-effector's weight is 152 kg so 101 kg is expected as a reaction force.

To check the accuracy and the repeatability, 5 different tests have been performed for this experiment to measure the ground reaction force from each test and the distance

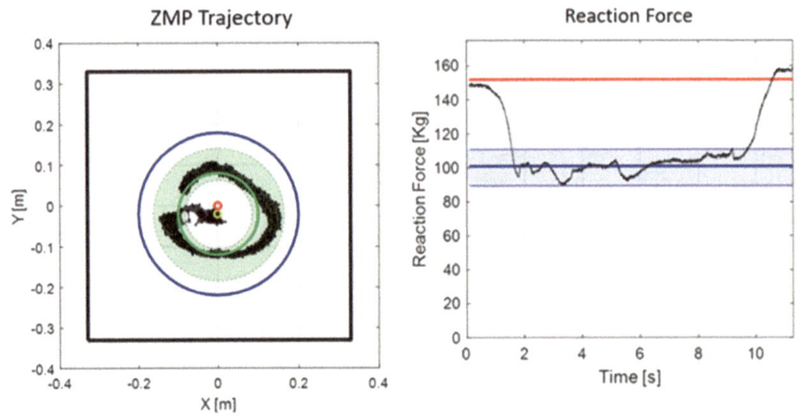

Fig. 7. ZMP and reaction force value of one random test from the experiment. The ZMP (black dots) begins in the CoM (yellow dot) and moves towards the desired initial buoyancy value in the first 2 s (solid green circle). After that the ZMP rotates 330° in 4 s and come to the initial buoyancy state in other 4 s. Finally, the end-effector returns to its initial state in 1 s. Blue circle represents the corresponding buoyancy force.

Table 1. Mean and standard deviation of each test.

Ground reaction force [kg]		ZMP deviation [mm]	
Mean	Std.	Mean	Std.
102.6	5.0	10.9	8.3
107.0	4.1	19.0	6.7
105.9	3.8	11.3	7.0
105.8	4.2	10.9	7.4
108.6	4.4	11.3	7.3

Table 2. Mean and standard deviation of the experiment.

	Ground reaction force [kg]		ZMP deviation [mm]	
	Mean	Std.	Mean	Std.
Measured	105.1	2.2	10.9	0.53
Expected	101.0	0	0	0
Abs. error	4.1	–	10.9	–
Rel. error	4.06%	–	–	–

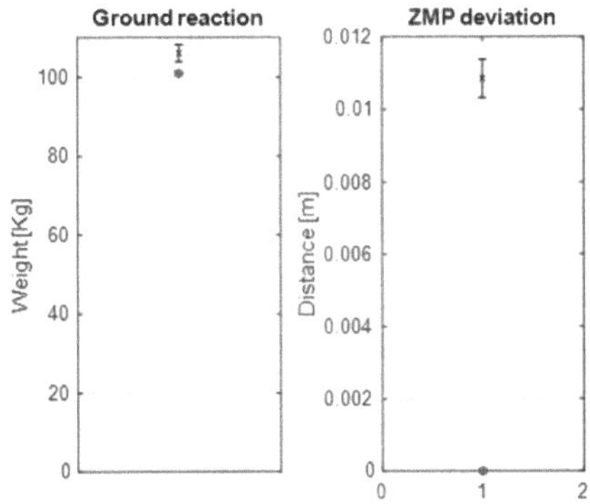

Fig. 8. Desired ground reaction and ZMP deviation expressed with red points. Measured data from the experiment showing the mean as a black cross and the standard deviation as a vertical line.

of the ZMP from the desired point of the circular trajectory. The results are shown in Table 1 while the accuracy and repeatability of the CDPR, measured with the mean and the standard deviation, are shown in Table 2. Figure 8 shows both graphs.

6 Discussion

The goal was to replicate the underwater buoyancy force exerted in a single body by using a CDPR. This rigid body acts as the main limb of an underwater robot and 6 DoF wrenches can be applied to it. This experiment allowed us to validate the exertion of the buoyancy force by measuring the expected reaction force in the ground. The application point of this force has been displaced along the base of support simulating the resultant force produced by moving different limbs of a multibody underwater robot emplacing on the seabed.

The position where the cables are attached must be known in order to calculate the Jacobian Matrix in real-time, as well as the volume of the limbs and their position to calculate the resultant buoyancy force and its position. For those cases where the robot is displacing, its position can be known by using external sensors like cameras, laser trackers, or internal measurements like the indirect provided by using forward kinematics algorithms and considering adequate cable models.

There are some discrepancies observed in Fig. 8 from one random test that shows that the upper force exerted is around 4 kg lower than expected. This could be related to the weight of the 8 cables that are not considered as the weight of the cable is 6.1 kg per 100 m. This issue could be solved by considering the cable weight of the section from the last pulley to the end-effector. On the other hand, the deviation in the position of the ZMP could be due to the different dynamic forces that act regarding the cable length deployed. This error could be minimized by considering the inertia of the cables and their dependence on the length deployed.

In addition, an appropriate analysis of the inertial and friction forces that act on the pulleys is needed to achieve better results. The winch dynamics has been overcome by applying a feed forward based on its inverse dynamic model like in [26]. Nonetheless, the mean error of 4.1 kg in the buoyancy force and 10.9 mm in the CoB is low enough to simulate the static underwater influence in heavy robots as the proposed and test their motion algorithms and stability control to maintain its ZMP inside the stability polygon. Also knowing that any wrench of 6DoF can be applied to the main body of the aquatic robot, underwater currents can be simulated as well as collisions, drag, or added mass forces.

Future works will improve the accuracy of the CDPR simulator, increasing its speed of motion of the CoB and working with mobile and autonomous underwater robots as well as making comparisons with real underwater environment.

References

1. Prats, M., et al.: Multipurpose autonomous underwater intervention: a systems integration perspective. In: 2012 20th Mediterranean Conference on Control & Automation (MED), pp. 1379–1384 (2012)
2. Saltaren, R., Barroso, A.R., Yakrangi, O.: Robotics for seabed teleoperation: part-1-conception and practical implementation of a hybrid seabed robot. IEEE Access **6**(c), 60559–60569 (2018). https://doi.org/10.1109/ACCESS.2018.2876040
3. Kim, J.-Y., Jun, B.-H.: Design of six-legged walking robot, Little Crabster for underwater walking and operation. Adv. Robot. **28**(2), 77–89 (2014)
4. Jun, B.-H., et al.: Development of seabed walking robot CR200. In: 2013 MTS/IEEE OCEANS-Bergen, pp. 1–5 (2013)
5. Milošević, Z., Fernandez, R.A.S., Dominguez, S., Rossi. C.: Guidance and navigation software for autonomous underwater explorer of flooded mines
6. Ishida, M., Drotman, D., Shih, B., Hermes, M., Luhar, M., Tolley, M.T.: Morphing structure for changing hydrodynamic characteristics of a soft underwater walking robot. IEEE Robot. Autom. Lett. **4**(4), 4163–4169 (2019)
7. Al-Homsy, A., Hartmann, J., Maehle, E.: Slippery and sandy ground detection for hexapod robots based on organic computing principles and somatosensory feedback. In: 2012 IEEE International Symposium on Robotic and Sensors Environments Proceedings, pp. 43–48 (2012)
8. Braginsky, B., Guterman, H.: Obstacle avoidance approaches for autonomous underwater vehicle: simulation and experimental results. IEEE J. Ocean. Eng. **41**(4), 882–892 (2016)
9. Pott, A., Bruckmann, T. (eds.): Cable-Driven Parallel Robots. MMS, vol. 32. Springer, Cham (2015). https://doi.org/10.1007/978-3-319-09489-2
10. Krishnamurthy, P., Khorrami, F., de Leeuw, J., Porter, M.E., Livingston, K., Long, J.H.: A multi-body approach for 6DOF modeling of biomimetic autonomous underwater vehicles with simulation and experimental results. In: 2009 IEEE Control Applications (CCA) & Intelligent Control (ISIC), pp. 1282–1287 (2009)
11. Nielsen, M.C., Eidsvik, O.A., Blanke, M., Schjølberg, I.: Validation of multi-body modelling methodology for reconfigurable underwater robots. In: OCEANS 2016 MTS/IEEE Monterey, pp. 1–8 (2016)
12. Antonelli, G.: Underwater Robots. STAR, vol. 96. Springer, Cham (2014). https://doi.org/10.1007/978-3-319-02877-4
13. Ridao, P., Battle, J., Amat, J., Carreras, M.: A distributed environment for virtual and/or real experiments for underwater robots. In: Proceedings 2001 ICRA. IEEE International Conference on Robotics and Automation (Cat. No. 01CH37164), vol. 4, pp. 3250–3255 (2001)

14. Surdilovic, D., Zhang, J., Bernhardt, R.: STRING-MAN: wire-robot technology for safe, flexible and human-friendly gait rehabilitation. In: 2007 IEEE 10th International Conference on Rehabilation Robotics (ICORR 2007), vol. 00, no. c, pp. 446–453 (2007). https://doi.org/10.1109/ICORR.2007.4428463
15. Khan, M.I., et al.: Enhancing Seated Stability Using Trunk Support Trainer (TruST). IEEE Robot. Autom. Lett. **2**(3), 1609–1616 (2017)
16. Zitzewitz, J.V., Rauter, G., Steiner, R., Brunschweiler, A., Riener, R.: A versatile wire robot concept as a haptic interface for sport simulation. In: Proceedings of IEEE International Conference on Robotics and Automation, pp. 313–318 (2009). https://doi.org/10.1109/ROBOT.2009.5152219
17. Otis, M.J.-D., Mokhtari, M., du Tremblay, C., Laurendeau, D., De Rainville, F.-M., Gosselin, C.M.: Hybrid control with multi-contact interactions for 6dof haptic foot platform on a cable-driven locomotion interface. In: 2008 Symposium on Haptic Interfaces for Virtual Environment and Teleoperator Systems, pp. 161–168 (2008)
18. Calisti, M., Corucci, F., Arienti, A., Laschi, C.: Dynamics of underwater legged locomotion: modeling and experiments on an octopus-inspired robot. Bioinspir. Biomim. **10**(4), 46012 (2015)
19. Ghassemi, S., Hong, D.: Feasibility study of a novel robotic system BALLU: Buoyancy assisted lightweight legged unit. In: 2016 IEEE-RAS 16th International Conference on Humanoid Robots (Humanoids), p. 144 (2016)
20. Rodriguez-Barroso, A., Saltaren, R., Portilla, G.A., Cely, J.S., Yakrangi, O.: Potential energy distribution of redundant cable-driven robot applied to compliant grippers: method and computational analysis. Sensors (Switzerland) **19**(15), 3403 (2019)
21. Jamshidifar, H., Khajepour, A., Fidan, B., Rushton, M.: Kinematically-constrained redundant cable-driven parallel robots: modeling, redundancy analysis, and stiffness optimization. IEEE/ASME Trans. Mechatron. **22**(2), 921–930 (2016)
22. Gouttefarde, M., Lamaury, J., Reichert, C., Bruckmann, T.: A versatile tension distribution algorithm for n-DOF parallel robots driven by n+2 cables. IEEE Trans. Robot. **31**(6), 1444–1457 (2015). https://doi.org/10.1109/TRO.2015.2495005
23. van Oort, G., Stramigioli, S.: Geometric interpretation of the zero-moment point. In: 2011 IEEE International Conference on Robotics and Automation, pp. 575–580 (2011)
24. Gouttefarde, M., Collard, J.F., Riehl, N., Baradat, C.: Geometry selection of a redundantly actuated cable-suspended parallel robot. IEEE Trans. Robot. **31**(2), 501–510 (2015). https://doi.org/10.1109/TRO.2015.2400253
25. Hernandez, Y., Kim, K.H., Benson, E., Jarvis, S., Meginnis, I., Rajulu, S.: Underwater space suit performance assessments part 1: motion capture system development and validation. Int. J. Ind. Ergon. **72**, 119–127 (2019)
26. Brablc, M., Sova, V., Grepl, R.: Adaptive feedforward controller for a DC motor drive based on inverse dynamic model with recursive least squares parameter estimation. In: 2016 17th International Conference on Mechatronics-Mechatronika (ME), pp. 1–5 (2016)

Author Index

M. Gouttefarde et al. (Eds.): CableCon 2021, MMS 104, pp. 427–428, 2021.
https://doi.org/10.1007/978-3-030-75789-2